THE BOOK OF SNAKES

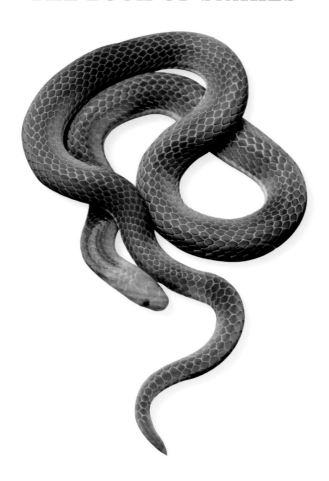

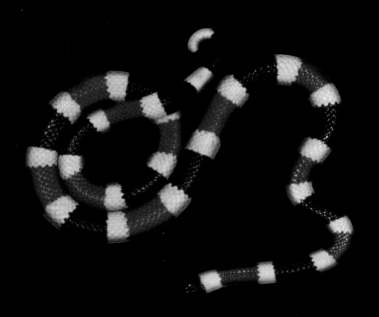

THE BOOK OF SNAKES

A LIFE-SIZE GUIDE TO SIX HUNDRED
SPECIES FROM AROUND THE WORLD

MARK O'SHEA

THE UNIVERSITY OF CHICAGO PRESS

MARK O'SHEA is a herpetologist, television presenter, zoologist, author, photographer, and lecturer. He is Professor of Herpetology at the University of Wolverhampton, UK and the Consultant Curator of Reptiles at West Midland Safari Park in the UK. He has been involved with snakes for more than five decades, initially keeping them in captivity, but gradually moving to become a professional herpetologist without a personal collection. Between 1999 and 2003 Mark presented four seasons of the internationally acclaimed *O'Shea's Big Adventure* for Animal Planet, co-produced with the UK's Channel 4 as *O'Shea's Dangerous Reptiles*. He has now presented almost forty documentaries including films for Discovery Channel, ITV, and the BBC. Mark has conducted herpetological fieldwork, or made films, in almost forty countries worldwide, on every continent except Antarctica. During 1987–88 he spent seven months as the herpetologist on the Royal Geographical Society Maracá Rainforest Project in Roraima, Brazil. He also ran herpetological projects on six expeditions for Operation Raleigh, Raleigh Executive, and Discovery Expeditions, during the 1980s–90s. He has made ten expeditions to Papua New Guinea (PNG) since 1986 and has worked on a long-term snakebite research project there, initially for Oxford University and the Liverpool School of Tropical Medicine, and since 2006 for the Australian Venom Research Unit, University of Melbourne. More recently he has worked on a snakebite project in Myanmar for the University of Adelaide. Since 2009 he has also worked as co-leader of a team working out of a Californian community college—the team has documented the first comprehensive herpetological survey of Timor-Leste, Asia's newest country. With ten phases of the project completed, they have documented around 70 species, 20–25 of them new to science. Mark's specialist area is the snake fauna of New Guinea and Wallacea, and his research takes him to natural history museums worldwide to examine snake specimens collected as far back as the nineteenth century. Mark has described a number of new snake species from PNG. In 2000 Mark received one of only eight Millennium Awards "for services to exploration" from the British Chapter of the Explorers' Club of New York, Mark's award being for zoology. In 2001 the University of Wolverhampton awarded him an honorary Doctor of Sciences degree for his "contributions to herpetology." He is the author of five books, including *A Guide to the Snakes of Papua New Guinea* (1996), of which he is currently working on the second edition. He has also written numerous articles and scientific papers. He lives in Shropshire, England, 15 miles from the birthplace of Charles Darwin.

The University of Chicago Press, Chicago 60637

© 2018 Quarto Publishing plc

Published 2018
Printed in China

27 26 25 24 23 22 21 20 19 18   1 2 3 4 5

ISBN-13: 978-0-226-45939-4 (cloth)
ISBN-13: 978-0-226-45942-4 (e-book)
DOI: https://doi.org/10.7208/chicago/9780226459424.001.0001

Library of Congress Cataloging-in-Publication Data

Names: O'Shea, Mark, author.
Title: The book of snakes : a life-size guide to six hundred species
  from around the world / Mark O'Shea.
Description: Chicago : The University of Chicago Press, 2018.
  | Includes bibliographical references and index.
Identifiers: LCCN 2018010069 | ISBN 9780226459394 (cloth)
  | ISBN 9780226459424 (e-book)
Subjects: LCSH: Snakes—Identification. | Snakes.
Classification: LCC QL666.O6 O74 2018 | DDC 597.96—dc23
LC record available at https://lccn.loc.gov/2018010069

This book was conceived and designed by
Ivy Press
An imprint of The Quarto Group
The Old Brewery, 6 Blundell Street
London N7 9BH, United Kingdom
T (0)20 7700 6700 F (0)20 7700 8066
www.QuartoKnows.com

Publisher SUSAN KELLY
Creative Director MICHAEL WHITEHEAD
Editorial Director TOM KITCH
Commissioning Editor KATE SHANAHAN
Senior Project Editor CAROLINE EARLE
Picture Researcher ALISON STEVENS
Editors LIZ DREWITT, SUSI BAILEY
Designer GINNY ZEAL
Illustrator DAVID ANSTEY

JACKET IMAGES
Matthieu Berroneau (*Atheris hispida*, *Vipera latastei*); Frank Cannon (*Xenopholis scalaris*);Adam G. Clause (*Contia tenuis*); Thor Håkonsen (*Protobothrops mangshanensis*); S. Blair Hedges (*Mitophis leptipileptus*, *Tetracheilostoma carlae*); Johan Marais (*Dipsina multimaculata*); Mark O'Shea (*Imantodes cenchoa*, *Lycodon capucinus*, *Micruroides euryxanthus*, *Myron richardsonii*, *Rhinotyphlops lalandei*); Mike Pingleton (*Pliocercus elapoides*); Nathan Rusli (*Trimeresurus insularis*); Shutterstock/Dr Morley Read (*Bothrops taeniatus*, *Helicops angulatus*); Laurie J. Vitt (*Siagonodon septemstriatus*).

LITHOCASE IMAGES
Matthieu Berroneau (*Vipera berus*); Omar Machado Entiauspe-Neto (*Xenodon dorbignyi*); Václav Gvoždík (*Philothamnus heterodermus*).

CONTENTS

RIGHT: **A venomous serpent** striking from a leafy bough probably epitomizes the worst nightmare for many people. Yet this Great Lakes Bushviper (*Atheris nitschei*) is actually a highly specialized and wonderfully adapted predator of mammals, lizards, or frogs that only uses its venom in defense when it feels truly threatened.

INTRODUCTION

BELOW: **A hooding Egyptian Cobra** (*Naje haje*) an iconic image, yet the purpose of the hood is to warn potential attackers that the cobra will defend itself if necessary; it is trying to avoid confrontation, not provoke it.

Snakes—almost everybody has an opinion about them, and often those opinions are extremely polarized, with people either fearing snakes or being fascinated by them. The sinuous movements of a snake's body, the oil-on-water effect of its iridescent scales, the hypnotic movements of its elevated head, the unblinking gaze, the continually flickering tongue, and often the serpent's totally unexpected and unannounced appearance, are all factors that play into its ability to convey both beauty and menace in a single tongue-flick.

It is no surprise that the snake has touched so many societies throughout history and been included in countless cultural and religious stories. By shedding its skin, it may be seen as a symbol of renewal and long life, but at the same time it is the bringer of death. It is likely that since humans first walked upon the Earth, the snake has held them in its thrall, an ever-present danger in the shadows, hidden in leaf litter, reaching out from a leafy bough, or lurking beneath dappled waters. So, why do so many people shudder at the very word "snake?" Obviously, one of the reasons people fear snakes is perfectly reasonable and natural—some snakes can, and do, kill humans. Worldwide, up to 125,000 lives are lost through snakebites every year. But, putting this into perspective, figures extrapolated from World Health Organization data suggest that 1.25 million people may have died in road-traffic accidents in 2015, ten times as many as were killed by snakes over a similar 12-month period.

There are a little over 3,700 living snake species known to science. They exhibit a truly amazing diversity of shape, size, color, pattern, and natural history. In *The Book of Snakes*, I will introduce the reader to 600 species, almost one in six of all snakes known. For those who are unfamiliar with snakes, I aim to dispel myths and bring enlightenment and understanding about one of the most maligned groups of animals on the planet. For those who are already snake aficionados, I hope to introduce rare or elusive species that may have previously passed beneath their radar.

In selecting which 600 species to feature in *The Book of Snakes*, I went for diversity, including many of the familiar names, both the popular, inoffensive snakes kept as pets, and the infamous, highly venomous species that claim lives. But I also wanted to illustrate less well-known species, from remote islands, cold mountains, arid deserts, verdant rainforests, and the open oceans—snakes with unique lifestyles or diets, and each with its own interesting story to tell. Some species are so rare that we struggled to find a single photograph to represent them. And that was the one hard-and-fast rule: if no image was available or of sufficient quality to represent the species at life size, then that species did not make the book. However, thanks to the many excellent photographers who have contributed images, we lost fewer than 30 species from the original selection.

I hope that *The Book of Snakes* will appeal to both armchair naturalists and experienced fieldworkers alike, but particularly that it will inspire budding herpetologists of future generations to respect, study, and protect snakes.

LEFT: **The little girl's expression,** her hand reaching out to explore, to make contact, both suggest awe and wonderment directed toward the sinuous serpent, a Burmese Rock Python (*Python bivittatus*), on the other side of the glass.

RIGHT: **This alligator lizard**, a member of the lizard suborder Anguimorpha, out of which the Serpentes are believed to have evolved, exhibits body elongation and limbed reduction on the road toward limblessness.

EVOLUTION & DIVERSITY OF SNAKES

Snakes are elongate animals with fragile skulls and skeletons, which may become disarticulated and separated post-mortem. It is therefore no surprise that relatively few complete snake fossils are available, most comprising a few vertebrae and skull fragments.

THE EVOLUTION OF SNAKES

There are two contrasting schools of thought regarding the evolutionary origin of snakes. One theory proposed that they evolved from a now extinct group of large marine reptiles known as mosasaurs, which dominated the Late Cretaceous oceans. The other theory holds that snakes have a terrestrial origin and evolved from within the Anguimorpha, a suborder of lizards that today contains the slow worms, alligator lizards, monitor lizards, and the venomous Gila monster, and beaded lizards. This latter theory is the more widely accepted, but there is still support for an aquatic mosasaur origin.

BELOW: **The earliest snakes** may have resembled this modern Boulenger's Pipesnake (*Cylindrophis boulengeri*) from Southeast Asia, being small species that preyed on cylindrical prey such as soft-bodied invertebrates or slender vertebrates. The macrostomatan snakes that could feed on broader mammalian prey probably evolved later.

The earliest snakes are now thought to date from the Middle Jurassic or Early Cretaceous period, 167–140 MYA (million years ago), with fossil examples discovered in England, Portugal, and Colorado, USA. These fossil examples comprise a few vertebrae and fragments of jawbones, but they can be readily identified as snakes by their strongly recurved teeth, a common characteristic of modern and ancestral snakes. Their discovery suggests a much earlier origin for snakes than the previously accepted Late Cretaceous, 95 MYA.

Early snakes are thought to have inhabited warm, wet, well-vegetated habitats, where they existed as terrestrial, nocturnal, wide-foraging, non-constricting stealth hunters, preying on soft-bodied invertebrates and vertebrates of lesser width than their own heads. A modern comparison might be the Asian pipesnakes (*Cylindrophis*). The greatest explosion in snake diversity appears to have occurred following the Cretaceous–Paleogene extinction event, 66 MYA. This led to the extinction of the dinosaurs, mosasaurs, and 75 percent of all life on Earth, but it also resulted in the rise of the mammals, a potential prey source of early snakes.

Some fossil snakes display hind limbs, including *Najash rionegrina*, from Late Cretaceous Patagonia, which has a well-developed pelvic girdle and what are believed to have been functional hind limbs. Three Middle Cretaceous marine species—*Pachyrhachis problematicus* and *Haasiophis terrasanctus* from Palestine, and *Eupodophis descouensi* from Lebanon—also had hind limbs. These species are grouped in the extinct family Simoliophiidae, but body elongation and loss of limbs does not necessarily separate snakes from lizards (see "Skeleton and Limbs," page 12).

As recently as 2016, an Early Cretaceous fossil from Brazil was described as *Tetrapodophis amplectus*. It had an extremely elongate body and four short pentadactyl limbs, and was reported worldwide as the first four-legged snake. But this discovery proved extremely controversial, and paleontologists now believe that the fossil is a dolichosaur, an extinct marine lizard-like reptile.

9

BELOW: **Early snakes were elongate** animals with vestigial hind limbs, such as this fossil (*Eupodophis descouensi*) from the Middle Cretaceous. Vestigial hind limbs are still present in extant boas and pythons.

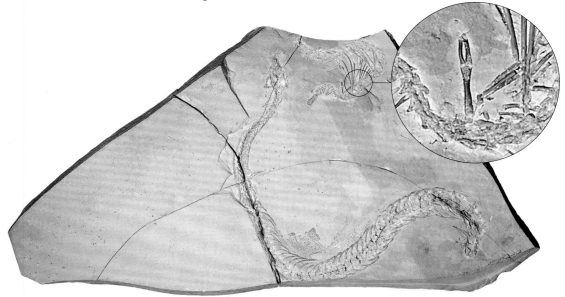

THE DIVERSITY OF MODERN SNAKES

The snakes (suborder Serpentes), along with the lizards (suborder Lacertilia) and the worm-lizards (suborder Amphisbaenia), comprise the order Squamata, the scaled reptiles. The sister clade (group) of the Squamata is the Rhynchocephalia, the beaked reptiles, a once diverse and widely distributed group of lizard-like reptiles that is now confined to New Zealand, where it is represented by a sole extant species, the Tuatara (*Sphenodon punctatus*). The Squamata and Rhynchocephalia together form the superorder Lepidosauria, the sister clade of the Archosauria, which contains crocodilians, birds, and extinct dinosaurs and pterosaurs.

Modern snakes are divided into two infraorders, the Scolecophidia (worm snakes) and the Alethinophidia (true snakes). The Scolecophidia comprises five families of small fossorial (burrowing) snakes. Although appearing primitive among living snakes, these are actually highly derived, having specialized considerably for their subterranean existence.

The Alethinophidia is divided into the Amerophidia, a small group that has not spread beyond Latin America, and the Afrophidia, which contains the majority of the true snakes. The Afrophidia are the "Out of Africa" clade, because the continent appears to be the group's evolutionary cradle, from where it radiated worldwide. The Afrophidia is further divided into the Henophidia ("old snakes"), which contains the boas, pythons, pipesnakes, shieldtails, and several smaller families of small-mouthed snakes, and the Caenophidia ("recent snakes").

The Caenophidia is divided into two superfamilies. The Acrochordoidea today contains just three species of aquatic filesnakes (*Acrochordus*), but once included the now extinct Nigerophiidae and Palaeophiidae. The sister clade to the Acrochordoidea is the huge Colubroidea, with its vast and diverse array of ratsnakes, watersnakes, treesnakes, cobras, seasnakes, and vipers. This superfamily comprises 11 families and more than 3,000 species—almost 82 percent of all living snakes.

A NOTE ON CLASSIFICATION AND SCIENTIFIC NAMES

Living organisms are classified using a hierarchical system. For snakes this would be: Kingdom: Animalia; Phylum: Chordata; Class: Reptilia; Order: Squamata; Suborder: Serpentes. Within the Serpentes, snakes are grouped into Superfamilies (ending –oidea), Families (ending –idae) and Subfamilies (ending –inae). Within the families and subfamilies are the Genera that contain the Species. A species name is a binomial, it comprises two words, and it is written in italics with only the generic part receiving a large case initial letter, i.e. *Natrix helvetica*. Names are not necessarily Latin, but if the name comes from another language it is latinized, i.e. the Sanskrit word "Naia," for the cobra genus, is latinized to *Naja*. A trinomial name indicates a subspecies. The name may be accompanied by the name of the describer, and the date of publication. If the name and date are contained in parentheses, this indicates that the name has changed since it was described, usually because it has been moved to another genus, i.e. the Indian Cobra, *Naja naja* (Linnaeus, 1758) was originally described by Linnaeus as *Coluber naja*.

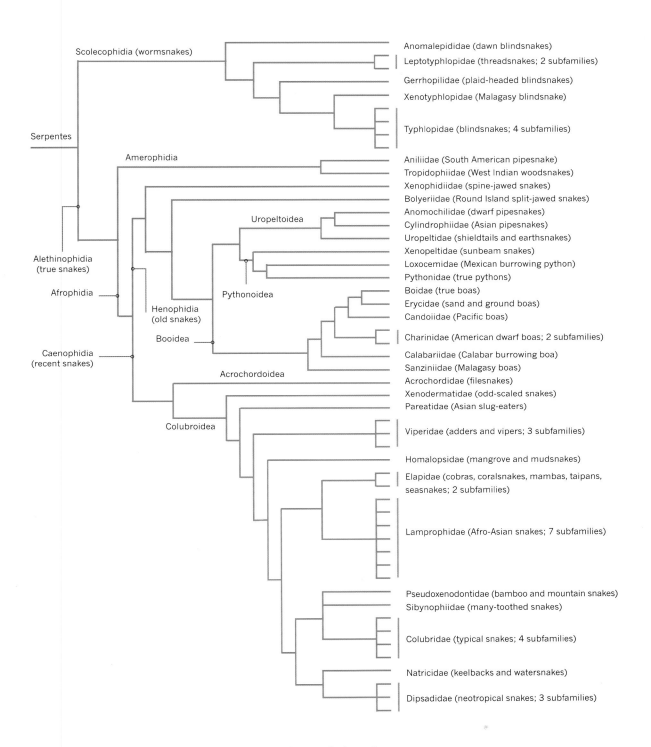

ABOVE: **A family tree of snakes** illustrating the divergence between the burrowing Scolecophidia (worm snakes) and the Alethinophidia (true snakes); the restricted Amerophidia and the much more successful Afrophidia; and the relatively primitive Henophidia (old snakes) and the more advanced Caenophidia (recent snakes). The Caenophidia is divided into the Acrochordoidea, and the diverse and widely distributed Colubroidea, which contains over 3,000 of the more than 3,700 living snake species. Thirty-three snake families are presented, 8 of which contain between 2 and 7 subfamilies, indicated by the vertical brown bars. Representatives from every family and subfamily are included in this book. This is a simplified family tree, the lengths of the arms are not intended to indicate the timelines since divergence between the various families or clades. Due to the constraints of space in the Acrochordoidea, entirely extinct snake families, such as the Nigerophiidae and Palaeophiidae, are omitted.

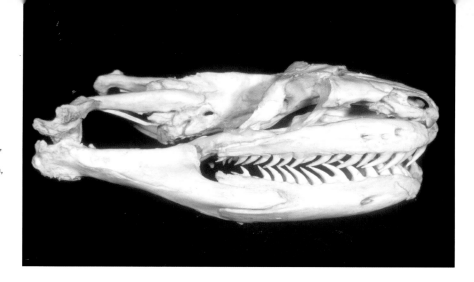

RIGHT: **The skull of a Green Anaconda** (*Eunectes murinus*) exhibiting the highly flexible bones of the skull, the three bones of the lower jaw (quadrate, compound, and dentary), and six rows of recurved teeth, which are typical of snakes and located on the maxillae and pterygoid-palatine (inner) bones of the upper jaw, and dentary bones of the lower jaw.

WHAT IS A SNAKE?

All amphibians, reptiles, birds, and mammals are pentadactyl tetrapods— vertebrates with four limbs, each with five digits. Snakes, as reptiles, are also pentadactyl tetrapods because their lizard ancestors were fully limbed.

SKELETON AND LIMBS

The snake skeleton comprises a skull and a spinal column. Because snakes possess extremely elongate, flexible bodies, they may have up to 500 vertebrae, although 120–240 is more common. Each vertebra is attached to a pair of ribs, which in the absence of a sternum are independent, being interconnected only by powerful intercostal muscles that enable the many modes of snake locomotion. The lack of a sternum allows the rib cage to expand outward so that the body can accommodate large meals, egg clutches, or litters of neonates. The outward expansion and mobility of the ribs is obvious in the dorsoventral flattening of a basking viper, the hooding of a cobra, and the lateral body compression of a swimming seasnake.

All snakes lack front limbs, but the vestiges of the pelvic girdle and hind limbs are present in the boas, pythons, and some other primitive snake groups. Externally, they are represented by a pair of curved horny spurs on either side of the cloaca (genital-excretory opening). Spurs are largest in males, which use them to court the female during copulation.

SKULLS AND TEETH

Unlike the skulls of mammals, turtles, or crocodilians, those of snakes exhibit kinesis, meaning that they are hugely flexible, and the individual bones are capable of the articulation required to manipulate and swallow prey. The large gape of a snake's mouth is achieved because the lower jaw comprises six separate, flexible bones. The tooth-bearing dentary bone is attached to a toothless compound bone, which in turn is attached to the skull via an elongate quadrate bone. This arrangement permits considerable mobility in all planes, further enhanced by the fact that the left and right dentary bones are not fused at the chin. Many snakes can expand their lower jaws extremely widely to accommodate large meals, and advance each side of the lower jaw independently as the prey is swallowed.

Most snakes have six rows of recurved solid teeth, arranged on the dentary bones of the lower jaw, and both the maxilla (outer) and pterygoid-palatine (inner) bones of the upper jaw. A few snakes lack teeth from some bones—for example, the blindsnakes (Typhlopidae) lack teeth from the

INTERNAL ORGANS

Snakes have the same internal organs as other vertebrates, but due to the elongation of their bodies they are arranged less symmetrically. Of two lungs, usually only the right lung is functional and it may run for a third the length of the body, while the left lung is small and vestigial. There are two elongate kidneys, also arranged asymmetrically, a liver, a pancreas, a gall bladder, and a heart that varies in its location depending on the snake's lifestyle—for example, whether a diving seasnake or climbing treesnake. The heart has three chambers, comprising two auricles and a ventricle, unlike the four chambers of mammalian and crocodilian hearts. The digestive system comprises an esophagus, a stomach, and small and large intestines. The sexual organs of a female consist of a pair of elongate ovaries, while males have a pair of testes and a paired hemipenis. This vulnerable organ is inverted inside the base of the tail until it is required.

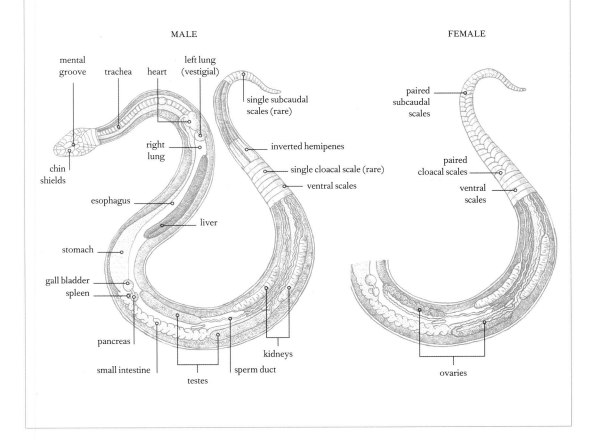

MALE

mental groove
trachea
heart
left lung (vestigial)
single subcaudal scales (rare)
right lung
chin shields
inverted hemipenes
single cloacal scale (rare)
ventral scales
esophagus
liver
stomach
gall bladder
spleen
pancreas
small intestine
testes
sperm duct
kidneys

FEMALE

paired subcaudal scales
paired cloacal scales
ventral scales
ovaries

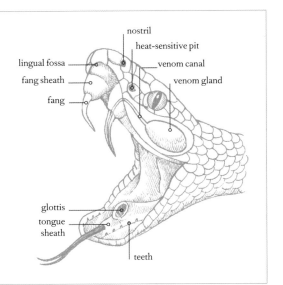

VENOM-DELIVERY MECHANISMS & FANGS

Venomous snakes have specialized venom-delivering mechanisms that culminate in their fangs. The most primitive are rear-fanged snakes, with enlarged, grooved teeth on the rear of the maxillae, down which venom trickles into a bite wound. In members of the venomous Elapidae and Viperidae families, the fangs are located on the front of the maxillae. In elapids, they are fixed in position, although the kinesis of the skull allows considerable movement. The vipers have short, toothless maxillae, to which are attached extremely long fangs that are hinged so that they can swing back horizontal to the skull when not in use. When a viper strikes, the flexibility of the skull and maxillae enables the fangs to swing forward like sabers, the highly kinetic skull absorbing the shock of the strike.

dentary bones, the threadsnakes (Leptotyphlopidae) lack teeth from the maxillae, and African egg-eating snakes (*Dasypeltis*) possess only a few teeth on the rear of the dentary and maxilla. The homologous nature of solid, ungrooved snake teeth makes it easy to distinguish snake fossils from those of lizards, which exhibit greater diversity of tooth type and shape.

SENSE ORGANS

Snakes are highly sensory animals whose sense organs differ from those of mammals. They lack an external ear or a tympanum, but they do possess a highly developed inner ear, with which they detect vibrations picked up by the columella bone, attached to the quadrate bone of the jaw. Snakes are not technically deaf; they just hear in a different way to other terrestrial vertebrates.

The snake's eyesight is also misunderstood. The retina of a vertebrate eye contains visual cells: rods, for night vision; and cones, for color vision and visual acuity. Fossorial blindsnakes may have eyes that are little more than photo-sensitive cells, warning them when they are exposed to daylight, but other snakes possess more elaborate vision. The pupils of diurnal snakes are round, whereas those of nocturnal or crepuscular snakes are vertically elliptical, or "catlike," providing the eye with more control over how much light reaches the retina. Many diurnal snakes also have dichromatic or trichromatic color vision.

The laterally positioned eyes of a snake provide it with 100–160-degree vision, but probably the best vision of any snake is that of the diurnal Asian treesnakes (*Ahaetulla*), which have horizontal keyhole-like pupils and a grooved snout, down which they can sight up their prey. This arrangement provides a 45-degree overlap in forward vision from both eyes, effectively providing binocular vision. These treesnakes also have a highly sensitive fovea centralis, a cone-heavy depression in the retina, enabling them to detect the slightest movements of a camouflaged lizard in the vegetation, and accurately judge distance to target.

In those species that spend their time underwater or buried in the sand, the eyes are often located in a more dorsolateral position, permitting vision without exposing the head. The Namib Sidewinding Adder (*Bitis peringueyi*, page 616) is one such sand-dweller with dorsally positioned eyes.

All snakes have a forked tongue. Located in the front of the lower jaw, this is often in continual movement, flicking in and out of the closed mouth through a small opening, the lingual fossa. Environmental molecules are transported on the tongue to a vomeronasal organ (olfactory sense organ) in the roof of the mouth, known as the Jacobson's organ, allowing snakes to track down either a mate or prey, and find their way around their home range. But they do not need to flick their tongues in order to smell—snake nostrils are also packed with sensitive olfactory tissues.

Many snakes that feed primarily on endothermic (warm-blooded) animals have evolved the ability to hunt in total darkness. Pythons, boas, and pitvipers (Crotalinae) all have thermosensory pits that detect the infrared body heat of their prey, enabling an accurate strike. In pythons and boas, there is a series of labial pits in the lip scales, whereas pitvipers

15

BELOW: **The Asian vinesnakes** (genus *Ahaetulla*) probably possess the best vision of any snake. They have horizontal pupils and can sight up their lizard prey down grooves on the snout, judging distance to target due to the considerable overlap in the vision from both eyes in front of the snout.

have a single loreal pit on either side of the head (located between the nasal and preocular scales—see scalation diagram opposite). More rudimentary structures, known as supranasal sacs, are present on the heads of Old World vipers (Viperinae), and may also function as infrared-sensitive receptors for hunting.

There are also several less-studied sensory receptors in snakes. The tentacles of the Tentacled Snake (*Erpeton tentaculum*, page 540), for example, are thought to detect vibrations in water that indicate the presence of fish. Similarly, the strange spinous, tuberculate scales of filesnakes (*Acrochordus*) are believed to detect swimming fish in cloudy water.

16

SEXUAL DIMORPHISM AND DICHROMATISM

Males and females of many snake species are almost indistinguishable, but there are clues to their gender. Males generally have longer tails than females, with a moderately bulbous basal area where the hemipenes are located. Females, meanwhile, may have shorter and more tapering tails, and often longer bodies than males. Females of some species are also much larger than males—for example, female Green Anacondas (*Eunectes murinus*, page 112) and Reticulated Pythons (*Malayopython reticulatus*, page 90) may reach around 20–23 ft (6–7 m) and 20–33 ft (6–10 m), respectively, while males are only around 10–13 ft (3–4 m) and 13–16 ft (4–5 m) in length. Larger females can carry more eggs or neonates, but in some species the sizes are reversed—female King Cobras (*Ophiophagus hannah*, page 480) reach only around 10 ft (3 m), while the largest recorded male was reportedly over 16 ft (5 m) in length.

Some species exhibit sexual dichromatism, whereby males and females have different coloration or patterns—for example, the male Northern Adder (*Vipera berus*, page 640) is silver-gray with black markings, while the female is brown with dark brown markings. Sexual dimorphism (differing body shape or size) is rarer in snakes than dichromatism.

BELOW: **The Malagasy Leafnose Snake** (*Langaha madagascariensis*, page 386) exhibits both sexual dimorphism and sexual dichromatism. The brown male bears a conical spike-shaped protuberance on his snout, while in the gray female this resembles a serrated spike

SCALATION

Snake scales are composed of keratin. The dorsal scales of the body are either smooth or keeled (ridged), and are usually arranged in imbricate (overlapping) rows. The ventral scales are usually broader, and also imbricate to permit locomotion on land, although many seasnakes have sacrificed their broad belly scales to enhance lateral body compression for swimming. The scales of the head are either a series of large "scutes," as in most colubrids or elapids, or are reduced in size to numerous undifferentiated granular scales, as in most vipers and some pythons. The number and arrangement of a snake's scales provide important clues for species identification.

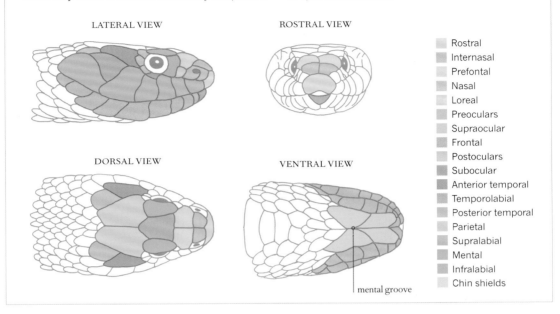

LATERAL VIEW

ROSTRAL VIEW

DORSAL VIEW

VENTRAL VIEW

mental groove

- Rostral
- Internasal
- Prefontal
- Nasal
- Loreal
- Preoculars
- Supraocular
- Frontal
- Postoculars
- Subocular
- Anterior temporal
- Temporolabial
- Posterior temporal
- Parietal
- Supralabial
- Mental
- Infralabial
- Chin shields

SHEDDING

As snakes grow, they need to shed their skins. A snake approaching a slough will exhibit "milky" eyes as the cells break down to separate the old and new skins, and when the eyes become clear, the snake is ready to shed. It rubs its snout on rough objects to begin the process, then crawls out of the old skin as it becomes snagged on rocks and twigs. Snakes do not possess eyelids, hence their unblinking gaze, but instead they have transparent coverings over the eyes known as "brilles" or "spectacles." These structures, which resemble contact lenses, are sloughed along with the rest of the skin, as is the skin on the forked tongue.

BELOW: **The dorsal scales** of snakes are usually smooth (A) but many aquatic or desert dwelling snakes (especially keelbacks and vipers) have keeled (ridged) scales (B), while the scales of the filesnakes (*Acrochordus*) are tuberculate (C).

A

B

C

PREY & HUNTING

All snakes are carnivorous, but as a group they hunt a wide diversity of prey types and sizes. Some snakes are generalist feeders, with a catholic diet, while others are specialists that concentrate on one type of prey.

INVERTEBRATE PREY

The smallest living snakes, the scolecophidian threadsnakes (Leptotyphlopidae) and blindsnakes (Typhlopidae), have tiny mouths and limited dentition. They feed on small, soft-bodied prey such as ant or termite larvae and eggs, although some of the larger species (*Acutotyphlops*) take earthworms.

Vermivory, or feeding on worms, is a common feature of the diets of snakes across many families. Earthworms are the prey of the Asian shieldtail snakes (Uropeltidae) and spine-jawed snakes (*Xenophidion*), but numerous advanced snakes also feed on earthworms. There are even a few venomous snakes that eat earthworms—the Fiji Snake (*Ogmodon vitianus*, page 520) and Papuan worm-eating snakes (*Toxicocalamus*) in the Elapidae, and the Udzungwa Mountain Viper (*Atheris barbouri*, page 604) in the Viperidae.

Biologists refer to snakes that feed on "slimy" prey—earthworms, slugs, and snails—as "goo-eaters." Molluscivorous snakes require specialized oral glands to neutralize the excessively sticky secretions produced by their prey, and have specially adapted jaws to enable them to extract the snails from their shells. Slugs and snails, and the snakes that eat them, are common in the tropics, with *Dipsas* and *Sibon* species found in tropical America, *Duberria* in Africa, and *Pareas* in Asia.

Specialized centipede-eaters are found in Central America (*Scolecophis*) and Africa (*Aparallactus*), while the American hooknose snakes (*Ficimia* and *Gyalopion*) prey on spiders and scorpions in addition to centipedes.

The Crab-eating Mangrove Snake (*Fordonia leucobalia*, page 541) and Gerard's Watersnake (*Gerarda prevostiana*, page 542), both found in Southeast Asia, feed on freshly molted crabs and mud lobsters. And in North America, crayfish snakes (*Liodytes*) have specially adapted skulls to enable them to feed on hard-shelled crayfish.

PISCINE (FISH) PREY

Fish feature in the diets of many snakes, especially the freshwater aquatic keelbacks and watersnakes (Natricidae), which hunt by sight and touch. The tuberculate skin of the Indo-Australian filesnakes (*Acrochordus*) enables them to grasp a slimy fish while they maneuver prey into the mouth, while the Tentacled Snake (*Erpeton tentaculatum*, page 540) detects fish using the curious tentacles on its head. Venomous freshwater piscivorous snakes include the Aquatic Coralsnake (*Micrurus surinamensis*, page 465) in Amazonia, the Banded Water Cobra (*Naja annulata*, page 466) in Africa, and the Cottonmouth (*Agkistrodon piscivorus*, page 554) in the USA. Seasnakes prey on a variety of gobies, moray eels, catfish, and pufferfish, while sea kraits (*Laticauda*) specialize in eels. A few seasnakes eat only fish eggs—the Mosaic Seasnake (*Aipysurus mosaicus*, page 489) takes the eggs of benthic gobies in their seabed burrows, while the Southern Turtle-headed Seasnake (*Emydocephalus annulatus*, page 500) uses its enlarged lateral lip scales to scrape the eggs of blennies and gobies from coral.

AMPHIBIAN PREY

Frogs and toads, including their tadpoles, also feature in the diets of many snakes—for example, the Western Grass Snake (*Natrix helvetica*, page 415), Mexican Hognose Snake (*Heterodon kennerlyi*, page 264), and Rinkhals (*Hemachatus haemachatus*, page 454). Cat-eyed snakes (*Leptodeira*) specialize in treefrog eggs laid on leaves. Salamanders and newts are eaten by gartersnakes (*Thamnophis*), while salamanders are the prey of the rare Oaxacan Dwarf Boa (*Exiliboa placata*, page 122). Rainbow snakes (*Farancia*) also prey on salamanders, including the fully aquatic sirens and amphiumas, while the South American Pipesnake (*Anilius scytale*, page 66) and the South American Coralsnake (*Micrurus lemniscatus*,

ABOVE: **The Iranian Spider-tailed Viper** (*Pseudocerastes urarachnoides*) lures birds within strike range with its spider-like tail tip.

20

ABOVE: **The African egg-eating snakes** (*Dasypeltis*) specialize in feeding on small birds' eggs, regurgitating the eggshell afterward.

page 460) eat caecilians. The Tiger Keelback (*Rhabdophis tigrinus*, page 424) even sequesters the powerful bufotoxins from its toad prey in its own skin, rendering it both venomous and poisonous.

REPTILIAN PREY

Many desert and grassland snakes—including sandsnakes (*Psammophis*), skaapstekers (*Psammophylax*), and small terrestrial vipers (*Bitis, Cerastes,* and *Echis*)—feed on lizards. The slender, rear-fanged vinesnakes of tropical America (*Oxybelis*), the Caribbean (*Uromacer*), Africa (*Thelotornis*), Madagascar (*Langaha*), and Asia (*Ahaetulla*) hunt agile, alert, camouflaged lizards by stealth. The Boomslang (*Dispholidus typus*, page 160) takes chameleons, and the Australian Black-headed Python (*Aspidites melanocephalus*, page 85) preys on agamid dragons and goannas. Amphisbaenians (worm-lizards) are preyed upon by fossorial and semi-fossorial snakes, such the Small-scaled Dawn Blindsnake (*Typhlophis squamosus*, page 41).

Some snakes also eat other snakes. This is called ophiophagy, a term that forms the basis of the generic name of the King Cobra (*Ophiophagus hannah*, page 480), which is capable of swallowing a Reticulated Python (*Malayopython reticulatus*, page 90) more than 6 ft (2 m) long. There are snake-eating snakes on most continents, including North American kingsnakes (*Lampropeltis*), Latin American mussuranas (e.g. *Mussurana* and *Clelia*), African filesnakes (*Gonionotophis*), Asian kraits (*Bungarus*) and coralsnakes (*Calliophis*), and the Australian bandy-bandys (*Vermicella*). Ophiophagy is not cannibalism unless a snake eats its own species, but there are plenty of examples of this behavior, too, especially within the Elapidae.

Snakes rarely prey on turtles, but there is a record of a Puff Adder (*Bitis arietans*, page 610) swallowing a small tortoise. Caiman are eaten by anacondas (*Eunectes*), while small crocodiles are taken by water pythons (*Liasis*). In Florida, introduced Burmese Pythons (*Python bivittatus*, page 95) have been documented preying on alligators.

AVIAN PREY

Birds are commonly included in the diets of arboreal snakes, such as the Boomslang, which will raid the pendulous nests of weaverbirds, and the Central American Puffing Snake (*Phrynonax poecilnotus*, page 213). One of the most unusual bird-eating snakes is the Iranian Spider-tailed Viper (*Pseudocerastes urarachnoides*, page 637), whose tail tip is shaped like a

spider to lure birds within strike range. Other habitual bird-eaters include the Golden Lancehead (*Bothrops insularis*, page 567) on Ilha da Queimada Grande, Brazil, and the Santa Catalina Island Rattlesnake (*Crotalus catalinensis*, page 574) from the Gulf of California, Mexico. Adult Tigersnakes (*Notechis scutatus*, page 519) on Australia's Mount Chappell Island gorge on muttonbird chicks once a year, while on Guam, the introduced Brown Treesnake (*Boiga irregularis*, page 146) has eaten most of the island's endemic flightless bird species into extinction.

ABOVE: **The Southern African Rock Python** (*Python natalensis*) can constrict and devour large mammals such as this nyala calf.

MAMMALIAN PREY

Rats, mice, rabbits, and similar small mammals constitute the primary prey of numerous nonvenomous and venomous snakes. Some even specialize in hunting bats, including the Borneo and Malaysian Cave Racers (*Elaphe taeniura grabowskyi* and *E. t. ridleyi*, page 172). Ratsnakes and boas employ constriction to kill their prey, but sit-and-wait ambushers like vipers and rattlesnakes use venom. Pythons, boas, and pitvipers can locate their warm-blooded mammalian prey in total darkness using their heat-sensitive facial pits (see page 15).

Larger mammals, such as deer, antelope, pigs, and monkeys, may be taken by adult pythons, boas, and anacondas. The largest mammals documented as snake prey were an adult female sun bear, eaten by a large Reticulated Python, and a puma, taken by a Green Anaconda (*Eunectes murinus*, page 112). Humans are also occasionally killed and eaten by giant pythons (see "Snakes & Humankind," pages 32–35).

ENEMIES & DEFENSE

Snakes have many natural enemies. Small snakes may fall foul of large venomous invertebrates, but it is among other vertebrates that most snake predators are to be found. To avoid being killed and eaten, snakes have evolved an entire armory of weapons and subterfuges.

ENEMIES

Probably the most well known of all "snake killers" is the mongoose, famously embodied as Rikki-Tikki-Tavi in Rudyard Kipling's *Jungle Book*. But these agile mammals are not confined to India; there are also mongooses throughout Southeast Asia and Africa. With their quick reactions and thick body fur, and a degree of immunity to cobra venom, mongooses are able to avoid, ward off, and even survive most snake strikes. Seen as a means of

RIGHT: **The large East African Gaboon Viper** (*Bitis gabonica*) has a body patterned like a pastel-colored Persian carpet, and a head like a huge leaf. This complex cryptic patterning serves to break up its outline when it is lying in woodland leaf litter.

controlling snakes and rats, mongooses were introduced to Okinawa, Jamaica, Hawaii, Fiji, Mauritius, and other islands, where they have caused considerable environmental damage and extinctions. Less well known than the mongoose story is the fact that the European hedgehog will also kill and eat snakes, as will many smaller carnivores such as cats.

Among the birds, snake-eagles, fish-eagles, hornbills, and secretary birds are all snake predators. In the reptiles, crocodiles, monitor lizards, and ophiophagous snakes—especially the King Cobra (*Ophiophagus hannah*, page 480), American kingsnakes (*Lampropeltis*), and African filesnakes (*Gonionotophis*)—all prey on snakes. After all, nothing fits inside a snake so well as another snake! Even the most venomous snakes, the seasnakes, are commonly found in the stomachs of tiger sharks. But humankind is the snake's worst enemy.

ABOVE: **The venomous Eastern Coralsnake** (*Micrurus fulvius*, top) can be distinguished from its harmless mimic, the Scarlet Milksnake (*Lampropeltis elapsoides*, bottom) by the order of its bands and the rhyme "Red to yellow, kill a fellow, Red to black, venom lack," but this rhyme does not work in South America.

DEFENSIVE STRATEGIES

Snakes have an array of defenses to avoid detection or warn off predators. Many snakes are highly camouflaged or cryptically patterned to enable them to hide in leaf litter or vegetation. Vinesnakes (*Oxybelis*, *Ahaetulla*, and *Thelotornis*) really do look like vines, while the head of a Gaboon Viper (*Bitis gabonica*, page 613) resembles a large dead leaf, and the rest of its Persian carpet patterning serves to break up its outline for both predator and prey alike. A few nocturnal snakes, such as rainbow boas (*Epicrates*), have iridescent body scales that shimmer in daylight and may also break up their outline when approached by a potential predator. American coralsnakes (*Micrurus* and *Micruroides*) are brightly banded red, yellow, and black, advertising that they are dangerous and should be left alone, a pattern mimicked by some harmless snakes to avoid interference.

BELOW: **The Cape Twigsnake** (*Thelotornus capensis*) inflates its throat to expose the interstitial skin, making itself look larger and more threatening to potential predators.

Some snakes go in for big visual threat displays. Examples include the cobras (*Naja*), with their hooding; the Boomslang (*Dispholidus typus*, page 160), which inflates its neck to expose the contrasting colors of its interstitial skin; and the Black Mamba (*Dendroaspis polylepis*, page 451), with its wide-mouth gaping, exposing the black interior. Other snakes rely on auditory warnings, such as the buzzing tail of a rattlesnake (*Crotalus*), the sawing

ABOVE **The Zebra Spitting Cobra** (*Naja nigricincta,* above left) spreads a hood as a warning that it is dangerous, and if that fails it will send twin jets of venom into the eyes of its perceived enemy. The Western Diamondback Rattlesnake (*Crotalus atrox,* above right) avoids confrontations with enemies by vibrating its rattle vigorously. Every time the rattlesnake sheds its skin it adds another interlocking link to the rattle as the cross-section above illustrates. The oldest links are at the tip.

sound made by a carpet or saw-scale viper (*Echis*) as its continual motion causes its serrated keeled dorsal scales to rub together, or the loud hissing of a Russell's Viper (*Daboia russelii*, page 623), Puff Adder (*Bitis arietans*, page 610), or Bullsnake (*Pituophis catenifer sayi*, page 215).

If escape is an option, then few snakes have a better method than the Paradise Flying Snake (*Chrysopelea paradisi*, page 132), which can flatten its entire body into a concave parachute, so that it simply glides away from the threat when it leaps from a tree. Still other snakes pretend to be something they are not, mimicking the body shape, patterning, or behavior of highly venomous coralsnakes, cobras, or vipers—the Aesculapian False Coralsnake (*Erythrolamprus aesculapii*, page 311), Large-eyed Mock Cobra (*Pseudoxenodon macrops*, page 441), and Northern False Lancehead (*Xenodon rabdocephalus*, page 346) all employ this defensive measure.

One of the strangest ways to avoid predation is to play dead, a behavior known as thanatosis. American hognose snakes (*Heterodon*), European grass snakes (*Natrix*), and the South African Rinkhals (*Hemachatus haemachatus*, page 454) all play dead when they feel threatened. The last of these also has another defense in its armory—it is a spitting cobra capable of sending jets of painful cytotoxic venom into the eyes of an enemy, effectively blinding it while the cobra makes good its escape. This is probably the only instance where a snake's venom has evolved for defense rather than hunting.

Another very strange defense measure is cloacal popping. Some coralsnakes (*Micruroides* and *Micrurus*) and hooknose snakes (*Ficimia* and *Gyalopion*) use this technique, forcefully and noisily expelling air from their cloacas.

HUMAN THREATS AND CONSERVATION

Humans have probably persecuted snakes for centuries, either killing them out of fear or harvesting them in unsustainable numbers for their skins, their meat, or their gall bladders, an apparently essential ingredient in an eastern tonic for failing libido in men. And of course snakes suffer from the same indirect threats to their survival as other animals, habitat destruction, fragmentation, and alteration. While some snake species have slipped into extinction through the agencies and actions of man, others have been saved by the intervention of conservation bodies, captive breeding, and educational campaigns aimed at teaching villagers or islanders to not only live alongside their serpentine neighbors, but also be proud of them and protect them.

American and European zoos have been actively engaged in a captive breeding program for the endangered Aruba Island Rattlesnake (*Crotalus durissus unicolor*, page 576), a small, pastel-colored subspecies of the large Tropical Rattlesnake, the aim being to bolster the population in the small desert center of its island. Two extremely primitive, endemic, egg-laying, split-jaw snakes, only distantly related to any other snakes on Earth, used to live on tiny (⅔ sq miles/1.69 km²) Round Island in the Indian Ocean, until seafarers introduced goats and rabbits to the island. The invasive mammals devoured the vegetation and then the rain washed the soil into the sea, and with it went the Round Island Burrowing Boa (*Bolyeria multocarinata*). The other species, the Keel-scaled Boa (*Casarea dussumieri*, page 79) was more fortunate, it was rescued by the Jersey Wildlife Preservation Trust (JWPT) who, working with the Government of Mauritius, have brought the species back from the brink so when the island has been restored it is hoped the species can go home. Sadly the Burrowing Boa was declared extinct by the IUCN in 1975.

The JWPT and other conservation organizations have been behind the recoveries of Caribbean snakes such as the Antiguan Racer (*Alsophis antiguae*), Great Bird Island Racer (*A. sajdaki*, page 298), and the Jamaican Boa (*Chilabothrus subflavus*, page 107). Snakes might not be everybody's "cup of tea" but that does not mean that they do not need conservation. And conserving snakes is actually beneficial to mankind, snakes are nature's rat catchers, eating the rodents that carry disease and devour our crops.

25

BELOW: **Snakes are rarely a vote winner** when it comes to popularity, but conservation of snakes is important. Endangered species such as the Jamaican Boa (*Chilabothrus subflavus*) been saved from extinction by conservation organizations.

RIGHT: **Every spring, thousands** of Red-sided Gartersnakes (*Thamnophis sirtalis parietalis*) emerge from communal dens in Manitoba, Canada. The males emerge first to await the females, whereupon they will compete with other males for the largest females. Some males mimic females to draw other males away and allow them to mate the females.

REPRODUCTIVE STRATEGIES

Unlike turtles, tortoises, crocodiles, alligators, and the tuatara, which all lay eggs, squamate reptiles employ both oviparity (egg-laying) and viviparity (live-bearing). There are advantages and disadvantages to each of these strategies.

COURTSHIP, MATING BALLS, AND SKEINS

Sexually receptive female snakes emit pheromones, leaving a trail for males to follow with their extremely sensitive forked tongues, sometimes multiple males pursuing a single female. At the famous Manitoba snake pits, thousands of male Red-sided Gartersnakes (*Thamnophis sirtalis parietalis*, page 432) emerge from hibernation, only to linger around the dens to await the later emergence of the females.

RIGHT: **Green Anacondas** (*Eunectes murinus*) may form great mating balls in the shallows or on land, one large female attracting up to a dozen smaller males, all jockeying for the best position to mate with her.

Multiple male Green Anacondas (*Eunectes murinus*, page 112) will often court a single female, forming a mating ball around her as they jockey for position with their tails in an attempt to mate. These mating balls are usually found in shallow water, and the female may mate with several males. A similar arrangement has been observed in more active species such as the Keel-bellied Whipsnake (*Dryophiops rufescens*, page 135) and the Paradise Flying Snake (*Chrysopelea paradisi*, page 132). Here, a single female is courted as she moves through vegetation by two or three males, which form a continually writhing entanglement, known as a "skein," about her body.

ABOVE: **Male pythons and boas** possess cloacal spurs (above right), the vestiges of their ancestral hind limbs, with which they will stroke the females during courtship; the spurs of females are small or absent. Similarly, male Southern Turtle-headed Seasnakes (*Emydocephalus annulatus*) have a spike on their snouts (above left), which they use during courtship to stroke the female's back.

MALE COMBAT

It is quite common for two male snakes, in pursuit of the same female, to engage in combat—this behavior has been observed in adders, rattlesnakes, the King Cobra (*Ophiophagus hannah*, page 480), and the Black Mamba (*Dendroaspis polylepis*, page 451). These combats are wrestling matches that involve the competitors entwining around one another, raising up and trying to force their opponent to the ground. Combats between venomous species do not usually lead to injuries, but those between large pythons may involve spurring and result in deep wounds.

THE MECHANICS OF MATING

Snakes reproduce sexually via copulation, the male snake inserting one of his paired hemipenes into the cloaca of the female, which will lift her tail to make penetration easier. The female's ova are fertilized internally with the male's sperm, although this may not happen immediately—

ABOVE: **A pair of male Dharman Ratsnakes** (*Ptyas mucosa*) writhe and wrestle with one another in an attempt to win the right to mate with a nearby female. Males of many species engage in this behavior, including King Cobras (*Ophiophagus hannah*) and Black Mambas (*Dendroaspis polylepis*).

RIGHT: **A female Osage Copperhead** (*Agkistrodon contortrix phaeogaster*) with her litter of neonates. The neonates are only days old but they will soon disperse into the surrounding undergrowth.

females of species that mate in the autumn are able to store sperm until more favorable conditions return in spring. Sperm is usually stored for only a few months, but much longer periods have been documented—six years in the case of the Banded Cat-eyed Snake (*Leptodeira annulata*, page 282).

Most tropical and subtropical snake species reproduce once a year, but in colder climates, where prey is only seasonally available and weather conditions are not ideal, the period during which a post-parturition female can recover her optimal body weight may be short. In these situations, species such as the Northern Adder (*Vipera berus*, page 640) in the Alps or the Carpet Python (*Morelia spilota*, page 93) in southern Australia may reproduce only biannually. Species in warm climates with abundant prey year-round, such as the Common Saw-scale Viper (*Echis carinatus*, page 624), may produce two litters each year.

OVIPARITY OR VIVIPARITY

Birds are oviparous (egg-laying), as are turtles, crocodilians, and monotreme mammals, while all remaining mammals are viviparous (live-bearing). But this clear-cut division of reproductive strategies becomes

LEFT: **A female Reticulated Python** (*Malayopython reticulatus*) with her clutch of leathery-shelled eggs. She will defend her eggs against egg thieves such as monitor lizards and incubate them by "shivering thermogenesis"—the continual twitching of her muscles which will elevate her body temperature by 13–23 °F (7–13 °C).

blurred in the Squamata, especially in the snakes. Oviparity is the primitive condition for snakes, and it is the sole strategy exhibited by 15 of the 33 snake families (see page 11), including many of the small, fossorial, and relatively primitive snakes. Live-bearing is the only strategy found in ten snake families, primarily in the Henophidia. The remaining eight families contain both oviparous and viviparous genera, and in some cases, genera that contain both oviparous and viviparous species—for example, the viviparous Smooth Snake (*Coronella austriaca*, page 155) and oviparous Southern Smooth Snake (*C. girondica*).

There are both advantages and disadvantages to live-bearing for snakes. Among the disadvantages is the fact that the gravid female must carry her offspring for the entire gestation period, which could be three months, during which time she is unlikely to feed, and if she is killed her entire reproductive investment is lost—she literally has all her "eggs" in one basket. In the tropics, it is probably advantageous for the female to lay her eggs so that she can begin hunting again, but in arctic–alpine climates live-bearing is a much better strategy. While eggs are vulnerable to cold temperatures, a live-bearing female can shelter below ground during the night or cold periods and then emerge to bask in the sun. When locomotion is combined with melanistic (black) coloration, the female snake becomes an efficient mobile incubator, seeking out the best basking spots.

Live-bearing is also a useful adaptation for aquatic snakes—the Colubrine Sea Krait (*Laticauda colubrina*, page 514), for example, is an oviparous marine species that must come onto land to lay its eggs, whereas viviparous true seasnakes simply give birth in the ocean and are not tied to land. Similarly, oviparous pythons must lay their eggs on land, while aquatic anacondas give birth directly into the shallow water.

ABOVE: **A hatchling Green Tree Python** (*Morelia viridis*) uses an egg-tooth on the tip of its upper lip to "pip" its egg. It will breath air for the first time but may wait some time before fully emerging. At approximately 15 months of age it will take on the green adult livery.

EGG-LAYING

Snake eggs are oval, with opaque white leathery shells. Some snakes lay their eggs soon after mating, when embryonic development has only just begun, while other species may retain the eggs until halfway through or near the end of the incubation period. The Persian False Horned Viper (*Pseudocerastes persicus*, page 636) has been recorded laying eggs that hatched 30–32 days later, half the usual 60–70-day incubation period, while the Sahara Sand Viper (*Cerastes vipera*, page 621) lays eggs only days before they are due to hatch. Most eggs are deposited when the embryo is at the 30 percent development stage.

Snakes may lay their eggs in a rocky crevice or an animal burrow, and leave them to incubate on their own. Western Grass Snakes (*Natrix helvetica*, page 415) often lay their eggs in garden compost heaps, sometimes communally, where the eggs benefit from heat generated by the decomposing vegetation. A few snake species remain with their eggs post-oviposition. Female pythons coil around their eggs, both to incubate and to protect them. Incubation is accomplished via a process of endogenous heat production known as shivering thermogenesis, which involves a prolonged period of rhythmic muscular contractions. Brooding Diamond Pythons (*Morelia spilota*, page 93) have been recorded shivering up to 50 times per minute throughout the two-month incubation period. Shivering thermogenesis will maintain an incubation temperature of 88–91 °F (31–33 °C), as much as 13–23 °F (7–13 °C) above the ambient air temperature. The pythons do not feed throughout this entire period.

A female King Cobra uses her body coils to build a nest of leaves, laying her eggs in the center. She will then remain at the nest site to protect the eggs from predators, such as the Water Monitor Lizard (*Varanus salvator*).

At the end of the incubation period, the young snakes hatch by making slits in the leathery shell of the egg. They use an egg-tooth on the front of the upper lip to achieve this, a process known as "pipping," although they may not actually emerge for some hours. Once out of the egg, the hatchling will shed its skin and become independent. Depending on the species and the size of the female, the clutch may vary from a single egg to more than 80 in the case of large pythons.

LEFT: **A female Copperhead** (*Agkistrodon contortrix*) giving birth to a litter of neonates which are born in membraneous packages from which they almost immediately escape to take their first breaths.

LIVE-BEARING

Just over 20 percent of snakes are live-bearers, a process that has evolved at least 35 times within the Serpentes. Two different forms of live-bearing have been defined—viviparity and ovoviviparity—but most recent authors do not differentiate between them. Boas, rattlesnakes, most vipers, American natricid watersnakes, true seasnakes, and the Red-bellied Blacksnake (*Pseudechis porphyriacus*, page 527) are all viviparous. Neonates are born coiled in a transparent egg membrane that ruptures soon after birth. Litter sizes for viviparous snakes vary depending on the species and the size of the female, from one to two in small species, to more than 100 in the Puff Adder (*Bitis arietans*, page 610).

VIRGIN BIRTH

Parthenogenesis is a form of reproduction whereby a female produces exact clones of herself without the need of a male. There are numerous parthenogenetic lizards, but only one truly obligate parthenogenetic snake, the Brahminy Blindsnake (*Indotyphlops braminus*, page 57), for which no males are known. There are also cases of females from normally bisexual species—including pythons, boas, filesnakes, gartersnakes, and pitvipers—producing offspring without first mating with a male. This is termed facultative parthenogenesis, and is the last-ditch attempt of a female to propagate when no males are available. The result is usually a small clutch or litter with a high mortality rate, and the offspring are always unisexual— all female in boas and pythons, and all male in rattlesnakes.

BELOW: **A two-headed Red-backed Ratsnake** (*Oocatochus rufodorsatus*). Two-headed or dicephalic snakes are effectively conjoined twins with two heads at the same end. Although most specimens are stillborn, or die soon after birth due to other defects, there have been many cases of dicephalic snakes being raised to adulthood.

SNAKES & HUMANKIND

There are few human cultures, through history or across the world, where humans have not, or do not, live alongside snakes. Probably no other animal group has had such an effect on human culture than the snake, which has become the symbol of life and longevity, and of sudden death.

SNAKES IN RELIGION

From the headdresses worn by ancient Egyptian pharaohs 3,000 years ago, to twentieth-century Appalachian snake handlers who follow the Gospel of St. Mark's "They shall take up serpents" literally, snakes have exerted a powerful grip on the human psyche, representing both good and evil.

ABOVE: **The cobra sheltered Buddha** from the rain by spreading its hood over his head. A grateful Buddha laid his two fingers (*Naja naja*) or his thumb (*N. kaouthia*) on the cobra's hood and left his mark.

In Judeo-Christian tradition, the serpent tempted Eve to pick the apple in the Garden of Eden, and for this act God condemned it to crawl on its belly and eat dust for the rest of its days. The snake reappears in the Bible as Aaron's rod, which swallows the rods of the Egyptian pharaoh's wise men, and as the staff of Moses, which parts the Red Sea during the Exodus. Aaron's "rod" was most likely an Egyptian Cobra (*Naja haje*, page 469), since cobras have an appetite for other snakes.

Cleopatra reputedly committed suicide by allowing herself to be bitten by an "asp," although the serpent was probably the Egyptian Cobra again. A queen such as she would have wished for a swift and painless death, and to remain looking beautiful after her passing, and the cobra is more likely to deliver on those wishes than what we now call an asp, a viper, or a side-stabbing snake (*Atractaspis*).

A cobra is said to have spread its hood to shelter Buddha from the rain, and as a sign of his thanks Buddha placed a mark on the hood. For Sri

Lankans, he placed two fingers and left the spectacle mark on the Indian Cobra (*Naja naja*, page 473), whereas for Thais he used his thumb, leaving the monoculate mark of the Thai Cobra (*N. kaouthia*).

SNAKES IN CULTURE

Snake symbolism is everywhere. The Hopi Indians of Arizona perform a rain dance with snakes gripped in their teeth, the serpents being seen as the guardians of water. Young Venda girls in South Africa perform the domba dance during their initiation into womanhood, when they mimic the movements of a large python. In Abruzzo, Italy, an annual snake festival sees the statue of St. Domenico being paraded through Cocullo draped with dozens of harmless Aesculapian snakes (*Zamenis longissimus*, page 251), while visitors to the Snake Temple on the Malaysian island of Penang marvel at hundreds of venomous Wagler's Temple Pitvipers (*Tropidolaemus wagleri*, page 603) lying languidly across the icons.

33

ABOVE: **The Egyptian Cobra** (*Naja haje*) is represented on the headdress of the Egyptian pharaohs.

Snakes appear in the art of many ancient cultures, from the Rainbow Serpent cave paintings and petroglyphs of Australian Aboriginal Dreamtime, to the Feathered Serpent, or Quetzalcoatl, of pre-Columbian Mayan and Aztec societies. At the ancient temple at Polonnaruwa, Sri Lanka, there is a huge viper carved into the rocks. Also at the site is an 800-year-old stone Ayurvedic "medicine boat," in which a dying snakebite victim would be placed, to be anointed with oils and herbs in the hope that he or she would survive.

Snakes feature in modern medicine, too. The Rod of Asclepius, the Greek god of healing, comprising a single Aesculapian snake curled around a staff, is used widely as the symbol for medicine. In the United States, it is sometimes exchanged for a caduceus, the symbol of the messenger god Hermes, which features two snakes coiled about a winged staff.

Today, even in our hectic commercial world, the representations of snakes are still all around us, as team mascots or in names of products as diverse as beer, candies, cement, condoms, and cars.

SNAKEBITE

Around the world, between 94,000 and 125,000 people die annually from snakebites. Most snakebite victims are poor rural farmers or children in the developing world. The countries with the highest incidences of snakebite deaths include India, Sri Lanka, Nepal, Myanmar, Nigeria, Mali, Togo, Benin, Senegal, and Papua New Guinea. No data exist for Indonesia, but death rates there are also assumed to be high.

Rather than going to hospital, victims will often visit a local shaman or medicine man in the vain hope that he can save them. Even for those people who survive a snakebite, the prognosis may not be good. Some snake venoms cause massive tissue destruction, leading to limb deformity or loss—each year, up to 400,000 snakebite victims may be disabled in this way. Snakebites are terrible to endure, but they are not without a cure.

Modern antivenoms are produced from the antibodies formed when horses and sheep are injected with increasing doses of snake venom, and are very effective for saving life and reducing the damage done by snake venoms, provided the victim gets to hospital quickly. Unfortunately, however, some Western drug companies are stopping the production of antivenoms, as they are less profitable than drugs for diseases like obesity, cancer, and heart disease. The world—and Africa in particular—may consequently be entering an antivenom crisis.

SNAKE VENOMS

Snake venoms are complex cocktails of different proteinaceous toxins that are designed to target prey. With the exception of spitting cobras, which spray jets of venom into the eyes of a perceived enemy and then effect an escape, snake venoms are not purposefully defensive. There are several different venom types, which are summarized below:

Neurotoxins

These paralyze the nervous system, preventing the passage of messages along nerves and leading to death through respiratory paralysis. There are two types: presynaptic neurotoxins, which destroy the transmitter sites on the "upstream" side of the synaptic gap; and post-synaptic neurotoxins, which block the receptor sites on the "downstream" side of the gap. These venoms are primarily found in the elapids—cobras (*Naja*), mambas (*Dendroaspis*), and taipan (*Oxyuranus*)—but also in some rattlesnakes, including the Mohave Rattlesnake (*Crotalus scutulatus*, page 580).

34

35

Hemotoxins

These toxins affect the blood and circulatory system. Anticoagulants cause prolonged bleeding by preventing blood coagulation, while procoagulants do the same by using up all the clotting factor in blood. Platelet inhibitors prevent normal blood clotting, and when combined with hemorrhagins (which puncture holes in the blood vessels), the result can be massive blood loss. Hemolytic toxins break down the red blood cells, causing blockage of the kidney tubules, leading to renal failure. Many vipers produce hemotoxins, but they are also found in some elapids such as taipans.

Myotoxins

These toxins affect the muscles, acting like neurotoxins by causing paralysis, or like hemotoxins by breaking down muscle tissue. They are primarily found in seasnakes.

Cytotoxins

These toxins digest protein, leading to massive tissue destruction. They are found in large vipers that need to digest bulky mammalian prey, and in the venoms of spitting cobras.

Other toxins

Sarafotoxins are cardiotoxins that cause a narrowing of the cardiac arteries; they are found in the venoms of the side-stabbing snakes (*Atractaspis*). The St. Lucia Lancehead (*Bothrops caribbaeus*, page 566) produces a cardiotoxin that causes arterial thrombosis, while the venom of the Gwardar (*Pseudonaja mengdeni*, page 528) contains a nephrotoxin that directly attacks the kidneys. Snake venoms are highly complex compounds.

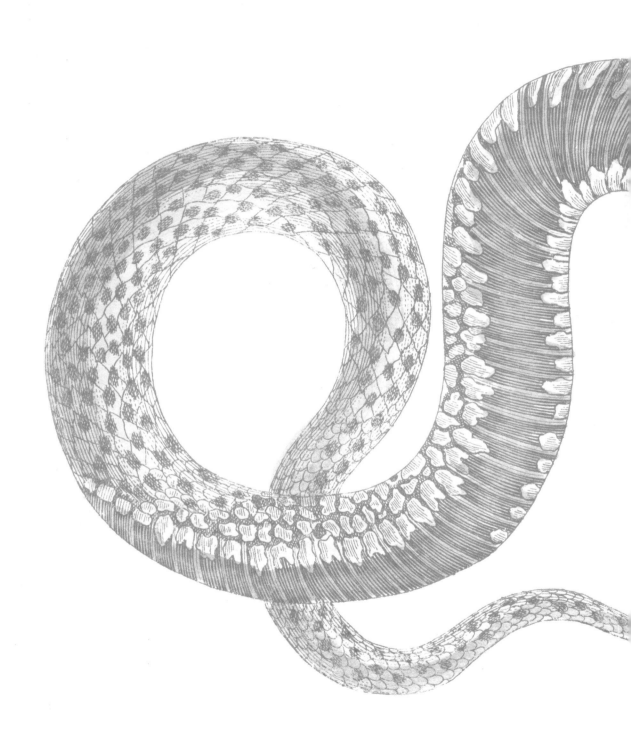

THE SNAKES

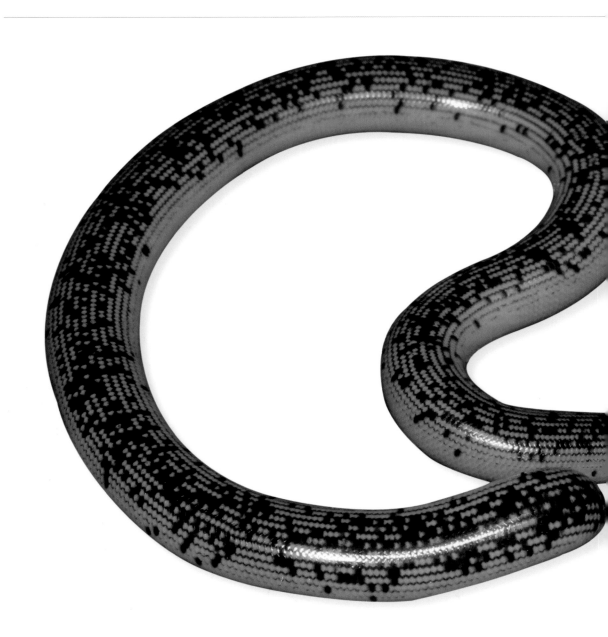

SCOLECOPHIDIA
The basal modern snakes

The Scolecophidia (*Scolec* = worm; *-ophidia* = snakes) are known as wormsnakes, blindsnakes, and threadsnakes. There are over 450 species, in five families: 12.3 percent of all living snakes. They are small and slender, with highly glossed, tight-fitting scales, yet a few achieve almost 3 ft 3 in (1 m). Fossorial in habit, blindsnakes are not actually blind. Their pigmented eyespots, under translucent scales, register sunlight and warn them to burrow if exposed. Scolecophidians are highly adapted subterranean snakes.

The Anomalepididae (dawn or early blindsnakes) are South American, 18 species in four genera, the most basal of all living snakes. They possess teeth on both the maxillary and dentary bones.

The largest family, the Typhlopidae (blindsnakes), with over 270 species, inhabits the tropics and subtropics. They only bear teeth on the maxillary bones. Two families were recently removed from the Typhlopidae: the Gerrhopilidae (plaid-headed blindsnakes), 21 species from India to New Guinea with subdermal glands on the head scales, and the Xenotyphlopidae, a monotypic family from Madagascar.

The Leptotyphlopidae (threadsnakes or slender blindsnakes) contain over 140 species from the Americas, Africa, and Asia. They differ from typhlopids in only possessing teeth on the dentary bone.

FAMILY	Anomalepididae
RISK FACTOR	Nonvenomous
DISTRIBUTION	South America: southern Brazil, southeast Paraguay, and northeastern Argentina
ELEVATION	510–3,000 ft (155–915 m) asl
HABITAT	Secondary Atlantic coastal forest, but also urbanized cities
DIET	Ant larvae and pupae (but not ant eggs), and occasionally termites
REPRODUCTION	Oviparous, with clutches of 2–24 eggs
CONSERVATION STATUS	IUCN Least Concern

ADULT LENGTH
4–15 in
(106–381 mm)

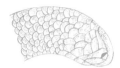

LIOTYPHLOPS BEUI
BEU'S DAWN BLINDSNAKE
(AMARAL, 1924)

Beu's Dawn Blindsnake is a smooth-scaled, glossy black, gray, or brown snake with a distinctive pale pink to yellow head and dorsal neck, and a pale patch over the cloaca. This patterning is common to many species of *Liotyphlops*. Species determination requires close examination of the head scalation. *Liotyphlops* species possess an enlarged rostral scale that extends dorsally to the level of the eyes.

The dawn blindsnakes, or early blindsnakes, comprise four genera and 18 species of neotropical blindsnakes. As the sister taxa to all other scolecophidian snakes, they represent the earliest divergence within extant snakes. Beu's Dawn Blindsnake inhabits southern Brazil, southeastern Paraguay, and northeastern Argentina, an area with considerable anomalepidid diversity. Originally a forest species, it is very common in highly urbanized São Paulo city. Fossorial in habit, it feeds almost exclusively on the larvae and pupae of small ants, especially aggressive fire ants, also occasionally eating termites, but never ant eggs. This species is named for Afrânio do Amaral's colleague T. Beu, who collected the holotype, which was destroyed, along with many other irreplaceable specimens, in the Instituto Butantan fire, on May 15th, 2010.

RELATED SPECIES

The closest relatives of *Liotyphlops beui* are probably Ternetz's Dawn Blindsnake (*L. ternetzii*) and the São Paulo Dawn Blindsnake (*L. schubarti*). All three species are found in southern Brazil, with Ternetz's Dawn Blindsnake also occurring in Paraguay and Argentina. At one time *L. beui* was included in the synonymy of *L. ternetzii*. Two other *Liotyphlops* species occur in southern Brazil, one in northeastern Brazil, and four in Colombia and neighboring countries.

Actual size

FAMILY	Anomalepididae
RISK FACTOR	Nonvenomous
DISTRIBUTION	Northeastern South America: eastern Venezuela, Guyana, Suriname, French Guiana, and northern Brazil
ELEVATION	0–2,130 ft (0–650 m) asl
HABITAT	Lowland rainforest
DIET	Ant and termite larvae and eggs, amphisbaenians (worm-lizards), legless lizards, and earthworms
REPRODUCTION	Oviparous, with clutches of 2–6 eggs
CONSERVATION STATUS	IUCN not listed

ADULT LENGTH
6–8 in,
occasionally 9 in
(150–200 mm,
occasionally 225 mm)

TYPHLOPHIS SQUAMOSUS
SMALL-SCALED DAWN BLINDSNAKE
(SCHLEGEL, 1839)

41

The Small-scaled Dawn Blindsnake is a common species in the Atlantic coastal rainforest of northeastern South America. It inhabits loose soil, ant or termite mounds, the rotten trunks of dead palms to 3 ft 3 in (1 m) above the ground, and even termite-infested man-made wooden pallets. It may also be found moving on the surface after heavy rain. It feeds on termite and ant larvae and eggs, but the Brazilian herpetologist Afrânio do Amaral claimed it also takes worm-lizards (amphisbaenians), legless lizards, and earthworms. When handled it will squirm vigorously, jabbing repeatedly with the terminal spine of its tail in an attempt to gain a purchase to escape. It can also retract its tail slightly, like an earthworm.

RELATED SPECIES

The genus *Typhlophis* is monotypic, but *T. squamosus* could be confused with the widely distributed White-nosed Dawn Blindsnake (*Liotyphlops albirostris*), Beu's Dawn Blindsnake (*L. beui*, page 40) from São Paulo, or the Reticulate Blindsnake (*Amerotyphlops reticulatus*, page 61) from Amazonia. However, the head of this species is covered in numerous small, undifferentiated scales, hence *squamosus*, in contrast to the heads of other blindsnakes.

Actual size

The Small-scaled Dawn Blindsnake is a shiny, smooth-scaled snake, with a black to dark brown dorsum, a uniform white venter, and a distinctive white or pink head. Some specimens have white spots on the short, curved tail.

FAMILY	Leptotyphlopidae: Epictinae
RISK FACTOR	Nonvenomous
DISTRIBUTION	Caribbean: Hispaniola (Haiti)
ELEVATION	1,200 ft (365 m) asl
HABITAT	Limestone hills, under pebbles or shade trees
DIET	Presumed termite and/or ant larvae and eggs
REPRODUCTION	Oviparous, clutch size unknown
CONSERVATION STATUS	IUCN not listed

ADULT LENGTH
7–8 in
(180–205 mm)

42

MITOPHIS LEPTIPILEPTUS
HAITIAN BORDER THREADSNAKE
(THOMAS, MCDIARMID & THOMPSON, 1985)

Actual size

The Haitian Border Threadsnake is a uniform silvery gray, although the head and neck may be a much paler gray color, and darker blotches may be present on the bodies of some specimens. There is also iridescence to the scales, a common feature among glossy-scaled leptotyphlopids.

All threadsnakes are slender, but this species from the southeastern Haitian border with the Dominican Republic is slender and elongate even by threadsnake standards, as evidenced by its name "*leptipileptus*," which translates as "thin on thin." Little is known of its natural history, although some specimens were collected from piles of stream-worn pebbles, under shady mango trees, and under rocks in a shaded ravine, all in an area dominated by limestone hills and a mosaic of cultivation and scrubland. An unusual feature of this species is the lack of any pelvic vestiges, whereas most other threadsnakes do possess the remnants of small pelvic bones. Nothing is known of its diet, although it is assumed to be similar to other threadsnakes and to feed on termite or ant eggs and larvae.

RELATED SPECIES

The genus *Mitophis* contains only four species, all from Hispaniola. Thomas' Threadsnake (*M. pyrites*) occurs in the south of Haiti and the Dominican Republic, whereas the other two species are from the Dominican Republic. The Sooty Threadsnake (*M. asbolepis*) occurs on the Sierra Martín García in the south, whereas the Samana Threadsnake (*M. calypso*) is from the Samana Peninsula in the north. All three species can be distinguished from the extra slender *M. leptepileptus* by the presence of four rather than three supralabial scales, while *M. pyrites* is also more stoutly built than the other species.

FAMILY	Leptotyphlopidae: Epictinae
RISK FACTOR	Nonvenomous
DISTRIBUTION	North America: USA and Mexico
ELEVATION	33–6,890 ft (10–2,100 m) asl
HABITAT	Arid grassland and semidesert, or oak–juniper woodland
DIET	Termites and ants, and other soft-bodied invertebrates
REPRODUCTION	Oviparous, with clutches of 1–8 eggs
CONSERVATION STATUS	IUCN Least Concern

ADULT LENGTH
5–11¾ in,
rarely 15 in
(130–300 mm,
rarely 380 mm)

RENA DULCIS
TEXAS THREADSNAKE
BAIRD & GIRARD, 1853

43

The Texas Threadsnake inhabits prairie grassland, oak–juniper woodland, and desert edges with yucca, cacti, or thornbush from Texas to Hidalgo and Veracruz. It lives under rocks or logs that retain moisture and occurs in towns and cities where soils contain sand or loam. It is only seen on the surface on cool, wet nights. It has been found in the nests of the Eastern screech owl, possibly having passed through the owl's gut unharmed. Prey consists of ants, termites, grasshopper nymphs, spiders, and solifuges. Specimens writhing in ant nests are not being stung but covering themselves in a pheromone to stop the ants attacking them. Males also follow female pheromonal trails and often form "mating balls" comprising several males and one female.

RELATED SPECIES
The leptotyphlopid genus *Rena* contains nine species throughout southwestern USA and Mexico, northern South America, and northern Argentina. The species closest to *R. dulcis* are probably the other two North American species: the New Mexico Threadsnake (*R. dissectus*), a former subspecies of *R. dulcis*, and the Western Threadsnake (*R. humilis*). *Rena dulcis* and *R. dissectus* possess supraocular scales but these are absent in *R. humilis*. *Rena dulcis* has a distinct, large anterior supralabial scale, while this is divided into two smaller scales in *R. dissectus*, hence its name.

The Texas Threadsnake is an unremarkable uniform red-brown throughout, without additional markings. Its scales are small and close-fitting and its eyes are reduced to pigmented areas under large, translucent head scales.

Actual size

FAMILY	Leptotyphlopidae: Epictinae
RISK FACTOR	Nonvenomous
DISTRIBUTION	West Africa: Senegal, The Gambia, southern Mali, Guinea-Bissau, and northern Guinea
ELEVATION	82–1,480 ft (25–450 m) asl
HABITAT	Paddy fields, gardens, streams, and ponds
DIET	Ant and termite larvae and eggs
REPRODUCTION	Oviparous, with clutches of 5–15 eggs
CONSERVATION STATUS	IUCN Least Concern

ADULT LENGTH
9½–18 in
(240–460 mm)

RHINOLEPTUS KONIAGUI
KONIAGUI THREADSNAKE
(VILLIERS, 1956)

44

The Koniagui Threadsnake is a shiny, smooth-scaled snake, entirely brown in color with yellow-orange iridescent hues. It has a sharply pointed snout, which protrudes well beyond its undercut lower jaw.

The Koniagui Threadsnake is one of the largest threadsnakes known. Its size was probably why it was originally included in the Typhlopidae, the family containing the generally larger, stouter blindsnakes. It is a very common species within its West African range, which encompasses the arid Sudanese climatic zone (the Koniagui are an ethnic Senegalese people). It is often uncovered during plowing or digging, and is found on the surface on rainy nights. The generic name *Rhinoleptus* means "slender-nosed" and is a reference to the narrow, pointed tip to the snout of this threadsnake. This species preys on small, soft-bodied invertebrates, such as ant or termite larvae and eggs.

RELATED SPECIES
Although *Rhinoleptus* is generally considered a monotypic genus, some authors include a second species, Parker's Threadsnake (*R. parkeri*) from the Ogaden of Ethiopia. Other authors include this species in the genus *Myriopholis*.

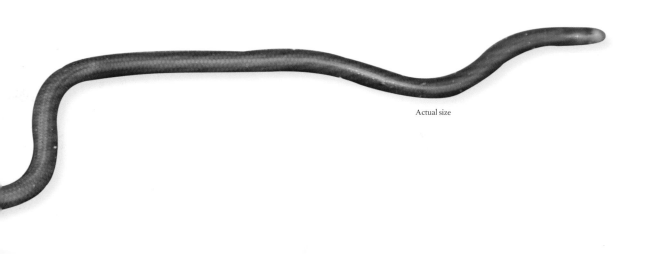

Actual size

FAMILY	Leptotyphlopidae: Epictinae
RISK FACTOR	Nonvenomous
DISTRIBUTION	Northern South America: southern Venezuela, the Guianas, and northern Brazil
ELEVATION	330–820 ft (100–250 m) asl
HABITAT	Rainforest soil and termite mounds
DIET	Termite larvae and eggs
REPRODUCTION	Oviparous, clutch size unknown
CONSERVATION STATUS	IUCN not listed

ADULT LENGTH
8–9¾ in
(200–250 mm)

SIAGONODON SEPTEMSTRIATUS
SEVEN-STRIPED THREADSNAKE
(SCHNEIDER, 1801)

45

The Seven-striped Threadsnake occurs from southern Venezuela to Guyana, Suriname, and French Guiana, and through the northern Brazilian states of Pará, Amazonas, and Roraima. It is a forest-floor-dwelling species, closely associated with termite mounds, inside which it both feeds on termite larvae and eggs, and lays its eggs. It also inhabits the rootstock of *Astrocaryum* palms. Although this species occurs in sympatry with several other threadsnakes, it is the only one within its range to be boldly striped. In common with many fossorial threadsnakes, it leads a secretive, nocturnal lifestyle, and is only encountered on the surface after heavy rain.

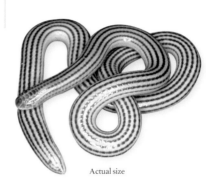

Actual size

The Seven-striped Threadsnake is yellow-brown above and paler yellow below, the markings consisting of seven dark brown longitudinal stripes along the body.

RELATED SPECIES
Siagonodon septemstriatus is closely related to three other threadsnake species: Ole Borch's Threadsnake (*S. borrichianus*) from northwestern Argentina; the Termite Threadsnake (*S. cupinensis*) from Suriname and the states of Amapa and Mato Grosso in Brazil; and the recently described Sharp-nosed Threadsnake (*S. acutirostris*) from Tocantins, Brazil.

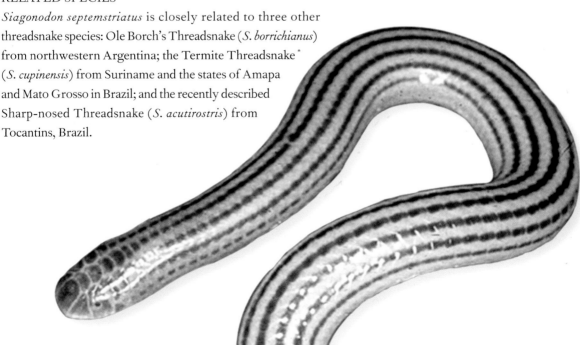

FAMILY	Leptotyphlopidae: Epictinae
RISK FACTOR	Nonvenomous
DISTRIBUTION	Lesser Antilles: Barbados
ELEVATION	330–920 ft (100–280 m) asl
HABITAT	Forest soil
DIET	Termites and ants; larvae, pupae, and eggs
REPRODUCTION	Oviparous, clutch size unknown
CONSERVATION STATUS	IUCN not listed

ADULT LENGTH
4–4⅛ in
(101–104 mm)

46

TETRACHEILOSTOMA CARLAE
BARBADOS THREADSNAKE
(HEDGES, 2008)

Actual size

The Barbados Threadsnake is the world's smallest snake species. It is known from only three specimens collected in 1889, 1963, and 2006, and was not formally described until 2008, when it was named for the describer's wife. Threadsnakes feed on the larvae, pupae, and eggs of ants and termites, and given that Barbados was once forested it can be assumed this is a fossorial forest-dwelling species. Barbados is one of the ten most densely populated countries in the world and most of the original forest has been cleared. Even secondary forest is limited to small plots, and the arrival of the parthenogenetic, colonizing Brahminy Blindsnake (*Indotyphlops braminus*, page 57) could threaten the survival of the tiny Barbados Threadsnake.

RELATED SPECIES

The smallest snake was once thought to be the Martinique Threadsnake (*Tetracheilostoma bilineatum*). The St. Lucia Threadsnake (*T. breuili*) is another diminutive species, described in the same paper as *T. carlae*.

The Barbados Threadsnake is dark brown to black with a pair of pale yellow-gray dorsolateral stripes, between which is a vertebral area of red-brown, while the undersides are grayish brown. There are whitish spots on the head and around the cloaca. The eyes are reduced to pigmented areas beneath large, translucent ocular scales.

FAMILY	Leptotyphlopidae: Leptotyphlopinae
RISK FACTOR	Nonvenomous
DISTRIBUTION	Southeastern Africa: southern Zambia, southern Malawi, eastern Zimbabwe, Mozambique, northeastern South Africa, and Swaziland
ELEVATION	655–5,250 ft (200–1,600 m) asl
HABITAT	Mesic savanna grasslands and termitaria
DIET	Termites
REPRODUCTION	Oviparous, with clutches of 3 eggs
CONSERVATION STATUS	IUCN Least Concern

ADULT LENGTH
6–7½ in
(150–193 mm)

LEPTOTYPHLOPS INCOGNITUS
INCOGNITO THREADSNAKE
BROADLEY & WATSON, 1976

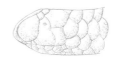

47

The Incognito Threadsnake occurs from southern Zambia to Swaziland and northeastern South Africa at low to medium elevations. Like other threadsnakes it is usually found sheltering under rocks or rotten logs, or inside termitaria, in savanna grassland habitats, although it may be forced onto the surface after rain at night. It feeds on small soft-bodied invertebrates, especially termites, and their larvae and eggs. Females may produce clutches of three eggs, which are extremely elongate, being five times as long as wide and often linked together like a string of small, pale sausages. These small snakes have many enemies, from ophiophagous snakes to meerkats and other small carnivorous mammals, birds, and scorpions. The epithet *incognitus* is a reference to how this widely distributed species remained unrecognized for so long.

RELATED SPECIES
One of 23 species remaining in the genus *Leptotyphlops*, following its revision and splitting into numerous genera, *L. incognitus* is a member of the southeastern African *L. scutifrons* species complex. This complex also includes Peters' Threadsnake (*L. scutifrons*), the Eastern Cape Threadsnake (*L. conjunctus*), the Pungwe Threadsnake (*L. pungwensis*), the Black-tipped Threadsnake (*L. nigroterminus*), and the Forest Threadsnake (*L. sylvicolus*).

Actual size

The Incognito Threadsnake is a slender black snake with smooth scales that present a very highly polished appearance. Its eyes are reduced to pigment patches, visible under translucent head scales.

FAMILY	Leptotyphlopidae: Leptotyphlopinae
RISK FACTOR	Nonvenomous
DISTRIBUTION	North Africa and western Asia: Egypt, Sudan, Somalia, and Kenya; Saudi Arabia, UAE, Yemen, Iran, Afghanistan, and Pakistan
ELEVATION	0–2,950 ft (0–900 m) asl
HABITAT	Soft, loose soil of agricultural fields, deciduous woodland, and thickets
DIET	Ants and their larvae, and possibly termites and other soft-bodied invertebrates
REPRODUCTION	Oviparous, clutch size unknown
CONSERVATION STATUS	IUCN not listed

ADULT LENGTH
8⅞–9 in
(225–229 mm)

48

MYRIOPHOLIS MACRORHYNCHA
HOOK-NOSED THREADSNAKE
(JAN, 1860)

This is one of the most widely distributed threadsnake species, although it is now thought that its true distribution may be limited to Africa, with Asian populations representing related, but currently undescribed, species. A typical leptotyphlopid, the Hook-nosed Threadsnake inhabits the soft soil of agricultural fields, but it is also found in arid deciduous woodland habitats, and it is thought to occur at higher elevations in grassland on volcanic soils in East Africa. In the Arabian Peninsula it is known to feed on ants and their larvae, but its diet in other parts of its range are undocumented, although termites and soft-bodied insect larvae are likely prey.

Actual size

The Hook-nosed Threadsnake appears in two forms. African specimens are uniform brown while many Asian specimens are translucent pink, so lacking in external pigment that their internal organs are visible. This is an extremely slender species with an elongate head that terminates in a downturned hook-nose.

RELATED SPECIES

The genus *Myriopholis* contains 21 species, but this number will increase should the situation with the *M. macrorhyncha* species complex be resolved. In the Arabian Peninsula it may be confused with the equally translucent Nurse's Threadsnake (*Leptotyphlops nursii*), named for a Lieutenant Colonel Nurse, an amateur entomologist in the Indian Army, rather than the profession. The most recently described species, in 2007, was Ionides' Threadsnake (*M. ionidesi*), named in honor of the famous Anglo-Greek snake man C. J. P. Ionides, who made Tanzania his home and collected the holotype.

FAMILY	Leptotyphlopidae: Leptotyphlopinae
RISK FACTOR	Nonvenomous
DISTRIBUTION	Southwest Africa: Angola and Namibia
ELEVATION	785–7,280 ft (240–2,220 m) asl
HABITAT	Namib Desert, Damaraland scrub and karoo semidesert
DIET	Not known, but presumed to be soft-bodied invertebrates, e.g. termites
REPRODUCTION	Presumed oviparous, clutch size unknown
CONSERVATION STATUS	IUCN not listed

ADULT LENGTH
6¾–11¾ in
(170–300 mm)

NAMIBIANA LABIALIS
DAMARA THREADSNAKE
(STERNFELD, 1908)

The Damara Threadsnake is named for the Damara people who inhabit an area once known as Damaraland in northern Namibia, although the Damara Threadsnake also occurs farther south into the Namib Desert, and north into Angola. It is adapted to live in an extremely arid, sandy, or rocky region where the only moisture available may be from sea fogs that move inland from the coast. As with many of the localized threadsnakes of Africa, the precise natural history of this species is relatively undocumented, but it is assumed to feed on soft-bodied invertebrates and lay eggs like other members of the Leptotyphlopidae.

Actual size

RELATED SPECIES

Namibiana labialis is one of five species of southwest African threadsnakes in the genus *Namibiana* (formerly included in *Leptotyphlops*), the others being Bocage's Threadsnake (*N. rostrata*) from western Angola, the Benguela Threadsnake (*N. latifrons*) from coastal southwest Angola, and both the Western Threadsnake (*N. occidentalis*) and Slender Threadsnake (*N. gracilior*) from Namibia and South Africa.

The Damara Threadsnake is a two-tone threadsnake, being gray-brown above, each scale edged with paler pigment, and paler also on the undersides. Its scales are smooth and shiny, and its eyes are minuscule areas of pigment under translucent head scales.

FAMILY	Gerrhopilidae
RISK FACTOR	Nonvenomous
DISTRIBUTION	Papua New Guinea: Milne Bay, Normanby Island
ELEVATION	2,035 ft (620 m) asl
HABITAT	Primary lowland rainforest
DIET	Presumed to comprise soft-bodied invertebrates, such as termites
REPRODUCTION	Presumed oviparous, clutch size unknown
CONSERVATION STATUS	IUCN not listed

ADULT LENGTH
10 in
(255 mm)

GERRHOPILUS PERSEPHONE
NORMANBY ISLAND BEAKED BLINDSNAKE
KRAUS, 2017

The Normanby Island Beaked Blindsnake is endemic to Normanby Island, in the d'Entercasteaux Archipelago in Milne Bay Province, southeast Papua New Guinea. Described in 2017, it is only known from a single specimen, which was collected climbing the lower trunk of a tree in primary lowland rainforest. The trees in this habitat are often home to the nests of arboreal termites and it is possible the snake was searching for such a nest. The reproductive strategy of this species is unknown but suspected to be oviparity, the normal process for scolecophidians. As a group the gerrhopilid blindsnakes are known as "plaid-headed blindsnakes" as their head scales exhibit a wickerwork appearance due to the presence of papillae-like sebaceous glands on every scale.

RELATED SPECIES

The genus *Gerrhopilus* contains 20 currently recognized species distributed across a wide geographical area: India (three), Sri Lanka (three), Andaman Islands (one), Thailand (one), Philippines (one), Java (one), Moluccas and West New Guinea (one), and Papua New Guinea (nine). The only other member of the *Gerrhopilidae* is the monotypic *Cathetorhinus melanocephalus*, which is thought to have originated from Mauritius. Other species from Milne Bay include the Panaeate Island Blindsnake (*G. addisoni*), Trobriand Island Blindsnake (*G. eurydice*), Rossel Island Blindsnake (*G. hades*), and the Plain Montane Blindsnake (*G. inornatus*) from the mainland.

Actual size

The Normanby Island Beaked Blindsnake is a slender snake, which is pale blue throughout, except for the head and neck, which is pinkish. It is unusual for a blindsnake to be anything other than dark brown, gray, or black. The head is rounded, with a downward sloping beak, distinct pale-gray eyes, and scales bearing numerous subcutaneous sebaceous glands.

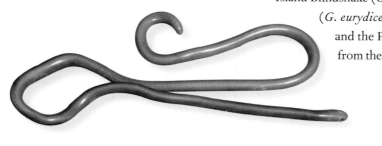

FAMILY	Typhlopidae: Afrotyphlopinae
RISK FACTOR	Nonvenomous
DISTRIBUTION	Sub-Saharan Africa: northeastern South Africa, Mozambique, Botswana, Namibia, and southern Angola
ELEVATION	0–3,850 ft (0–1,175 m) asl
HABITAT	Savanna, grassland, and coastal forest
DIET	Termites and their larvae
REPRODUCTION	Oviparous, with clutches of 8–60 eggs
CONSERVATION STATUS	IUCN not listed

ADULT LENGTH
23¾–35½ in
(600–900 mm)

AFROTYPHLOPS SCHLEGELII
SCHLEGEL'S GIANT BLINDSNAKE
(BIANCONI, 1847)

51

Schlegel's Giant Blindsnake, also known as Schlegel's Beaked Blindsnake, is probably the largest African blindsnake. The nominate subspecies (*Afrotyphlops s. schlegelii*) occurs in northeastern South Africa and southern Mozambique, while Peter's Giant Blindsnake (*A. s. petersii*) is found in Botswana, northern Namibia, and southern Angola. These large blindsnakes possess beaks that enable them to burrow into termitaria or compacted soil. The giant blindsnakes live much deeper underground than their smaller relatives, and they are only seen on the surface at night after a prolonged period of heavy rain. They feed on termites and their larvae and store fat reserves in the posterior half of the body. Large females may lay up to 60 eggs.

RELATED SPECIES
The genus *Afrotyphlops* contains 26 African species. The Zambesi Blindsnake (*A. mucruso*) was once a subspecies of *A. schlegelii*. The Somali Giant Blindsnake (*A. brevis*) and the Angolan Giant Blindsnake (*A. anomalus*) are also related species. Some authors place these four largest species in the genus *Megatyphlops*.

Actual size

Schlegel's Giant Blindsnake is a large snake with clearly visible, smooth scales that in some pale specimens appear almost as a mosaic pattern. The dorsum is usually yellow or yellow-brown with darker speckling or blotches, while the venter is uniform pale yellow. The head bears a downward-projecting beak and the tail has a terminal spine used to force the snake forward when burrowing. Large specimens may be 1 in (25 mm) wide.

FAMILY	Typhlopidae: Afrotyphlopinae
RISK FACTOR	Nonvenomous
DISTRIBUTION	South Asia: India
ELEVATION	0–985 ft (0–300 m) asl
HABITAT	Habitat preferences unknown, but range includes arid and wet habitats
DIET	Soft-bodied invertebrates and earthworms
REPRODUCTION	Oviparous, clutch size unknown
CONSERVATION STATUS	IUCN not listed

ADULT LENGTH
11¾–23¾ in
(300–600 mm)

GRYPOTYPHLOPS ACUTUS
INDIAN BEAKED BLINDSNAKE
(DUMÉRIL & BIBRON, 1844)

The Indian Beaked Blindsnake is a uniform shiny brown blindsnake, with a paler venter, a rounded head with a projecting beak, and distinct dark eyes.

Actual size

Sometimes also called the Beaked Wormsnake, this is the largest Asian blindsnake known. The Beaked Blindsnake is endemic to peninsular India, where it is found from the Ganges Valley in the north, through the dry Deccan Plateau and the wet Western Ghats, to the southern tip of India, although it is far rarer in the south than in the north. Its natural history and biology are poorly known. It is believed to feed on soft-bodied invertebrates such as termites and their larvae, and also earthworms, but although it is presumed to be oviparous, like other blindsnakes, its clutch size is unknown. The *Grypo* in the generic name means hooked, while *acutus* means sharply.

RELATED SPECIES

Grypotyphlops acutus is the only member of the typhlopid subfamily Afrotyphlopinae known to occur in Asia. Its relationships are therefore unknown, but it is thought to represent a link back to the Gondwanan origins of both Africa and India.

FAMILY	Typhlopidae: Afrotyphlopinae
RISK FACTOR	Nonvenomous
DISTRIBUTION	Southern Africa: South Africa, Swaziland, Lesotho, Mozambique, Zimbabwe, and Botswana
ELEVATION	0–5,380 ft (0–1,640 m) asl
HABITAT	Savanna, grassland, semidesert, and coastal fynbos
DIET	Termites and their larvae
REPRODUCTION	Oviparous, with clutches of 2–8 eggs
CONSERVATION STATUS	IUCN not listed

ADULT LENGTH
11¾–13¾ in
(300–350 mm)

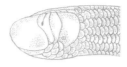

RHINOTYPHLOPS LALANDEI
DELALANDE'S BEAKED BLINDSNAKE
(SCHLEGEL, 1839)

53

Delelande's Beaked Blindsnake is widely distributed across southern Africa, from the Cape, north to Namaqualand, and northeast through Lesotho, Swaziland, Mpumalanga, and Limpopo, to eastern Botswana, western Mozambique, and Zimbabwe. It inhabits savanna, grasslands, semidesert, and coastal fynbos heathland, where it is most often discovered under rocks or logs during the day, or on the surface at night, after heavy rain. This species bears a distinct beak for excavating termite mounds, where it feeds on termites and their larvae, and lays up to eight eggs. Pierre Antoine Delalande (1787–1823) was a French naturalist based at the Muséum National d'Histoire Naturelle, Paris. He made collections in the Cape of South Africa.

RELATED SPECIES

The genus *Rhinotyphlops* contains seven species of small, beaked blindsnakes, three species in southern Africa, and four in East Africa. Some authors place the East African species in the genus *Letheobia*, a genus containing 18 other African blindsnakes. The other two southern African *Rhinotyphlops* are Boyle's Beaked Blindsnake (*R. boylei*), and Schinz's Beaked Blindsnake (*R. schinzi*), from Botswana, Namibia, and Namaqualand, South Africa.

Delalande's Beaked Blindsnake is smooth and shiny. The dorsum of the body is pink-brown with every scale edged with paler pigment to present a checkerboard effect. The undersides and lower flanks are pink and the snout and tail tips are paler still. On the pale, rounded head the dark eyes are very visible, while the tail bears a terminal spine.

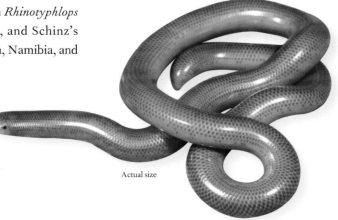

Actual size

FAMILY	Typhlopidae: Asiatyphlopinae
RISK FACTOR	Nonvenomous
DISTRIBUTION	Melanesia: Solomon Islands and Bougainville (Papua New Guinea)
ELEVATION	50–805 ft (15–245 m) asl
HABITAT	Rainforests
DIET	Earthworms
REPRODUCTION	Oviparous, clutch size unknown
CONSERVATION STATUS	IUCN not listed

ADULT LENGTH
9¾–14½ in
(250–366 mm)

ACUTOTYPHLOPS INFRALABIALIS
RED SHARP-NOSED BLINDSNAKE
(WAITE, 1918)

The Red Sharp-nosed Blindsnake occurs on Bougainville, (Papua New Guinea), and in the Solomon Islands, on Malaita, Guadalcanal, New Georgia, and the Nggela Islands. A relatively stocky blindsnake, it inhabits rainforest leaf litter and loose subsoil, and preys primarily on earthworms, rather than the termite and ant larvae and pupae that constitute the diet of smaller, more slender blindsnakes. The sharp snout of this species serves as a burrowing aid, whether in pursuit of its earthworm prey, or escaping desiccation or predation. Female Red Sharp-nosed Blindsnakes lay eggs, but their clutch size is unknown. The related Kunua Sharp-nosed Blindsnake (*A. kunuaensis*) lays single or pairs of eggs, and it is probable that the Red Sharp-nosed Blindsnake produces similar-sized clutches.

RELATED SPECIES

The genus *Acutotyphlops* contains four other species of sharp-nosed blindsnakes. The Kunua Sharp-nosed Blindsnake and Bougainville Sharp-nosed Blindsnake (*A. solomonis*) are endemic to the island of Bougainville, while the Bismarck Sharp-nosed Blindsnake (*A. subocularis*) is endemic to the Bismarck Archipelago, on the islands of New Britain, New Ireland, Duke of York and Umboi. One species (*A. banaorum*) occurs on Luzon in the Philippines.

The Red Sharp-nosed Blindsnake has shiny, smooth scales, a moderately stout body and a sharply pointed head, which is used for burrowing. It is pinkish red dorsally and pale yellow ventrally, with every dorsal scale pale-edged to present a reticulate appearance. The head is pale like the venter and the eyes are visible as two small dark eyespots.

Actual size

FAMILY	Typhlopidae: Asiatyphlopinae
RISK FACTOR	Nonvenomous
DISTRIBUTION	South and Southeast Asia: Pakistan, northeastern India, Nepal, Bangladesh, Myanmar, Thailand, Laos, Vietnam, and China
ELEVATION	460–5,000 ft (140–1,525 m) asl
HABITAT	Low and mid-montane forests
DIET	Termites and ants and their larvae, and earthworms
REPRODUCTION	Oviparous, with clutches of 4–14 eggs
CONSERVATION STATUS	IUCN Least Concern

ADULT LENGTH
13¾–17 in
(350–430 mm)

ARGYOPHIS DIARDI
DIARD'S BLINDSNAKE
(SCHLEGEL, 1839)

55

Diard's Blindsnake is a common inhabitant of low to mid-montane forest, such as exists in the Himalayan Terai of southern Nepal. Fossorial, nocturnal, and only encountered on the surface after rain, it may be found during the day sheltering under rotten logs or large stones. Being stout-bodied with a broader head than many other blindsnakes, Diard's Blindsnake can probably ingest larger prey than some of its smaller congeners, taking not only termites and ants, their larvae and pupae, but also earthworms and potentially other soft-bodied invertebrates. Pierre-Medard Diard (1794–1863) was a French explorer and naturalist who collected in Southeast Asia.

RELATED SPECIES

The genus *Argyophis* contains 12–13 South and Southeast Asian blindsnake species. Two subspecies of Diard's Blindsnake are recognized: the nominate form (*A. d. diardi*) occurs through most of the range, while a second form (*A. d. platyventris*) is found in Pakistan. Müller's Blindsnake (*A. muelleri*) from eastern Indonesia is a former subspecies of Diard's Blindsnake. Some authors do not recognize *Argyophis*, preferring to retain these species within *Typhlops*.

Actual size

Diard's Blindsnake is a moderately stout-bodied blindsnake with smooth, shiny mid-brown scales and a rounded snout. The eyes are just visible as small dark eyespots, and the tail terminates in a short spine.

FAMILY	Typhlopidae: Asiatyphlopinae
RISK FACTOR	Nonvenomous
DISTRIBUTION	Indonesia: southern Sulawesi and Buton
ELEVATION	1,640 ft (500 m) asl
HABITAT	Grassland with sparse scrub
DIET	Prey preferences unknown
REPRODUCTION	Reproductive strategy unknown, presumed oviparous
CONSERVATION STATUS	IUCN not listed

ADULT LENGTH
6 in
(150 mm)

CYCLOTYPHLOPS DEHARVENGI
DEHARVENG'S BLINDSNAKE
IN DEN BOSCH & INEICH, 1994

Actual size

Deharveng's Blindsnake is smooth-scaled, dark brown dorsally and laterally, and light brown ventrally, becoming yellowish on the anterior undersides. The head is light brown with dark spots on every scale. This species is unique in possessing a single round frontal scale on the dorsum of the head.

So far only recorded from southern Sulawesi and the island of Buton, Indonesia, Deharveng's Blindsnake is a minuscule species characterized by a single round frontal scale on the dorsum of the head. It inhabits cool, moist meadows with sparse scrubby vegetation, and is probably nocturnal. One of the few specimens known was found under a basaltic rock. Nothing is known of the natural history and biology of this rare species, although given its small size it is presumed to feed on small soft-bodied invertebrates such as termite larvae. Since all other blindsnakes are oviparous it is presumed this species is also. It is named in honor of the collector of the holotype, Louis Deharveng, the research director of the French Centre National de la Recherche Scientifique.

RELATED SPECIES

Cyclotyphlops is a monotypic genus with no close relatives. The Australasian genus *Anilios* has been recognized as the closest genus to *Cyclotyphlops*.

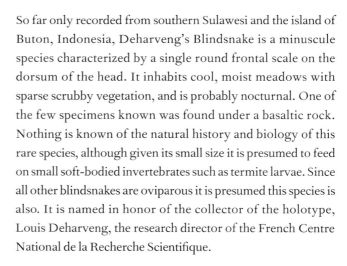

FAMILY	Typhlopidae: Asiatyphlopinae
RISK FACTOR	Nonvenomous
DISTRIBUTION	Worldwide: from India to all continents except Antarctica, and even remote islands
ELEVATION	0–6,560 ft (0–2,000 m) asl
HABITAT	Most habitats but especially gardens, nurseries, and coastal ports
DIET	Termite and ant larvae and eggs, and also possibly earthworms and caterpillars
REPRODUCTION	Oviparous, with clutches of 1–6 eggs
CONSERVATION STATUS	IUCN not listed

ADULT LENGTH
6–7 in
(150–180 mm)

INDOTYPHLOPS BRAMINUS

BRAHMINY BLINDSNAKE

(DAUDIN, 1803)

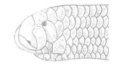

57

The Brahminy Blindsnake originates from India, but today it is the most widely distributed snake in the world, albeit by introduction. This is the only obligate parthenogenetic snake species; only female specimens are known. A small snake that hides in the root balls of pot plants, the Brahminy Blindsnake is easily transported internationally with exotic plants or agricultural crops such as oil-palm seedlings, earning it the alternative name of Flowerpot Snake. Females produce up to six eggs containing viable clones; it is an excellent colonizer requiring only one specimen to initiate a new colony. It is also often found underneath oil drums or stones. Prey comprises termite or ant eggs and larvae, and possibly small earthworms or caterpillars.

Actual size

The Brahminy Blindsnake is a smooth, shiny black or dark brown snake with a rounded head and small eyes that are barely visible, and a short spine at the terminus of the tail.

RELATED SPECIES

The genus *Indotyphlops* contains 23 Asian blindsnake species. One species occurs in Hong Kong (*I. laz̧elli*), and another on Komodo and Flores, Indonesia (*I. schmutz̧i*), but the remainder, with the exception of *I. braminus*, are confined to India, Sri Lanka, Nepal, Myanmar, or Thailand. Some authors do not recognize *Indotyphlops* and retain these species in *Ramphotyphlops*. While the Brahminy Blindsnake is established worldwide, one related species, the Christmas Island Blindsnake (*R. exocoeti*), is in danger of extinction due to the aggression of introduced crazy ants (*Paratrechina longicornis*).

FAMILY	Typhlopidae: Asiatyphlopinae
RISK FACTOR	Nonvenomous
DISTRIBUTION	Indonesia: Lesser Sunda Islands—Lombok, Sumbawa, Komodo, Moyo, Flores, Sumba, and Timor
ELEVATION	0–3,940 ft (0–1,200 m) asl
HABITAT	Wet forest, riverine forest, and lushly vegetated mountainsides
DIET	Probably soft-bodied invertebrates
REPRODUCTION	Oviparous, with clutches of up to 9 eggs
CONSERVATION STATUS	IUCN not listed

ADULT LENGTH
9¾–16½ in
(250–420 mm)

SUNDATYPHLOPS POLYGRAMMICUS
LESSER SUNDA BLINDSNAKE
(SCHLEGEL, 1839)

The Lesser Sunda Blindsnake is smooth-scaled and shiny with a rounded head and a terminal tail spine. The dorsum may be uniform brown, gray, or almost black, or marked by a series of fine, longitudinal stripes, 11 brown stripes alternating with ten yellow stripes. The undersides are uniform white, yellow, or yellow-brown, although the throat may be dark.

The Lesser Sunda Blindsnake has been recorded from a number of islands in Indonesia's southeasternmost archipelago. This species is associated with wet forest habitats, from wet montane forests to densely vegetated riverine valleys. A fossorial and nocturnal species, it is only seen on the surface during or after heavy rain, which forces it from its burrow and onto the surface. It may otherwise be found under rotten logs or large stones, but it is generally an inhabitant of montane country and is less common near the coast. Little is known about its biology and natural history, though it is thought to feed on soft-bodied invertebrates, and a clutch of nine eggs is on record.

RELATED SPECIES

Sundatyphlops is a monotypic genus that some authors still include in the Australo-Papuan genus *Anilios*, which contains 47 species. The Torres Blindsnake (*A. torresianus*) was once a subspecies of *S. polygrammicus* and the two genera are closely related. *Sundatyphlops polygrammicus* currently contains five island-specific subspecies from Timor (*S. p. polygrammicus*), Sumba (*S. p. brongersmai*), Lombok (*S. p. elberti*), Flores (*S. p. florensis*), and Komodo, Moyo, and Sumbawa (*S. p. undecimlineatus*). These subspecies may actually be valid species, or even species complexes in the case of the Timorese nominate form.

Actual size

FAMILY	Typhlopidae: Asiatyphlopinae
RISK FACTOR	Nonvenomous
DISTRIBUTION	Eurasia: Greece, Albania, Serbia, Macedonia, Bulgaria, Turkey, Israel, Jordan, Lebanon, Syria, Egypt, Caucasus, Iran, and Afghanistan
ELEVATION	0–6,230 ft (0–1,900 m) asl
HABITAT	Sandy habitats with sparse vegetation
DIET	Ants, their eggs, larvae, and pupae, earthworms, and insect larvae
REPRODUCTION	Oviparous, with clutches of up to 9 eggs
CONSERVATION STATUS	IUCN not listed

ADULT LENGTH
8–11¾ in,
rarely 15¾ in
(200–300 mm,
rarely 400 mm)

XEROTYPHLOPS VERMICULARIS
EURASIAN BLINDSNAKE
(MERREM, 1820)

59

Also known as the Vermiculate Blindsnake, the Eurasian or European Blindsnake is the most widely distributed *Xerotyphlops*. Its preferred habitat is sparsely vegetated hillsides and valleys on a loose, sandy soil substrate. It is fossorial, either using existing burrows or excavating its own, and is active at night or at dusk, but it may be found on the surface during the day after heavy rain. Specimens are most frequently encountered under stones or logs. The preferred prey of the Eurasian Blindsnake includes ants, their eggs, larvae, and pupae, other insect larvae, and small earthworms. Growth rates are slow, with adult specimens sloughing their skins only once a year.

RELATED SPECIES

The genus *Xerotyphlops* contains four other species with extremely limited distributions: Etheridge's Blindsnake (*X. etheridgei*) from western Mauritania, the Socotra Blindsnake (*X. socotranus*) from Socotra Island, the Lorestan Blindsnake (*X. luristanicus*) from Iran, and Wilson's Blindsnake (*X. wilsoni*) from southwestern Iran. Some authors do not recognize this genus and retain these species in *Typhlops*.

Actual size

The Eurasian Blindsnake is smooth-scaled and shiny. It is uniform light brown or pink in color, slightly darker dorsally than ventrally, and has a rounded head with a pale snout. The eyes are dark and clearly visible under the translucent scales of the head, and the tail bears a short terminal spine to aid locomotion.

FAMILY	Typhlopidae: Madatyphlopinae
RISK FACTOR	Nonvenomous
DISTRIBUTION	Madagascar: southwestern and western Madagascar
ELEVATION	0–3,230 ft (0–985 m) asl
HABITAT	Coastal sand dunes and thornbush savanna
DIET	Probably soft-bodied invertebrates
REPRODUCTION	Probably oviparous, clutch size unknown
CONSERVATION STATUS	IUCN Critically Endangered

ADULT LENGTH
8¾ in
(220 mm)

MADATYPHLOPS ARENARIUS
MALAGASY SAND BLINDSNAKE
(GRANDIDIER, 1872)

Actual size

The Malagasy Sand Blindsnake is smooth-scaled and glossy. The body and rounded head are uniform unpigmented pink, against which the dark eyes are clearly visible.

The Malagasy Sand Blindsnake is primarily found in the southwest of Madagascar, although there is a single record from the northwest. It is an inhabitant of coastal sand dunes and thornbush savanna. Specimens may be discovered under stones, but one specimen was located climbing the bark of a tree at night. Arboreal habits are not uncommon in blindsnakes, several species having been recorded aloft where they feed on termites found in termitaria on the branches. Little is known about the natural history or biology of the Malagasy Sand Blindsnake, though it is assumed to feed on soft-bodied invertebrates such as termites and/or their larvae, and lay eggs like other typhlopids. The word *arenarius* means sand-dwelling.

RELATED SPECIES

Madatyphlops contains 13 species, eight from Madagascar, one from Mayotte in the Comoros Islands (*M. comorensis*), three from Somalia (*M. calabresii*, *M. cuneirostris*, and *M. leucocephalus*), and one from Tanzania (*M. platyrhynchus*). In southwestern Madagascar, *M. arenarius* occurs in sympatry with Boettger's Sand Blindsnake (*M. boettgeri*).

FAMILY	Typhlopidae: Typhlopinae
RISK FACTOR	Nonvenomous
DISTRIBUTION	Amazonian South America: Colombia, Venezuela, the Guianas, northern Brazil, western Peru, and northern Bolivia
ELEVATION	0–2,460 ft (0–750 m) asl
HABITAT	Primary and secondary forest
DIET	Leaf-cutter ants and termites, and their larvae
REPRODUCTION	Oviparous, with clutches of up to 10 eggs
CONSERVATION STATUS	IUCN not listed

ADULT LENGTH
9¾–15¾ in
(250–400 mm)

AMEROTYPHLOPS RETICULATUS
RETICULATE BLINDSNAKE
(LINNAEUS, 1758)

61

The largest blindsnake of the Amazonian and Guianan region, the Reticulate Blindsnake is often found in sympatry with threadsnakes of the family Leptotyphlopidae. The stout-bodied Reticulate Blindsnake is able to excavate its own burrow, rather than having to utilize those made by termites and other fossorial organisms. It is active nocturnally but is also encountered abroad in the early morning, and it is a common species. Habitats include both primary and secondary rainforest, with the Reticulate Blindsnake inhabiting leaf litter, decaying logs, termitaria, and ants' nests. Prey includes ants and ant larvae, and probably also the larvae and adults of termites. Up to ten eggs are laid in aggressive leaf-cutter ant nests for protection from predators.

RELATED SPECIES

The genus *Amerotyphlops* contains 15 species of Central and South American blindsnakes, 13 mainland species, one species endemic to Trinidad (*A. trinitatus*), and another endemic to Grenada (*A. tasymicris*). Only two other species occur in the Amazon–Guiana region, Brongersma's Blindsnake (*A. brongersmianus*) and the Amazon Basin Blindsnake (*A. minuisquamus*). The patterning of *A. reticulatus* distinguishes it from these two species.

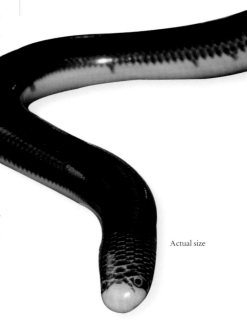

Actual size

The Reticulate Blindsnake has smooth, shiny scales, a body as thick as a finger, and a very distinctive pattern. The dorsum of the body is light, mid-, or dark brown, while the venter is immaculate white or cream, the division between the two colors occurring in the midlateral region. The tip of the snout is also colored pale like the venter and a band of similar pale pigment encircles the tail anterior to its terminus.

FAMILY	Typhlopidae: Typhlopinae
RISK FACTOR	Nonvenomous
DISTRIBUTION	West Indies: eastern Cuba (Guantánamo)
ELEVATION	Sea level
HABITAT	Coastal habitats
DIET	Probably soft-bodied invertebrates
REPRODUCTION	Probably oviparous, clutch size unknown
CONSERVATION STATUS	IUCN not listed

ADULT LENGTH
11–11¾ in
(282–301 mm)

CUBATYPHLOPS NOTORACHIUS
IMIAS BLINDSNAKE
THOMAS & HEDGES, 2007

Actual size

The Imias Blindsnake is smooth-scaled, and glossy light brown, being darker on the dorsum than on the venter. The dark eyes are clearly visible.

The Imias Blindsnake occurs in extreme eastern Cuba, close to the town of Imias, in Guantánamo Province. The name *notorachius* means "southern shore," a reflection of where this species occurs, at sea level on the southeastern corner of Cuba. No natural history data are available for this species, but in common with other typhlopid blindsnakes it may be expected to be nocturnal, fossorial but active on the surface after rain, to feed on soft-bodied invertebrates such as termites, and to lay a small number of soft-shelled eggs. The pale coloration and lack of pigmentation in this species suggests it is more an inhabitant of sandy soils than dark soils, and its coastal distribution would indicate this.

RELATED SPECIES

The genus *Cubatyphlops* contains 12 species of western Caribbean blindsnakes, mostly from Cuba but also from the Cayman Islands and Bahamas. *Cubatyphlops notorachius* belongs to the *C. biminiensis* species group, which also includes four other Guantánamo Province species, the Maisi Blindsnake (*C. anchaurus*), Cuban Pallid Blindsnake (*C. anousius*), Cuban Short-nosed Blindsnake (*C. contorhinus*), and Guantánamo Bay Blindsnake (*C. perimychus*). Two other American genera are found in the West Indies, the eastern Caribbean genus *Antillotyphlops* and the widespread genus *Typhlops*.

FAMILY	Xenotyphlopidae
RISK FACTOR	Nonvenomous
DISTRIBUTION	Madagascar: northern Madagascar
ELEVATION	0–165 ft (0–50 m) asl
HABITAT	Forested and scrubby coastal sand dunes
DIET	Probably soft-bodied invertebrates
REPRODUCTION	Probably oviparous, clutch size unknown
CONSERVATION STATUS	IUCN Critically Endangered

ADULT LENGTH
9½–10¼ in
(240–263 mm)

XENOTYPHLOPS GRANDIDIERI
GRANDIDIER'S MALAGASY BLINDSNAKE
(MOCQUARD, 1905)

63

Grandidier's Malagasy Blindsnake can be recognized by its broad, flat rostral scale, giving a "bulldozer" appearance. The type locality is unknown but other specimens have been collected under stones on the Baie de Sakalava on the northern tip of Madagascar in forested and scrubby coastal sand-dune habitats. The natural history and biology of this species is almost unknown. It probably feeds on soft-bodied invertebrates like termites and/or their larvae, and it probably lays eggs. Its small range is being deforested in order to provide wood for the charcoal-burning industry, and it is also Critically Endangered by mining in the area. Alfred Grandidier (1836–1921) was a French ornithologist who worked in Madagascar and discovered the bones of the elephant bird.

RELATED SPECIES

The endemic Madagascan Xenotyphlopidae is a monotypic blindsnake family, distinct and separate from all other scolecophidian snakes, although it shares some characteristics with members of the threadsnake family Leptotyphlopidae, such as the Koniagui Threadsnake (*Rhinoleptus koniagui*, page 44). Some authors recognize a second species, Mocquard's Malagasy Blindsnake (*Xenotyphlops mocquardi*), but others consider this species to be a synonym. *Xenotyphlops grandidieri* has no other close relatives.

Actual size

Grandidier's Malagasy Blindsnake is smooth-scaled and glossy. The body is dark pink anteriorly and pale pink posteriorly. The head is blunt, covered in small papillae, and terminates acutely in a large rostral scale with a small hook at the base. The eyespots are not visible on the head.

ALETHINOPHIDIA: AMEROPHIDIA

Out of America

The Alethinophidia (*Alethin* = real, true; *-ophidia* = snakes) are the True Snakes, a clade comprising snakes more advanced than the Scolecophidia (page 39). Some alethinophidian snakes are fossorial, with small mouths that limit their prey size but the majority of snakes are macrostomatan (big-mouthed) species with highly adapted jaws that permit them to swallow prey much wider than their own heads.

During the mid-Cretaceous, 116–97 MYA, the Alethinophidia diverged to form two clades—the Amerophidia and Afrophidia—as a result of the break-up of West Gondwana, the supercontinent comprising modern-day South America and Africa. Almost 99 percent of all living alethinophidian snakes belong to the Afrophidia (page 127).

The Amerophidia contains two families, three genera, and 35 species. The monotypic Aniliidae is a semi-fossorial, small-mouthed snake which exhibits ecological similarities to the Asian pipesnakes (Cylindrophiidae) of the Afrophidia, while the Tropidophiidae contains 34 species of macrostomatan woodsnakes or dwarf boas, which are convergent with similar-sized dwarf boas (Charinidae) in the Afrophidia. Amerophidian snakes are confined to Central and South America, and the West Indies. They are nonvenomous constrictors.

FAMILY	Aniliidae
RISK FACTOR	Nonvenomous
DISTRIBUTION	South America: Brazil, Venezuela, Trinidad, Guyana, French Guiana, Suriname, Colombia, Bolivia, Ecuador, and Peru
ELEVATION	98–2,300 ft (30–700 m) asl
HABITAT	Tropical rainforest and cultivated habitats
DIET	Small snakes, amphisbaenians, and fish, including eels
REPRODUCTION	Viviparous, with litters of 7–15 neonates
CONSERVATION STATUS	IUCN not listed

ADULT LENGTH
23¾–35½ in,
rarely 3 ft 3 in
(600–900 mm,
rarely 1.0 m)

66

ANILIUS SCYTALE
SOUTH AMERICAN PIPESNAKE
(LINNAEUS, 1758)

The South American Pipesnake is a cylindrical-bodied snake with smooth scales, a short tail, a rounded head and small eyes. It has alternating red and black transverse bands, the red bands being wider than the black bands, and the first black band crossing the posterior head. The bands are never yellow or white like a coralsnake, although pale yellow replaces red on the underside, with black blotches that may coincide or alternate with the dorsal bands.

The South American Pipesnake is a common inhabitant of primary and secondary rainforest leaf litter and cultivated habitats throughout the Amazonian countries and the Guianas. It is nocturnal or crepuscular, and semi-fossorial in habit. The South American Pipesnake preys on small snakes, from blindsnakes (*Typhlops* and *Amerotyphlops*, page 61) to groundsnakes (*Atractus*, pages 268–269), as well as amphisbaenians (worm-lizards), swamp eels, and elongate fish. It may also take caecilians (legless amphibians) and elongate lizards. Constriction is used to restrain and subdue prey prior to swallowing. South American Pipesnakes are inoffensive and do not bite, even when handled. When disturbed they mirror the behavior of Asian pipesnakes (*Cylindrophis*, pages 73–75) by burying their head in the coils and elevating and inverting the tail, either to deflect attention from the real head or to intimidate the enemy.

RELATED SPECIES

Two subspecies of *Anilius scytale* are recognized, the nominate form (*A. s. scytale*), distributed throughout most of Amazonia and the Guianas, and a second form (*A. s. phelpsorum*) from Bolivar and Amacuro states, Venezuela. This second form, which some authors raise to species level, can be distinguished from the nominate form by wider black than red bands and a higher ventral scale count. This pipesnake could be confused with highly venomous coralsnakes (*Micrurus*, pages 457–465) and mildly venomous false coralsnakes (*Erythrolamprus aesculapii*, page 311).

Actual size

FAMILY	Tropidophiidae
RISK FACTOR	Nonvenomous
DISTRIBUTION	Central and South America: Pacific coastal Panama, Colombia, and Ecuador
ELEVATION	0–2,460 ft (0–750 m) asl
HABITAT	Lowland tropical rainforest
DIET	Prey preferences unknown
REPRODUCTION	Viviparous, with litters of 2–6 neonates
CONSERVATION STATUS	IUCN not listed

ADULT LENGTH
15¾ in
(400 mm)

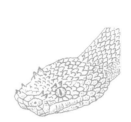

TRACHYBOA BOULENGERI

NORTHERN EYELASH BOA

PERACCA, 1910

67

The Northern Eyelash Boa is a strange snake, and rarely encountered in the field. It inhabits the Pacific lowland Chocó rainforests of Panama, Colombia, and Ecuador, where it adopts a secretive existence in the leaf litter, often close to water sources. The natural history and biology of these snakes is poorly known. They are known to be live-bearers, producing small litters, but their prey preferences are a mystery, captive specimens refusing all food offered except fish and the occasional small mouse. No records are available regarding their diet in the wild. These cryptozoic snakes exhibit an unusual defense—when handled, they freeze. This is neither defensive "balling" nor thanatosis ("playing dead"), but it may be a precursor to one or both of those defensive strategies.

RELATED SPECIES

A second species of *Trachyboa* is known, the Southern Eyelash Boa (*T. gularis*) from Pacific coastal Ecuador, although this second species does not possess "eyelashes." The term "boa" is applied loosely with respect to these snakes because they are not closely related to the Boidae (pages 104–113), instead only being related to the West Indian woodsnakes (*Tropidophis*, pages 68–69) in the Tropidophiidae. At first glance *T. boulengeri* has the appearance of the African Rough-scaled Bushviper (*Atheris hispida*, page 607) or an Eyelash Palm-pitviper (*Bothriechis schlegelii*, page 558).

The Northern Eyelash Boa is a small snake with strongly keeled scales, a short tapering tail, and a bulbous head with small eyes, with vertically elliptical pupils, over which rise a series of small fleshy supraciliary ("eyelash") scales. Further elevated scales are present on the snout. Coloration is brown with darker saddles, and a pale yellow or white tail.

Actual size

FAMILY	Tropidophiidae
RISK FACTOR	Nonvenomous, constrictor
DISTRIBUTION	West Indies: Hispaniola (Haiti and Dominican Republic)
ELEVATION	0–2,690 ft (0–820 m) asl
HABITAT	Lowland rainforests, plantations, rocky areas, and creeks
DIET	Frogs, lizards, and small mammals
REPRODUCTION	Viviparous, with litters of 4–9 neonates
CONSERVATION STATUS	IUCN not listed

ADULT LENGTH
19¾–28 in
(500–712 mm)

TROPIDOPHIS HAETIANUS
HAITIAN WOODSNAKE
(COPE, 1879)

Actual size

Although often called "dwarf boas," the snakes of the genus *Tropidophis* are not closely related to the Boidae and should possibly be known by their alternative name of "woodsnakes." Twenty-seven species occur in the West Indies, but the Haitian Woodsnake is the only species of *Tropidophis* to occur on Hispaniola, the island comprising Haiti and the Dominican Republic. It inhabits lowland rainforest, especially alongside creeks, but is also found in plantations where it hides inside piles of discarded coconut husks or other debris. At night it hunts and constricts small mammals, frogs, and lizards. It is both terrestrial and arboreal, being found sheltering inside epiphytic bromeliads on palms and other trees. Woodsnakes are completely harmless, there being no dangerous snakes in Haiti or the Dominican Republic.

RELATED SPECIES

Three subspecies of *Tropidophis haetianus* are recognized, the widespread nominate form (*T. h. haetianus*), an eastern Dominican subspecies (*T. h. hemerus*), and a Tiburon Peninsula subspecies (*T. h. tiburonensis*) in Haiti. A record from northeastern Cuba may be an introduction. Three related Jamaican species were once also subspecies: Southern Jamaican Woodsnake (*T. jamaicensis*), Northern Jamaican Woodsnake (*T. stejnegeri*), and the Portland Point Woodsnake (*T. stullae*).

The Haitian Woodsnake is a small, moderately muscular snake with a laterally compressed body, and an angular head with small eyes and vertically elliptical pupils. The body may be pale gray to dark brown or tan, with patterning absent or consisting of two irregular parallel rows of dark brown dorsal blotches.

FAMILY	Tropidophiidae
RISK FACTOR	Nonvenomous constrictor
DISTRIBUTION	West Indies: Cuba
ELEVATION	33–2,620 ft (10–800 m) asl
HABITAT	Moist woodland, rainforest, gardens, pastures, and rocky outcrops
DIET	Frogs, lizards, birds, and small mammals
REPRODUCTION	Viviparous, with litters of 8–36 neonates
CONSERVATION STATUS	IUCN not listed

ADULT LENGTH
3 ft 6 in
(1.06 m)

TROPIDOPHIS MELANURUS
CUBAN WOODSNAKE
(SCHLEGEL, 1837)

69

The most widely distributed of 17 Cuban *Tropidophis*, the Cuban Woodsnake is found in moist woodland, tropical rainforest, open habitats, rocky habitats, and even human-altered habitats throughout Cuba, and on the satellite Isla de la Juventud. It is also one of the largest species in the genus. Nocturnal, terrestrial, and arboreal, it is a predator of frogs, lizards, birds, and small mammals, which are killed by constriction. This snake demonstrates a number of defensive strategies if handled. It will "ball," with its head in the center of its coils, it may exude a noxious white secretion from its cloacal glands, or it may autohemorrhage from its mouth and eyes. Otherwise, this snake is completely harmless, there being no dangerous venomous snakes on Cuba.

RELATED SPECIES

Thirty-two species of *Tropidophis* woodsnakes are recognized, of which three are Brazilian, two Ecuadorian–Peruvian, and the remainder West Indian. Of those 17 occur on Cuba, all but two endemic. Three subspecies of *T. melanurus* are recognized, the widespread nominate form (*T. m. melanurus*), a northern race (*T. m. dysodes*) from Pinar del Rio Province, and an island endemic from Isla de la Juventud (*T. m. ericksoni*). The Navassa Woodsnake (*T. bucculentus*) is a former subspecies and probably endangered due to the introduction of goats.

The Cuban Woodsnake is a slender, muscular snake with keeled or smooth scales, an angular head, and small eyes with vertically elliptical pupils. The dorsal color may be gray, brown, buff, orange, tan, or red, with a darker pattern of longitudinal stripes, transverse blotches, spots, or zigzags, which vary from subspecies to subspecies and locality to locality. The tail is black or white, and the undersides are buff with darker markings.

Actual size

ALETHINOPHIDIA: AFROPHIDIA: HENOPHIDIA

Out of Africa: Old Snakes

The Afrophidia, snakes of African evolutionary origin, contains almost 99 percent of all alethinophidian snakes, the majority belonging in the Caenophidia (pages 126–645). The remaining 185 species are the Henophidia (*Heno* = old, true; *-ophidia* = snakes), the Old Snakes. The Henophidia contains three families of fossorial, small-mouthed Southeast Asian snakes— Anomochilidae, Cylindrophiidae, and Uropeltidae, the pipesnakes and shieldtails from Southeast Asia, often grouped as the superfamily Uropeltoidea—and two *incertae sedis* (of unknown position) families: the Indian Ocean Bolyeriidae and Southeast Asian Xenophidiidae.

The Henophidia also contains two major macrostomatan (big-mouthed) superfamilies. Widely distributed and well known, these clades include the world's largest snakes capable of taking large vertebrate prey. Superfamily Pythonoidea contains the Afro-Asian-Australasian Pythonidae, the Southeast Asian Xenopeltidae, and the Mexican Loxocemidae. Superfamily Booidea contains the American families Boidae and Charinidae, the African Calabariidae, Afro-Asian Erycidae, Malagasy Sanzinidae, and the Pacific-New Guinea Candoiidae. Afrophidian snakes are distributed worldwide, including back in the Americas where they occur alongside amerophidian snakes. All are nonvenomous and many are constrictors.

FAMILY	Anomochilidae
RISK FACTOR	Nonvenomous
DISTRIBUTION	Southeast Asia: Malaysian Borneo (Sabah)
ELEVATION	4,760–4,970 ft (1,450–1,515 m) asl
HABITAT	Low to medium elevation montane rainforests
DIET	Not known, but presumed to comprise small or slender invertebrates, or possibly vertebrates
REPRODUCTION	Possibly oviparous, clutch size unknown
CONSERVATION STATUS	IUCN Data Deficient

ADULT LENGTH
20½ in
(520 mm)

ANOMOCHILUS MONTICOLA
KINABALU LESSER PIPESNAKE
DAS, LAKIM, LIM & HUI, 2008

72

The Kinabalu Lesser Pipesnake is a cylindrical-bodied snake with a short, rounded head, indistinct from the neck, and a short tail. Its scales are smooth and glossy. The snake is iridescent blue-black in color with pale yellow spots on the flanks, larger blotches on the venter, a broken band of pale yellow across the snout, and an orange band around the tail.

The Kinabalu Lesser Pipesnake is the largest known species of *Anomochilus*. Endemic to the Malaysian state of Sabah, Borneo, it is only known from a few specimens collected on Mt. Kinabalu, at elevations around 4,920 ft (1,500 m asl). It is an inhabitant of rainforest leaf litter where it adopts a semi-fossorial existence. Lesser pipesnakes lack a mental groove under the chin, the result being that they have a narrow mouth gape and can only swallow small or slender prey, presumably invertebrates but possibly also elongate vertebrates. Their reproductive strategy is also very poorly known. While the Asian pipesnakes (Cylindrophiidae) and shieldtails (Uropeltidae) are viviparous, it is thought that *Anomochilus* may be oviparous, the only known gravid female (*A. leonardi*) containing four soft-shelled eggs.

RELATED SPECIES
The genus and family contain two other localized species, the Malayan Lesser Pipesnake (*Anomochilus leonardi*) from Peninsular Malaysia and Sabah, Borneo, and the Sumatran Lesser Pipesnake (*A. weberi*) from Sumatra and Kalimantan, Borneo. This family of three species is considered more primitive than the Asian pipesnakes of genus *Cylindrophis*, but recent research suggests they are close to the Sri Lankan Pipesnake (*C. maculatus*, page 74).

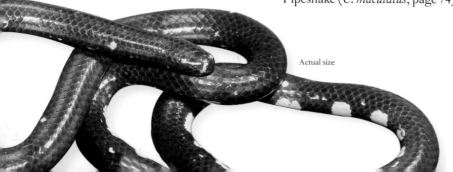

Actual size

FAMILY	Cylindrophiidae
RISK FACTOR	Nonvenomous
DISTRIBUTION	Southeast Asia: Indonesia and Timor-Leste
ELEVATION	395–1,280 ft (120–390 m) asl
HABITAT	Lowland forest, riverine forest, banana groves, and bamboo stands
DIET	Presumed slender vertebrates such as blindsnakes
REPRODUCTION	Viviparous, litter size unknown
CONSERVATION STATUS	IUCN not listed

ADULT LENGTH
13 in
(330 mm)

CYLINDROPHIS BOULENGERI

BOULENGER'S PIPESNAKE

ROUX, 1911

Boulenger's Pipesnake is only known from approximately a dozen specimens collected in Indonesian West Timor, independent Timor-Leste (East Timor), and the Indonesian islands of Wetar to the northeast and Babar to the east. These small snakes are infrequently encountered, usually in lowland forest or riverine forest, but also in bamboo stands and cultivated areas such as banana groves. They are semi-fossorial to fossorial, specimens also being found under large rocks near rivers. Although there are no natural history records for this species, it probably feeds on blindsnakes and other slender vertebrates, and in common with the much better documented Red-tailed Pipesnake (*Cylindrophis ruffus*, page 75) it is probably viviparous. The true taxonomic status of the populations from different islands has yet to be fully determined.

RELATED SPECIES

Several other poorly known Asian pipesnakes occur on islands in southeastern Indonesia: the Lesser Sunda Pipesnake (*Cylindrophis opisthorhodus*) on Lombok, Sumbawa, Komodo, and Flores, the Tanahjampea Island Pipesnake (*C. isolepis*), the Yamdena Island Pipesnake (*C. yamdena*), and the Aru Island Pipesnake (*C. aruensis*).

Actual size

Boulenger's Pipesnake is smooth-scaled and glossy, with a rounded head and short tail. It is black above and cream on the venter, with a series of irregular upward extensions of this pale pigment onto the flanks, each with an orange spot at its terminus, those of the neck, anterior body, and tail meeting or almost meeting to form orange rings. Cream bars also extend onto the sides of the head from the chin region. The pale venter is black-spotted.

FAMILY	Cylindrophiidae
RISK FACTOR	Nonvenomous
DISTRIBUTION	South Asia: Sri Lanka
ELEVATION	0–3,940 ft (0–1,200 m) asl
HABITAT	Lowland and low montane forested hills and plains, gardens, and cultivated areas
DIET	Small snakes, earthworms, and insects
REPRODUCTION	Viviparous, with litters of 1–15 neonates
CONSERVATION STATUS	IUCN not listed

ADULT LENGTH
28 in
(715 mm)

CYLINDROPHIS MACULATUS
SRI LANKAN PIPESNAKE
(LINNAEUS, 1758)

74

The Sri Lankan Pipesnake is a smooth-scaled snake with a slightly pointed and flattened head, only just distinct from the neck, and a very short tail. Dorsally it is black with a double series of brick red spots that reduce the black pigment to fine cross-bars. Ventrally it is white with black cross-bars, including under the tail.

The Sri Lankan Pipesnake is a nocturnal and semi-fossorial inhabitant of lowland plains and low montane rainforests, but it may also be found in cultivated areas such as gardens or paddy fields. It is often found under large stones or logs, or in piles of leaf litter. It hunts small snakes such as shieldtails (Uropeltidae), blindsnakes (Asiatyphlopinae), or roughsides (*Aspidura*, page 399), but is also reported to feed on earthworms and insects. It is a live-bearing species, and females may produce up to 15 neonates. When it feels threatened by a potential predator the Sri Lankan Pipesnake will flatten its body, exposing its red spots to best advantage, hide its head, and elevate its tail like a cobra, the tip inverting to expose the contrasting black and white undersides.

RELATED SPECIES

Asian pipesnakes are primarily a Southeast Asian group, occurring from Myanmar to Sulawesi and the Tanimbar and Aru islands in Indonesia, with *Cylindrophis maculatus* being the only South Asian member of the genus. They occur on Sri Lanka along with members of the related shieldtail family Uropeltidae.

Actual size

FAMILY	Cylindrophiidae
RISK FACTOR	Nonvenomous
DISTRIBUTION	Southeast Asia: south China, Myanmar, Thailand, Laos, Vietnam, Cambodia, Indonesia, and Malaysia
ELEVATION	0–5,500 ft (0–1,675 m) asl
HABITAT	Lowland rainforest, swamps, rice paddies, and saltwater lagoons
DIET	Snakes and eels
REPRODUCTION	Viviparous, with litters of 5–13 neonates
CONSERVATION STATUS	IUCN Least Concern

ADULT LENGTH
27½–35½ in
(700–900 mm)

CYLINDROPHIS RUFFUS
RED-TAILED PIPESNAKE
(LAURENTI, 1768)

75

The Red-tailed or Common Asian Pipesnake is a widely distributed and frequently encountered member of the genus *Cylindrophis*, being recorded from south China and mainland Southeast Asia, to Sumatra, Java, Borneo, Sulawesi, and their satellite archipelagos. It is found in a wide variety of lowland habitats. The prey of the Red-tailed Pipesnake consists of cylindrical vertebrates such as snakes or eels. Pipesnakes are live-bearers, a large Red-tailed Pipesnake producing litters of up to 13 neonates. When threatened the pipesnake will bury its head in its coils and elevate its tail, inverting the tip to expose the red pigment beneath. This may be intended to mimic a head and distract attention from the real head, or to intimidate the predator.

Actual size

The Red-tailed Pipesnake is a smooth-scaled, cylindrical snake with a rounded head, only slightly distinct from the neck, very small eyes, and a short tail. It is glossy black or brown above with a series of white or cream transverse bars, which may coalesce dorsally as a complete or broken collar and series of rings. The venter is black and white, except for the underside of the tail, which is bright red, this pigment continuing onto the dorsum of the tail as a red ring.

RELATED SPECIES

The genus *Cylindrophis* contains 14 species, 13 of which occur in Southeast Asia, only the Sri Lankan Pipesnake (*C. maculatus*, page 74) occurs in South Asia. *Cylindrophis ruffus* occurs in sympatry with the Burmese Pipesnake (*C. burmanus*) in Myanmar, the Lined Pipesnake (*C. lineatus*) and Engkari Pipesnake (*C. engkariensis*) in Sarawak, Borneo, Mirza's Pipesnake (*C. mirzae*) on Singapore, Jodi's Pipesnake (*C. jodiae*) in Vietnam, the Suboculate Pipesnake (*C. subocularis*) on Java, and the Sulawesi Black Pipesnake (*C. melanotus*).

FAMILY	Uropeltidae
RISK FACTOR	Nonvenomous
DISTRIBUTION	South Asia: southwestern India
ELEVATION	3,490–7,380 ft (1,065–2,250 m) asl
HABITAT	Rainforest
DIET	Presumed earthworms
REPRODUCTION	Viviparous, with litters of 3–6 neonates
CONSERVATION STATUS	IUCN not listed

ADULT LENGTH
17¼ in
(440 mm)

PLECTRURUS PERROTETII
NILGIRI EARTHSNAKE
DUMÉRIL & BIBRON, 1854

The Nilgiri Earthsnake is a cylindrical snake with smooth, glossy scales, a narrow, pointed head, small eyes, and a short, blunt tail that tapers and terminates as two small points, one above the other. The general coloration is light or dark brown but the scales may be marked with a red or yellow central streak, and the underside of the tail is often orange.

The Nilgiri Earthsnake is endemic to southwestern India, where it inhabits the Nilgiri Mountains of western Tamil Nadu, Karnataka, and Kerala, in the Western Ghats. It spends the majority of its time underground, only moving onto the surface after heavy rain, which is quite a common occurrence in this part of India. It is a rainforest floor species and reputedly common where it occurs, but it is also found in cultivated fields, especially those manured with horse dung. Such habitats will accumulate large populations of earthworms, the primary prey of uropeltid snakes. At night the Nilgiri Earthsnake is said to move out of the ground and under the horse dung as the air and ground temperature drops. Gustave Samuel Perrotet (1793–1867) was a French explorer and naturalist collector.

RELATED SPECIES
The genus *Plectrurus* contains two other species, both endemic to south India: the Kerala Earthsnake (*P. aureus*) and Günther's Earthsnake (*P. guentheri*).

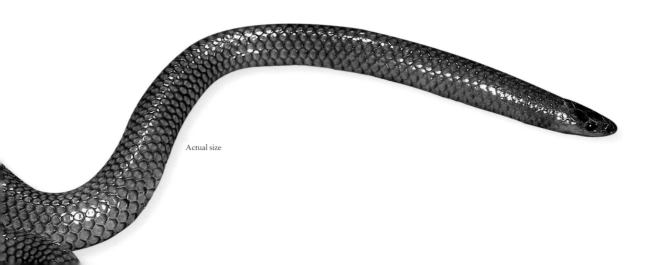

Actual size

FAMILY	Uropeltidae
RISK FACTOR	Nonvenomous
DISTRIBUTION	South Asia: Sri Lanka
ELEVATION	2,460–3,120 ft (750–950 m) asl
HABITAT	Forest edges, paddy fields, farms, and gardens
DIET	Earthworms
REPRODUCTION	Viviparous, with litters of 2–4 neonates
CONSERVATION STATUS	IUCN not listed

ADULT LENGTH
8–11¾ in
(200–300 mm)

RHINOPHIS HOMOLEPIS
TREVELYAN'S SHIELDTAIL
HEMPRICH, 1820

77

Trevelyan's Shieldtail occurs at low elevations in central Sri Lanka, where it inhabits hillside forest, forest edges, paddy fields, gardens, farms, and cattle pens. Often small colonies are found together in agricultural habitats, especially those with soft soil and an abundant earthworm fauna, their primary prey. This species is also common near watercourses and swampy areas. Like all shieldtails for which reproductive strategy is known, this species is a live-bearer. At first glance it is easy to mistake which is the head and which is the tail. The head is the sharp, pointed end of the snake while the bulbous, bright yellow end with a pink center is the tail.

Trevelyan's Shieldtail is a small, cylindrical snake with a long, pointed head and very small eyes. The tail terminates as a slightly bulbous pink and yellow shield, which contrasts with the glossy blue-black coloration of most of the body. Small yellow triangles or spots are distributed along the flanks and a yellow collar may encircle the neck, while the lower flanks may be white with dark flecks.

RELATED SPECIES

The genus *Rhinophis* contains 19 species, 14 endemic to Sri Lanka, four endemic to India, and one (*R. oxyrhynchus*) occurring in both countries. This genus includes most of the Sri Lankan shieldtails. This species was also known as *R. trevelyanus* but the older name *R. homolepis* has priority. Similar species include Drummond-Hay's Shieldtail (*R. drummondhayi*), a species considered Near Threatened by the IUCN.

Actual size

FAMILY	Uropeltidae
RISK FACTOR	Nonvenomous
DISTRIBUTION	South Asia: western India
ELEVATION	0–4,490 ft (0–1,370 m) asl
HABITAT	Forests, paddy fields, and agricultural fields
DIET	Earthworms and insects
REPRODUCTION	Presumed viviparous, litter size unknown
CONSERVATION STATUS	IUCN Least Concern

ADULT LENGTH
11¾–12¾ in
(300–323 mm)

UROPELTIS MACROLEPIS
LARGE-SCALED SHIELDTAIL
(PETERS, 1862)

78

The Large-scaled Shieldtail is stout-bodied, with smooth, iridescent scales and a rounded head with relatively large eyes. The tail looks as if it has been cut with a knife, the shield being covered in strongly bicarinate and tricarinate scales. Coloration is black above, while the flanks are often marked with yellow spots or bars that may form a longitudinal lateral stripe, extending under the throat, or onto the dorsum.

Shieldtails are also known as rough-tails or thorntails. Some species have gradually tapering tails, but most look as if they have been cut acutely through the tail with a sharp knife and the wound has healed with a rough scab or scar. This is the shield that earns them their common name. They are expert burrowers and the terminal shield effectively plugs the burrow behind them, even collecting dirt on its rough surface. The Large-scaled Shieldtail is found in southern Gujarat and Maharashtra states, western India, where it is reputedly relatively common. It feeds on earthworms and soft-bodied invertebrates and is encountered on the surface in the monsoon season when its burrows are flooded. Shieldtails are inoffensive snakes that have many enemies and only one defense: to burrow.

RELATED SPECIES

Two subspecies are recognized, the Bombay Large-scaled Shieldtail (*Uropeltis macrolepis macrolepis*) and the Mahabaleshwar Shieldtail (*U. m. mahableshwarensis*), which occurs farther south in the northern Western Ghats. *Uropeltis* contains 23 of the 55 known species of shieldtails and earthsnakes. The genus is endemic to south India. Similar species include Elliot's Shieldtail (*U. ellioti*).

Actual size

FAMILY	Bolyeriidae
RISK FACTOR	Nonvenomous, constrictor
DISTRIBUTION	Indian Ocean: Round Island (Mauritius)
ELEVATION	920 ft (280 m) asl
HABITAT	Dry forest
DIET	Lizards
REPRODUCTION	Oviparous, with clutches of 3–11 eggs
CONSERVATION STATUS	IUCN Endangered, CITES Appendix I

ADULT LENGTH
3 ft 3 in–4 ft 2 in
(1.0–1.28 m)

CASAREA DUSSUMIERI
ROUND ISLAND KEEL-SCALED BOA
(SCHLEGEL, 1837)

79

The Round Island Keel-scaled Boa belongs to the Bolyeriidae, the primitive split-jawed snakes, with a split maxillary bone that hinges into anterior and posterior halves. It lacks the vestigial pelvic spurs of true boas and lays eggs rather than giving birth to neonates. It hunts sleeping day geckos and skinks at night, killing them by constriction, and shelters under palm fronds or in seabird burrows. Extirpated from Mauritius, possibly by rats, it now only occurs on tiny Round Island 14 miles (22.5 km) to the north. With its highest point only 920 ft (280 m) asl and a total area of just ⅔ sq miles (1.69 sq km), it is a protected area with a unique and highly endemic fauna and flora, but in the past it was devastated by introduced goats and rabbits, causing habitat loss and soil erosion on a large scale.

The Round Island Keel-scaled Boa is a slender snake with small, keeled scales and an angular, pointed head with small eyes and vertically elliptical pupils. Its coloration is gray-brown or orange, with or without a dorsal stripe of darker brown pigment.

RELATED SPECIES

Casarea dussumieri has only one close relative, the Round Island Burrowing Boa (*Bolyeria multocarinata*), a fossorial species that suffered more from the erosion of the island's soil than did *C. dussumieri*. This second species has not been seen since 1975 and is officially listed as Extinct by the IUCN. *Casarea dussumieri* has been saved from extinction by a captive breeding program run by the Jersey Wildlife Conservation Trust, working in collaboration with the Government of Mauritius.

Actual size

FAMILY	Xenophidiidae
RISK FACTOR	Nonvenomous
DISTRIBUTION	Southeast Asia: Peninsular Malaysia
ELEVATION	330 ft (100 m) asl
HABITAT	Lowland rainforest
DIET	Believed to comprise earthworms, and possibly insect larvae
REPRODUCTION	Probably oviparous, clutch size unknown
CONSERVATION STATUS	IUCN Data Deficient

ADULT LENGTH
10¼ in
(263 mm)

XENOPHIDION SCHAEFERI
MALAYAN
SPINE-JAWED SNAKE
GÜNTHER & MANTHAY, 1995

The Malayan Spine-jawed Snake is known only from its holotype, collected in the 4⅔ sq mile (12.14 sq km) Templer Park, Selangor, 13½ miles (22 km) from Kuala Lumpur. Its habitat is lowland primary rainforest. This small snake is nocturnal, terrestrial, and thought to feed on earthworms and insect larvae, although small lizards may also feature in its diet (see below). Its reproductive strategy is unknown, but it is probably oviparous. The name "spine-jawed snake" originates from the presence of a unique elongate palatine process on the maxilla. The IUCN list this species as Data Deficient, but it is perhaps severely endangered, locally extirpated, or even extinct, as its type locality, 1¼ miles (2 km) south of the park entrance, was cleared and planted with bananas in 1990.

The Malayan Spine-jawed Snake is a small snake with a compressed body, weakly keeled scales, a short tail, and an elongate squarish head with small round-pupilled eyes and large prefrontal scales. Coloration and patterning consist of a dark brown dorsum with a broad grayish-white zigzagging dorsal stripe, itself overlain by a finer vertebral dark brown stripe. The venter is pale gray.

RELATED SPECIES

The only other species in the Xenophidiidae, also only known from its holotype, is the Borneo Spine-jawed Snake (*Xenophidion acanthognathus*) from Mt. Kinabalu at 1,970 ft (600 m) asl in Sabah, Borneo. Slightly larger than its Malayan congener, this species feeds on skinks and lays eggs, but its type locality has also been severely damaged.

Actual size

FAMILY	Xenopeltidae
RISK FACTOR	Nonvenomous, constrictor
DISTRIBUTION	Southeast Asia: Myanmar to southern China, south to Sumatra, Java, and Borneo, and east to the Philippines
ELEVATION	0–4,590 ft (0–1,400 m) asl
HABITAT	Lowland rainforest, swamps, and rice paddies
DIET	Small mammals, birds, lizards, snakes, and frogs
REPRODUCTION	Oviparous, with clutches of 3–17 eggs
CONSERVATION STATUS	IUCN Least Concern

ADULT LENGTH
2 ft 7 in–3ft 7 in
(0.8–1.1 m)

XENOPELTIS UNICOLOR
SUNBEAM SNAKE
REINWARDT, 1827

81

Also known as the Iridescent Earthsnake, the Sunbeam Snake is found throughout mainland Southeast Asia, Sumatra, Java, Borneo, and the Philippines. An inhabitant of lowland rainforest, freshwater swamps, and cultivated habitats such as rice paddies, it is a terrestrial to semi-fossorial species that lives in leaf litter and animal burrows, and may even become fossorial, burrowing into soft mud. Secretive and nocturnal by nature, Sunbeam Snakes are usually only encountered abroad, crossing tracks or roads, after monsoonal rain. The Sunbeam Snake has a rather catholic diet, preying on rodents, birds, lizards, snakes, and frogs, all of which are killed by constriction.

Actual size

RELATED SPECIES
The only species in the family Xenopeltidae are *Xenopeltis unicolor* and its close relative the Hainan Sunbeam Snake (*X. hainanensis*) from China's Hainan Island. The nearest relatives to the sunbeam snakes are the pythons and the Mexican Burrowing Python (*Loxocemus bicolor*, page 82).

The Sunbeam Snake is well named, its glossy-scaled body having the iridescence of a rainbow or sunbeam with every color in the spectrum represented in daylight. It has a cylindrical body, a short tail, and a flattened, compressed head with small eyes. Juveniles often bear a white or yellow collar, but this disappears with increased maturity.

FAMILY	Loxocemidae
RISK FACTOR	Nonvenomous, constrictor
DISTRIBUTION	North and Central America: central Mexico to Costa Rica
ELEVATION	66–1,970 ft (20–600 m) asl
HABITAT	Arid lowland seasonal and deciduous forests, thorn forest, and lower montane forests
DIET	Small mammals, lizards, frogs, and reptile eggs
REPRODUCTION	Oviparous, with clutches of 4–12 eggs
CONSERVATION STATUS	IUCN Least Concern, CITES Appendix II

ADULT LENGTH
3–5 ft
(0.9–1.5 m)

82

LOXOCEMUS BICOLOR
MEXICAN BURROWING PYTHON
COPE, 1861

The Mexican Burrowing Python does not look like a boa or a python. It is uniform gray above and white below, with the change of color strongly demarcated on the flanks. The head is pointed, with small eyes and dorsal scutes. The tail is non-prehensile.

The Mexican Burrowing Python is found in Central America, from central Mexico to Guatemala, Honduras, Nicaragua, and Costa Rica. It is particularly common on the Pacific versant but there are also isolated records from the Atlantic versant. It is nocturnal and semi-fossorial, hiding in leaf litter during the day. The Mexican Burrowing Python feeds on small mammals, lizards, and frogs, which are killed by constriction, but also appears to be a reptile egg specialist. It actively searches out the nests of iguanas and sea turtles, devouring most of the eggs whole by pressing them against its body coils and working its jaws over the egg. During the breeding season males will combat and bite one another for access to available females. Unlike the boas of the Americas, the Mexican Burrowing Python lays eggs.

RELATED SPECIES
Loxocemus bicolor has no close relatives in the Americas, although it shares some characteristics with the boas. The closest relatives of *Loxocemus* are the Old World pythons and the Sunbeam Snake (*Xenopeltis unicolor*, page 81), hence this New World "python" is placed in its own monotypic family, Loxocemidae.

Actual size

FAMILY	Pythonidae
RISK FACTOR	Nonvenomous, constrictor
DISTRIBUTION	Australia: northern Australia
ELEVATION	0–195 ft (0–60 m) asl
HABITAT	Arid rocky escarpments, especially around caves, dry savanna woodland, coastal forest, and urban environments
DIET	Small mammals, especially bats, and birds, lizards, and frogs
REPRODUCTION	Oviparous, with clutches of 7–20 eggs
CONSERVATION STATUS	IUCN not listed, CITES Appendix II

ADULT LENGTH
2 ft 7 in–3 ft 7 in
(0.8–1.1 m)

ANTARESIA CHILDRENI
CHILDREN'S PYTHON
(GRAY, 1842)

83

Children's Python is a small, inoffensive species. However, it is not so named because it makes a suitable pet species for children, but rather because it was named in honor of the British naturalist John George Children (1777–1852). Children's Python is found across northern Australia, west of the Great Dividing Range from the Kimberley of Western Australia to Mt. Isa in Queensland, and is usually associated with arid rocky escarpments, where it may be found hunting on roosting bats. It is also frequently encountered in wooded, forested, or grassland habitats, and even in urban areas, such as Darwin, where it hunts small terrestrial mammals, and a wide variety of other small vertebrates, from birds to lizards and frogs.

RELATED SPECIES

The genus *Antaresia* contains three other species: the Spotted Python (*A. maculosa*) from Queensland and southern New Guinea, Stimson's Python (*A. stimsoni*), which has western and eastern subspecies across Australia, and the Anthill Python (*A. perthensis*), the smallest python in the world at a maximum length of 24 in (610 mm).

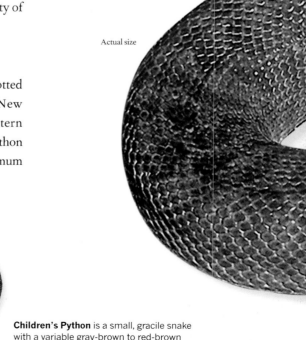

Actual size

Children's Python is a small, gracile snake with a variable gray-brown to red-brown body, speckled with darker pigment, and a small head, small, light-colored eyes, and a non-prehensile tail.

FAMILY	Pythonidae
RISK FACTOR	Nonvenomous, powerful constrictor
DISTRIBUTION	Australasia: New Guinea (mainland and many offshore islands)
ELEVATION	0–985 ft (0–300 m) asl
HABITAT	Lowland rainforest, monsoon forest, riverine forest, and also gardens
DIET	Mammals and reptiles
REPRODUCTION	Oviparous, with clutches of up to 22 eggs
CONSERVATION STATUS	IUCN not listed, CITES Appendix II

ADULT LENGTH
6 ft 7 in–14 ft 9 in,
rarely > 16 ft 5 in
(2.0–4.5 m,
rarely > 5.0 m)

APODORA PAPUANA
PAPUAN PYTHON

(PETERS & DORIA, 1878)

The Papuan Python is a stout-bodied snake with a broad, chunky head. It is usually two shades of brown, the back being darker than the flanks, with a gray head, every scale demarcated by black interstitial skin. The iris of the eye is also light gray. The python is soft and velvety to the touch, due to the small size of the body scales.

The Papuan Python is the heaviest and largest of the New Guinea pythons, with rare specimens exceeding 13 ft (4 m), even 16 ft 5 in (5 m) in length, and weighing over 59½ lb (27 kg). It is a stoutly built snake capable of overpowering and constricting wallabies of its own weight, but it is also a noted predator of reptiles, especially snakes, even taking pythons of similar lengths to itself. Although widely distributed throughout New Guinea and its satellite islands, this large python is one of the least frequently encountered species, being usually associated with heavily forested areas near large rivers, where it may be found crossing roads on wet nights. However, it also occurs in gardens, where it hunts bandicoots and rats, which are killed by constriction.

RELATED SPECIES

This python is sometimes placed in the genus *Liasis* with its closest relatives, the Australian Olive Python (*Liasis olivaceus*, page 89), the Brown Water Python (*L. fuscus*), and Macklot's Water Python (*L. mackloti*, page 88).

Actual size

FAMILY	Pythonidae
RISK FACTOR	Nonvenomous, powerful constrictor
DISTRIBUTION	Australasia: northern Australia (Western Australia to Queensland)
ELEVATION	50–195 ft (15–60 m) asl
HABITAT	Humid coastal forest, dry woodland, and arid rocky desert
DIET	Mammals, ground-dwelling birds, and reptiles
REPRODUCTION	Oviparous, with clutches of 6–18 eggs
CONSERVATION STATUS	IUCN not listed, CITES Appendix II

ADULT LENGTH
5–10 ft
(1.5–3.0 m)

ASPIDITES MELANOCEPHALUS
BLACK-HEADED PYTHON
(KREFFT, 1864)

85

The Black-headed Python, and its relative the Woma (see below), are considered the most basal of Australasian pythons. These are the only pythons to lack heat-sensitive pits in their labial and rostral scales, a sign that they do not primarily hunt warm-blooded prey. Although the Black-headed Python will take mammals and birds, it is much more attuned to preying on reptiles, including "goannas" (monitor lizards) and snakes, even venomous species. Across its range, the Black-headed Python may be found in wooded habitats and coastal forests, but it is especially common on rocky escarpments or in semidesert habitats, although it gives way to its congener in true desert. Black-headed Pythons are recorded from Port Headland, Western Australia, to Rockhampton in Queensland.

The Black-headed Python is instantly recognizable from the jet-black head and neck that earn it both its common and scientific names (*melano* = black, *-cephalus* = headed). The lips and snout do not bear the thermosensitive pits found in other pythons.

RELATED SPECIES
The only close relative of *Aspidites melanocephalus* is the Woma (*A. ramsayi*), an arid habitat specialist found across the center of Australia. It is banded brown and lacks the black head of *A. melanocephalus*.

Actual size

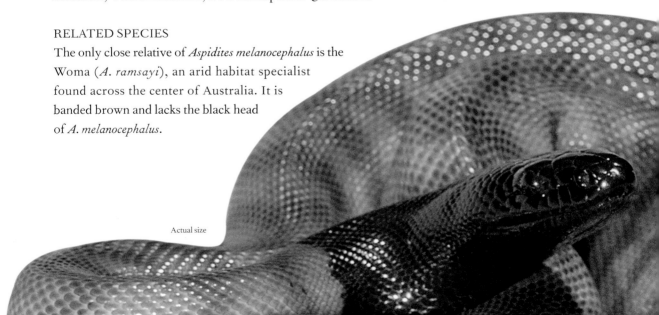

FAMILY	Pythonidae
RISK FACTOR	Nonvenomous, constrictor
DISTRIBUTION	Australasia: Bismarck Archipelago (New Britain, New Ireland, and satellite islands)
ELEVATION	0–985 ft (0–300 m) asl
HABITAT	Primary and secondary forest, and also oil-palm and coconut plantations
DIET	Small mammals and reptiles
REPRODUCTION	Oviparous, with clutches of up to 9 eggs
CONSERVATION STATUS	IUCN Least Concern, CITES Appendix II

ADULT LENGTH
2 ft 7 in–4 ft 3 in
(0.8–1.3 m)

BOTHROCHILUS BOA
BISMARCK RINGED PYTHON
(SCHLEGEL, 1837)

The Bismarck Ringed Python is often boldly patterned with orange or brown rings that alternate with black rings. The elongate head is often glossy black, like that of the white-lipped pythons. The patterning is most evident in juvenile specimens and may fade in adults.

The Bismarck Ringed Python is one of only two pythons found in the Bismarck Archipelago, Papua New Guinea, the other being the much larger Amethystine Python (*Simalia amethistina*, page 100). Although primarily a rainforest species, the Bismarck Python is also commonly encountered in oil-palm and coconut plantations, where it hunts small mammals, lizards such as skinks, and smaller snakes, and it is not above the act of cannibalism. This is probably the most secretive of the New Guinea pythons, being semi-fossorial, often hiding inside coconut husk piles or oil-palm frond rows, and rarely seen in the open, except after heavy rain, when it may be found on roads at night.

RELATED SPECIES

The closest relatives to *Bothrochilus boa* are the white-lipped pythons of genus *Leiopython* (page 87). In the past all these species have been included in one or other of these genera. This is one of only two New Guinea python species not found on mainland New Guinea, the other being the Biak Island White-lipped Python (*L. biakensis*).

Actual size

FAMILY	Pythonidae
RISK FACTOR	Nonvenomous, constrictor
DISTRIBUTION	Australasia: New Guinea (northern mainland and St. Matthias Archipelago)
ELEVATION	0–5,410 ft (0–1,650 m) asl
HABITAT	Lowland rainforest, riverine forest, swamps, and plantations
DIET	Small mammals
REPRODUCTION	Oviparous, with clutches of up to 17 eggs
CONSERVATION STATUS	IUCN not listed, CITES Appendix II

ADULT LENGTH
3 ft 3 in–5 ft 7 in
(1.0–1.7 m)

LEIOPYTHON ALBERTISII
NORTHERN WHITE-LIPPED PYTHON
(PETERS & DORIA, 1878)

87

Often called the D'Albertis Python, having being named in honor of the Italian naturalist Luigi Maria d'Albertis (1841–1901) who collected in New Guinea, the Northern White-lipped Python inhabits the lowland forests and plantations of northern New Guinea. It does not occur in the Bismarck Archipelago to the east, which is inhabited by the related Bismarck Ringed Python (*Bothrochilus boa*, page 86), but there is an isolated island population in the St. Matthias Islands to the north of New Ireland. Where it occurs, this python may be the most frequently encountered python species, especially on roads at night after heavy rain when it is out hunting bandicoots and rats. It is an inoffensive species that rarely bites, but like all pythons is capable of delivering a harmless but bloody bite.

The Northern White-lipped Python is usually chestnut brown on the body, merging to yellow-brown on the flanks and white below, with a strongly contrasting glossy black head and distinctive black and white "piano-key" markings along the lips. It is much more stunningly marked than its larger, maroon, southern relative *L. meridionalis*.

RELATED SPECIES

Leiopython albertisii is closely related to the other white-lipped pythons. These include the Biak Island White-lipped Python (*L. biakensis*) from the Schouten Islands, west New Guinea; Karimui White-lipped Python (*L. fredparkeri*) from Simbu, Papua New Guinea; Huon White-lipped Python (*L. huonensis*) from Morobe, Papua New Guinea, Wau White-lipped Python (*L. montanus*) from Eastern Highlands Province, and the much larger (8 ft 2 in / 2.5 m) Southern White-lipped Python (*L. meridionalis*) from the southern coastal lowlands. This genus is most closely related to the Bismarck Python (*Bothrochilus boa*).

Actual size

FAMILY	Pythonidae
RISK FACTOR	Nonvenomous, constrictor
DISTRIBUTION	Indo-Australia: Indonesia (eastern Lesser Sunda Islands), Timor-Leste, and possibly northwestern Australia
ELEVATION	0–1,480 ft (0–450 m) asl
HABITAT	Lowland forest and wetlands, creeks, plantations, and rice paddies
DIET	Small mammals, birds and their eggs, and possibly lizards, frogs, or fish
REPRODUCTION	Oviparous, with clutches of 8–20 eggs
CONSERVATION STATUS	IUCN not listed, CITES Appendix II

ADULT LENGTH
3 ft 3 in–7 ft 3 in
(1.0–2.2 m)

LIASIS MACKLOTI
MACKLOT'S WATER PYTHON
DUMÉRIL & BIBRON, 1844

Macklot's Water Python, also known as the Freckled Python, is found on the islands of the eastern Lesser Sundas and there is evidence to suggest it may also occur in northwestern Australia, which was very close to the island of Timor during glacial times when sea-levels were lower. Macklot's Water Pythons are associated with low-lying aquatic habitats such as seasonally flooded grasslands, wet forest, and rice paddies. They may venture into towns using creeks and culverts, and they are most often encountered in the wet season when many of these slow-moving snakes are killed on the roads. Prey probably ranges from rats to birds, and possibly lizards or frogs, but these are understudied pythons in nature. Heinrich Christian Macklot (1799–1832) was a German naturalist who was killed in an uprising in Java.

Actual size

RELATED SPECIES

Liasis mackloti comprises three subspecies, which some authors treat as species. Two are island endemics, the 5 ft (1.5 m) Sawu Python (*L. m. savuensis*) and the 7 ft 3 in (2.2 m) Wetar Python (*L. m. dunni*). The islands between Sawu and Wetar are inhabited by the 5 ft 2 in (1.6 m) nominate form. The Brown Water Python (*L. fuscus*) of northern Australia and southern New Guinea is its closest relative. Northwestern Australian water pythons may also belong to *L. mackloti*.

Macklot's Water Python is gray-brown with a scattering of darker scales, presenting a freckled appearance. It is off-white or yellowish below and on the lips. The head is moderately elongate and the iris of the eye may contrast with the body color, especially so in white-eyed Sawu Pythons.

FAMILY	Pythonidae
RISK FACTOR	Nonvenomous, powerful constrictor
DISTRIBUTION	Australia: northern Australia (Western Australia to Queensland)
ELEVATION	0–2,130 ft (0–650 m) asl
HABITAT	Lowland rainforest, riverine forest, coastal lowlands, and rocky outcrops with permanent water
DIET	Mammals, birds, and reptiles
REPRODUCTION	Oviparous, with clutches of 7–31 eggs
CONSERVATION STATUS	IUCN not listed, CITES Appendix II

ADULT LENGTH
10 ft–21 ft 3 in
(3.0–6.5 m)

LIASIS OLIVACEUS
AUSTRALIAN OLIVE PYTHON
GRAY, 1842

89

The Australian Olive Python is Australia's second largest python, after the Scrub Python (*Morelia kinghorni*) of Cape York Peninsula, Queensland. It is a powerful constrictor, capable of subduing and swallowing large mammals such as wallabies, but it also preys on birds, lizards such as frilled lizards and goannas (monitor lizards), and snakes, including other pythons. Australian Olive Pythons are particularly common in rocky habitats, especially those with permanent water sources, where it lingers in ambush for animals coming to drink, but it may also be encountered in dry tropical forests and wooded savannas. There are two main ranges for this species, with a northern population from the Kimberley of Western Australia, through the Northern Territory to western Cape York Peninsula, Queensland, and a separate, more southern population in the Pilbara of Western Australia.

Actual size

RELATED SPECIES
Liasis olivaceus is related to the water pythons (*L. fuscus* and *L. mackloti*, page 88) and the Papuan Python (*Apodora papuana*, page 84). There are two subspecies, the nominate Northern Olive Python and the isolated Western Olive Python (*L. o. barroni*) in the Pilbara, which may be a full species.

The Australian Olive Python is uniform brown above and white or yellow below. It has a robust body, and an elongate head with large scutes rather than granular scales on the dorsum.

FAMILY	Pythonidae
RISK FACTOR	Nonvenomous, powerful constrictor
DISTRIBUTION	Southeast Asia: Bangladesh to the Philippines, south to Timor-Leste
ELEVATION	0–4,270 ft (0–1,300 m) asl
HABITAT	Riverine forest, rainforest, and mangrove swamp
DIET	Mammals from rodents to deer; rare accounts of sun bears or humans being taken as prey
REPRODUCTION	Oviparous, with large clutches of 50–100 eggs
CONSERVATION STATUS	IUCN not listed, CITES Appendix II

ADULT LENGTH
Male
20–23 ft (6.0–7.0 m)

Female
20–33 ft (6.0–10 m)

MALAYOPYTHON RETICULATUS
RETICULATED PYTHON
(SCHNEIDER, 1801)

90

The Reticulated Python exhibits an instantly recognizable pattern of oranges, blacks, whites, and buffs, arranged in a reticulate or netlike design. Some specimens also have bright yellow heads. The head is more elongate than in other pythons, and bears a black and orange postocular stripe. The iris of the eye is orange with a black, vertically elliptical pupil.

The Reticulated Python is the longest snake species alive today, outsized females achieving 33 ft (10 m) and 165 lb (75 kg), although giant specimens are now rare. Primarily a mammal-predator, it has highly visible, heat-sensitive pits along the lips and on the tip of the snout, to enable it to hunt warm-blooded prey. It can take large prey, such as an adult female sun bear in Borneo, and humans are also occasionally taken. Young pythons sleep on branches over rivers, choosing sites above deep pools into which they plunge to escape predators. The female lays up to 100 leathery-shelled eggs that adhere together as a pile. She incubates them for 65–105 days and defends her nest against predation by egg-thieves. Annually, thousands of Reticulated Pythons are collected and slaughtered for their skins and meat.

RELATED SPECIES

Two small island populations are recognized as subspecies of *Malayopython reticulatus*, from Selayar (*M. r. saputrai*) and Tanahjampea (*M. r. jampeanus*), south of Sulawesi, Indonesia. The closest relative of this species is the Lesser Sunda Python (*M. timoriensis*, page 91), while recent research demonstrates a closer relationship to Australo-Papuan pythons (*Simalia*, pages 100–102), rather than Afro-Asian pythons (*Python*, pages 95–99).

Actual size

FAMILY	Pythonidae
RISK FACTOR	Nonvenomous constrictor
DISTRIBUTION	Southeast Asia: Lesser Sunda Islands from Lombok to Alor, Indonesia, but not Timor
ELEVATION	0–1,640 ft (0–500 m) asl
HABITAT	Dry and wet deciduous forest, montane forest
DIET	Small mammals, possibly also birds
REPRODUCTION	Oviparous, with clutches of 4–6 eggs
CONSERVATION STATUS	IUCN not listed, CITES Appendix II

ADULT LENGTH
3 ft 3 in–4 ft 3 in
(1.0–1.3 m)

MALAYOPYTHON TIMORIENSIS
LESSER SUNDA PYTHON
(PETERS, 1876)

This species is often called the "Timor Python" due to its scientific name, but this is an error based on a lack of collection data. The first specimen sent to the Berlin Museum, where it was described by Wilhelm Carl Hartwig Peters (1815–83), was shipped from Kupang in West Timor, leading Peters to believe it originated from Timor. However, Kupang was then an important Dutch East Indian trading and shipping port, and it is now believed that this specimen originated from elsewhere in the Lesser Sundas, but not Timor, where it does not appear to occur and where locals are unfamiliar with photographs of this species. However, they do recognize the other two Timorese pythons, Macklot's Water Python (*Liasis mackloti*, page 88) and the Reticulated Python (*Malayopython reticulatus*, page 90). The Lesser Sunda Python occurs in a variety of forest habitats where it uses its heat-sensitive pits to hunt warm-blooded mammals, and possibly also birds.

RELATED SPECIES

The only closely related species is the much larger and more widespread Reticulated Python, *Malayopython reticulatus* (page 90). Recent research has demonstrated that these two pythons are more closely related to Australo-Papuan pythons of the genus *Simalia* than they are to Afro-Asian pythons of the genus *Python*.

The Lesser Sunda Python exhibits the same body shape and elongate head of its much larger relative, the Reticulated Python. The body has a brown or yellow-brown background, with black reticulations, at least anteriorly, and a pale yellow underside. The iris of the eye is brown, and less contrasting than in the Reticulated Python, and the postocular stripe is faint or absent.

Actual size

FAMILY	Pythonidae
RISK FACTOR	Nonvenomous, constrictor
DISTRIBUTION	Australia: Western Australia (Kimberley)
ELEVATION	33–670 ft (10–205 m) asl
HABITAT	Monsoon forest in rocky riverine gorges
DIET	Probably small mammals and/or birds
REPRODUCTION	Oviparous, with clutches of 10–14 eggs
CONSERVATION STATUS	IUCN not listed, CITES Appendix II

ADULT LENGTH
5 ft–6 ft 7 in
(1.5–2.0 m)

MORELIA CARINATA
ROUGH-SCALED PYTHON
(SMITH, 1981)

92

The Rough-scaled Python is easily recognized by its strongly keeled scales, and the single round scale in the center of its head. In coloration and patterning it rather resembles the widespread Carpet Python (*Morelia spilota*).

The Rough-scaled Python is the only python with strongly keeled scales. It is also characterized by a single large, round scale in the center of the head, surrounded by small granular scales. It has the smallest range of any Australian python, being confined to the lower reaches of the Mitchell, Hunter, and Moran rivers, in the Kimberley, Western Australia, where it inhabits stands of monsoon forest in the steep rocky, riverine gorges. This is an understudied python in nature; until 2000 only six specimens were known. It could have been threatened by poaching for the reptile trade, or a local natural disaster, but thanks to captive breeding efforts by the Australian Reptile Park in New South Wales, its future seems secure. Its wild prey preferences and biology are still unknown.

RELATED SPECIES

The closest relatives of *Morelia carinata* are the widespread Carpet Python (*M. spilota*, page 93) and the Centralian Python (*M. bredli*) of central Australia.

Actual size

FAMILY	Pythonidae
RISK FACTOR	Nonvenomous, constrictor
DISTRIBUTION	Australasia: Australia and southern New Guinea
ELEVATION	0–3,690 ft (0–1,125 m) asl
HABITAT	Tropical rainforest, dry woodland, savanna woodland, rocky outcrops, and urban environments
DIET	Small mammals, birds, and lizards
REPRODUCTION	Oviparous, with clutches of 9–52 eggs
CONSERVATION STATUS	IUCN Least Concern, CITES Appendix II

ADULT LENGTH
5 ft–8 ft 2 in
(1.5–2.5 m)

MORELIA SPILOTA
CARPET PYTHON
(LACÉPÈDE, 1804)

93

Across Australia, except the arid center, Carpet Pythons exist as six subspecies. The Diamond Python, the nominate subspecies, inhabits the Hawkesbury sandstone outcrops of New South Wales. Carpet Pythons are also common in southern New Guinea. Habitats include wet and dry forests, savanna-woodland, rocky habitats, and even cities. The IUCN lists it as Least Concern, but some subspecies receive Australian state protection.

RELATED SPECIES
Close relatives are the Centralian Python (*Morelia bredli*) and the Rough-scaled Python (*M. carinata*, page 92). The other subspecies are the Eastern Carpet Python (*M. s. mcdowelli*) and Inland Carpet Python (*M. s. metcalfei*), east and west of the Great Dividing Range; Jungle Carpet Python (*M. s. cheynei*), Atherton Tablelands, Queensland; Top End Carpet Python (*M. s. variegata*), Arnhem Land; and Western Carpet Python (*M. s. imbricata*), south-western Australia. The status of Papuan populations is undetermined.

Actual size

The Carpet Python is generally reddish or brown with dark-edged yellow cross-bars and blotches, but the Diamond Python (*Morelia spilota spilota*) is yellow or gray with every scale edged with black and a series of dark-edged cross-bars, while the Jungle Carpet Python (*M. s. cheynei*) exhibits a bold pattern of black and yellow cross-bands or stripes.

FAMILY	Pythonidae
RISK FACTOR	Nonvenomous, constrictor
DISTRIBUTION	Australasia: southern New Guinea, Indonesia (Aru Islands), and Australia (Queensland)
ELEVATION	0–5,910 ft (0–1,800 m) asl
HABITAT	Rainforest, monsoon forest, and oil-palm plantations
DIET	Small mammals, birds, and lizards
REPRODUCTION	Oviparous, with large clutches of 8–30 eggs
CONSERVATION STATUS	IUCN Least Concern, CITES Appendix II

ADULT LENGTH
3 ft 3 in–5 ft
(1.0–1.5 m)

MORELIA VIRIDIS
SOUTHERN GREEN TREE PYTHON
SCHLEGEL, 1872

The Southern Green Tree Python is a stunning snake, being vivid green with markings of yellow, white, or blue, with some specimens almost entirely blue or blue-green. Hatchlings start life bright yellow or orange, with markings of black and white, changing to the adult livery at 15 months of age.

The Southern Green Tree Python's bright green livery makes it one of the most iconic and attractive snake species in the world, although the yellow or orange juveniles are even more stunning. This highly arboreal python has a slender, muscular body and a prehensile tail. It sleeps looped over a branch like a coiled rope. The head is large, the mouth can gape widely, and long teeth will secure struggling prey. Prey consists of small mammals such as mice, rats, and small bandicoots, but birds and lizards are also taken. It is nocturnal and may be encountered hunting on the ground, after rain. Although a tropical forest species, it has adapted to live in oil-palm plantations.

RELATED SPECIES

Morelia viridis was split into two species separated by the central cordillera of New Guinea. They appear morphologically identical, but are genetically different. The Northern Green Tree Python (*M. azurea*) inhabits northern New Guinea and nearby islands. To the layman, green tree pythons also closely resemble the emerald tree boas (*Corallus caninus*, page 108, and *C. batesii*) of South America, providing a classic example of convergent evolution.

Actual size

FAMILY	Pythonidae
RISK FACTOR	Nonvenomous, powerful constrictor
DISTRIBUTION	Indo-China: Myanmar to Thailand and China, west to Nepal, with isolated populations in Java, Bali, and Sulawesi; introduced into Florida, USA
ELEVATION	0–3,940 ft (0–1,200 m) asl
HABITAT	Tropical dry forest, and riverine grassland and woodland
DIET	Mammals from rodents to deer, and sometimes birds
REPRODUCTION	Oviparous, with large clutches of 30–100 eggs
CONSERVATION STATUS	IUCN Vulnerable, CITES Appendix II

ADULT LENGTH
10–22 ft
(3.0–6.7 m)

PYTHON BIVITTATUS
BURMESE PYTHON
KUHL, 1820

95

The Burmese Python is the second-largest snake species in Asia, after the Reticulated Python (*Malayopython reticulatus*, page 90), but it achieves a greater girth than Reticulated Pythons of the same length. It inhabits tropical dry forests and riverine grasslands in Indo-China and along the Ganges Valley of southern Nepal, but is absent from the rainforests of Malaysia, Borneo, and Sumatra. Isolated populations occur in Java, Bali, and Sulawesi, which are remnants of a wider population from glacial times, when sea-levels were lower and dry forest more extensive. Mammals are the main prey, up to the size of deer, and are ambushed and constricted. Burmese Pythons were accidentally introduced into the Florida Everglades during a hurricane, where they thrive in habitat similar to their native Asian grasslands. The population has grown despite thousands being captured and culled.

The Burmese Python's patterning comprises brown saddles on a yellowish-brown background, with similarly colored blotches on the flanks and a pale brown V-shape on the dorsum of the head. The best character to distinguish it from the similar Indian Python (*Python molurus*) is the presence of a subocular scale under the eye.

RELATED SPECIES
Python bivittatus is closely related to the Indian Python (*Python molurus*, page 97) and was treated as a subspecies for many years. The isolated population on Sulawesi is now recognized as a separate subspecies, *P. b. progschai*.

Actual size

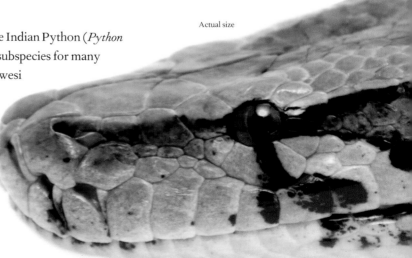

FAMILY	Pythonidae
RISK FACTOR	Nonvenomous, constrictor
DISTRIBUTION	Australasia: Borneo (Sarawak, Sabah, Brunei, and Kalimantan)
ELEVATION	0–3,280 ft (0–1,000 m) asl
HABITAT	Lowland and mid-montane rainforest, riverine forest, swamps, and plantations
DIET	Small mammals and birds
REPRODUCTION	Oviparous, with clutches of 10–15 eggs
CONSERVATION STATUS	IUCN Least Concern, CITES Appendix II

ADULT LENGTH
3 ft 3 in–6 ft 7 in
(1.0–2.0 m)

PYTHON BREITENSTEINI
BORNEO SHORT-TAILED PYTHON
STEINDACHNER, 1881

The Borneo Short-tailed Python is a relatively short but very stout-bodied snake with a short tail. It is generally tan in color with a patterning of irregular dark brown saddles and blotches. The dorsum of the head is tan while the sides of the head are much darker.

Actual size

The Borneo Short-tailed Python is endemic to the island of Borneo, where it is widely distributed and relatively common. This is a heavy-bodied terrestrial python that is not built for climbing into vegetation, but rather leads a secretive existence in the leaf litter. Although primarily an inhabitant of low montane and lowland rainforest and wetland areas, it has adapted well to living in oil-palm plantations, where it feeds on small mammals and birds, especially the abundant rodents. Rat damage is one of the most serious threats faced by the oil-palm industry and snakes like the python are some of the best biological controls available, yet this does not save the python, which is often killed on sight by workers.

RELATED SPECIES

Python breitensteini was formerly one of three subspecies of the Short-tailed or Blood Python (*P. curtus*). *Python curtus* is from western and southern Sumatra and known as the Sumatran Short-tailed Python, while the other former subspecies, the Malayan Short-tailed or Blood Python (*P. brongersmai*), occurs on the Malay Peninsula and eastern Sumatra. This last species is well named, being red in color and willing to draw blood from anybody who tries to handle it. A new and related species has been described from Thailand (*P. kyaiktiyo*).

FAMILY	Pythonidae
RISK FACTOR	Nonvenomous, powerful constrictor
DISTRIBUTION	Asia: India, Pakistan, and Sri Lanka
ELEVATION	0–8,200 ft (0–2,500 m) asl
HABITAT	Dry forest, rainforest, grasslands, and scrubland
DIET	Mammals, from rodents to deer
REPRODUCTION	Oviparous, with large clutches of up to 107 eggs
CONSERVATION STATUS	IUCN not listed, CITES Appendix I

ADULT LENGTH
10–22 ft
(3.0–6.7 m)

PYTHON MOLURUS
INDIAN PYTHON
(LINNAEUS, 1758)

97

The Indian Python inhabits dry forests and grasslands and spends the day sheltering in porcupine burrows, emerging at night to hunt mammals, from bats and rats to jackals, civets, wild boar, and deer, which are ambushed and killed by constriction. Although large, they do not hunt humans. The IUCN does not currently list the Indian Python, but it is placed on Appendix I (most protected) of CITES, and India also protects the species domestically. The Indian Python occurs throughout Sri Lanka, India, and Pakistan, but gives way to its close relative, the Burmese Python (*Python bivittatus*, page 95), to the north in Nepal and in Assam, northeastern India.

The Indian Python is similar to the Burmese Python (*Python bivittatus*), but it has a pale gray rather than yellowish ground color, and a less distinct arrowhead on the dorsum of its head.

Actual size

RELATED SPECIES

Python molurus is related to its former subspecies, the Burmese Python (*P. bivittatus*). The Sri Lankan population was also once treated as a subspecies (*P. m. pimbura*), but the subspecies is generally not recognized today. *Python molurus* and *P. bivittatus* can be distinguished by a subocular scale between the eye and the supralabials, present in *P. bivittatus* but absent in *P. molurus*.

FAMILY	Pythonidae
RISK FACTOR	Nonvenomous, powerful constrictor
DISTRIBUTION	Southern and East Africa: Namibia to southern Kenya, Mozambique, and eastern and northeastern South Africa
ELEVATION	0–5,910 ft (0–1,800 m) asl
HABITAT	Lowland grassland and woodlands, coastal thickets, and rocky outcrops
DIET	Mammals and birds
REPRODUCTION	Oviparous, with large clutches of up to 100 eggs
CONSERVATION STATUS	IUCN not listed, CITES Appendix II

ADULT LENGTH
13 ft–16 ft 5 in
(4.0–5.0 m)

PYTHON NATALENSIS
SOUTHERN AFRICAN PYTHON
SMITH, 1840

The Southern African Python is a stoutly built python with a body pattern of dark brown irregular saddles on a light brown background, and a brown arrowhead marking on the dorsum of its head.

The Southern African Python is Africa's second-longest snake, after the Central African Python (*Python sebae*), which may achieve 24 ft 7 in (7.5 m). It is an inhabitant of low-lying coastal thickets, savanna woodland, and rocky outcrops, often in close proximity to water, but does not occur in arid deserts or cool montane habitats. In South Africa it is found in the north and east, but despite its scientific name, ironically, it is rare in KwaZulu-Natal. A large and powerful constrictor, it preys on mammals from rats to antelopes and monkeys, and on rare occasions has even been documented preying on humans. Recently fed pythons are themselves vulnerable to attack by humans or carnivores such as African wild dogs or hyenas. They will seek a burrow in which to digest their meals in safety.

RELATED SPECIES

The closest relative of *Python natalensis* is the Central African Python (*P. sebae*), from West, Central, and East Africa, of which *P. natalensis* was formerly a subspecies. The two can be distinguished by the granular dorsal head scalation of *P. natalensis*, compared to the large head scutes of *P. sebae*.

Actual size

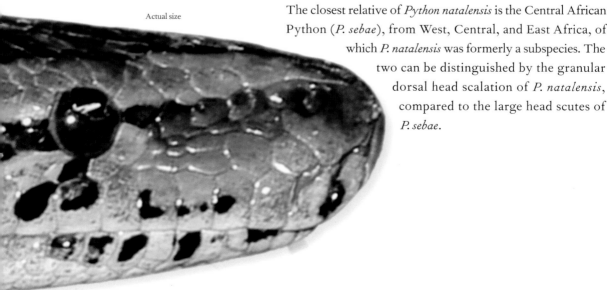

FAMILY	Pythonidae
RISK FACTOR	Nonvenomous, constrictor
DISTRIBUTION	West and Central Africa: Senegal to Sudan
ELEVATION	0–3,280 ft (0–1,000 m) asl
HABITAT	Open savannas and savanna woodland or scrub
DIET	Small mammals
REPRODUCTION	Oviparous, with clutches of 5–15 eggs
CONSERVATION STATUS	IUCN Least Concern, CITES Appendix II

ADULT LENGTH
3 ft 3 in–5 ft
(1.0–1.5 m)

PYTHON REGIUS
BALL PYTHON
(SHAW, 1802)

99

Although referred to as the Ball Python in the United States, this species is called the Royal Python in the UK, a literal translation of its specific name. The name Ball Python is derived from the defensive posture these pythons adopt, of coiling into a ball with their heads buried in the center. Ball Pythons are inoffensive snakes and rarely bite. This species is found across a wide swath of West and Central Africa, and spend much of their time in animal burrows, emerging at night to hunt. Hundreds of thousands of Ball Pythons were imported into the West for the pet trade, but hopefully this unsustainable harvest has declined due to the massive captive breeding market, which has also seen the development of many color morphs known as "designer morphs." If the snake-keeping public prefer to acquire the unusual albino color morphs, this may take the pressure off the wild populations.

Actual size

RELATED SPECIES
Python regius is one of Africa's two smallest python species, the other species being the Angolan Python (*P. anchietae*) of Angola and Namibia.

The Ball Python is a stout but relatively short python with a distinctive pattern of pale brown saddles and stripes on a dark brown to black background, and a dark brown head with a pale stripe through each eye.

FAMILY	Pythonidae
RISK FACTOR	Nonvenomous, powerful constrictor
DISTRIBUTION	New Guinea: west New Guinea and Papua New Guinea, including Bismarck Archipelago (New Britain, New Ireland)
ELEVATION	0–5,580 ft (0–1,700 m) asl
HABITAT	Rainforest, riverine forest, mangrove and freshwater swamps, savannas, and plantations
DIET	Mammals and birds; also lizards as juveniles
REPRODUCTION	Oviparous, with clutches of 10–11 eggs
CONSERVATION STATUS	IUCN not listed, CITES Appendix II

ADULT LENGTH
16 ft 5 in–19 ft 8 in
(5.0–6.0 m)

SIMALIA AMETHISTINA
AMETHYSTINE PYTHON
(SCHNEIDER, 1801)

The Amethystine Python is fairly variable in coloration, ranging from dark brown to straw yellow, with or without darker markings, but all specimens exhibit the iridescent sheen in daylight. Amethystine pythons are slender snakes with elongate heads and numerous thermosensitive labial pits.

Actual size

The Amethystine Python was a very widely distributed species, until the Queensland subspecies was elevated to species status and the Indonesian island populations were described as separate species. The remaining New Guinea–Bismarck Archipelago *Simalia amethistina* may still be a species complex. This is a slender inhabitant of both closed and open habitats, and it feeds on a variety of mammals, from rats and bandicoots to small wallabies, as well as birds, although juveniles may also take skinks. Adults are often found in the vicinity of flying fox rookeries where the pythons find easy prey. The common and scientific names of this python originate from the "oil-on-water" iridescent sheen that is visible on the scales in daylight, and which may serve to camouflage the outline of a sleeping python.

RELATED SPECIES

The genus *Simalia* contains those species formerly in *Morelia* that exhibit regular dorsal head scutes, rather than granular scales. The closest relatives of *S. amethistina* are the former subspecies, the Scrub Python (*S. kinghorni*) from Queensland, and the former island populations, the Southern Moluccan Python (*S. clastolepis*), Halmahera Python (*S. tracyae*), and Tanimbar Python (*S. nauta*), all from Indonesia.

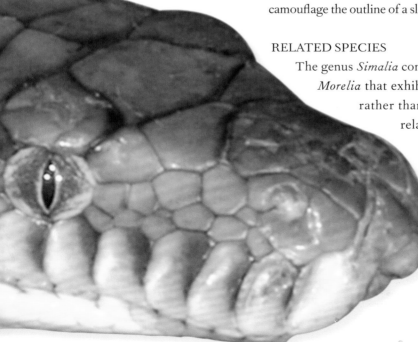

FAMILY	Pythonidae
RISK FACTOR	Nonvenomous, constrictor
DISTRIBUTION	New Guinea: west New Guinea and Papua New Guinea
ELEVATION	4,270–9,840 ft (1,300–3,000 m) asl
HABITAT	Montane rainforest
DIET	Poorly known; mammals, and possibly birds
REPRODUCTION	Oviparous, with clutches of 14–20 eggs
CONSERVATION STATUS	IUCN not listed, CITES Appendix II

ADULT LENGTH
6 ft 7 in–19 ft 8 in
(2.0–3.0 m)

SIMALIA BOELENI
BOELEN'S PYTHON
(BRONGERSMA, 1953)

101

Boelen's Python, named for the Dutch surgeon who collected
the holotype, is a stunning species also known as the Black
Python due to its unique coloration. Unlike other New
Guinea pythons, it only occurs in montane rainforest
above 4,270 ft (1,300 m) asl, in cool, humid conditions.
Endemic to New Guinea, it is distributed along the
length of the Central Cordillera, but is nowhere
common. Sought after by the pet trade, it is also
threatened by the wholesale clearance of montane
forest for timber, slash-and-burn agriculture, oil
prospecting, and gold mining. Its natural history is
poorly documented, with even its prey preferences
an open question, although it probably feeds on rats,
bandicoots, or ground-nesting birds.

RELATED SPECIES
The closest relative of *Simalia boeleni* is probably
the Amethystine Python (*S. amethistina*, page 100).

Actual size

Boelen's Python is iridescent black or blue-black with white
or yellow markings on the lips, throat, and anterior underside,
with fingers of white extending upward onto the flanks. The
iris of the eye is light gray. It is a stockily built snake with a
broad, chunky head. Juveniles are red-brown in color.

FAMILY	Pythonidae
RISK FACTOR	Nonvenomous, constrictor
DISTRIBUTION	Australia: Northern Territory (west Arnhem Land)
ELEVATION	0–950 ft (0–290 m) asl
HABITAT	Sandstone escarpments
DIET	Small mammals, and possibly lizards
REPRODUCTION	Oviparous, with clutches of 6–10 eggs
CONSERVATION STATUS	IUCN not listed, CITES Appendix II

ADULT LENGTH
10 ft–16 ft 5 in
(3.0–5.0 m)

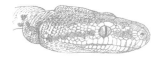

SIMALIA OENPELLIENSIS
OENPELLI PYTHON
(GOW, 1977)

The Oenpelli Python is very slender with an elongate head and almost forward-facing eyes. It is silver-gray with markings confined to irregular, rather subdued mottling of darker pigment, although this species can change color depending on the prevailing conditions, from grayish at night to reddish in the daytime.

Named for the small town of Oenpelli on the East Alligator River, the iridescent Oenpelli Python is believed by some to be the "Rainbow Serpent" of Aboriginal Dreamtime. It certainly inhabits the right habitat—sandstone escarpments in the Kakadu National Park and surrounding Arnhem Land, where it is known as "Nawaran." This is a very rarely encountered species that hunts bandicoots, possums, rock rats, and flying foxes in the crevices and caves of the escarpments at night, although lizards may feature in the diets of juveniles. As with any species that inhabits small, limited ranges, there are concerns for the future of this python, and it is listed as Threatened in the Northern Territory. The Oenpelli Python is the subject of a captive breeding and conservation project involving local people.

RELATED SPECIES

The Australian python that is closest to *Simalia oenpelliensis* is probably the much larger Scrub Python (*S. kinghorni*) of Queensland.

Actual size

FAMILY	Calabariidae
RISK FACTOR	Nonvenomous, constrictor
DISTRIBUTION	West and Central Africa: Guinea, Sierra Leone, Liberia, Côte d'Ivoire, Ghana, Togo, Benin, Nigeria, Cameroon, Central African Republic, Equatorial Guinea, Gabon, Congo, and DRC
ELEVATION	10–3,440 ft (3–1,050 m) asl
HABITAT	Rainforests, evergreen and deciduous forests, savanna woodland, and plantations
DIET	Small mammals, shrews, and rodents
REPRODUCTION	Oviparous, with clutches of 1–4 elongate eggs
CONSERVATION STATUS	IUCN not listed

ADULT LENGTH
2 ft–3 ft 3 in
(0.6–1.03 m)

CALABARIA REINHARDTII

CALABAR GROUND BOA

(SCHLEGEL, 1851)

103

The Calabar Ground Boa, named for a region of Nigeria, was previously thought to be a python because it lays elongate eggs while boas are live-bearers. It is found in the rainforests of West and Central Africa, and also inhabits savanna woodland and human-mediated habitats, such as oil-palm plantations. Nocturnal and fossorial, the Calabar Ground Boa is usually only seen after heavy rain, when it emerges to hunt. Prey comprises mainly small mammals, such as rodents and shrews. When it feels threatened the Calabar Ground Boa will coil into a tight ball, its head in the center and its tail acting as a pseudo-head, distracting attention away from the real head. The scientific name honors Johannes Theodor Reinhardt (1816–82), a Danish naturalist.

RELATED SPECIES

As the only member of the Calabariidae, *Calabaria reinhardtii* does not have any close relatives. In the past it has been linked with the pythons (Pythonidae, pages 83–102), the rosy and rubber boas (Charinidae, pages 120–121) of western USA, Canada, and Mexico, and the Afro-Asian sand boas (Erycidae, pages 117–119), but current opinion is that the true boas (Boidae, pages 104–114) are its closest relatives and it may represent a basal element within the boas.

The Calabar Ground Boa is a cylindrical-bodied snake with a rounded head, indistinct from the neck, small eyes, and a short stumpy tail. Its scales are smooth and tight-fitting, and its coloration and patterning comprise orange and brown blotches and scattered scales on a dark brown to dark gray background.

Actual size

FAMILY	Boidae
RISK FACTOR	Nonvenomous, powerful constrictor
DISTRIBUTION	South America: Colombia, Venezuela, Trinidad, the Guianas, Brazil, Peru, Ecuador, Bolivia, Paraguay, Uruguay, and Argentina
ELEVATION	0–3,280 ft (0–1,000 m) asl
HABITAT	Rainforest, grassland, semidesert, islands, cultivated land, and human settlements
DIET	Mammals, birds, and lizards
REPRODUCTION	Viviparous, with litters of 10–64 neonates
CONSERVATION STATUS	IUCN not listed, CITES Appendix II (except *B. c. occidentalis* Appendix I)

ADULT LENGTH
6 ft 7 in–10 ft,
rarely 16 ft 5 in
(2.0–3.0 m,
rarely 5.0 m)

BOA CONSTRICTOR
COMMON BOA
LINNAEUS, 1758

104

The Common Boa, or Boa Constrictor, inhabits tropical rainforests, woodland, grassland, semidesert, cultivated land, and human settlements. Nocturnal, it hunts aloft or on the ground, feeding mostly on mammals, birds, and lizards. It has been known to take vampire bats, village dogs, porcupines, coatimundis, green iguanas, deer, and even an ocelot. In Manaus, Brazil it is an important control on the disfiguring human disease leishmaniasis, because it preys on the common opossum that acts as a disease reservoir. It is not large enough to endanger humans.

RELATED SPECIES

Amazonian and Guianan boas belong to the nominate subspecies, and three other subspecies are recognized: the Long-tailed Boa (*B. c. longicauda*), the Peruvian Red-tail Boa (*B. c. ortonii*), and Argentine *Boa* (*B. c. occidentalis*). Four other Boa species also exist: the Central American or Imperial Boa (*B. imperator*), and Western Mexico Boa (*B. sigma*), and from the Lesser Antilles, the St. Lucia Boa (*B. orophias*) and Dominican Clouded Boa (*B. nebulosa*).

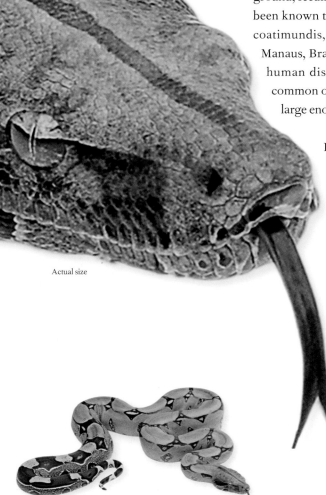

Actual size

The Common Boa is a powerfully built snake with a prehensile tail, angular head, and relatively small eyes. It is gray or brown with a series of distinctive dark brown dorsolateral saddles. The head is gray-brown with a faint brown midline and a dark brown postocular stripe. The tail is ringed red in juveniles, but usually turns brown with maturity.

FAMILY	Boidae
RISK FACTOR	Nonvenomous, powerful constrictor
DISTRIBUTION	West Indies: Cuba and Isla de la Juventud
ELEVATION	0–1,070 ft (0–325 m) asl
HABITAT	Moist and dry woodland, rocky outcrops, and caves
DIET	Mammals, birds, lizards, and snakes
REPRODUCTION	Viviparous, with litters of 1–7 neonates
CONSERVATION STATUS	IUCN Near Threatened, CITES Appendix II

ADULT LENGTH
6 ft 7 in–10 ft,
rarely 13 ft
(2.0–3.0 m,
rarely 4.0 m)

CHILABOTHRUS ANGULIFER
CUBAN BOA
(COCTEAU & BIBRON, 1840)

The Cuban Boa is the largest snake in the Caribbean, excluding
Trinidad, which is home to the Green Anaconda (*Eunectes
murinus*, page 112) and Common Boa (*Boa constrictor*, page
104). It is found throughout the island of Cuba, including
Guantánamo Bay, and on the Isla de la Juventud. It inhabits
woodland habitats, but is also associated with rocky
outcrops, and especially caves where it hunts bats.
Adults also hunt rodents, village chickens, iguanas,
and woodsnakes (*Tropidophis*, pages 68–69),
while juveniles take anole lizards and mice.
Most specimens are less than 10 ft (3 m) in total
length, but one specimen collected at
Guantánamo Bay was reported to measure
16 ft (4.85 m). Cuban Boas are noted for their
irascible temperament, biting easily, but they are
not dangerous to humans.

Actual size

RELATED SPECIES
The West Indian boas were all once contained in the South
American genus *Epicrates* but were relatively recently
transferred to the resurrected endemic West Indian genus
Chilabothrus. *Chilabothrus angulifer* is the largest of the 12
species. It has no subspecies. Other species that reach over
6 ft 7 in (2 m) include the Haitian Boa (*C. striatus*) from
Hispaniola (Haiti and Dominican Republic), the widely
distributed Bahamian Boa (*C. strigilatus*), and the recently
described Silver Boa (*C. argenteum*), from the Conception
Island Bank, in the Bahamas.

The Cuban Boa is a muscular snake with a
prehensile tail and a broad, rounded head with
moderately large eyes. It is usually brown above
and paler gray on the lower flanks, with variable
patterning, western specimens being boldly marked
with a black or dark brown angular pattern, while
eastern specimens exhibit indistinct or no markings.

FAMILY	Boidae
RISK FACTOR	Nonvenomous, constrictor
DISTRIBUTION	West Indies: Puerto Rico (Isla Mona)
ELEVATION	0–165 ft (0–50 m) asl
HABITAT	Dry subtropical forest or rocky outcrops
DIET	Small lizards, rarely mammals
REPRODUCTION	Viviparous, with litters of 4 neonates
CONSERVATION STATUS	IUCN Endangered, CITES Appendix I

ADULT LENGTH
31½–35½ in,
rarely 3 ft 3 in
(800–900 mm,
rarely 1.0 m)

CHILABOTHRUS MONENSIS
MONA ISLAND BOA
(ZENNECK, 1898)

106

The Mona Island Boa is a slender snake with a distinct head, moderately large eyes, and a prehensile tail. It is dorsally gray-brown with a pattern comprising transverse dark brown angular blotches, which may be broken at the midline. The head is patternless, apart from a postocular stripe in some specimens, and the undersides are immaculate white or stippled with brown.

The Mona Island Boa is endemic to the tiny Puerto Rican island of Isla Mona, with an area of 22 sq miles (57 sq km), located in the Mona Passage, between Puerto Rico and the Dominican Republic. Habitat loss and the introduction of feral cats, which kill the small boas, are among the reasons for its decline. The Mona Island Boa is an inhabitant of dry subtropical forest but it is also found on rocky outcrops and it may enter caves. It is also said to utilize termitaria and even the rafters of houses. It is a nocturnal predator of sleeping anole lizards and occasionally takes large ameiva lizards, rodents, and small bats, although these may also be declining on Isla Mona.

RELATED SPECIES

Chilabothrus monensis from the west of Puerto Rico is most closely related to the Virgin Islands Boa (*C. granti*), a former subspecies occurring on the Isla Culebra and Cayo Diablo, and the US and British Virgin Islands, to the east of Puerto Rico.

Actual size

FAMILY	Boidae
RISK FACTOR	Nonvenomous, constrictor
DISTRIBUTION	Caribbean: Jamaica and Great Goat Island
ELEVATION	0–130 ft (0–40 m) asl
HABITAT	Woodland, forest, and rocky outcrops, including caves
DIET	Small mammals, birds, and lizards
REPRODUCTION	Viviparous, with litters of 3–39 neonates
CONSERVATION STATUS	IUCN Vulnerable, CITES Appendix I

ADULT LENGTH
5 ft–6 ft 7 in
(1.5–2.0 m)

CHILABOTHRUS SUBFLAVUS
JAMAICAN BOA
(STEJNEGER, 1901)

107

Known locally as "yellow snake," the Jamaican Boa is listed as Appendix I of CITES, a status level only applied to ten other snake species. Once found throughout Jamaica, it has been virtually extirpated in the wild. Causes for this decline include active persecution, habitat loss to housing, agriculture and mining, and the introduction of pigs, dogs, cats, and the small Indian mongoose. Today it may only survive in the wild on Great Goat Island, off the south of Jamaica. However, captive breeding of this species by zoos has created a large captive population and prevented its ultimate extinction. It is a woodland inhabitant that also inhabits rocky outcrops and caves, preying on rodents, bats, and lizards, but there is also a record of a parrot being eaten. Jamaica has no dangerous snakes.

Actual size

RELATED SPECIES
Many of the West Indian island-dwelling boas are also Endangered like *Chilabothrus subflavus*. The Puerto Rican Boa (*C. inornatus*) and the Mona Island Boa (*C. monensis*, page 106) are also on CITES Appendix I.

The Jamaican Boa is an agile-bodied snake with a long, angular head, large eyes, and a prehensile tail. The anterior body and head are olive to yellow with black scale tips, with black transverse bands appearing by the midbody, which increase in intensity until the posterior of the body and the tail are virtually black. All scales are iridescent. Juveniles are often yellowish or pinkish with dark bands.

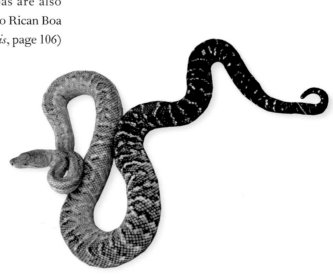

FAMILY	Boidae
RISK FACTOR	Nonvenomous, constrictor
DISTRIBUTION	Northern South America: Venezuela, Guyana, Suriname, French Guiana, and Brazil
ELEVATION	0–3,280 ft (0–1,000 m) asl
HABITAT	Lowland tropical rainforest
DIET	Small mammals, and lizards
REPRODUCTION	Viviparous, with litters of 6–15 neonates
CONSERVATION STATUS	IUCN not listed, CITES Appendix II

ADULT LENGTH
4 ft 7 in–6 ft,
rarely 6 ft 7 in
(1.4–1.8 m,
rarely 2.0 m)

108

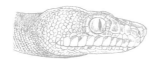

CORALLUS CANINUS
GUIANAN EMERALD TREEBOA
(LINNAEUS, 1758)

Until recently the Emerald Treeboa was a single species distributed throughout northern South America, east of the Andes, but today this species has been split into two separate species, with *Corallus caninus* the name applied to emerald treeboas from north of the Amazon River and east of the Rio Negro, in northeastern Brazil, eastern Venezuela, and the Guianas. Emerald treeboas are extremely similar in appearance to the green tree pythons (*Morelia azurea* and *M. viridis*, page 94) of New Guinea, an example of convergent evolution where two unrelated and distant species resemble one another. Emerald treeboas are nocturnal, arboreal lowland rainforest snakes that hunt and constrict rodents and lizards. Their long front teeth maintain a grip on the struggling prey following the strike.

RELATED SPECIES

The Amazonian Emerald Tree Boa (*Corallus batesii*) inhabits the remainder of the original range, in Colombia, western and central Brazil, Peru, and northern Bolivia. The species most closely related to the two emerald treeboas are the Annulated Treeboa (*C. annulatus*) of Central America and northwestern Colombia, and Blomberg's Treeboa (*C. blombergi*) of western Ecuador.

The Guianan Emerald Treeboa is a muscular snake with a laterally compressed body, a long head, vertically elliptical pupils, and a prehensile tail. The adult livery is bright green with irregular white transverse bars across the back, while juveniles may be yellow, green, or orange. Emerald treeboas can be distinguished from green tree pythons by their longer heads and the presence of heat-sensitive pits on all the supralabials rather than just a few.

Actual size

FAMILY	Boidae
RISK FACTOR	Nonvenomous, constrictor
DISTRIBUTION	Northern South and southern Central America: Costa Rica, Panama, Colombia, Venezuela, and Trinidad
ELEVATION	0–3,280 ft (0–1,000 m) asl
HABITAT	Lowland rainforest, wet and dry forests, flooded grassland, and mangrove swamps
DIET	Small mammals, birds, and lizards
REPRODUCTION	Viviparous, with litters of 9–15 neonates
CONSERVATION STATUS	IUCN not listed, CITES Appendix II

ADULT LENGTH
4 ft 7 in–6 ft 7 in,
occasionally 7 ft 7 in
(1.4–2 m,
occasionally 2.3 m)

CORALLUS RUSCHENBERGERII
CARIBBEAN COASTAL TREEBOA
(COPE, 1875)

109

The Caribbean Coastal Treeboa is found on the American mainland from Pacific coastal Costa Rica to Caribbean coastal Venezuela, Trinidad, Tobago, and many smaller offshore islands. It is the largest member of the genus *Corallus*. Treeboas are nocturnal and, curiously, they are the only snakes known that can be located by their eyeshine, reflecting a flashlight beam from hundreds of feet away. Although largely arboreal, treeboas often hunt terrestrial prey from a head-down posture, dangling close to the ground, ready to strike anything running underneath. The prey of the Caribbean Coastal Treeboa comprises bats, birds, rodents, mouse opossums, and lizards. Often large numbers of treeboas may be found in a relatively small area such as a mangrove swamp. Commodore Dr. William Samuel Waithman Ruschenberger (1807–95) was a U.S. Navy and Army physician.

RELATED SPECIES

Corallus ruschenbergerii is most closely related to the Amazonian Treeboa (*C. hortulanus*) of northern South America, and the Grenadian Treeboa (*C. grenadensis*) and Cook's Treeboa (*C. cookii*), both from the Lesser Antilles. A more distant relative is the extremely rare Cropani's Treeboa (*C. cropanii*) from the Brazilian Atlantic Forests. Not seen alive since 1953, it was believed extinct until a specimen was captured in 2017. This snake is now being radio-tracked in the wild.

The Caribbean Coastal Treeboa is a large, slender but powerful snake with a long head, with distinctive labial pits, and a long prehensile tail. Specimens may be uniform buff, orange-brown, or gray, with or without black-tipped scales, or they may be patterned with hollow, brown diamond markings along the flanks. The undersides are off-white while the sides of the head are pale.

Actual size

FAMILY	Boidae
RISK FACTOR	Nonvenomous, constrictor
DISTRIBUTION	Amazonian South America: Colombia, Venezuela, the Guianas, Brazil, Peru, Ecuador, and Bolivia
ELEVATION	0–9,020 ft (0–2,750 m) asl
HABITAT	Rainforest, flooded forest, and savanna woodland
DIET	Small mammals and birds
REPRODUCTION	Viviparous, with litters of 6–28 neonates
CONSERVATION STATUS	IUCN not listed, CITES Appendix II

ADULT LENGTH
5 ft–6 ft 7 in
(1.5–2.0 m)

EPICRATES CENCHRIA
BRAZILIAN RAINBOW BOA
(LINNAEUS, 1758)

The name "Rainbow Boa" originates from the highly visible iridescent sheen on the scales of this species, which presents an "oil-on-water" appearance. This may be a defensive adaptation to aid the crypsis of a nocturnal snake in daylight, because it is reflected in the names of other, unrelated nocturnal snakes, such as the Sunbeam Snake (*Xenopeltis unicolor*, page 81) and Amethystine Python (*Simalia amethistina*, page 100). The Brazilian Rainbow Boa inhabits the Amazon Basin and the Guiana Shield. It achieves a relatively large size and can capture and constrict a wide variety of small to medium-sized mammals and birds. The presence of heat-sensitive pits on its lips enables it to efficiently hunt endotherms at night. Brazilian Rainbow Boas shelter in holes or under logs during the day.

Actual size

RELATED SPECIES

Epicrates cenchria was previously viewed as a single species with nine subspecies, but now three of these are recognized as full species, while the others are no longer recognized. These are the Argentine Rainbow Boa (*E. alvarezi*), Paraguayan Rainbow Boa (*E. crassus*), and Caatinga Rainbow Boa (*E. assisi*) from northeastern Brazil.

The Brazilian Rainbow Boa has a powerful body, long, arrow-shaped head, and prehensile tail. It is orange or red above, with a pattern of hollow black rings along the back, and yellow-centered black spots along the flanks. The head is orange with black stripes, while the undersides are white with black spots.

FAMILY	Boidae
RISK FACTOR	Nonvenomous, powerful constrictor
DISTRIBUTION	Northern South and southern Central America: Costa Rica, Panama, Colombia, Venezuela, Trinidad, the Guianas, and northeastern Brazil
ELEVATION	0–8,630 ft (0–2,630 m) asl
HABITAT	Lowland deciduous woodland, gallery forests, palm groves, marshes, non-flooded forests, savannas, and thorn forest
DIET	Small mammals, birds, and lizards
REPRODUCTION	Viviparous, with litters of 6–20 neonates
CONSERVATION STATUS	IUCN not listed, CITES Appendix II

ADULT LENGTH
5 ft–6 ft 7 in
(1.5–2.0 m)

EPICRATES MAURUS
NORTHERN RAINBOW BOA
GRAY, 1849

111

Also known as the Brown Rainbow Boa or Colombian Rainbow Boa, the northernmost member of the genus *Epicrates* inhabits a wide variety of habitats, from dry woodland to thorn forest, savanna woodland, and swamps. Where it occurs in sympatry with the Brazilian Rainbow Boa (*E. cenchria*, page 110) it inhabits more open habitats or forest-savanna edge, leaving the rainforest proper to the other species. Prey comprises a variety of small mammals, birds, and lizards, which are captured at night and killed by constriction. Females may produce up to 20 neonates and facultative parthenogenesis has been reported in this species. Females have also been known to ingest undeveloped eggs or stillborn embryos as a means of recovering resources lost to reproduction.

RELATED SPECIES

Of all the forms of rainbow boa, *Epicrates maurus* was elevated from subspecific status long before the most recent revisions of the *E. cenchria* group, which resulted in three new species, and no subspecies. The former subspecies *E. c. barbouri*, from Ilha de Marajó, Pará, Brazil, is now synonymized with *E. maurus*.

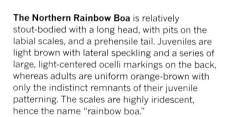

Actual size

The Northern Rainbow Boa is relatively stout-bodied with a long head, with pits on the labial scales, and a prehensile tail. Juveniles are light brown with lateral speckling and a series of large, light-centered ocelli markings on the back, whereas adults are uniform orange-brown with only the indistinct remnants of their juvenile patterning. The scales are highly iridescent, hence the name "rainbow boa."

FAMILY	Boidae
RISK FACTOR	Nonvenomous, powerful constrictor
DISTRIBUTION	Northern South America: Colombia, Venezuela, Trinidad, the Guianas, Brazil, Peru, Ecuador, and Bolivia
ELEVATION	0–785 ft (0–240 m) asl
HABITAT	Rivers in rainforest, lakes, seasonally flooded savannas
DIET	Mammals from agouti to tapir, waterbirds, and caimans
REPRODUCTION	Viviparous, with large litters of 20–40 neonates
CONSERVATION STATUS	IUCN not listed, CITES Appendix II

ADULT LENGTH
Male
10–13 ft
(3.0–4.0 m)

Female
23 ft–26 ft 3 in
(7.0–8.0 m)

EUNECTES MURINUS
GREEN ANACONDA
(LINNAEUS, 1758)

The Green Anaconda is characteristically patterned dark green with large black ocelli markings. The head exhibits a pair of curved, black-edged, orange postocular stripes. In large specimens these stripes move more dorsally, as the head broadens with size and age, and in clear sunlit water they may resemble a pair of horns, their outlines "drop-shadowed" by the black edges.

This is the heaviest snake in the world, with female specimens over 220 lb (100 kg) recorded. A large female can hunt, eat, mate, and give birth in water, without needing to venture onto land. Her weight may make crawling difficult but does not affect swimming. Females grow much larger than males. In the breeding season, "mating balls" may form with several smaller males courting a single female. Green Anacondas prey on mammals as large as brocket deer, capybaras, and young tapirs, and on waterbirds and caimans. Prey is constricted to death. Cannibalism is also documented. There are no confirmed records of humans being eaten. Green Anacondas in some habitats will estivate through the dry season, while those in rainforest rivers may be active all year round.

RELATED SPECIES

Three other anacondas are recognized: the Yellow Anaconda (*Eunectes notaeus*, page 113) from southern Brazil, Paraguay, and northern Argentina; De Schauensee's Anaconda (*E. deschauenseei*) from the mouth of the Amazon; and the recently described Beni Anaconda (*E. beniensis*) from Bolivia.

Actual size

FAMILY	Boidae
RISK FACTOR	Nonvenomous, constrictor
DISTRIBUTION	Central South America: Paraguay, Bolivia, southwestern Brazil, northeastern Argentina, and Uruguay
ELEVATION	260–490 ft (80–150 m) asl
HABITAT	Seasonally flooded savannas, and riverine gallery forest
DIET	Mammals, waterbirds, caimans, turtles, and possibly fish
REPRODUCTION	Viviparous, with litters of 10–40 neonates
CONSERVATION STATUS	IUCN not listed, CITES Appendix II

ADULT LENGTH
Male
6 ft 7 in–8 ft
(2.0–2.4 m)

Female
10–13 ft
(3.0–4.0 m)

EUNECTES NOTAEUS
YELLOW ANACONDA
COPE, 1862

Actual size

Also known as the Paraguayan Anaconda, the Yellow Anaconda inhabits the vast seasonally flooded grasslands of the Pantanal in southwestern Brazil and neighboring countries, sometimes in sympatry with the Green Anaconda (*Eunectes murinus*, page 112). It hunts mammals, waterbirds, and Broad-snouted Caiman, which are killed by constriction. Turtles and fish are also occasionally taken. The largest prey animals are probably young peccaries and brocket deer. Although not found in Amazonia, the Yellow Anaconda may occur in closed-canopy riverine forest within its range. The larger adult females attract a number of smaller males as suitors, and form "mating balls" during the breeding season. Although not as large as its northern relative, this species may still achieve weights of 110–120 lb (50–55 kg).

RELATED SPECIES

Eunectes notaeus is closely related to two similarly patterned species, De Schauensee's Anaconda (*E. deschauenseei*) from Ilha de Marajó, in the mouth of the Amazon, and French Guiana, and the Beni Anaconda (*E. beniensis*) from Beni River, Bolivia, which was originally thought to be the result of hybridization between Yellow and Green Anacondas.

The Yellow Anaconda's patterning is less distinctive than that of the Green Anaconda. The background color is yellow to brown, with black spots coalescing to form irregular zigzag or dumbbell patterns on the back. The head bears a black arrowhead marking, and a dark postocular stripe is present from the eye to the angle of the jaw. The undersides are yellow.

FAMILY	Candoiidae
RISK FACTOR	Nonvenomous, constrictor
DISTRIBUTION	Melanesia: eastern Indonesia and New Guinea
ELEVATION	0–3,280 ft (0–1,000 m) asl
HABITAT	Rainforest, and coconut and oil-palm plantations
DIET	Small mammals, frogs, and lizards
REPRODUCTION	Viviparous, with litters of 5–48 neonates
CONSERVATION STATUS	IUCN not listed, CITES Appendix II

ADULT LENGTH
Male
15¾–17¾ in
(400–450 mm)

Female
19¾–27½ in,
rarely 36½ in
(500–700 mm,
rarely 930 mm)

114

CANDOIA ASPERA
NEW GUINEA GROUND BOA
(GÜNTHER, 1877)

The New Guinea Ground Boa is an extremely stocky snake with heavily keeled scales and a short, non-prehensile tail. The head is angular and viper-like and covered in small granular scales, and the eyes have vertically elliptical pupils. Cloacal spurs are present in males, but reduced in size or absent in females. Coloration is generally dark or light brown, or yellow, with a pattern of darker squarish saddles. The undersides may be dark or light and are often speckled with brown or red.

Colloquially known as the "Viper Boa," the New Guinea Ground Boa is found throughout most of the island of New Guinea, the Bismarck Archipelago to the east, and the northern Moluccas of Indonesia to the west. It inhabits both rainforest and man-made environments such as coconut or oil-palm plantations, where it has a preference for dark, cool, damp locations, such as overgrown creeks or inside piles of discarded coconut husks. It is a sedentary, terrestrial sit-and-wait ambusher of rodents, bandicoots, lizards, and frogs, and it is itself preyed upon by the New Guinea Small-eyed Snake (*Micropechis ikaheka*, page 518). It is so static in its habits that Papuans often refer to it as the "sleepy snake." However, large specimens can deliver painful, bloody but otherwise harmless bites if handled.

RELATED SPECIES

Most of the range is occupied by *Candoia aspera schmidti*, while the nominate subspecies, *C. a. aspera*, is found in New Ireland, at the eastern extreme of the range. *Candoia aspera* may be confused with the Solomon Islands Ground Boa (*C. paulsoni*) and the two do occur together in parts of its range. It is also sometimes mistaken for the venomous Smooth-scaled Death Adder (*Acanthophis laevis*, page 484), but it more closely resembles the Rough-scaled Death Adder (*A. rugosus*) of the Trans-Fly region, though *C. aspera* is absent from that area.

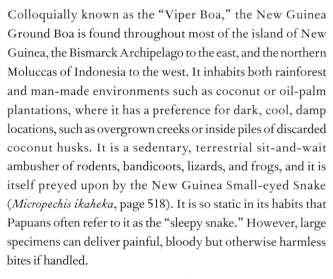

Actual size

FAMILY	Candoiidae
RISK FACTOR	Nonvenomous, constrictor
DISTRIBUTION	Melanesia: eastern Indonesia and New Guinea
ELEVATION	0–5,000 ft (0–1,525 m) asl
HABITAT	Rainforest, and coconut and cocoa plantations
DIET	Lizards and frogs
REPRODUCTION	Viviparous, with litters of 5–6 neonates
CONSERVATION STATUS	IUCN not listed, CITES Appendix II

CANDOIA CARINATA

NEW GUINEA TREEBOA

(SCHNEIDER, 1801)

ADULT LENGTH
Male
15¾–22¾ in
(400–575 mm)

Female
23¾–28 in
(600–715 mm)

115

The New Guinea Treeboa, also known as a "bevel-nosed boa" due to its overhung snout, occurs throughout New Guinea, the Bismarck Archipelago to the east, and in eastern Indonesia, the Moluccas, and Sulawesi. Its pattern is so similar to the lichen-like colors of the tree bark that if it remains motionless it is virtually invisible. Unlike the terrestrial New Guinea Ground Boa (*Candoia aspera*, page 114), with which it occurs in sympatry, the New Guinea Treeboa is arboreal, where it hunts lizards, either tree-dwelling species or terrestrial species that are ambushed from above. Prey includes skinks, geckos, and possibly tree frogs. It is an extremely docile snake. Females produce small litters of neonates as thick as matches.

The New Guinea Treeboa is an extremely slender snake with a long, prehensile tail, and a long, angular head, covered in granular scales, with very small eyes. It may be either light gray with dark gray saddles, or light brown with dark brown saddles or stripes. There is often a yellow-brown saddle over the cloaca and a large white spot posterior to the cloaca. Males have large cloacal spurs; those of females are tiny.

RELATED SPECIES

Two subspecies are recognized, the nominate form (*Candoia carinata carinata*) occurring through most of the range, with a newly described subspecies (*C. c. tepedeleni*) in the Bismarck Archipelago, east of New Guinea. *Candoia carinata* is most similar to the equally slender Palau Treeboa (*C. superciliosa*), which was once included in this species. The Solomon Islands Ground Boa (*C. paulsoni*, page 116) was also treated as a subspecies of *C. carinata*, despite the physical differences in their appearances.

Actual size

FAMILY	Candoiidae
RISK FACTOR	Nonvenomous, constrictor
DISTRIBUTION	Melanesia: eastern Indonesia, New Guinea, and the Solomon Islands
ELEVATION	0–6,000 ft (0–1,830 m) asl
HABITAT	Rainforest and oil-palm plantations
DIET	Small mammals, frogs, and lizards
REPRODUCTION	Viviparous, with litters of 16–48, rarely 60 neonates
CONSERVATION STATUS	IUCN not listed, CITES Appendix II

ADULT LENGTH
Male
27½–33 in
(700–840 mm)

Female
3 ft 3 in–4 ft 3 in
(1.0–1.3 m)

CANDOIA PAULSONI
SOLOMON ISLANDS GROUND BOA
(STULL, 1956)

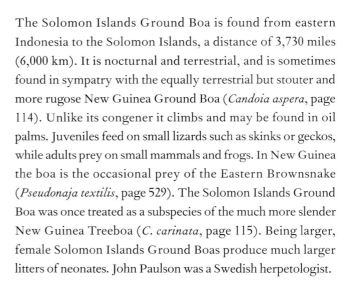

The Solomon Islands Ground Boa is a moderately stout snake, midway between *Candoia aspera* and *C. carinata*. Its long angular head is covered by granular scales, it has small eyes with vertical pupils, and its moderate tail is shorter than that of *C. carinata* though much longer than that of *C. aspera*. Coloration and patterning are extremely variable, with browns or grays predominating and a dark zigzag vertebral line common.

The Solomon Islands Ground Boa is found from eastern Indonesia to the Solomon Islands, a distance of 3,730 miles (6,000 km). It is nocturnal and terrestrial, and is sometimes found in sympatry with the equally terrestrial but stouter and more rugose New Guinea Ground Boa (*Candoia aspera*, page 114). Unlike its congener it climbs and may be found in oil palms. Juveniles feed on small lizards such as skinks or geckos, while adults prey on small mammals and frogs. In New Guinea the boa is the occasional prey of the Eastern Brownsnake (*Pseudonaja textilis*, page 529). The Solomon Islands Ground Boa was once treated as a subspecies of the much more slender New Guinea Treeboa (*C. carinata*, page 115). Being larger, female Solomon Islands Ground Boas produce much larger litters of neonates. John Paulson was a Swedish herpetologist.

RELATED SPECIES

The nominate subspecies (*Candoia paulsoni paulsoni*) inhabits the Solomon Islands and the islands east of the Bismarck Archipelago, while another subspecies (*C. p. mcdowelli*) occurs in Papua New Guinea and the Milne Bay Islands. Island endemics are found on Bougainville (*C. p. vindumi*), Misima (*C. p. rosadoi*), and Woodlark (*C. p. sadlieri*). This species appears to be absent from Indonesian west New Guinea, but a sixth subspecies (*C. p. tasmai*) occurs further west, in the north Moluccas. *Candoia paulsoni* resembles the Pacific Boa (*C. bibroni*), which occurs from the Solomons to Fiji.

Actual size

FAMILY	Erycidae
RISK FACTOR	Nonvenomous, constrictor
DISTRIBUTION	Southeastern Europe, North Africa and Middle East: Balkans to Greece, Turkey to northern Saudi Arabia and Iran, Egypt to Morocco
ELEVATION	0–4,920 ft (0–1,500 m) asl
HABITAT	Beaches, arable land, dry valleys, rocky scrub, and semidesert
DIET	Small mammals, birds, reptile eggs, lizards, and some invertebrates
REPRODUCTION	Viviparous, with litters of 6–20 neonates
CONSERVATION STATUS	IUCN not listed, CITES Appendix II

ADULT LENGTH
15¾–23¾ in,
rarely 31½ in
(400–600 mm,
rarely 800 mm)

ERYX JACULUS
JAVELIN SAND BOA
(LINNAEUS, 1758)

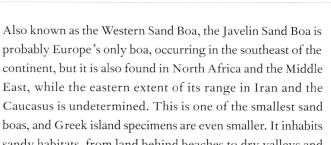

117

Also known as the Western Sand Boa, the Javelin Sand Boa is probably Europe's only boa, occurring in the southeast of the continent, but it is also found in North Africa and the Middle East, while the eastern extent of its range in Iran and the Caucasus is undetermined. This is one of the smallest sand boas, and Greek island specimens are even smaller. It inhabits sandy habitats, from land behind beaches to dry valleys and semidesert, but is also found in arable land. It likes rocky, scrubby land and avoids entirely sandy soils. Hiding under cover during the day, it lies in ambush under the sand at night, waiting for small rodents, lizards, and crickets, or forages for nestling birds or reptile eggs. This is a live-bearing species.

RELATED SPECIES

Three subspecies may be recognized, a nominate form (*Eryx jaculus jaculus*) in North Africa, a Eurasian subspecies (*E. j. turcicus*) from the Balkans to Syria, and a Caucasian subspecies (*E. j. familiaris*) from Armenia and Iran. Some authorities do not recognize any subspecies. The related Dwarf Desert Sand Boa (*E. miliaris*) occurs from Afghanistan to the northern Caucasus and may just enter Europe. *Eryx jaculus* also occurs in sympatry with the Central Asian Sand Boa (*E. elegans*) in Iran.

The Javelin Sand Boa is moderately stout with a short tail, smooth scales, and a head that terminates in a pointed rostral scale for digging. The eyes are small with vertical pupils. It is generally gray to gray-brown, darker above than on the flanks, with an irregular pattern of darker brown or orange transverse bars, which may coalesce to form a zigzag pattern or vertebral stripe. Males have larger cloacal spurs than females.

Actual size

FAMILY	Erycidae
RISK FACTOR	Nonvenomous, constrictor
DISTRIBUTION	Arabia: Saudi Arabia, Kuwait, Yemen, Oman, and UAE
ELEVATION	0–3,610 ft (0–1,100 m) asl
HABITAT	Sandy desert
DIET	Lizards, small mammals, and invertebrates
REPRODUCTION	Oviparous, with clutches of up to 4 eggs
CONSERVATION STATUS	IUCN Least Concern, CITES Appendix II

ADULT LENGTH
11¾–17¾ in,
rarely 25¼ in
(300–450 mm,
rarely 640 mm)

ERYX JAYAKARI
ARABIAN SAND BOA
(BOULENGER, 1888)

118

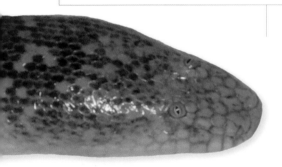

The Arabian Sand Boa is a true desert-adapted snake, occurring in the Arabian Peninsula from Kuwait to Oman and Yemen. It inhabits sandy deserts and is marvelously adapted for life below the surface. It does not inhabit rocky or montane desert. Its head is shaped for digging and the almost dorsal eyes enable it to keep watch without exposing its whole head. Juveniles take soft-bodied invertebrates and small geckos while adults take adult geckos and small mammals like shrews or mice. This species may be fairly common, although it is nocturnal and rarely seen on the surface. The Arabian Sand Boa is an egg-laying species, producing a small number of large eggs. Colonel Atmaram Jayakar (1844–1911) was an Indian Army surgeon who collected the holotype in Oman.

The Arabian Sand Boa is a moderately stout snake, with a short tail and a flattened head that terminates with a rounded flat rostral, which is used for excavation in the sand. The eyes are extremely small, with vertically elliptical pupils, and more dorsally positioned than in other sand boas. Cloacal spurs are present, larger in males than females. Coloration is tan, orange, or yellow with irregular darker transverse markings.

RELATED SPECIES

No subspecies of *Eryx jayakari* are recognized. Its range may overlap that of the Javelin Sand Boa (*E. jaculus*, page 117) in the north. There is also a record of the East African or Egyptian Sand Boa (*E. colubrinus*) in Yemen. At night, in the dark, sand boas also bear a superficial resemblance to the dangerously venomous carpet or saw-scale vipers (*Echis*, pages 624–628).

Actual size

FAMILY	Erycidae
RISK FACTOR	Nonvenomous, constrictor
DISTRIBUTION	South Asia: eastern Iran and Afghanistan, Pakistan, India, and Nepal
ELEVATION	0–3,150 ft (0–960 m) asl
HABITAT	Sandy soil habitats, semidesert, and agricultural habitats
DIET	Small mammals, birds, lizards, snakes, and invertebrates
REPRODUCTION	Viviparous, with litters of 6–8 neonates
CONSERVATION STATUS	IUCN not listed, CITES Appendix II

ADULT LENGTH
2 ft 5 in–3 ft 3 in
(0.75–1.0 m)

ERYX JOHNII
RED SAND BOA
(RUSSELL, 1802)

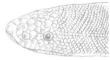

119

The Red Sand Boa, also known as John's Sand Boa, is one of the largest sand boas. It may be encountered in farmland. It is fossorial and nocturnal, ambushing rodents, birds, lizards, large invertebrates, and even other snakes, from a position just below the surface. Its stumpy tail resembles its head, leading to the colloquial but mistaken name of "two-headed snake." This is a live-bearing species. Sand boas are popular with snake charmers, being large and impressive, but safe and docile to handle. The Red Sand Boa was described by the Honourable East India Company physician-naturalist and snake expert Patrick Russell (1726–1805), in honor of the missionary-medic and herpetologist Rev. Christoph John (1747–1813).

RELATED SPECIES

Eryx johnii occurs in sympatry with the Common Indian or Rough-scaled Sand Boa (*E. conicus*), an almost equally large species but with extremely rough, keeled scales. The Rough-scaled Sand Boa is also found in Sri Lanka, where *E. johnii* is absent. The Iranian–Afghan population of *E. johnii* may warrant subspecific status.

The Red Sand Boa is an extremely stout-bodied snake with a rounded head, indistinct from the neck, small eyes with vertical pupils, an enlarged shovel-shaped rostral for burrowing, and a short, stumpy tail, which resembles the head. It is smooth-scaled throughout and usually red or reddish-brown in color, although some specimens are gray or yellow and others have broad black bands that increase in definition on the posterior of the body and the tail.

Actual size

FAMILY	Charinidae: Charininae
RISK FACTOR	Nonvenomous, constrictor
DISTRIBUTION	Western North America: northwestern USA and British Columbia (Canada)
ELEVATION	1,640–10,000 ft (500–3,060 m) asl
HABITAT	Upland coniferous or pine–oak woodlands, grassland, and desert edges
DIET	Small mammals, birds, lizards, salamanders, and reptile eggs
REPRODUCTION	Viviparous, with litters of 1–10 neonates
CONSERVATION STATUS	IUCN Least Concern, CITES Appendix II

ADULT LENGTH
19¾–23¾ in,
rarely 32¾ in
(500–600 mm,
rarely 830 mm)

CHARINA BOTTAE
NORTHERN RUBBER BOA
(BLAINVILLE, 1835)

The Northern Rubber Boa is a small snake with a cylindrical body, rounded head, small eyes with vertical pupils, and stumpy tail. Coloration is usually mid-brown to olive-brown or green, without any markings, but lighter on the flanks and underbelly.

The Northern Rubber Boa may be the northernmost boa in the world, since it occurs from California, throughout the northwestern United States, to British Columbia, Canada. It often shelters beneath logs or boulders. Juveniles feed on reptile eggs and lizards, and adults on rodents, moles, birds, lizards, or salamanders. It stalks its prey, often with its mouth agape, strikes, and kills it by constriction. One defense involves coiling into a ball, head in the center, and the blunt tail elevated as a pseudo-head. While raiding mouse nests it has been observed to make mock strikes at the adult mouse with its tail, while it eats the young. Northern Rubber Boas may live for 20 years, and their tails are often scarred, possibly due to the above activity. Paulo Botta (1802–70) was an Italian physician and archeologist who visited California.

RELATED SPECIES

Until recently *Charina bottae* was treated as a single species, with two subspecies. The isolated southern Californian populations have now been elevated to specific status as the Southern Rubber Boa (*C. umbratica*). Rubber boas are unlikely to be confused with the only other boas within their range, the rosy boas (*Lichanura*, page 121).

Actual size

FAMILY	Charinidae: Charininae
RISK FACTOR	Nonvenomous, constrictor
DISTRIBUTION	North America: USA (California and Arizona) and Mexico (Baja California and northwestern Sonora)
ELEVATION	0–6,560 ft (0–2,000 m) asl
HABITAT	Semidesert, rocky slopes, and talus slopes, usually near water
DIET	Small mammals, birds, lizards, and snakes
REPRODUCTION	Viviparous, with litters of 1–12 neonates
CONSERVATION STATUS	IUCN Least Concern, CITES Appendix II

ADULT LENGTH
31½–35½ in,
rarely 3 ft 7 in
(800–900,
rarely 1.1 m)

LICHANURA TRIVIRGATA
ROSY BOA
(COPE, 1861)

121

The Rosy Boa is popular in captivity because of its docility and small size. It occurs in a variety of subspecies and forms and inhabits semidesert habitats, usually where there is an abundance of rocks for shelter and hunting, such as talus slopes or rocky valleys. It is most common near water sources. Prey consists of small mammals, small birds, lizards, and other snakes; even a small Sidewinder Rattlesnake (*Crotalus cerastes*, page 575) has been reported as prey. Only males possess cloacal spurs, with which they stroke the female during courtship, a common feature of boa and python behavior. Despite their popularity in captivity, Rosy Boas are little studied in nature.

RELATED SPECIES

The status of the subspecies of *Lichanura trivirgata* has been the subject of considerable discussion and change. Some authorities recognize three subspecies: Mexican Rosy Boa (*L. t. trivirgata*) on the Baja California Peninsula, and the Desert Rosy Boa (*L. t. gracia*) and Coastal Rosy Boa (*L. t. roseofusca*), both from California. The northern Baja population is also sometimes treated as a separate subspecies (*L. t. saslowi*). Some authorities do not recognize any subspecies. The Northern Striped Rosy Boa from California is a separate species (*L. orcutti*).

The Rosy Boa is a moderately stout-bodied snake, with smooth scales, a longish head that is only slightly distinct from the neck, small eyes, and a relatively short but prehensile tail. Coloration is variable and is the basis of some of the subspecies definitions. The body may be gray, brown, orange, or pink, overlain by three broad longitudinal stripes, which may be orange, rose-red, brown, or black, the vertebral stripe extending onto the dorsum of the head.

Actual size

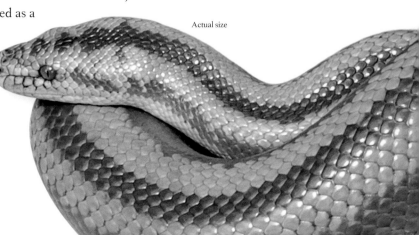

FAMILY	Charinidae: Ungaliophiinae
RISK FACTOR	Nonvenomous, constrictor
DISTRIBUTION	North America: Oaxaca (Mexico)
ELEVATION	6,560–8,040 ft (2,000–2,450 m) asl
HABITAT	Cool, moist cloud forest
DIET	Small frogs, their eggs, and salamanders
REPRODUCTION	Viviparous, with litters of 8–16 neonates
CONSERVATION STATUS	IUCN Vulnerable, CITES Appendix II

ADULT LENGTH
15¾–18½ in
(400–470 mm)

EXILIBOA PLACATA
OAXACAN DWARF BOA
BOGERT, 1968

The Oaxacan Dwarf Boa is a small snake with glossy black scales, a head just distinct from the neck, covered by enlarged scutes rather than granular scales, slightly protruding eyes, and a short tail. The only break in the black pigment is a white cloacal spot.

Only recorded from the Sierra de Juárez and Sierra Mixe in northern Oaxaca state, Mexico, the Oaxacan Dwarf Boa inhabits cloud forest. The holotype was found under a large boulder and subsequent specimens have been found under large, flat stones, often the same stone yielding specimens over several years. Boas have also occasionally been seen moving around in light rain. The natural history of the Oaxacan Dwarf Boa in nature is poorly known but captive specimens fed on small frogs, though only in darkness, suggesting nocturnal activity. A museum specimen contained a salamander so it is assumed that amphibians are important in the diet. A docile species, it resorts to defensive balling when handled, or emptying the noxious contents of its cloacal glands.

RELATED SPECIES

Exiliboa placata differs from all other neotropical boas in the possession of a single large internasal scale. It does not appear to be closely related to any other species, but is closer to *Ungaliophis* (page 123) than other neotropical boas. It looks more like a semi-fossorial colubroid snake than a boa, but both sexes possess small cloacal spurs, something not seen in advanced snakes.

Actual size

FAMILY	Charinidae: Ungaliophiinae
RISK FACTOR	Nonvenomous, constrictor
DISTRIBUTION	North and Central America: southern Mexico, Guatemala, Honduras, and Nicaragua
ELEVATION	215–7,550 ft (65–2,300 m) asl
HABITAT	Lowland tropical rainforest, montane pine woodlands, and cloud forest
DIET	Birds, bats, frogs, and lizards
REPRODUCTION	Viviparous, with litters of 2–10 neonates
CONSERVATION STATUS	IUCN not listed, CITES Appendix II

ADULT LENGTH
19¾–30 in
(500–760 mm)

UNGALIOPHIS CONTINENTALIS
ISTHMIAN BROMELIAD BOA
MÜLLER, 1880

123

The Isthmian Bromeliad Boa, also known as the Northern Bromeliad or Isthmian Dwarf Boa, is a rare species, only known from a few specimens. It is known to inhabit lowland rainforest and upland pine forests, specimens being found on boulders above a stream, under rotten pine logs, and on a bromeliad-laden tree. Such a microhabitat would provide shelter, moisture, and potential prey. Virtually everything known regarding the natural history of this snake comes from captive specimens. Captive specimens have taken neonate mice, lizards, and frogs; in nature they also take birds and bats. The male uses his cloacal spurs to court the female, but he also engages in copulatory biting of her tail as he coils his body around hers.

RELATED SPECIES

Ungaliophis species possess a single large prefrontal scale on their heads, a feature not seen in other boas. A second species, the Panamanian Bromeliad Boa (*U. panamensis*), occurs from southern Nicaragua to Colombia. The two species exhibit a number of scalation and dental differences, and *U. panamensis* has triangular rather than ovoid markings.

The Isthmian Bromeliad Boa is a small, slender snake with smooth scales, a slightly pointed head, just distinct from the neck, and a short, prehensile tail. It is generally gray, the ground color heavily flecked with black, and with orange on the lower flanks and a pale-edged black arrowhead marking on the dorsum of the head, which continues down the back as a double row of distinctive light-edged, ovoid black ocelli.

Actual size

FAMILY	Sanziniidae
RISK FACTOR	Nonvenomous, constrictor
DISTRIBUTION	Indian Ocean: southern and central Madagascar
ELEVATION	0–4,350 ft (0–1,325 m) asl
HABITAT	Thornbush savanna, dry forest, and cultivated habitats
DIET	Mammals and birds
REPRODUCTION	Viviparous, with litters of 6–13 neonates
CONSERVATION STATUS	IUCN Least Concern, CITES Appendix I

ADULT LENGTH
4–5 ft,
rarely 10 ft
(1.25–1.5 m,
rarely 3.0 m)

124

ACRANTOPHIS DUMERILI
DUMÉRIL'S BOA
JAN, 1860

Duméril's Boa is the second largest snake in Madagascar after its congener, the Madagascan Ground Boa (*Acrantophis madagascariensis*), which may achieve 10 ft 6 in (3.2 m). Duméril's Boa, which was named in honor of the French herpetologist André Marie Constant Duméril (1774–1860), is a common species. Terrestrial and nocturnal, it is often killed on roads. Prey includes mammals, from rats and bats to tenrecs and lemurs, but birds including domestic fowl are also taken. While the Madagascan Ground Boa produces small litters (two to six) of large neonates, Duméril's Boa produces large litters (6–13) of small neonates. Curiously, males are larger than females, while in most other boas the female is the larger. Both species are referred to in Malagasy as "do."

RELATED SPECIES

While *Acrantophis dumerili* occurs in southern and central Madagascar, its closest relative, the Madagascan Ground Boa (*A. madagascariensis*), inhabits the north of the island, although there are locations on the western coast where both species may occur in sympatry. The two can be distinguished by the enlarged scutes on the anterior head of *A. madagascariensis*, which are absent in *A. dumerili*.

Actual size

Duméril's Boa is a stout snake with a powerful, muscular body, a broad, angular head that is distinct from the neck, and a prehensile tail. It is generally gray or brown, with irregular black or dark brown dumbbell-shaped saddles that narrow considerably to cross the back, and a dark postocular stripe following the eye. *Acrantophis dumerili* and *A. madagascariensis* bear a striking resemblance to the Common Boa (*Boa constrictor*, page 104) of the Neotropics.

FAMILY	Sanziniidae
RISK FACTOR	Nonvenomous, constrictor
DISTRIBUTION	Indian Ocean: eastern Madagascar
ELEVATION	0–5,250 ft (0–1,600 m) asl
HABITAT	Primary and secondary rainforest, plantations, and cultivated and inhabited areas
DIET	Mammals, birds, and frogs
REPRODUCTION	Viviparous, with litters of 1–19 neonates
CONSERVATION STATUS	IUCN Least Concern, CITES Appendix I

ADULT LENGTH
5–6 ft,
rarely > 6 ft 7 in
(1.5–1.85 m,
rarely > 2.0 m)

SANZINIA MADAGASCARIENSIS
MADAGASCAN TREEBOA
(DUMÉRIL & BIBRON, 1844)

125

The Madagascan Treeboa, also known as the "Sanzinia," is a very common snake in eastern Madagascar. It is arboreal during the day, but terrestrial during the night, and found in a wide variety of habitats, ranging from primary rainforest to plantations and even village gardens. As a relatively large constricting snake it can take a wide array of prey, from rodents and tenrecs, to small lemurs, birds, and frogs. Warm-blooded prey may be located using the treeboa's heat-sensitive labial pits, captured, and constricted. The Madagascan Treeboa has long teeth for securing prey and it will bite if disturbed. Despite this snake being nonvenomous, the teeth can deliver a painful, bloody bite. Stories of 8 ft 2 in–13 ft (2.5–4 m) specimens are not thought credible. The local name for this snake is "mandrita."

RELATED SPECIES

The name *Boa mandrita* was used for this species for a period during the late twentieth century because when *Sanzinia* and the two *Acrantophis* species were sunk into genus *Boa* there would have been two species called *Boa madagascariensis*. Today a second species is recognized (*S. volontany*), inhabiting the drier western forests. Although the two species look similar they differ genetically.

The Madagascan Treeboa has an elongate, muscular body, prehensile tail, and a broad chunky head with obvious labial pits. Coloration is variable but most specimens are brown or olive with a series of transverse, light-edged dumbbell markings, and a black postocular stripe from the eye to the angle of the jaw. Males and some females have cloacal spurs.

Actual size

ALETHINOPHIDIA: AFROPHIDIA: CAENOPHIDIA

Out of Africa: Recent Snakes

The Caenophidia (*Caeno* = new, recent; *-ophidia* = snakes) are the Recent Snakes, containing almost 3,000 species, 82 percent of all living snakes. The Caenophidia contains two superfamilies: the Acrochordoidea, containing only the three species in the family Acrochordidae, genus *Acrochordus*, and the Colubroidea, containing eleven families.

The largest family in the Colubroidea is the Colubridae (> 860 species), which dominates North America, Europe, and Asia but is less well represented in the southern continents. The Natricidae (> 220 species) are similarly distributed. The Dipsadidae (> 750 species) dominate Central and South America, with representatives in North America and Tibet, while the Lamprophiidae (> 300 species) inhabit Africa and Madagascar, with representative species in Europe and Asia. Endemic Southeast Asia families include Pseudoxenodontidae (11 species), Pareatidae (21 species), and Xenodermatidae (18 species). The Homalopsidae (55 species) inhabit Asia and Australasia, and the Sibynophiidae (11 species) are Asian and American. The Elapidae (> 360 species) inhabit the Americas, Africa, and Asia, and they are the dominant family of Australasia and the Pacific and Indian Oceans. The Viperidae (> 330 species) inhabit the Americas, Europe, Africa, and Asia.

FAMILY	Acrochordidae
RISK FACTOR	Nonvenomous
DISTRIBUTION	Australasia: southern New Guinea and northern Australia
ELEVATION	0–98 ft (0–30 m) asl
HABITAT	Freshwater lagoons, billabongs, creeks, slow-moving rivers, and swamps
DIET	Freshwater fish
REPRODUCTION	Viviparous, with litters of 11–25 neonates
CONSERVATION STATUS	IUCN Least Concern

ADULT LENGTH
Male
3 ft 3 in–4 ft
(1.0–1.2 m)

Female
4 ft 7 in–5 ft 7 in,
rarely 8 ft 2 in–10 ft
1.4–1.7 m,
rarely 2.5–3.0 m

128

ACROCHORDUS ARAFURAE
ARAFURA FILESNAKE
MCDOWELL, 1979

The Arafura Filesnake is a bulky, flabby, tuberculate-skinned aquatic snake with a large head, small eyes, dorsally positioned, valvular nostrils, and a long prehensile tail. Coloration is generally yellow-brown or red-brown with a pattern of dark brown or black reticulations and pale spots ventrolaterally, although the patterning becomes obscured in adult specimens. The undersides are off-white or light brown.

The large, freshwater-dwelling Arafura Filesnake is found in northern Australia, particularly Kakadu, and southern New Guinea, in the Trans-Fly. It is fully aquatic and helpless on land. The loose, flabby skin of the Arafura Filesnake is covered in sensory tubercles that also assist the snake in maintaining a grip on its captured prey while it repositions it for swallowing. Arafura Filesnakes can swallow large fish such as barramundi or catfish. They swim well but they are sluggish, spending much time hunting or sheltering in the submerged roots of screw palms or other aquatic trees. In northern Australia, aboriginal women hunt filesnakes for food, feeling for them with their feet, while in southern New Guinea their skins are used on traditional kundu drums.

RELATED SPECIES
The closest relative of *Acrochordus arafurae* is the marine-adapted Little Filesnake (*A. granulatus*, page 129), although it more closely resembles the other freshwater species, the Javan Filesnake (*A. javanicus*) of Southeast Asia. A 10 ft (3 m) extinct freshwater fossil species (*A. dehmi*) is also known to have inhabited Asia during the Miocene, 6.35 million years ago.

Actual size

FAMILY	Acrochordidae
RISK FACTOR	Nonvenomous
DISTRIBUTION	Asia and Australasia: Pakistan to China, Malaysia, Indonesia, the Philippines, New Guinea, northern Australia, and Solomon Islands
ELEVATION	295 ft (90 m) asl to −66 ft (−20 m) bsl
HABITAT	Estuaries, river mouths, tidal rivers, mudflats, mangrove swamps, and coral reefs
DIET	Marine fish
REPRODUCTION	Viviparous, with litters of 1–12 neonates
CONSERVATION STATUS	IUCN Least Concern

ACROCHORDUS GRANULATUS
LITTLE FILESNAKE
(SCHNEIDER, 1799)

ADULT LENGTH
Male
2 ft 7 in–3 ft 3 in
(0.8–1.0 m)

Female
3–4 ft,
rarely 5 ft 2 in
(0.9–1.2 m,
rarely 1.6 m)

129

The Little Filesnake is the smallest and only marine member of the genus *Acrochordus*, and the most widely distributed. It occurs along the Asian coast from Pakistan to China, and the Indo-Australian Archipelago to New Guinea, northern Australia, and the Solomon Islands. It inhabits turbid river mouths, estuaries, mudflats, and mangrove swamps, but it is also found on coral reefs and it travels many miles upstream in tidal rivers, to enter freshwater rivers and lakes. It preys on small fish, mostly gobies such as mudskippers. The flabby, loose-skinned body and lack of ventral scales make this snake helpless on land, but in the ocean it flattens its body like a ribbon and swims effortlessly. The tubercles on the skin are sensory and also help the Little Filesnake grip its slippery prey.

The Little Filesnake is instantly recognizable due to its loose-fitting, rough, tuberculate skin. The head is small, the eyes are small, and the nostrils are dorsally located for breathing at the surface and valvular to prevent water entry. The tail is prehensile for gripping vegetation. Coloration may be uniform brown but many specimens are banded gray, brown, black, or reddish.

RELATED SPECIES
The only other living relatives of *Acrochordus granulatus* are the two freshwater species, the Arafura Filesnake (*A. arafurae*, page 128) and the Javan Filesnake (*A. javanicus*), the Arafura Filesnake being its closest relative.

Actual size

FAMILY	Colubridae: Ahaetullinae
RISK FACTOR	Rear-fanged, mildly venomous; harmless to humans
DISTRIBUTION	South and Southeast Asia: India, Sri Lanka, Nepal, Bangladesh, Myanmar, Thailand, Cambodia, Laos, and Vietnam
ELEVATION	0–6,890 ft (0–2,100 m) asl
HABITAT	Lowland and low montane forests, gardens, and secondary growth
DIET	Amphibians, lizards, birds, and small mammals
REPRODUCTION	Viviparous, with litters of 3–23 neonates
CONSERVATION STATUS	IUCN not listed

ADULT LENGTH
5 ft–6 ft 7 in
(1.5–2.0 m)

AHAETULLA NASUTA
LONG-NOSED VINESNAKE
(BONNATERRE, 1790)

The Long-nosed Vinesnake is an extremely slender snake with a long tail and an elongate, pointed head with a protruding nasal extension. The eyes have horizontal pupils, which are sighted down a groove on the side of the snout. Coloration ranges from green to brown, often marked with oblique dark stripes on the dorsum and a fine yellow longitudinal stripe on the lower flanks. The undersides are green, yellow, or gray.

The Long-nosed Vinesnake is diurnal and highly arboreal, hunting in the tangled vegetation of rainforests or gardens. Prey includes frogs, lizards, small birds, and small mammals in the case of larger specimens. Hunting by stealth, this camouflaged vinesnake stalks its prey with punctuated movements resembling the movements of the vegetation, and judging distance by sighting up the prey with its horizontal pupils down "gunsight-like" grooves on the elongate snout. *Ahaetulla* vinesnakes might possess the best vision of any snake. Prey is killed with venom from the rear fangs, but Asian vinesnakes are not dangerous to humans. This species is found in India, Sri Lanka, and mainland Southeast Asia.

RELATED SPECIES

The genus *Ahaetulla* contains eight species distributed from India to Indonesia and the Philippines. *Ahaetulla nasuta* has a nasal protuberance, which distinguishes it from all other species, except perhaps the Brown-speckled Vinesnake (*A. pulverulenta*) of India, Sri Lanka, and Bangladesh. In the Western Ghats these two species occur in sympatry with *A. nasuta*: Günther's Vinesnake (*A. dispar*) and the Western Ghats Vinesnake (*A. perroteti*).

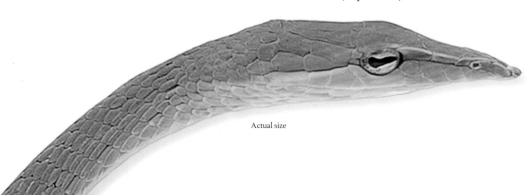

Actual size

FAMILY	Colubridae: Ahaetullinae
RISK FACTOR	Rear-fanged, mildly venomous; harmless to humans
DISTRIBUTION	South and Southeast Asia: northeast India, Bhutan, Bangladesh, southern China, Myanmar, Thailand, Cambodia, Laos, Vietnam, Malaysia, Indonesia, and the Philippines
ELEVATION	0–4,530 ft (0–1,380 m) asl
HABITAT	Lowland and low montane forest edges, regrowth, and gardens
DIET	Lizards and birds
REPRODUCTION	Viviparous, with litters of 4–10 neonates
CONSERVATION STATUS	IUCN Least Concern

ADULT LENGTH
5 ft–6 ft 5 in
(1.5–1.95 m)

AHAETULLA PRASINA
ORIENTAL VINESNAKE
(BOIE, 1827)

131

The Oriental Vinesnake is a common diurnal snake of forest edges and gardens where it hunts lizards and small birds, and possibly frogs or mice. Camouflaged, and possessing excellent vision with its horizontal pupils, it will jerkily stalk its prey. Its venom kills small vertebrates but is harmless to humans. The Oriental Vinesnake occurs from northeast India and Bhutan, to Southeast Asia, southern China, Indonesia, and the Philippines. *Ahaetulla* is a Singhalese word meaning "eye-plucker," a reference to the elongate head and body shape.

RELATED SPECIES

Four subspecies are recognized, with the nominate form occupying most of the range, and subspecies in China (*Ahaetulla p. medioxima*) and the Philippines (*A. p. suluensis* and *A. p. preocularis*). This species may be confused with the River Vinesnake (*A. fronticincta*) of Myanmar, Speckled-headed Vinesnake (*A. fasciolata*) of Malaysia and Indonesia, or the Malaysian Green Vinesnake (*A. mycterizans*).

The Oriental Vinesnake is an extremely slender snake with a long tail and an elongate head that terminates in a point. The eyes have horizontal pupils, which are directed down a groove on the side of the snout, enabling the Oriental Vinesnake to see and stalk highly camouflaged prey. Coloration is variable, from green or yellow to brown. Often the lower flanks bear a fine yellow longitudinal stripe. The venter is green, yellow, or gray.

Actual size

FAMILY	Colubridae: Ahaetullinae
RISK FACTOR	Rear-fanged, mildly venomous; harmless to humans
DISTRIBUTION	Southeast Asia: Myanmar, Thailand, Singapore, Andaman Islands, Sumatra, Borneo, Java, Sulawesi, and the Philippines
ELEVATION	0–5,000 ft (0–1,525 m) asl
HABITAT	Lowland and low montane rainforest, dry forest
DIET	Lizards
REPRODUCTION	Oviparous, with clutches of 5–8 eggs
CONSERVATION STATUS	IUCN Least Concern

ADULT LENGTH
3 ft 3 in–5 ft
(1.0–1.5 m)

CHRYSOPELEA PARADISI
PARADISE FLYING SNAKE
BOIE, 1827

The Paradise Flying Snake is slender-bodied with obliquely arranged smooth scales, a long tail, an elongate head, large eyes, and round pupils. Every dorsal scale is emerald green or yellow, but black-edged, the overall result being a stunning reticulate pattern of black and green or yellow. The green or yellow venter is also sutured with black. A vivid series of red or orange spots is often present, forming a punctuated longitudinal stripe. The head may bear four pale, black-edged transverse bars, and a black stripe passes through the eye and along the supralabials.

Flying snakes do not actually fly, they glide, but how they achieve this is not well understood. Snakes lack a sternum (breastbone) so they have more mobility in their ribs than mammals, and just as a cobra can expand the ribs of the anterior body to form a hood, the slender and lightweight flying snakes can expand the ribs along the length of the body to form a concave cavity that catches the air when they leap into space and enables them to glide to safety. The Paradise Flying Snake is found throughout Southeast Asia and is one of only three snake species to have recolonized Krakatau since the 1883 eruption. Its weak venom is only powerful enough to kill geckos.

RELATED SPECIES

Chrysopelea paradisi contains three subspecies, the nominate form (*C. p. paradisi*) throughout most of the range, a Philippine subspecies (*C. p. variabilis*), and a Sulawesi subspecies (*C. p. celebensis*). The genus *Chrysopelea* also contains four other species of flying snakes, the Golden Flying Snake (*C. ornata*) from India and China to the Philippines; the Twin-barred Flying Snake (*C. pelias*) from Malaysia and Indonesia; the Moluccan Flying Snake (*C. rhodopleuron*); and the Sri Lankan Flying Snake (*C. taprobanica*), which also occurs in India.

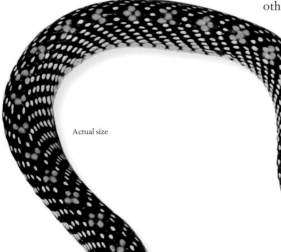

Actual size

FAMILY	Colubridae: Ahaetullinae
RISK FACTOR	Nonvenomous
DISTRIBUTION	Australasia: eastern Indonesia, New Guinea, and the Solomon Islands
ELEVATION	0–3,770 ft (0–1,150 m) asl
HABITAT	Coastal bush, coconut plantations, gardens, and islands
DIET	Frogs and lizards
REPRODUCTION	Oviparous, with clutches of 5–8 eggs
CONSERVATION STATUS	IUCN Least Concern

ADULT LENGTH
3 ft 3 in–4 ft 3 in
(1.0–1.3 m)

DENDRELAPHIS CALLIGASTRA
COCONUT TREESNAKE
(GÜNTHER, 1867)

133

Also called the Northern Treesnake in Australia, the Coconut Treesnake is most frequently encountered in coastal or island bush, or coconut plantations, but it also occurs in highland gardens. It has a preference for open sunny habitats rather than closed-canopy forest. The keels on the outer edges of its ventral scales enable it to scale vertical coconut palms with ease, and it uses its long tail as an anchorage when attempting to bridge gaps. This is a widely distributed species found in New Guinea, the Solomon Islands, northeastern Cape York Peninsula, Queensland, and the Moluccas of Indonesia. A relatively gracile snake, the alert, diurnal Coconut Treesnake feeds on frogs, skinks, geckos, reptile eggs, and small agamid lizards, which are actively hunted and chased down, or captured while they sleep.

RELATED SPECIES

(*Dendrelaphis calligastra*) currently includes the former Solomons Treesnake (*D. solomonis*), but it may be a species-complex containing several cryptic geographical species. Several other treesnake species occur in sympatry with *D. calligastra* in New Guinea, the Montane treesnake (*D. gastrostictus*); Side-striped Treesnake (*D. lineolatus*); Lorentz's Treesnake (*D. lorentzi*), and Big-eyed Treesnake (*D. macrops*), while it occurs alongside the much larger Common Treesnake (*D. punctulatus*) in Queensland.

The Coconut Treesnake is a slender snake with obliquely arranged, smooth scales, a long prehensile tail, an elongate head that is only slightly distinct from the neck, large eyes, and round pupils. It is bronze-brown to olive above with light blue interstitial skin, which shows between the scales when the snake moves. It is pale yellow to white below anteriorly, light gray posteriorly, and a black stripe passes from the snout, through the eye and back onto the anterior body, before fading out.

Actual size

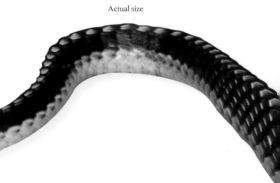

FAMILY	Colubridae: Ahaetuliinae
RISK FACTOR	Nonvenomous
DISTRIBUTION	South Asia: India, Sri Lanka, Pakistan, Nepal, Bangladesh, and Myanmar
ELEVATION	0–6,560 ft (0–2,000 m) asl
HABITAT	Most wooded habitats, plantations, riverine forest, etc.
DIET	Frogs, lizards, and birds
REPRODUCTION	Oviparous, with clutches of 6–8 eggs
CONSERVATION STATUS	IUCN not listed

ADULT LENGTH
2 ft 7 in–3 ft 3 in,
rarely 5 ft 2 in
(0.8–1.0 m,
rarely 1.6 m)

134

DENDRELAPHIS TRISTIS
COMMON INDIAN BRONZEBACK
(DAUDIN, 1803)

The Common Indian Bronzeback is a slender snake with obliquely arranged, smooth scales, a long whiplike tail, and an angular, elongate head, only slightly distinct from the neck, with large eyes and round pupils. It is rich red or bronze-brown above and brownish gray below, the two colors separated by a distinctive white lateral stripe that extends forward through the supralabials, under the eye, to the snout. The interstitial skin of the neck is pale blue, exposed when the snake inflates its throat in a defensive display.

Asian members of the diurnal treesnake genus *Dendrelaphis* are usually known as bronzebacks, as many have brown dorsums. The Common Indian Bronzeback can be found throughout the Indian subcontinent from Sri Lanka to Nepal, and Pakistan to Myanmar. It occurs in a wide variety of wooded or forested habitats, both native rainforest and man-made plantations, and is also found in gardens or on lone-standing trees. It is a common sight, hunting primarily frogs, but also small lizards and birds, during the day, but if it stops moving it blends into its surroundings perfectly with its slender body and cryptic patterning. Bronzebacks climb well due to a pair of ridged keels along either side of the ventral scales. They are nonvenomous and rarely bite, even if handled.

RELATED SPECIES

Dendrelaphis contains 45 species distributed throughout South and Southeast Asia, and also Australasia. The Asian species are called bronzebacks but the Australasian species are usually just called treesnakes; for example, the Coconut Treesnake (*D. calligastra*, page 133). The species most closely related to *D. tristis* are the Karnataka Bronzeback (*D. chairecacos*), from the Western Ghats, and Schokar's Bronzeback (*D. schokari*), from southern India and Sri Lanka.

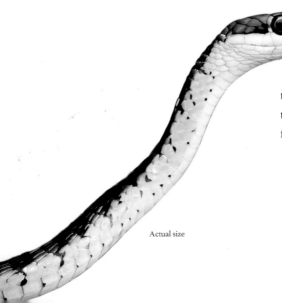

Actual size

FAMILY	Colubridae: Ahaetullinae
RISK FACTOR	Rear-fanged, mildly venomous
DISTRIBUTION	Southeast Asia: southern Thailand, Cambodia, Peninsular Malaysia, Singapore, Sumatra, Java, and Borneo
ELEVATION	0–1,640 ft (0–500 m) asl
HABITAT	Lowland forests and forest edges
DIET	Lizards
REPRODUCTION	Oviparous, with clutches of 2–3 eggs
CONSERVATION STATUS	IUCN Least Concern

ADULT LENGTH
2 ft 5 in–3 ft 3 in,
occasionally 4 ft
(0.75–1.0 m,
occasionally 1.2 m)

DRYOPHIOPS RUBESCENS
KEEL-BELLIED WHIPSNAKE
(GRAY, 1835)

135

The Keel-bellied Whipsnake is an extremely slender, vine-like snake that inhabits lowland rainforest and forest-edge situations. It occurs in both primary and secondary rainforest and also enters gardens. It is diurnally active and highly arboreal in behavior, although it is often found in bushes relatively low to the ground. Its prey consists of small lizards; it uses acute vision, due to its horizontal pupils, to locate them, and venom, injected via its rear fangs, to kill them. The Keel-bellied Whipsnake is docile when handled and its venom is harmless to humans. Often several males will court a single female, the males being smaller with darker heads. Being a slender-bodied snake the female only lays small clutches of two to three eggs.

RELATED SPECIES

A second species, the Philippine Whipsnake (*Dryophiops philippina*), occurs in the central and northern Philippines. Member of this genus are similar in appearance to the Asian vinesnakes (*Ahaetulla*, pages 130–131), although they do not possess the long pointed snout of the vinesnakes.

Actual size

The Keel-bellied Whipsnake is an extremely elongate and slender snake with smooth dorsal scales, keeled ventral scales, a long tail, and an elongate head, distinct from the neck, with moderately large eyes and horizontal pupils. The head and body are red-brown to gray-brown, patterned with numerous small light and dark flecks, while the head bears dark streak markings, and a dark stripe passes through the eye to the postocular area. The venter is yellow or olive-brown.

FAMILY	Colubridae: Calamariinae
RISK FACTOR	Nonvenomous
DISTRIBUTION	Southeast Asia: Thailand, Peninsular Malaysia, Singapore, Sumatra, Java, Borneo, and the Philippines
ELEVATION	655–5,500 ft (200–1,676 m) asl
HABITAT	Lowland and low montane rainforest
DIET	Earthworms and insect larvae
REPRODUCTION	Oviparous, clutch size unknown
CONSERVATION STATUS	IUCN Least Concern

ADULT LENGTH
23¾–25¼ in
(600–640 mm)

CALAMARIA LUMBRICOIDEA
VARIABLE REEDSNAKE
BOIE, 1827

The Variable Reedsnake is a small, fairly robust snake with smooth scales, a narrow head, and small eyes. The dorsum is black with narrow white or yellow rings, while the venter is the reverse, being yellow with black bands. The juvenile has a red or orange head but this becomes darker brown with increased maturity.

The genus *Calamaria* contains 61 species and is the largest genus in the endemic Southeast Asian subfamily Calamariinae, which also contains six other genera. At up to 25¼ in (640 mm) in length, the Variable Reedsnake is one of the largest species of *Calamaria*, most others being under 15¾ in (400 mm), and many under 7¾ in (200 mm). A terrestrial snake, it inhabits leaf litter in lowland and low montane rainforest, and is also sometimes found in gardens, provided they are close to less disturbed habitats. Prey recorded for this species ranges from earthworms to insect larvae, but it achieves a size where it may also prey upon smaller snakes. It is oviparous, like all reedsnakes, but its clutch size is unknown.

RELATED SPECIES

Calamaria lumbricoidea is, as its common name suggests, a very variable species across its considerable range, which overlaps with the ranges of many other reedsnakes with which it may be confused. The juvenile patterning may also mimic the highly venomous Red-headed Krait (*Bungarus flaviceps*, page 446), or the Blue Long-glanded Coralsnake (*Calliophis bivirgata*, page 448). At one time the juvenile morphotype was described as a separate species, *Calamaria bungaroides*, due to this close resemblance.

Actual size

FAMILY	Colubridae: Calamariinae
RISK FACTOR	Nonvenomous
DISTRIBUTION	Southeast Asia: Thailand, Peninsular Malaysia, Singapore, Sumatra, Java, Bali, and Borneo
ELEVATION	0–5,250 ft (0–1,600 m) asl
HABITAT	Lowland rainforest
DIET	Frogs and slugs
REPRODUCTION	Oviparous, clutch size unknown
CONSERVATION STATUS	IUCN Least Concern

ADULT LENGTH
15¾–17¾ in
(400–450 mm)

CALAMARIA SCHLEGELI
RED-HEADED REEDSNAKE
DUMÉRIL, BIBRON & DUMÉRIL, 1854

137

This reedsnake is also sometimes called the Pink-headed or White-headed Reedsnake, given its variable head coloration. It is just one of many reedsnake species that inhabit the lowland rainforests of peninsular Thailand and Malaysia, Singapore, and the islands of Sundaland (Sumatra, Borneo, Java, and Bali). Reedsnakes are terrestrial leaf-litter dwellers that feed on invertebrates and small vertebrates, and the Red-headed Reedsnake has been recorded to feed on small frogs and slugs. These small snakes are also the prey of many larger, ophiophagous snakes such as kraits (*Bungarus*, pages 444–447) and coralsnakes (*Calliophis*, pages 448–449). This species was named in honor of the noted German naturalist and herpetologist Hermann Schlegel (1804–84), who worked for the Rijksmuseum van Natuurlijke Historie in Leiden at the time when large collections were being made in the Dutch East Indies.

The Red-headed Reedsnake is a small, slender snake with smooth scales, a head no wider than the neck, and small eyes. The dorsum is black or dark brown, the venter is yellow or white without other markings, and the head and neck are red, pink, white, or brown.

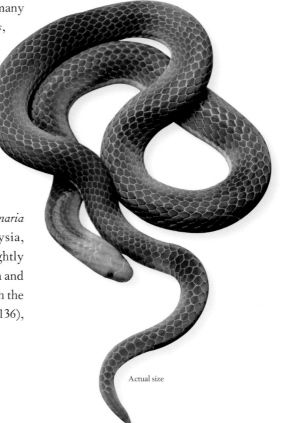

RELATED SPECIES

Two subspecies are recognized, the nominate form (*Calamaria schlegeli schlegeli*) from Thailand, Peninsular Malaysia, Singapore, Sumatra, and Borneo, which exhibits a brightly colored head, and a second form (*C. s. cuvieri*) from Java and Bali that has a dark brown head. It may be confused with the juvenile of the Variable Reedsnake (*C. lumbricoidea*, page 136), which also has a red head.

Actual size

FAMILY	Colubridae: Calamariinae
RISK FACTOR	Nonvenomous
DISTRIBUTION	Southeast Asia: Peninsular Malaysia
ELEVATION	4,920–6,460 ft (1,500–1,970 m) asl
HABITAT	Low montane rainforest
DIET	Unknown, possibly earthworms, mollusks, insects, or geckos
REPRODUCTION	Presumed oviparous, clutch size unknown
CONSERVATION STATUS	IUCN Least Concern

ADULT LENGTH
17¾–19¾ in
(450–500 mm)

MACROCALAMUS TWEEDIEI
TWEEDIE'S MOUNTAIN REEDSNAKE
LIM, 1963

138

The montane *Macrocalamus* tend to be more localized in their distribution than their numerous and widely distributed relatives in the genus *Calamaria* (pages 136–137). All seven species are endemic to the Malay Peninsula. Tweedie's Mountain Reedsnake inhabits the leaf litter of low montane rainforest in the Cameron and Genting Highlands of Pahang state, Malaysia, as do most of its congeners. Its diet in the wild is unknown, but in captivity it takes geckos, while Chanard's Mountain Reedsnake (*M. chanardi*) feeds on earthworms, slugs, and insect larvae. The reproductive strategy of Tweedie's Reedsnake is also undocumented, but presumed to be oviparous. Michael Tweedie (1907–93) was a Singapore-based British naturalist and Raffles Museum director, who specialized in the reptiles, fish, and crabs of Malaysia.

RELATED SPECIES

Macrocalamus contains seven species of montane reedsnakes, of which *M. tweediei* is the second largest at 19¾ in (500 mm), exceeded only by Jason's Reedsnake (*M. jasoni*) at 29½ in (750 mm). *Macrocalamus tweediei* appears similar to Schultz's Mountain Reedsnake (*M. schultzi*), which is brown rather than black, or the black and yellow Genting Highlands Reedsnake (*M. gentingensis*).

Actual size

Tweedie's Mountain Reedsnake is a robustly built snake with a small head and small eyes, smooth scales, and a short tail. It is shiny black above and checkerboard black and yellow below, the yellow pigment of the throat and neck extending onto the lower flanks of the anterior body, and onto the supralabials.

FAMILY	Colubridae: Calamariinae
RISK FACTOR	Nonvenomous
DISTRIBUTION	Southeast Asia: southern Thailand, Peninsular Malaysia, Sumatra, Borneo, Singapore, and Nias, Mentawai, and Riau archipelagos
ELEVATION	0–1,640 ft (0–500 m) asl
HABITAT	Lowland rainforest, rice paddies, and plantations
DIET	Earthworms, insects, and insect larvae
REPRODUCTION	Oviparous, with clutches of 2–3 eggs
CONSERVATION STATUS	IUCN Least Concern

ADULT LENGTH
7¾–9 in
(200–230 mm)

PSEUDORABDION LONGICEPS
SHARP-NOSED DWARF REEDSNAKE
(CANTOR, 1847)

139

Actual size

The Dwarf Reedsnakes of genus *Pseudorabdion* are small snakes of less than 11¾ in (300 mm) in length, the Sharp-nosed Dwarf Reedsnake achieving up to 9 in (230 mm). This species is found on the Malay Peninsula, and on Sumatra and Borneo and several other archipelagos, but reports of its occurrence on Sulawesi are thought to be erroneous. It is an inhabitant of lowland rainforest with specimens also collected in rice paddies and plantations, although it has been suggested they may have been carried there with floodwater and are not naturally occurring in these man-made microhabitats. The Sharp-nosed Dwarf Reedsnake feeds on earthworms, insects, and insect larvae, which it finds in the leaf litter or subsoil. It is a semi-fossorial species, which is generally active at night. It lays two or three elongate eggs.

RELATED SPECIES

The genus *Pseudorabdion* contains 15 species distributed across Southeast Asia, with the related genus *Rabdion*, which contains two species, endemic to Sulawesi. Most of the species of *Pseudorabdion* have a similar body and head shape and most have narrow or broad collars of white, yellow, or red, but some species are more localized in their distribution, several being endemic to Borneo or smaller island groups from Thailand to the Philippines.

The Sharp-nosed Dwarf Reedsnake is a small, slender snake with a long, pointed head, indistinct from the neck, and smooth scales. It is iridescent black above and brown below, the only markings being a thin white or yellow collar around the neck.

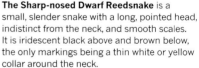

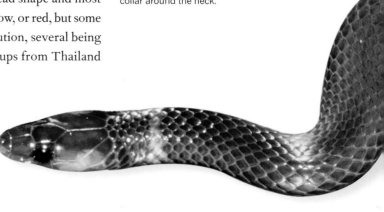

FAMILY	Colubridae: Colubrinae
RISK FACTOR	Nonvenomous, constrictor
DISTRIBUTION	Southeast Asia: northeast India, southern China, Myanmar, Laos, and Vietnam
ELEVATION	3,280–9,840 ft (1,000–3,000 m) asl
HABITAT	Low to mid-montane forests
DIET	Not known
REPRODUCTION	Oviparous, with clutches of 6 eggs
CONSERVATION STATUS	IUCN not listed

ADULT LENGTH
2 ft 7 in–3 ft 3 in
(0.8–1.0 m)

ARCHELAPHE BELLA
ELEGANT RATSNAKE
(STANLEY, 1917)

140

The Elegant Ratsnake is a very poorly known species. It inhabits low to mid-montane rainforest habitats and is rare and secretive. Very little is known regarding its natural history, although captive specimens have been maintained and bred in Russia. Captive specimens fed on mice, but the Elegant Ratsnake's prey preferences in nature are unknown. Even its taxonomic position has been the source of confusion because it was included in the racer genus *Coluber*, the ratsnake genus *Elaphe*, the smooth snake genus *Coronella*, the kukri snake genus *Oligodon*, and the genus *Maculophis*, before *Archelaphe*. *Arch* means early and -*elaphe* means ratsnake, an indication that this species is considered to represent a primitive form of ratsnake.

The Elegant Ratsnake is a smooth-scaled, cylindrical snake with a dorsoventrally compressed, blunt-snouted head, relatively small eyes and round pupils. The overall coloration is red to pink with a series of pale to bright yellow bands around the body, some of which fork on the flanks to form a chain-link pattern, and all of which are edged with black. The head is pale at the front, with an elongate red, dorsal, V-shaped marking surrounded by pale pigment, and red postocular stripes. The undersides are marked with a checkerboard of black and yellow.

RELATED SPECIES

Two subspecies are recognized, the nominate form (*Archelaphe bella bella*) from India, China, and Myanmar, and a Vietnamese subspecies (*A. b. chapaensis*), which may also occur in Laos. The closest relatives of *A. bella* are thought to be green ratsnakes (*Gonyosoma*, pages 175–176), trinket snakes (*Elaphe*, pages 169–172), and East Asian ratsnakes (*Euprepiophis*, page 173).

Actual size

FAMILY	Colubridae: Colubrinae
RISK FACTOR	Nonvenomous, constrictor
DISTRIBUTION	South Asia: India, Pakistan, Sri Lanka, and Bangladesh
ELEVATION	0–82 ft (0–25 m) asl
HABITAT	Lowland plains, forests, parks, and gardens, with dense brush, rock piles, or rodent burrows
DIET	Insects, frogs, lizards, and small mammals
REPRODUCTION	Oviparous, with clutches of 2–7 eggs
CONSERVATION STATUS	IUCN not listed

ADULT LENGTH
25½–29½ in,
rarely 4 ft 3 in
(650–750 mm,
rarely 1.3 m)

ARGYROGENA FASCIOLATA
BANDED RACER
(SHAW, 1802)

141

The Banded Racer is usually only banded as a juvenile, the adults being more uniform in color. This is a common and fast-moving inhabitant of plains and low hills, being found in woodlands, parks, and gardens. It prefers dense brush, rock piles, and areas with many rodent burrows, where it hunts mice and rats as an adult, killing them by constriction, although juveniles feed more on insects, lizards, and frogs. Shrews and bats are also eaten. The Banded Racer is found throughout peninsular India (except the extreme southeast), Pakistan, Sri Lanka, and Bangladesh, but reports from Nepal are unproven. When it feels threatened it will raise its body and flatten its neck into a narrow hood, leading to it being mistaken for an Indian Cobra (*Naja naja*, page 473).

The Banded Racer is a relatively muscular snake with smooth scales, a head slightly distinct from the neck, moderately large eyes with round pupils, and a rounded snout. Juveniles are brown with distinctive fine black and white bands, whereas adults are uniform brown to red-brown with a white or yellowish venter.

RELATED SPECIES
The genus *Argyrogena* only contains one other species, the little-known Stripe-tailed Racer *A. vittacaudata* from Darjeeling and West Bengal. Apart from the Indian Cobra (*Naja naja*), *A. fasciolata* also resembles a small Dharman Ratsnake (*Ptyas mucosa*, page 221).

Actual size

FAMILY	Colubridae: Colubrinae
RISK FACTOR	Nonvenomous, constrictor
DISTRIBUTION	North America: southwestern USA and northern Mexico
ELEVATION	0–6,000 ft (0–1,830 m) asl
HABITAT	Desert, thornbush scrub, chaparral, creosote–mesquite scrub, grasslands, oak–hickory woodland, and rocky valleys, with loose sandy or loamy substrates
DIET	Lizards and small mammals, occasionally snakes, birds, or insects
REPRODUCTION	Oviparous, with clutches of 3–23 eggs
CONSERVATION STATUS	IUCN Least Concern, protected in Kansas and Utah

ADULT LENGTH
2 ft 7 in–3 ft 3 in,
rarely 5 ft 7 in
(0.8–1.0 m,
rarely 1.7 m)

142

ARIZONA ELEGANS
GLOSSY SNAKE
KENNICOTT, 1859

The Glossy Snake is a moderately slender snake with a head just distinct from the neck, and moderately large eyes with round pupils. Its scales are smooth with a highly glossy appearance. Patterning shows geographical variation but is usually gray or brown above, with a series of paler or darker middorsal rhomboid blotches or transdorsal cross-bars, while the undersides and lower flanks are off-white to pale gray or brown, without any markings. A black stripe passes over the head, through the eyes, and back to the angle of the jaw.

The Glossy Snake has a very patchy distribution with numerous subspecies isolated from one another over a large swath of southwestern USA and northern Mexico. The northernmost population just enters southern Nebraska, while the southernmost extent of its range is Aguascalientes and northern Jalisco, Mexico. This is a snake of arid habitats, ranging from desert to thorny scrubland, chaparral, grassland, and rocky habitats, often in extremely hot locations, provided the substrate is loose sand or loam. It primarily hunts lizards but will also take rodents, small snakes, or birds. Prey captured on the surface is constricted but the Glossy Snake also spends a lot of time exploring subterranean burrows, where it kills prey by squeezing it against the burrow walls, there being insufficient room to apply its constricting coils.

RELATED SPECIES

Arizona elegans comprises eight subspecies, the Texas (*A. e. arenicola*), Mohave (*A. e. candida*), Desert (*A. e. eburnata*), Kansas (*A. e. elegans*), Chihuahuan (*A. e. expolita*), Arizona (*A. e. noctivaga*), California (*A. e. occidentalis*), and Painted Desert Glossy Snake (*A. e. philipi*), while a second species, the Peninsular Glossy Snake (*A. pacata*), a former subspecies, is found in southern Baja California.

Actual size

FAMILY	Colubridae: Colubrinae
RISK FACTOR	Nonvenomous, constrictor
DISTRIBUTION	North America: southwestern USA and northern Mexico
ELEVATION	1,480–5,910 ft (450–1,800 m) asl
HABITAT	Arid and semiarid rocky habitats, from sandy grassland and creosote bush, mesquite or cacti desert lowlands, to arid oak or cedar woodland at higher elevations
DIET	Lizards, birds, and small mammals
REPRODUCTION	Oviparous, with clutches of 3–14 eggs
CONSERVATION STATUS	IUCN Least Concern

ADULT LENGTH
4 ft 7 in–5 ft 7 in,
rarely 6 ft
(1.4–1.7,
rarely 1.8 m)

BOGERTOPHIS SUBOCULARIS
TRANS-PECOS RATSNAKE
(BROWN, 1901)

143

The Trans-Pecos Ratsnake is primarily a north Mexican species, from the Chihuahuan Desert in Durango, north to southern Texas and New Mexico in the United States. This is a large, nocturnal, desert-adapted species that is rarely seen. The Trans-Pecos Ratsnake was a holy grail for herpetologists visiting southwestern Texas, where most specimens are encountered on desert roads at night. A powerful constrictor, it preys on lizards, birds, and rodents. Male Trans-Pecos Ratsnakes will pursue a female and bite her repeatedly during courtship, and even during mating. This species is unique in that it has a host–parasite relationship with one particular species of tick, known only from this ratsnake, that congregates in large numbers on the dorsum of the snake's tail. Charles Mitchell Bogert (1908–92) was an American herpetologist.

The Trans-Pecos Ratsnake is a slender, muscular snake with a long tail, weakly keeled scales, and an angular, slightly pointed head. The eyes have vertically elliptical pupils. Coloration is buff or tan with a pair of dark, longitudinal paravertebral stripes, which may be broken, and which are conjoined at intervals by H-shaped transverse blotches. The genus *Bogertophis* is characterized by a row of subocular scales under the eyes.

RELATED SPECIES

A second species is recognized in the genus *Bogertophis*, the Baja California Ratsnake (*B. rosaliae*), which lacks the dark patterning of its congener. Two subspecies are recognized, a northern subspecies (*Bogertophis subocularis subocularis*) and a southern subspecies (*B. s. amplinotus*). *Bogertophis subocularis* is related to the glossy snakes (*Arizona*, page 142), king- and milksnakes (*Lampropeltis*, pages 182–187), and ratsnakes (*Pantherophis*, pages 207–210, *Pseudelaphe*, page 219, and *Senticolis*, page 230).

Actual size

FAMILY	Colubridae: Colubrinae
RISK FACTOR	Rear-fanged, mildly venomous; harmless to humans
DISTRIBUTION	South and Southeast Asia: northeastern India, Bangladesh, Bhutan, Myanmar, Thailand, Peninsular Malaysia, Nicobar Islands, Cambodia, Laos, Vietnam, and southern China
ELEVATION	490–6,890 ft (150–2,100 m) asl
HABITAT	Lowland to low montane rainforest, including secondary growth
DIET	Frogs, lizards, birds and their eggs, small mammals, and other snakes
REPRODUCTION	Oviparous, with clutches of 4–10 eggs
CONSERVATION STATUS	IUCN not listed

ADULT LENGTH
5–6 ft
(1.5–1.87 m)

BOIGA CYANEA
GREEN CATSNAKE
(DUMÉRIL, BIBRON & DUMÉRIL, 1854)

The Green Catsnake is a slender snake with a laterally compressed, muscular body, a long prehensile tail, and a large, rounded head with large bulging eyes and vertically elliptical, catlike pupils. The scales are smooth and arranged in oblique rows. It is bright emerald green, contrasting with the black of the interstitial skin and the pale gray of the iris. The undersides are yellow-green while the throat may be white or pale blue. Juveniles are red-brown except for the head, which is green.

One of the most spectacular members of a spectacular genus, the Green Catsnake is a vividly colored snake. When it threatens, it opens its mouth widely to gape, exposing the black lining, which contrasts with its emerald green body and gray eyes. This is a widely distributed species found from the southern Himalayas of Bhutan to the Malay Peninsula and the Nicobar Islands, in the Andaman Sea. It is highly arboreal and nocturnal, inhabiting pristine and disturbed rainforest habitat, both lowland and low montane. Prey preferences include most vertebrates, from frogs and lizards to birds and their eggs, rodents, and other snakes. The Green Catsnake is a rear-fanged venomous species that uses its venom to subdue prey, combined with constriction, but it is not considered dangerous to humans.

RELATED SPECIES

The genus *Boiga* contains 33–35 species, mostly distributed through South and Southeast Asia, but one species, the Brown Treesnake (*B. irregularis*, page 146) occurs in New Guinea and northern Australia. *Boiga cyanea* is most similar in appearance to the Banded Green Catsnake (*B. saengsomi*) from southern Thailand.

Actual size

FAMILY	Colubridae: Colubrinae
RISK FACTOR	Rear-fanged, mildly venomous
DISTRIBUTION	Southeast Asia: southern Thailand, Cambodia, Vietnam, Malaysia, Singapore, Indonesia (including Borneo), Sumatra, Java, Sulawesi, and the Philippines
ELEVATION	0–1,970 ft (0–600 m) asl
HABITAT	Lowland rainforest, mangrove swamps, and mixed dipterocarp forest
DIET	Frogs, lizards, birds and their eggs, small mammals, and other snakes
REPRODUCTION	Oviparous, with clutches of 4–15 eggs
CONSERVATION STATUS	IUCN not listed

BOIGA DENDROPHILA

MANGROVE SNAKE

(BOIE, 1827)

ADULT LENGTH
5 ft–8 ft 2 in
(1.5–2.5 m)

Also known as the Black-and-Gold Treesnake, the Mangrove Snake is the third largest *Boiga*, after the 9 ft (2.75 m) Dog-toothed Catsnake (*B. cynodon*) and the 10 ft (3 m) Brown Treesnake (*B. irregularis*, page 146). A frequent inhabitant of mangrove swamps, and an excellent swimmer, it should not be confused with the unrelated mangrove mudsnakes (Homalopsidae, pages 536–545). This large, highly arboreal, nocturnal treesnake also inhabits lowland rainforest and dipterocarp forest. Prey preferences include frogs, lizards, birds, snakes, and mammals, from mice to shrews and mouse-deer. Prey is subdued through a combination of venom, injected with a chewing bite by its rear fangs, and constriction. This species has its own enemies, the King Cobra (*Ophiophagus hannah*, page 480) being a frequent predator. This is a large snake and bites should be avoided.

RELATED SPECIES

Nine subspecies exist, the nominate subspecies being Javan. Other subspecies occur in Borneo (*B. d. annectans*), the Malay Peninsula and Sumatra (*B. d. melanota*), Sumatra and Nias (*B. d. occidentalis*), Sulawesi (*B. d. gemmicincta*), Palawan (*B. d. levitoni*), north Philippines (*B. d. divergens*), central Philippines (*B. d. multicincta*), and south Philippines (*B. d. latifasciata*).

Actual size

The Mangrove Snake is a large and powerful snake with smooth scales, a long prehensile tail and a large head, distinct from its neck, with large bulbous eyes, and vertically elliptical, catlike pupils. The various subspecies vary considerably, but are generally glossy black, with numerous broad or narrow yellow (occasionally white) rings around the body and tail, and a yellow throat and supralabials, sutured with black. The undersides are dark gray to black and may include the edges of the yellow dorsal markings.

FAMILY	Colubridae: Colubrinae
RISK FACTOR	Rear-fanged, venomous
DISTRIBUTION	Southeast Asia and Australasia: eastern Indonesia, northern Australia and New Guinea, the Solomon Islands, and neighboring archipelagos; introduced onto Guam
ELEVATION	0–7,500 ft (0–2,286 m) asl
HABITAT	Rainforest, coastal forest, gardens and other agricultural areas, scrubland, and around human habitations
DIET	Frogs, lizards, birds, and their eggs, small mammals, and other snakes
REPRODUCTION	Oviparous, with clutches of 2–11 eggs
CONSERVATION STATUS	IUCN not listed

ADULT LENGTH
Male
6 ft 7 in–8 ft,
rarely 10 ft
(2.0–2.4 m,
rarely 3 m)

Female
6 ft 7 in–7 ft 7 in
(2.0–2.3 m)

BOIGA IRREGULARIS
BROWN TREESNAKE
(BOIE, 1827)

146

The Brown Treesnake is a powerful but relatively slender snake with a long, prehensile tail, smooth, obliquely arranged scales, a large head with bulbous eyes, and vertically elliptical, catlike pupils. It may be yellow, brown, orange, red, or gray, with or without irregular transverse chevron markings and a speckling of darker pigment. Northwestern Australian specimens are boldly banded brown and white. The undersides are immaculate yellow or white, or speckled with darker pigment. A fine brown postocular stripe is present.

It is not the native populations of the Brown Treesnake that are under scrutiny, but the introduced population on Guam, in the Mariana Islands. After World War II the Brown Treesnake was accidentally introduced to Guam with military equipment, en route to the continental USA. Today more than one million snakes inhabit the island, and as voracious predators on a previously snake-free island they have eaten several endemic flightless birds into extinction; invaded houses, and inflicted serious (but so far nonfatal) bites to human babies; and caused numerous power outages by entering electrical installations. In its natural range, this species is very common but not a problem, feeding on a range of vertebrate prey and inhabiting many habitats. This is one of the most studied snakes in the world.

RELATED SPECIES

Boiga irregularis cannot be confused with any other species within its range, although the boldly banded specimens from northwestern Australia are sometimes treated as a separate species, the Tiger Catsnake (*Boiga fusca*). It occurs in sympatry on Sulawesi with the Mangrove Snake (*Boiga dendrophila gemmicincta*, page 145).

Actual size

FAMILY	Colubridae: Colubrinae
RISK FACTOR	Nonvenomous
DISTRIBUTION	North America: eastern and southeastern USA
ELEVATION	0–2,460 ft (0–750 m) asl
HABITAT	Pine, hardwood, or pine–oak woodland with a wiregrass understory
DIET	Reptile eggs, also lizards, small snakes, salamanders, small frogs, insects, mollusks, and neonate mice
REPRODUCTION	Oviparous, with clutches of 7–19 eggs
CONSERVATION STATUS	IUCN not listed, threatened in Indiana and Texas, rare in Missouri

ADULT LENGTH
19¾–21¾ in,
rarely 32½ in
(500–550 mm,
rarely 828 mm)

CEMOPHORA COCCINEA
SCARLETSNAKE
(BLUMENBACH, 1788)

147

The Scarletsnake is endemic to the United States, being found in every Atlantic state from New Jersey to Florida and west to Oklahoma and Texas. It is found in woodland habitats, from pine to oak and including mixed woodland, usually on loamy soil and with an understory of wiregrass. The principal diet of the Scarletsnake comprises reptile eggs, of both lizards and snakes, including its own species. Small eggs are swallowed whole while large eggs are chewed until the enlarged posterior maxillary teeth can slit the egg open, the snake then feeding on the contents, squeezing them along with its coils. The Scarletsnake cannot break into birds' eggs but occasionally it will prey on lizards, small snakes, salamanders, small frogs, invertebrates, or even neonate mice.

The Scarletsnake is a small, slender snake with a narrow, pointed head, only slightly distinct from the neck, small eyes with round pupils, and a short tail. The patterning varies geographically and between subspecies, but is generally pale yellow to off-white with numerous broad red bands, saddles, or patches, sandwiched between narrow black bands. The head is red, followed by a black band and then by yellow and black nape bands. The undersides are white.

RELATED SPECIES

Three subspecies are recognized, the Florida Scarletsnake (*Cemophora coccinea coccinea*), Northern Scarletsnake (*C. c. copei*), and Texas Scarletsnake (*C. c. lineri*). The monotypic genus *Cemophora* is most closely related to the milksnakes and kingsnakes (*Lampropeltis*, pages 182–187). It may also be confused with the Eastern Coralsnake (*Micrurus fulvius*, page 458) or the Texas Coralsnake (*M. tener*), but the order of the bands should distinguish it; North American coralsnakes have the red and yellow bands in contact while *C. coccinea* has the red and black bands in contact.

Actual size

FAMILY	Colubridae: Colubrinae
RISK FACTOR	Nonvenomous
DISTRIBUTION	North America: southwestern USA and northwestern Mexico
ELEVATION	0–3,000 ft (0–915 m) asl
HABITAT	Desert arroyos and washes, and rocky uplands including habitats dominated by saguaro cactus, mesquite, creosote, and thornbush
DIET	Cockroaches, ant and termite larvae, and centipedes
REPRODUCTION	Oviparous, with clutches of 2–4 eggs
CONSERVATION STATUS	IUCN not listed

ADULT LENGTH
6½–9½ in,
occasionally 11¼ in
(165–240 mm,
occasionally 285 mm)

148

CHILOMENISCUS STRAMINEUS
VARIABLE SANDSNAKE
COPE, 1860

Actual size

The Variable Sandsnake is a small snake with mostly smooth, glossy scales, a cylindrical body, a short tail, and a narrow, pointed head, indistinct from the neck. The eyes are small. Patterning is variable but commonly this snake is banded orange, yellow, or reddish, alternating with dark brown or black. The bands may completely ring the body, be broken on the pale yellow or white venter, or even be absent. The snout is pale yellow, with a broad black hood over the posterior head and neck.

This species was previously known as *Chilomeniscus cinctus* but the older name *C. stramineus* takes precedence. It occurs in southern Arizona, USA, and northwestern Sonora, Mexico, and also on the Baja California Peninsula and many of the islands of the Gulf of California, including Tiburon Island. The Variable Sandsnake is a small snake that inhabits arid habitats from desert arroyos and washes to areas dominated by saguaro cactus or thorn scrub. It burrows easily in loose sand and may feed on the surface or underground. It is an accomplished "sand-swimmer," being able to move quickly for long distances just beneath the surface, leaving S-shaped tracks in its wake. It is most active on wet nights. Prey consists mainly of insects, from ants to cockroaches, but dangerous centipedes are also eaten.

RELATED SPECIES

A second species of *Chilomeniscus*, the Isla Cerralvo Sandsnake (*C. savagei*), is endemic to Cerralvo Island, in the southern Gulf of California, Mexico. The closest relatives to *Chilomeniscus* are the American groundsnakes (*Sonora*, page 232) and the shovelnose snakes (*Chionactis*, page 149), both of which occur in the same region.

FAMILY	Colubridae: Colubrinae
RISK FACTOR	Nonvenomous
DISTRIBUTION	North America: southwestern USA and northwestern Mexico
ELEVATION	0–2,490 ft (0–760 m) asl
HABITAT	Upland desert and arroyos with saguaro cactus or mesquite–creosote bush cover
DIET	Insects, centipedes, and spiders
REPRODUCTION	Oviparous, with clutches of 3–5 eggs
CONSERVATION STATUS	IUCN not listed

ADULT LENGTH
9¾–11¾ in,
rarely 17 in
(250–300 mm,
rarely 430 mm)

CHIONACTIS PALAROSTRIS
SONORAN SHOVELNOSE SNAKE
(KLAUBER, 1937)

149

The Sonoran Shovelnose Snake is poorly known. It occurs from extreme southern Arizona, USA, into Sonora, Mexico, and it inhabits upland desert with vegetation ranging from saguaro cacti to mesquite or creosote bushes, but its natural history is poorly documented. Sonoran Shovelnose Snakes are believed to spend the daylight hours under flat rocks, in rocky crevices or in animal burrows, emerging at night to hunt. The Sonoran Shovelnose Snake is believed to be a less proficient burrower than its relative, the Western Shovelnose Snake (*Chionactis occipitalis*). The prey of the Sonoran Shovelnose Snake includes spiders, small centipedes, and insects and their larvae. This is a very secretive snake that, if uncovered, will adopt an elevated S-shaped posture and launch repeated strikes at its perceived enemy.

RELATED SPECIES

Two subspecies are recognized, the nominate Sonoran Shovelnose Snake (*Chionactis palarostris palarostris*) in the south of the range, and the Organ Pipe Shovelnose Snake (*C. p. organica*) in the north, which includes the US population. The second species, the Western Shovelnose Snake (*C. occipitalis*), occurs from Arizona to southern Nevada, California, northern Baja California, and Sonora. Shovelnose snakes are closely related to the American sandsnakes (*Chilomeniscus*, page 148) and the American groundsnakes (*Sonora*, page 232).

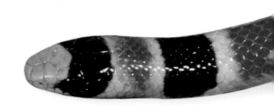

Actual size

The Sonoran Shovelnose Snake is a small, cylindrical snake with smooth scales and a short tail, a narrow, pointed head, and small eyes. Patterning consists of broad black bands with broader white or pale yellow interspaces, each with a squarish, red dorsal saddle marking that does not extend below the midline on the flank. The anterior of the head is pale yellow, the posterior being covered by the first black band.

FAMILY	Colubridae: Colubrinae
RISK FACTOR	Nonvenomous
DISTRIBUTION	Central and South America: Costa Rica, Panama, Venezuela, the Guianas, Brazil, Ecuador, Peru, Bolivia, and northern Argentina
ELEVATION	0–7,870 ft (0–2,400 m) asl
HABITAT	Primary and secondary rainforest, forest edge situations, gallery forest, forested creeks, and forest clearings
DIET	Frogs, lizards, and salamanders
REPRODUCTION	Oviparous, with clutches of 4–12 eggs
CONSERVATION STATUS	IUCN not listed

ADULT LENGTH
3 ft 3 in–5 ft
(1.0–1.5 m)

150

CHIRONIUS EXOLETUS
COMMON SIPO
(LINNAEUS, 1758)

Actual size

The sipos of genus *Chironius* are some of the few neotropical colubrid snakes that have an even dorsal scale count, with odd scale counts that include a strengthening vertebral row being the norm. The others are tiger ratsnakes (*Spilotes*, page 234), which have 14 or more dorsal scale rows, compared to *Chironius* which has 12 or fewer. The Common Sipo occurs across most of the range of the genus and is sympatric with several congeners. This species inhabits many forest habitats and also cleared areas. It is a common snake, diurnal, and both arboreal and terrestrial. The Common Sipo preys on frogs and lizards, and takes salamanders in Central America. Although this species is nonvenomous it will bite vigorously if handled, and will draw blood. The word "sipo" may be from the Tupi word *cipó*, for vine.

RELATED SPECIES

The genus *Chironius* contains 22 species distributed through northern and central South America and lower Central America, with an endangered species, the St. Vincent Racer (*C. vincenti*), in the Lesser Antilles. In South America *C. exoletus* may be confused with the Northern Sipo (*C. carinatus*), the Atlantic Coastal Sipo (*C. bicarinatus*), or the Yellow-striped Sipo (*C. flavolineatus*).

The Common Sipo is a relatively large but slender snake with a slightly laterally compressed body, a long tail, smooth scales (except the paravertebral rows, which are keeled), a moderately broad head, distinct from the neck, and large eyes with round pupils. It is generally uniform green, brown, olive, gray, or bluish, without spots or stripes, although juveniles may be banded. The venter is paler than the dorsum and the throat and neck may be white or yellow. A black postocular stripe may be present on the side of the head.

FAMILY	Colubridae: Colubrinae
RISK FACTOR	Nonvenomous, constrictor
DISTRIBUTION	Southeast Asia: Lesser Sunda Islands (Indonesia and Timor-Leste)
ELEVATION	0–3,940 ft (0–1,200 m) asl
HABITAT	Coastal forest, montane rainforest, and the outskirts of towns
DIET	Frogs, small mammals, and birds
REPRODUCTION	Oviparous, with clutches of up to 6 eggs
CONSERVATION STATUS	IUCN not listed

ADULT LENGTH
5 ft–5 ft 7 in,
occasionally 6 ft 7 in
(1.5–1.7 m,
occasionally 2.0 m)

COELOGNATHUS SUBRADIATUS

LESSER SUNDA RATSNAKE

(SCHLEGEL, 1837)

151

The Lesser Sunda Ratsnake is sometimes called the Lesser Sunda Racer, the names being interchangeable. It is the only ratsnake or racer occurring in the Lesser Sundas, from Lombok to Wetar in the Inner Banda Arc, and Sumba to Timor in the Outer Banda Arc. It is known to prey on frogs in Timor-Leste but is also thought to take small mammals and birds, which are killed by constriction. The Lesser Sunda Ratsnake is found in many habitats, from coastal forests to montane rainforest and the outskirts of towns. When confronting a potential threat it raises its body into an S-shape and makes rapid, open-mouthed strikes, while vibrating its tail on the leaf litter, and it may bite freely but is nonvenomous.

RELATED SPECIES

The genus *Coelognathus* contains seven species distributed from Pakistan to the Philippines. The Enggano Island Ratsnake (*C. enganensis*), from west of Sumatra, was once a subspecies of *C. subradiatus*. Another similar species is the Radiated Ratsnake (*C. radiatus*), occurring from Nepal to Java. Other species are the Indian Trinket Snake (*C. helena*), also occurring in Sri Lanka; Palawan Trinket Snake (*C. philippinus*); Red-tailed Trinket Snake (*C. erythrurus*), from the Philippines and Sulawesi; and Yellow-striped Trinket Snake (*C. flavolineatus*), from Thailand, Malaysia, Sumatra, Java, and Borneo.

The Lesser Sunda Ratsnake is a large, powerful, muscular snake with a long tail and elongate head, with large eyes and round pupils. It varies in coloration from yellow or buff to deep reddish brown, with an immaculate white or yellowish venter. Dorsal markings, if present, comprise a pair of dark dorsolateral, longitudinal stripes, although these may be fragmented, and sometimes a finer lateral stripe or broader vertebral stripe. The intervening scales may be flecked with dark pigment, and a dark postocular stripe is usually present running onto the neck.

Actual size

FAMILY	Colubridae: Colubrinae
RISK FACTOR	Nonvenomous
DISTRIBUTION	North and Central America: southern Canada, USA, Mexico, Guatemala, and Belize
ELEVATION	0–8,010 ft (0–2,440 m) asl
HABITAT	Woodland, forests, grasslands, prairies, swamps, and urban habitats
DIET	Small mammals, lizards, small snakes, small turtles, reptile eggs, fish eggs, frogs, salamanders, spiders, and insects
REPRODUCTION	Oviparous, with clutches of 1–36 eggs
CONSERVATION STATUS	IUCN Least Concern

ADULT LENGTH
5 ft–6 ft 3 in
(1.5–1.9 m)

COLUBER CONSTRICTOR
NORTH AMERICAN RACER
LINNAEUS, 1758

152

The North American Racer is one of the most widely distributed snakes in North America, occurring on both seaboards and throughout most of the USA, north into British Columbia, Canada, and south through Mexico to Guatemala and Belize. It is found in many habitats, and is often associated with watercourses. It does not inhabit deserts, and is less widely distributed south of the US–Mexican border. North American Racers prey on many organisms, from insects and spiders to amphibians, lizards, small snakes, and even small turtles. Small mammals are also taken, from mice to shrews and squirrels. Despite its scientific name, the North American Racer does not constrict its prey, but chews it vigorously until it is dead.

RELATED SPECIES

The genus *Coluber* is now confined to Nearctic racers, but while some authors only recognize *C. constrictor* in the genus, others include 11 species that may be placed in the genus *Masticophis*, such as the Coachwhip (*M. flagellum*, page 194). The Common Racer consists of 11 subspecies, including the Everglades Racer (*C. c. paludicola*) and the Mexican Racer (*C. c. oaxaca*).

The North American Racer is a slender snake with smooth scales, a long tail, a head broader than the neck, and large eyes with round pupils under a shelved supraocular scale, giving a scowling impression. Coloration is variable depending on subspecies, with clues provided in the common names; for example, the Northern and Southern Black Racers (*Coluber constrictor constrictor* and *C. c. priapus*), and Yellow-bellied and Western Yellow-bellied Racers (*C. c. flaviventris* and *C. c. mormon*).

Actual size

FAMILY	Colubridae: Colubrinae
RISK FACTOR	Nonvenomous
DISTRIBUTION	Southeast Asia: Vietnam
ELEVATION	2,360 ft (720 m) asl
HABITAT	Secondary evergreen forest
DIET	Not known
REPRODUCTION	Not known
CONSERVATION STATUS	IUCN Data Deficient

ADULT LENGTH
19¾ in
(500 mm)

COLUBROELAPS NGUYENVANSANGI

NGUYEN VAN SANG'S SNAKE

ORLOV, KHARIN, ANANJEVA, THIEN TAO & QUANG TRUONG, 2009

153

Nguyen Van Sang is an eminent Vietnamese herpetologist with a specialist interest in the snakes of his country. His Russian and Vietnamese colleagues named this unusual snake, which he collected, in his honor. Nguyen Van Sang's Snake was until recently only known from the female holotype, which was collected during the 2003 dry season in the Loc Bac Forest Enterprise, Lam Dong Province, southern Vietnam. It was found in the leaf litter of a secondary evergreen forest. A second specimen was later obtained in Bu Gia Map National Park, Binh Phuoc Province, also in the south. Nothing is known of its prey preferences or its reproductive strategy, but given its extremely slender body shape it is unlikely to be able to feed on anything larger than earthworms or insects.

RELATED SPECIES

Although *Colubroelaps nguyenvansangi* is included in the Colubridae it bears a strong resemblance to the elapid coralsnakes of *Calliophis* (pages 448–449) and *Sinomicrurus* (page 482). However, it lacks fangs or venom glands, which is why its describers coined the generic name *Colubroelaps*. It is unlike any other known colubrine species, despite displaying a number of colubrine characteristics. Its future taxonomic position has yet to be determined, probably by molecular analysis.

Nguyen Van Sang's Snake is an extremely slender, worm-like snake with smooth, iridescent scales, a long, cylindrical tail that terminates abruptly, and a small, rounded head, with small eyes, round pupils, and large, regular scutes. The flanks are blue-black, and the dorsum is orange-brown with a narrow blue-black vertebral stripe running the length of the body and tail. The venter is immaculate white. The head is yellow and black, while the throat and chin are white, with black markings.

Actual size

FAMILY	Colubridae: Colubrinae
RISK FACTOR	Nonvenomous
DISTRIBUTION	North America: south-central Mexico
ELEVATION	5,740–10,200 ft (1,750–3,100 m) asl
HABITAT	Pine-oak forest, thorn-forest, savanna-grasslands, cloud forest, agricultural lands, and around human habitations
DIET	Insects and insect larvae
REPRODUCTION	Viviparous, with litters of 2–7 neonates
CONSERVATION STATUS	IUCN Least Concern

ADULT LENGTH
9¾– 12½ in
(250–320 mm)

CONOPSIS LINEATA
LINED TOLUCAN EARTHSNAKE
(KENNICOTT, 1859)

The Lined Tolucan Earthsnake is small with smooth scales and a slightly pointed head that is only slightly distinct from the neck, and a short tail. It may be gray, brown, or orange, either unpatterned or with a single vertebral stripe or a series of three or five longitudinal stripes. Its eyes protrude slightly and have round pupils.

The Lined Tolucan Earthsnake is widely distributed above 5,740 ft (1,750 m) asl. It has been recorded across the Mexican Plateau and in the central Mexican highlands, from Zacatecas in the north to Oaxaca in the south. It is found in a wide variety of elevated habitats, from mesquite grassland, to thorn or pine-oak forest, in high-elevation cloud forest, and also in cultivated plots, where it shelters under debris or small rocks. It may also be found close to human dwellings. This is a small but brave snake that may attempt to mimic one of the many small montane rattlesnakes if uncovered. Totally harmless and all bluff, its prey consists of insects and their larvae, and females give birth to between two and seven neonates. *Conopsis* is the only live-bearing colubrid genus in the Americas.

RELATED SPECIES

Some authors recognize three subspecies of *Conopsis lineata*, depending on the presence or absence of stripes. Genus *Conopsis* also contains a further five species. The most widely distributed is the Large-nosed Earthsnake (*C. nasus*), which occurs through most of the western Mexico highlands, as far north as Sonora and Chihuahua. The Two-lined Mexican Earthsnake (*C. biserialis*) occurs from Jalisco to Hidalgo, while the Twin-spotted Mexican Earthsnake (*C. amphistcha*), Spotted Mexican Earthsnake (*C. acutus*), and Carlos San Filip Earthsnake (*C. megalodon*) are found in Oaxaca and Guerrero. *Conopsis* may be most closely related to the hook-nosed snakes (*Ficimia*, page 174 and *Gyalopion*, page 177).

Actual size

FAMILY	Colubridae: Colubrinae
RISK FACTOR	Nonvenomous, constrictor
DISTRIBUTION	Eurasia: southern England to Kazakhstan and northern Iran, Sicily, and Spain to southern Sweden
ELEVATION	0–7,380 ft (0–2,250 m) asl
HABITAT	Southern heathlands, railway embankments (England), open forest and woodland, scree slopes, dry areas of marshes, vineyards, dry stone walls, and montane heathland
DIET	Lizards, snakes, small mammals, and birds and their eggs
REPRODUCTION	Viviparous, with litters of 2–16 neonates
CONSERVATION STATUS	IUCN not listed, protected in the United Kingdom

CORONELLA AUSTRIACA

SMOOTH SNAKE

LAURENTI, 1768

ADULT LENGTH
23¾–29½ in,
occasionally 31½ in
(600–750 mm,
occasionally 800 mm)

155

Britain's rarest reptile, the Smooth Snake is confined to the southern heaths of England, but in Europe it is distributed east to Kazakhstan and Iran, west to Galicia, Spain, north to Scandinavia, and south to Sicily. It inhabits woodland, rocky scree slopes, marshes, vineyards, dry-stone walls, and montane regions. The Smooth Snake feeds on reptiles, primarily small lacertid lizards, but slow worms and snakes are also taken, as are mice and birds. The Smooth Snake uses constriction to subdue its prey. In the United Kingdom this species is totally protected.

RELATED SPECIES

Three subspecies may be recognized. The nominate form (*Coronella austriaca austriaca*) occupies most of the range while the western Iberian and southern Italian and Sicilian populations are recognized as separate subspecies by some authors (*C. a. acutirostris* and *C. a. fitzingeri*, respectively). The other two smooth snakes are the oviparous Southern Smooth Snake (*C. girondica*), from Iberia and Morocco, and the Indian Smooth Snake (*C. brachyura*). The Smooth Snake is related to the Red-backed Ratsnake (*Oocatochus rufodorsatus*, page 201), the only other viviparous Eurasian colubrine snake.

The Smooth Snake is a muscular snake with smooth scales, a moderately long tail, a head just distinct from the neck, a rounded snout, and moderately large eyes with round pupils. It ranges from silver or dark gray to yellow-brown, with a pair of broad, dorsolateral, longitudinal stripes or a series of short vertebral cross-bars, of darker gray or brown than the ground color. The head bears a dark postocular stripe and a bilobed nape marking that may run into the first dorsal markings. The venter is gray or brown, reddish in juveniles.

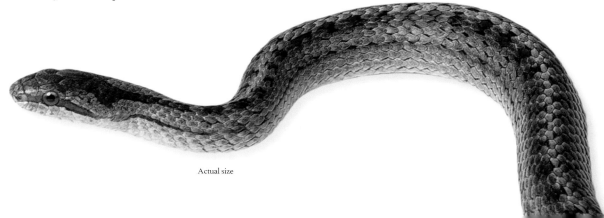

Actual size

FAMILY	Colubridae: Colubrinae
RISK FACTOR	Rear-fanged, mildly venomous; harmless to humans
DISTRIBUTION	Sub-Saharan Africa: Senegal to Eritrea, south to the Cape, South Africa
ELEVATION	0–8,200 ft (0–2,500 m) asl
HABITAT	Wet savanna and woodland, especially in association with water
DIET	Frogs
REPRODUCTION	Oviparous, with clutches of 6–19 eggs
CONSERVATION STATUS	IUCN not listed

ADULT LENGTH
23¾–28 in,
rarely 31½ in
(600–710 mm,
rarely 800 mm)

156

CROTAPHOPELTIS HOTAMBOEIA
RED-LIPPED HERALD SNAKE
(LAURENTI, 1768)

The Red-lipped Herald Snake is a small snake with smooth scales anteriorly, keeled scales posteriorly, a short tail, and a broad head with small eyes and vertically elliptical pupils. It is usually uniform gray, brown, or olive with an iridescent, shiny black head and either red or white supralabials.

The Red-lipped Herald Snake, also known as the White-lipped Herald Snake, is a slow-moving nocturnal and terrestrial predator of frogs. It is widely distributed through Sub-Saharan Africa, from Senegal to Eritrea, and south down the eastern side of the continent to the Cape, South Africa. It occurs in savanna and woodland habitats and demonstrates a preference for wetland areas, where its prey is abundant. If disturbed it will elevate its anterior body, flatten its head to display its contrastingly colored lips, hiss loudly, and launch open-mouthed strikes. If given the opportunity this little snake will deliver a chewing bite with its long, bladelike rear fangs, but although its venom is effective against frogs it is believed harmless to humans.

RELATED SPECIES

The genus *Crotaphopeltis* contains six species. *Crotaphopeltis hotamboeia* occurs in sympatry with the Barotse Watersnake (*C. barotseensis*) in the Okavango Swamp and upper Zambezi River in Botswana. In Kenya it occurs in sympatry with the Tana Herald Snake (*C. braestrupi*) and the Yellow-flanked Herald Snake (*C. degeni*), and in Tanzania with Tornier's Catsnake (*C. tornieri*). The West African Herald Snake is *C. hippocrepis*.

Actual size

FAMILY	Colubridae: Colubrinae
RISK FACTOR	Nonvenomous
DISTRIBUTION	Sub-Saharan Africa: Somalia, Kenya, Tanzania (including Zanzibar and Mafia Island), Zimbabwe, Mozambique, Malawi, and eastern South Africa
ELEVATION	0–3,280 ft (0–1,000 m) asl
HABITAT	Coastal scrub, moist savannas, and lowland evergreen forest
DIET	Birds' eggs
REPRODUCTION	Oviparous, with clutches of 6–28 eggs
CONSERVATION STATUS	IUCN not listed

DASYPELTIS MEDICI
EAST AFRICAN EGG-EATER
BIANCONI, 1859

ADULT LENGTH
Male
19¾–23¾ in
(500–600 mm)
Female
23¾–29½ in,
rarely 3 ft 3 in
(600–750 mm,
rarely 1.0 m)

157

Also known as the Rufous Egg-eater, the East African Egg-eater occurs on the coast of Somalia, Kenya, and Tanzania, and Zanzibar, Lamu, and Mafia islands. Farther south it inhabits Malawi, eastern Zimbabwe, Mozambique, and KwaZulu-Natal, South Africa. It feeds exclusively on small birds' eggs, which are "tongue-flicked" for freshness and sized up before the swallowing process begins, the egg then being slowly engulfed by the articulable jaws. When it reaches the throat a forward-facing spinal process pierces the shell, the contents are swallowed, and the compressed remains of the shell are regurgitated. The teeth are minute or absent, so in defense, egg-eaters mimic the highly venomous carpet vipers (*Echis*, pages 624–628) in patterning and display. Michele Medici (1782–1859) was an Italian physician and naturalist.

The East African Egg-eater is a small snake with strongly keeled, serrated scales, a long tail, and a rounded head, indistinct from the neck, with large eyes and vertically elliptical pupils. It may be brown, red, orange, gray, or pink in color, either uniform or with a dark brown vertebral stripe with evenly spaced white spots. Several chevron markings are present on the back of the head and neck. The iris is distinctive, being yellow, orange, or gray.

RELATED SPECIES

There are two subspecies, a nominate southern form (*Dasypeltis medici medici*) and a northern form (*D. m. lamuensis*), which has Lamu Island as its type locality. The genus *Dasypeltis* contains 13 species, all found in Africa. Other egg-eaters occurring in sympatry with *D. medici* include the widespread Common Egg-eater (*D. scabra*), which also inhabits the Arabian Peninsula, and the Southern Brown Egg-eater (*D. inornata*).

Actual size

FAMILY	Colubridae: Colubrinae
RISK FACTOR	Nonvenomous
DISTRIBUTION	Central America and northern South America: Costa Rica, Panama, western Colombia, and western Ecuador
ELEVATION	0–5,250 ft (0–1,600 m) asl
HABITAT	Lowland and low montane rainforest, evergreen forest, riverine forest, and open cleared areas
DIET	Frogs, lizards, and small mammals
REPRODUCTION	Oviparous, with clutches of up to 7 eggs
CONSERVATION STATUS	IUCN not listed

ADULT LENGTH
3 ft 3 in–5 ft
(1.0–1.5 m)

DENDROPHIDION CLARKII
RAINBOW FOREST RACER
DUNN, 1933

158

The Rainbow Forest Racer is a slender snake with a long, whiplike tail, weakly keeled scales, and a distinctive head with very large eyes, round pupils, and a squarish snout. It is bright green anteriorly, except the head, which is brown, becoming olive-green and then olive-brown by the midbody, this section also exhibiting a lateral pattern of light-centered black ocelli or cross-bands, and finally red-brown on the tail. The underside of the head and neck is yellow, while the venter of the body is gray-brown with dark and light spotting.

The Rainbow Forest Racer, also known as Clark's Forest Racer, inhabits lowland and montane rainforest, evergreen forest, riverine forest, and open areas, such as treefall gaps or cleared gardens. An alert, diurnal species, it is primarily terrestrial but is also at home aloft. It preys mainly on frogs, but lizards and small rodents are also taken. If disturbed it will attempt to flee, but if cornered it will make mock strikes and inflate its neck to display the contrastingly colored interstitial skin. If grasped by the tail it may demonstrate caudal pseudautotomy, deliberately shedding its tail, but unlike lizards it cannot regenerate a new tail. Former United Fruit employee Herbert Clark (1877–1960) organized the annual Panamanian snake census from 1929 to 1953.

RELATED SPECIES

The South and Central American forest racers are a complicated genus containing 15 species. *Dendrophidion clarkii* was resurrected from synonymy in *D. nuchale*, the Red-headed Forest Racer, at the same time as the Red-tailed Forest Racer (*D. rufiterminorum*) was described.

Actual size

FAMILY	Colubridae: Colubrinae
RISK FACTOR	Rear-fanged, mildly venomous; harmless to humans
DISTRIBUTION	West and Central Africa: Guinea-Bissau to DRC, and western Uganda, Rwanda, and Tanzania
ELEVATION	4,920–9,840 ft (1,500–3,000 m) asl
HABITAT	Montane and mid-montane rainforest, and plantations
DIET	Frogs and tadpoles
REPRODUCTION	Presumed oviparous, clutches size unknown
CONSERVATION STATUS	IUCN not listed

ADULT LENGTH
2 ft 4 in–3 ft 3 in,
rarely 4 ft 2 in
(0.7–1.0 m,
rarely 1.28 m)

DIPSADOBOA UNICOLOR
GÜNTHER'S GREEN TREESNAKE
GÜNTHER, 1858

159

Günther's Green Treesnake is a nocturnal West and Central African species, found in mid- to high-elevation rainforest from Guinea-Bissau to western Uganda and south to Rwanda, Burundi, and the Democratic Republic of the Congo. Although it occurs in rainforest habitats and plantations, and has the demeanor of a treesnake, it is not especially adapted to arboreal life because it lacks the ridged ventral scales that give many other treesnakes their climbing abilities. This species is actually often seen on the ground, but little is known regarding its natural history. It is believed to feed exclusively on frogs and their tadpoles, while its congeners are known to take skinks, geckos, and chameleons, which are subdued by venom injected by the rear fangs. Its reproductive strategy is probably oviparity, like other *Dipsadoboa* species.

Günther's Green Treesnake is a slender snake with smooth scales, a long tail, a broad head, and protruding eyes with vertically elliptical pupils. It is a highly variable species. Juveniles are gray or brown, paler anteriorly, darker posteriorly. At maturity the body becomes green, blue, or black with a blue tail, while the interstitial skin becomes black or gray.

RELATED SPECIES

The genus *Dipsadoboa* has the appearance of the smaller species of Asian catsnakes (*Boiga*, pages 144–146), and along with *Toxicodryas* (page 247) and *Telescopus* (pages 243–244) it may occupy the nocturnal cat-eyed snake niche in tropical Africa. There are ten species distributed through Sub-Saharan Africa with at least five other species in the West African rainforest, including the Green Treesnake (*D. viridis*), Duchesne's Treesnake (*D. duchesnii*), and Underwood's Treesnake (*D. underwoodi*).

Actual size

FAMILY	Colubridae: Colubrinae
RISK FACTOR	Rear-fanged, highly venomous: procoagulants and hemorrhagins, and possibly anticoagulants
DISTRIBUTION	Sub-Saharan Africa: Senegal to Eritrea, south to the Cape, South Africa
ELEVATION	0–7,870 ft (0–2,400 m) asl
HABITAT	Savanna woodland, mopane woodland, coastal bush, and gardens, excluding deserts and rainforests
DIET	Lizards and birds, occasionally frogs or rodents
REPRODUCTION	Oviparous, with clutches of 10–25 eggs
CONSERVATION STATUS	IUCN not listed

ADULT LENGTH
Male
3 ft 3 in–4 ft 3 in
(1.0–1.29 m)

Female
3 ft 3 in–4 ft,
rarely 6 ft 7 in
(1.0–1.26 m,
rarely 2.0 m)

160

DISPHOLIDUS TYPUS
BOOMSLANG
(SMITH, 1828)

The Boomslang, which means "treesnake," is the most dangerous rear-fanged snake in the world. A juvenile specimen claimed the life of eminent Field Museum herpetologist Karl Patterson Schmidt in 1957. Other deaths are on record, usually of snake handlers, and specific Boomslang antivenom is now produced in South Africa. Small quantities of venom are injected from the large rear fangs, located under the eyes in the large mouth, but it may cause massive internal hemorrhage and renal failure 24–48 hours later. The Boomslang is found throughout sub-Saharan Africa, in savanna woodland and coastal scrub, but it also occupies gardens and hedges. It is an alert and agile diurnal predator of chameleons, other lizards, and birds, being able to enter suspended weaverbird nests. Defensive Boomslangs inflate their throats to expose the contrasting interstitial skin color.

The Boomslang is a slender snake with obliquely arranged, keeled scales, a long tail and a short, relatively large head, dominated by very large eyes with round pupils. Juveniles are brown above with blue interstitial skin, and white below, with a yellow throat and emerald green irises. Adult females are brown or olive above and white or brown below, while males range from light blue or green above, with black interstitial skin and pale green undersides, to yellow-green with black scale edging, red-brown with pink undersides, or even black above and gray below, with black scale edging.

Actual size

RELATED SPECIES

The genus *Dispholidus* is monotypic, although three subspecies of *D. typus* are recognized by some authors. Most of the range is occupied by the nominate subspecies (*D. t. typus*), with separate subspecies in the Rift Valley, from Kenya to Zambia (*D. t. kivuensis*), and Angola, Republic of the Congo, Democratic Republic of the Congo, and northwestern Zambia (*D. t. punctatus*). *Dispholidus* is most closely related to the African black treesnakes (*Thrasops*).

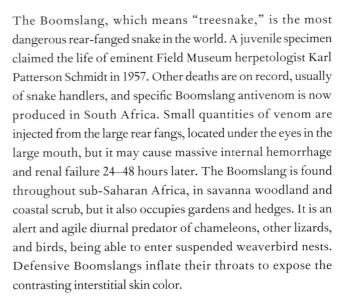

FAMILY	Colubridae: Colubrinae
RISK FACTOR	Nonvenomous, constrictor
DISTRIBUTION	Southeastern Europe: Greece, Albania, European Turkey, southern Bulgaria, Romania, Moldova, Ukraine, and southwestern Russia; small populations in Hungary, Serbia, Montenegro, and Croatia
ELEVATION	0–5,250 ft (0–1,600 m) asl
HABITAT	Rocky hillsides, vineyards, hedges, dry stone walls, forests, steppes, and semidesert
DIET	Lizards, birds, small mammals, occasionally other snakes
REPRODUCTION	Oviparous, with clutches of 5–12 eggs
CONSERVATION STATUS	IUCN not listed

ADULT LENGTH
6 ft 7 in–8 ft 2 in
(2.0–2.5 m)

DOLICHOPHIS CASPIUS
CASPIAN WHIPSNAKE
(GMELIN, 1789)

161

The Caspian Whipsnake is probably Europe's largest snake. It is widely distributed through southeastern Europe, from Albania to the northern shores of the Caspian Sea. This is a relatively common, diurnal, and primarily terrestrial species, which is most frequently encountered on rocky hillsides or in open sandy valleys, but also occurs in vineyards, forests, steppe, and semidesert habitats. When encountered the Caspian Whipsnake will seek to flee, but if cornered it will defend itself vigorously, striking high and biting. Although wounds bleed profusely, this is a nonvenomous snake of no danger to humans. The preferred prey comprises lizards, especially for juvenile Caspian Whipsnakes, but birds, small mammals, and other snakes are also included in the diets of adults.

RELATED SPECIES

The closest relative of *Dolichophis caspius* is Schmidt's Whipsnake (*D. schmidti*) from Turkey, Iran, and Armenia. Three other *Dolichophis* species are the Cypriot Whipsnake (*D. cypriensis*), Gyaros Island Whipsnake (*D. gyarosensis*), from the Cyclades, and the Large Whipsnake (*D. jugularis*), from Greece, Turkey, the Middle East, and Iran.

The Caspian Whipsnake is a moderately slender snake with smooth glossy scales, a long tail, and a head just distinct from the neck. It has large eyes and round pupils under shelved supraocular scales that present a slightly scowling expression in adults. In adults the dorsum is gray, olive-brown, or olive-green, with every scale black-edged, giving a reticulate appearance, and the venter is yellow to pale green. Juveniles are gray to brown with short transverse black bars.

Actual size

FAMILY	Colubridae: Colubrinae
RISK FACTOR	Nonvenomous
DISTRIBUTION	North America: southeastern USA
ELEVATION	0–490 ft (0–150 m) asl
HABITAT	Coastal scrub, mangrove swamp, pine flatlands, prairie and hardwood hammocks, and sawgrass plains
DIET	Frogs, rodents, other snakes, and turtles
REPRODUCTION	Oviparous, with clutches of 4–11 eggs
CONSERVATION STATUS	IUCN Least Concern, threatened and federally protected in the USA

ADULT LENGTH
5–6 ft,
rarely 8 ft 6 in
(1.5–1.8 m,
rarely 2.6 m)

DRYMARCHON KOLPOBASILEUS
GULF COAST INDIGO SNAKE
(HOLBROOK, 1842)

The Gulf Coast Indigo Snake is a powerful, stout-bodied snake with smooth, iridescent scales, a long tail, and a large head with large eyes and round pupils. Unlike some cribos from South America, the indigo snakes are glossy blue-black throughout, with a reddish throat and occasional white patches, while the venter is orange or blue-gray.

The Gulf Coast Indigo Snake was recently split from the Eastern Indigo Snake (*Drymarchon couperi*). The largest native nonvenomous snake found north of Mexico, it occurs from the Florida Keys to the bayous of Louisiana, in a variety of habitats from mangrove swamps to pine flatlands, seasonally flooded sawgrass plains, and drier hardwood hammocks. It shares the burrows of gopher tortoises and nine-banded armadillos. Once abundant, Gulf Coast Indigo Snakes are now federally protected, but may still be declining due to commercial collecting and to harvesting of the stumps they use for winter shelter. They hunt a variety of vertebrates, from rodents to turtles, as well as other snakes, from Cornsnakes (*Pantherophis guttatus*, page 209) to Eastern Diamondback Rattlesnakes (*Crotalus adamanteus*, page 572). Indigo snakes are immune to the venom of their rattlesnake prey.

RELATED SPECIES

The closest relative is the Eastern Indigo Snake (*Drymarchon couperi*) from eastern Florida, Alabama, and Georgia, which was once a subspecies of *D. corais*, now known as the Yellow-tailed Cribo, which occurs from Venezuela to Argentina. The other former subspecies of *D. corais* are now subspecies of the Mexican Cribo (*D. melanurus*, page 163), although the Isla Margarita Cribo (*D. margaritae*) was elevated to specific status. The Falcon Cribo (*D. caudomaculatus*) is another recently described species from northern Venezuela.

Actual size

FAMILY	Colubridae: Colubrinae
RISK FACTOR	Nonvenomous
DISTRIBUTION	North, Central, and South America: Texas, through Mexico and Central America, to Venezuela and South America west of the Andes
ELEVATION	0–5,250 ft (0–1,600 m) asl
HABITAT	Lowland and montane rainforest, and dry forest, savanna woodland, rocky arroyos, and mangrove forest
DIET	Frogs, lizards, birds and their eggs, small mammals, and snakes
REPRODUCTION	Oviparous, with clutches of 4–11 eggs, rarely up to 25
CONSERVATION STATUS	IUCN Least Concern

ADULT LENGTH
5 ft–6 ft 7 in,
occasionally 9 ft 8 in
(1.5–2.0 m,
occasionally 2.95 m)

DRYMARCHON MELANURUS
MEXICAN CRIBO
(DUMÉRIL, BIBRON & DUMÉRIL, 1854)

163

The Mexican Cribo inhabits wet and dry forests, savanna woodlands, and rocky arroyos from Texas to Peru, west of the Andes, and Venezuela, east of the Andes. Central America's largest snake after the Common Boa (*Boa constrictor*, page 104) and bushmasters (*Lachesis*, pages 588–589), it is a formidable predator of lizards, birds, rodents, and particularly snakes. Cribos pursue other snakes, even large venomous species, moving alongside their quarry and grasping the head in their jaws, preventing retaliatory bites, before chewing vigorously to kill it, then swallowing it headfirst. They are immune to bites from venomous snakes. Rodents are killed underground by being crushed against the burrow walls; one cribo may kill all the occupants of an entire rat's nest at the same time.

The Mexican Cribo is a large, stout-bodied snake with smooth scales, a long tail, and a large head, distinct from the neck, with moderately large eyes and round pupils. Coloration varies with subspecies, but generally the anterior body and head are light to olive-brown with markings confined to a horizontal bar on the neck and dark supralabial suturing; the midbody and posterior body are dark brown with darker scale suturing, while the tail varies with location from red to black.

RELATED SPECIES

Genus *Drymarchon* was once one species and eight subspecies. Today it comprises six species but only *D. melanurus* has subspecies: the Black-Tailed Cribo (*D. m. melanurus*) of northwestern South America; Texas Indigo Snake (*D. m. erebennus*) from Texas and northeastern Mexico; Red-tailed Cribo (*D. m. rubidus*) from northwest Mexico; Orizaba Cribo (*D. m. orizabensis*) from Veracruz, Mexico; and Central American Cribo (*D. c. unicolor*) from Chiapas, Mexico to Costa Rica.

Actual size

FAMILY	Colubridae: Colubrinae
RISK FACTOR	Nonvenomous
DISTRIBUTION	North, Central, and South America: Texas, through Mexico and Central America, to Colombia
ELEVATION	0–6,000 ft (0–1,830 m) asl
HABITAT	Forest edges, clearings, lowland and low montane wet and dry forests, and arid arroyos
DIET	Frogs, lizards, birds, small mammals, and snakes; sometimes insects
REPRODUCTION	Oviparous, with clutches of 4–8 eggs
CONSERVATION STATUS	IUCN Least Concern

ADULT LENGTH
2 ft 4 in–3 ft 3 in,
rarely 4 ft 3 in
(0.7–1.0 m,
rarely 1.3 m)

DRYMOBIUS MARGARITIFERUS
SPECKLED RACER
(SCHLEGEL, 1837)

The Speckled Racer is a slender, smooth-scaled species with a long tail and a head slightly broader than the neck, with moderately large eyes and round pupils. The dorsum may be pale or bright green, yellow, or reddish, with every scale black-edged to present a speckled or reticulate pattern, and the undersides are yellow.

A fast-moving, diurnal inhabitant of forest edge and clearings, the Speckled Racer may also be found in lowland and low montane forests, and inhabits rocky, arid, overgrown arroyos, too. It is primarily a terrestrial snake, but is able to climb well and may be encountered in low vegetation. It is a widely distributed species, being found from Texas, down both versants of Mexico and Central America, to Colombia. Its prey consists largely of frogs and toads, but lizards, reptile eggs, small mammals, and occasionally small snakes are also taken by adults. There are reports of fish being on the menu, and juveniles take small vertebrates but also feed on insects. Speckled Racers are themselves preyed upon by the ophiophagous Mexican Cribo (*Drymarchon melanurus*, page 163).

RELATED SPECIES

There are four subspecies: the Northern Speckled Racer (*Drymobius margaritiferus margaritiferus*) from Texas to northern South America; West Mexican Speckled Racer (*D. m. fistulosus*) from Sonora to Oaxaca; Central American Speckled Racer (*D. m. occidentalis*) from Chiapas, Mexico to El Salvador; and Big Corn Island Speckled Racer (*D. m. maydis*) from Nicaragua. There are three other *Drymobius* species: the Green Highland Racer (*D. chloroticus*) and Black Forest Racer (*D. melanotropis*) from Central America, and Esmarald or Rhombic Racer (*D. rhombifer*) from Central and South America.

Actual size

FAMILY	Colubridae: Colubrinae
RISK FACTOR	Nonvenomous
DISTRIBUTION	South America: Colombia, Venezuela, Guyana, Suriname, French Guiana, Brazil, Ecuador, eastern Peru, and Bolivia
ELEVATION	0–11,500 ft (0–3,500 m) asl
HABITAT	Primary and secondary rainforest
DIET	Lizards, frogs, reptile eggs, and small snakes
REPRODUCTION	Oviparous, with clutches of 2–6 eggs
CONSERVATION STATUS	IUCN Least Concern

ADULT LENGTH
2 ft 4 in–4 ft
(0.7–1.2 m)

DRYMOLUBER DICHROUS
NORTHERN WOODLAND RACER
(PETERS, 1863)

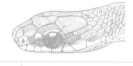

165

The Northern Woodland Racer is a widespread and common species that inhabits primary and secondary rainforest in the Guianas and the Amazon Basin countries. It is a diurnally active, terrestrial or semi-arboreal species that hunts a wide variety of lizards and frogs, but also takes reptile eggs, and other snakes, and has been known to be cannibalistic. This is an ontogenetically patterned species, juveniles exhibiting a pattern distinctly different from that of the adults. The Northern Woodland Racer is also able to pseudautotomize its tail in response to attack by a potential predator, but unlike many lizards, which autotomize (shed) their tails, snakes that practice pseudautotomy are unable to regenerate a new tail.

RELATED SPECIES

The generic name *Drymoluber* was coined to demonstrate a perceived close relationship with two other racer genera, *Drymobius* (page 164) and *Coluber* (page 152), but *Drymoluber* is actually more closely related to the tropical racers (*Mastigodryas*, page 195). Two other species are included in the genus *Drymoluber*, the high-elevation Apurímac Woodland Racer (*D. apurimacensis*) from Peru, and Vital Brazil's Woodland Racer (*D. brazili*) from Brazil and Paraguay.

The Northern Woodland Racer is a moderately robust snake with smooth scales, a long tail, and a head just distinct from the neck, with large eyes and round pupils. The head may be black, and the body leaf-green on the flanks to dark green or olive on the dorsum, with a yellow venter and white on the throat, neck, and labials. Juveniles are light brown with darker brown bands and these may also be faintly visible on adult snakes.

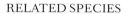
Actual size

FAMILY	Colubridae: Colubrinae
RISK FACTOR	Nonvenomous
DISTRIBUTION	South Asia: southern and northeastern India, northern Sri Lanka, and False Island, Myanmar
ELEVATION	0–655 ft (0–200 m) asl
HABITAT	Lowland habitats, poorly known
DIET	Dietary preferences unknown, presumed lizards
REPRODUCTION	Presumed oviparous, clutch size unknown
CONSERVATION STATUS	IUCN Least Concern

ADULT LENGTH
15¾–20½ in
(400–520 mm)

DRYOCALAMUS GRACILIS
SCARCE BRIDAL SNAKE
(GÜNTHER, 1864)

The Scarce Bridal Snake is well named, as it is remarkably rare. It is recorded from northeast India, the Eastern and Western Ghats of southern India, and from northern Sri Lanka, where only three specimens have ever been found. There is also a record from False Island, on the Rakhine coast of Myanmar. It inhabits lowland habitats, and has been found near habitations or crossing roads in rice-growing districts. It is generally a very poorly known species, but probably mirrors its congener, the Common Bridal Snake (*Dryocalamus nympha*), in being oviparous and feeding on small lizards. The Common Bridal Snake also adopts a defensive strategy that involves tying itself in knots and it is possible the Scarce Bridal Snake has the same defense.

RELATED SPECIES

The closest relative to *Dryocalamus gracilis* is the Common Bridal Snake (*D. nympha*) from India, and some authors consider *D. gracilis* to be a synonym of *D. nympha*. The genus contains four Southeast Asian species: Davison's Bridal Snake (*D. davisonii*), the Three-striped Bridal Snake (*D. tristrigatus*), the Half-banded Bridal Snake (*D. subannulatus*), and the Palawan Bridal Snake (*D. philippinus*). Bridal snakes are closely related to the wolfsnakes (*Lycodon*, page 190).

The Scarce Bridal Snake is a small snake with a slender, cylindrical body and a slightly elongate, rounded head, just distinct from the neck. The eyes are small and the pupils are elliptical. It is a cream-colored snake, immaculate below but dorsally marked with long, oval brown saddles, the pale interspaces being spotted with brown, and with a light brown cap over the dorsum of the head.

Actual size

FAMILY	Colubridae: Colubrinae
RISK FACTOR	Nonvenomous
DISTRIBUTION	Europe and Southeast Asia: Greece, Turkey, Georgia, Armenia, Azerbaijan, Nagorno-Karabakh, Dagestan, and Iran
ELEVATION	0–6,560 ft (0–2,000 m) asl
HABITAT	Arid rocky hills with sparse vegetation, stony steppe, clearings in oak forest
DIET	Centipedes, scorpions, other invertebrates, and small lizards
REPRODUCTION	Oviparous, with clutches of 3–8 eggs
CONSERVATION STATUS	IUCN Least Concern

ADULT LENGTH
19¾–24½ in
(500–620 mm)

EIRENIS MODESTUS

ASIA MINOR DWARF SNAKE

(MARTIN, 1838)

167

The Asia Minor Dwarf Snake, also known as the Ring-headed Dwarf Racer, is the only member of this relatively large genus to occur in European Turkey, from where it occurs east to the Caucasus (Dagestan, Georgia, Armenia, and Azerbaijan). It also occurs on several Greek islands in the Aegean, which really constitute part of Asia Minor, where it is widely distributed, through Turkey and into Iran. It is a small snake that frequents arid open habitats, such as sparsely vegetated rocky hillsides, stony steppe, and sunny clearings in woodland, but it avoids the direct sun, being active under or around rocks, boulders, or vegetation. Its prey comprises mostly invertebrates, including centipedes, millipedes, spiders, scorpions, cockroaches, beetles, woodlice, earthworms, and snails, but small lizards are also taken.

RELATED SPECIES

The genus *Eirenis* contains 20 species in several subgenera. *Eirenis modestus* is contained in the subgenus *Eirenis*, with the Gold-striped Dwarf Snake (*E. aurolineatus*) from Turkey. These two sister species are believed to be basal (primitive) members of the genus. *Eirenis modestus* contains three subspecies, with the nominate form (*E. m. modestus*), which may be a species complex, in the Caucasus and Turkey, a southern Turkish subspecies (*E. m. cilicius*), and a northwestern subspecies (*E. m. semimaculatus*) from the Aegean and Bosphorus.

The Asia Minor Dwarf Snake is delicate, being little thicker than a pencil, with smooth scales, a relatively long tail, a head only slightly distinct from the neck, small eyes, and round pupils. It is gray, brown, or olive-green without body markings, apart from a broad crescent marking on the nape. The head is dorsally brown with transverse yellow cross-bars posterior to the eyes and anterior to the nape marking. There are dark markings on the pale supralabials. The venter is yellow, pale gray, or off-white.

Actual size

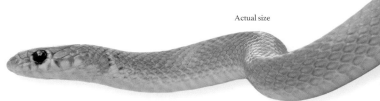

FAMILY	Colubridae: Colubrinae
RISK FACTOR	Rear-fanged, mildly venomous; harmless to humans
DISTRIBUTION	South Asia: India, Bangladesh, and Nepal
ELEVATION	820–1,640 ft (250–500 m) asl
HABITAT	Low-lying deciduous woodland on plains or low foothills, and near water
DIET	Birds' eggs, and possibly birds or small mammals
REPRODUCTION	Believed oviparous, with clutches of up to 7 eggs
CONSERVATION STATUS	IUCN Least Concern

ADULT LENGTH
27½–31½ in
(700–800 mm)

ELACHISTODON WESTERMANNI
INDIAN EGG-EATING SNAKE
REINHARDT, 1863

The Indian Egg-eating Snake is small and slender with smooth scales, a relatively long tail, a head that is a little broader than the neck, moderately large eyes, and vertically elliptical pupils. It is generally dark brown to black with cream markings, in the form of a checkerboard pattern or random blotches. A pale longitudinal stripe runs the length of the body on the enlarged vertebral scales, and the venter is off-white with brown speckling. The head is brown with a dorsal black arrowhead marking.

Unlike the virtually toothless African egg-eaters (*Dasypeltis*, page 157), the rare Indian Egg-eating Snake has teeth and rear fangs. It is nocturnal and arboreal. Once only known from Bengal, it has now been recorded more widely in India, Bangladesh, and Nepal. The Indian Egg-eating Snake may be found near the nests of small birds such as sparrows. It enters the nest and selects an egg by smell, before swallowing it whole. Vigorous movements force the egg against sharp vertebral processes; the contents are swallowed, and the shell regurgitated. The presence of venom and a thermosensitive pit in the nostril raises an interesting question as to whether birds or small mammals are also hunted at night. Gerardus Frederik Westermann (1807–90) was a zoologist who founded Amsterdam Zoo in 1838.

RELATED SPECIES

Elachistodon westermanni is contained in a monotypic genus. The fact that both it and the African egg-eaters feed on birds' eggs in the same way is probably more as a result of convergent evolution than any close relationship between the two genera, so it is difficult to determine which genera are most closely related to *Elachistodon*. However, they are likely to be Asian, rear-fanged, and arboreal, with an enlarged vertebral scale row.

Actual size

FAMILY	Colubridae: Colubrinae
RISK FACTOR	Nonvenomous, constrictor
DISTRIBUTION	East Asia: Japan (Ryukyu Islands, Bonin Islands), and Kurile Islands (Russia)
ELEVATION	0–4,350 ft (0–1,325 m) asl
HABITAT	Grassland, rice paddies, bamboo thickets, forests, and around human habitations
DIET	Small mammals, birds, and frogs
REPRODUCTION	Oviparous, with clutches of 7–12 eggs
CONSERVATION STATUS	IUCN not listed

ELAPHE CLIMACOPHORA
AODAISHO
(BOIE, 1826)

ADULT LENGTH
3 ft 3 in–6 ft 7 in,
rarely 7 ft 7 in
(1.0–2.0 m,
rarely 2.3 m)

169

Also called the Japanese Ratsnake, the Aodaisho is found on Japan's four main islands, in the northern Ryukyu and northern Bonin Islands, and on Kunashir in the southern Kurile Islands, Russia. It is one of the few snake species for which albino individuals are commonly encountered in nature, and a leucistic (pigmentless) population is also known. The Aodaisho inhabits forests, bamboo thickets, and agricultural areas, and enters houses to hunt rodents, although birds, squirrels, and frogs are also taken. Birds' eggs are ruptured by a vertebral process, as in the Indian Egg-eating Snake (*Elachistodon*, page 168) or African egg-eaters (*Dasypeltis*, page 157), although an Aodaisho is large enough to swallow eggs whole. Its main defense is to empty the foul-smelling contents of its cloacal glands, a tactic also used by its relative, the aptly-named Stinking Goddess (*E. carinata*).

RELATED SPECIES

The much-revised genus *Elaphe* contains 11 species distributed from Japan, Korea, and Amur, Russia, to western Europe. Four other ratsnakes occur within Japanese territories, the Stinking Goddess (*E. carinata*), Japanese Four-lined Ratsnake (*E. quadrivirgata*), Japanese Burrowing Ratsnake (*Euprepiophis conspillata*), and Beauty Ratsnake (*Orthriophis taeniurus*).

The Aodaisho is a large, muscular but slender snake with smooth to weakly keeled scales, a moderately long tail, a large head, and large eyes with round pupils. Patterned specimens are dark olive-green dorsally, with or without darker mottling, or four longitudinal stripes on the body, but many specimens are albino or leucistic. Juveniles exhibit a blotched pattern.

Actual size

FAMILY	Colubridae: Colubrinae
RISK FACTOR	Nonvenomous, constrictor
DISTRIBUTION	Southeast Asia: southern China and northern Vietnam
ELEVATION	165–1,640 ft (50–500 m) asl
HABITAT	Deciduous forest on karst limestone, also bamboo thickets, and meadows near water
DIET	Small mammals and birds
REPRODUCTION	Oviparous, with clutches of 6–12 eggs
CONSERVATION STATUS	IUCN Vulnerable

ADULT LENGTH
5 ft 2 in–6 ft,
rarely 8 ft 2 in
(1.6–1.8 m,
rarely 2.5 m)

ELAPHE MOELLENDORFFI
MÖLLENDORFF'S RATSNAKE
(BOETTGER, 1886)

Möllendorff's Ratsnake is a large snake with an elongate, laterally compressed body, a dorsally red, gray-lipped head, and small eyes with red irises. Its body is light gray, patterned with irregular, light-centered, dark-bordered, rusty-brown blotches arranged vertebrally and laterally, which continue onto the posterior body and tail, where the ground color may be partially obscured by a wash of red pigment. Some specimens lack any patterning except the red head.

Möllendorff's Ratsnake, also known as the Hundred-flower Snake, or the Red-headed Ratsnake or Trinket Snake, was named in honor of Otto Franz von Möllendorff (1848–1903), a German malacologist (mollusk expert) who spent many years in China. It is confined to extreme southern China and northern Vietnam, where it inhabits deciduous forest, particularly on karst limestone outcrops, and also bamboo thickets and meadows near water. It has been little studied in nature, but like all other ratsnakes (except the Red-backed Ratsnake, *Oocatochus rufodorsatus*, page 201) it is oviparous. Its prey preferences appear to be for warm-blooded animals; it feeds on small mammals such as mice, rats, and bats, but also takes birds. The IUCN list Möllendorff's Ratsnake as Vulnerable due to its harvesting for meat, skins, and potions, and its popularity in the pet trade.

RELATED SPECIES
Elaphe moellendorffi is one of four ratsnakes that were until recently included in the genus *Orthriophis*, but which are now returned to *Elaphe*, the others being the Eastern Trinket Snake (*E. cantoris*) from Nepal, northeast India, Bhutan, and Myanmar; Hodgson's Ratsnake (*E. hodgsoni*) from northern India, Nepal, and Tibet; and the widely distributed Beauty Ratsnake (*E. taeniura*, page 172).

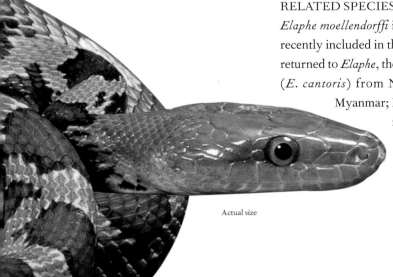

Actual size

FAMILY	Colubridae: Colubrinae
RISK FACTOR	Nonvenomous, constrictor
DISTRIBUTION	Southeastern Europe: Italy, Slovenia, Croatia, Bosnia, Montenegro, Macedonia, Albania, and Greece
ELEVATION	0–8,200 ft (0–2,500 m) asl
HABITAT	Arid habitats, clearings in deciduous woodland, old quarries, abandoned buildings, vegetated karst outcrops, and hedges and dry stone walls around agricultural land
DIET	Small mammals, birds, eggs, and lizards
REPRODUCTION	Oviparous, with clutches of 3–18 eggs
CONSERVATION STATUS	IUCN Near Threatened

ADULT LENGTH
4 ft 3 in–5 ft 2 in,
rarely 6 ft 7 in
(1.3–1.6 m,
rarely 2.0 m)

ELAPHE QUATUORLINEATA
FOUR-LINED RATSNAKE
(BONNATERRE, 1790)

171

The Four-lined Ratsnake is a southern and southeastern European species, occurring in Italy, the Balkan countries along the Adriatic coast, Greece, and the islands of the Aegean Sea. It demonstrates a preference for arid habitats, such as clearings in deciduous woodland, old quarries, broken stone walls, and karst outcrops, but it is also found near water, and around pastures, under hedges, and around old buildings. It is diurnal, and a terrestrial predator of a range of small mammals, from mice to young rabbits, as well as birds and their eggs, and lizards. It kills its prey by constriction. It is also an adept climber that may be found aloft. A proficient rat-killer, this snake should be encouraged in farmland because it performs an invaluable service, ridding the area of vermin.

The Four-lined Ratsnake is stout-bodied, with smooth or weakly keeled scales, a relatively long tail, a head broader than the neck, large eyes, and round pupils. Juveniles are pale gray with bold dark blotches, but adults are generally mid- or dark brown above, paler brown or yellow on the flanks, and off-white below, with four broad, black longitudinal stripes, and a black postocular stripe on the side of the head. Amorgos specimens are almost patternless, and some mainland specimens are almost melanistic.

RELATED SPECIES
The Western Four-lined Ratsnake (*Elaphe quatuorlineata quatuorlineata*) is found on the mainland and many islands, and three Aegean island subspecies are also recognized: the Cyclades Four-lined Ratsnake (*E. q. muenteri*), on Naxos, Amorgos, and neighboring islands; Paros Ratsnake (*E. q. parensis*); and Skyros Ratsnake (*E. q. scyrensis*). The Amorgos Ratsnake (*E. rechingeri*) is now a synonym of *E. q. muenteri*, while the Eastern Four-lined Snake (*E. sauromates*), a former subspecies, inhabits Bulgaria, Romania, Moldova, Ukraine, Turkey, and the Caucasus.

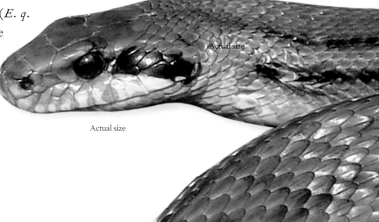

Actual size

Actual size

FAMILY	Colubridae: Colubrinae
RISK FACTOR	Nonvenomous, constrictor
DISTRIBUTION	East and Southeast Asia: Russia, China, Ryukyu Islands, Taiwan, northeast India, Bhutan, Myanmar, Thailand, Vietnam, Laos, Peninsular Malaysia, Borneo, and Sumatra
ELEVATION	0–10,200 ft (0–3,100 m) asl
HABITAT	Deciduous forests, rainforests, montane forests, rice paddies, around human habitations, and in limestone caves
DIET	Small mammals, bats, and birds
REPRODUCTION	Oviparous, with clutches of 5–25 eggs, depending on population
CONSERVATION STATUS	IUCN not listed

ADULT LENGTH
4 ft 3 in–8 ft 2 in
(1.3–2.5 m)

172

ELAPHE TAENIURA
BEAUTY RATSNAKE
(COPE, 1861)

The Beauty Ratsnake has smooth scales, a long tail, and an elongate head. It is variably patterned, being olive, yellow-brown, or gray-brown, with a dorsal black chain-link pattern and a bold stripe on the tail. The flanks may be heavily flecked with dark pigment. The head is brown above, pale on the lips, with a broad black postocular stripe. The body of the Malaysian Cave Racer (*E. t. ridleyi*) is pale anteriorly, darker posteriorly and on the tail, with a distinctive yellow vertebral stripe.

The Beauty Ratsnake occurs in a wide variety of geographical locations and habitats, from deciduous montane woodland in China to tropical rainforest in Southeast Asia. It may also be found in rice paddies, and around human habitations, entering roof spaces in search of prey. Some populations are associated with karst limestone caves, in Peninsular Malaysia, Sumatra, and Borneo, and are known as Cave Racers. Cave Racers have been found up to 2 miles (3 km) deep in caves, hunting bats and cave swiftlets, while Beauty Ratsnakes above ground feed on a variety of small mammals and birds, which are killed by constriction.

RELATED SPECIES

Nine subspecies are recognized: from eastern China (*Orthriophis taeniurus taeniura*); western China, northeast India, Myanmar, Thailand, and Laos (*Elaphe. t. yunnanensis*); Myanmar and Thailand (*E. t. helfenbergeri*); Vietnam, Cambodia, and Thailand (*E. t. callicyanous*); Taiwan (*E. t. friesi*); the Ryukyu Islands (*E. t. schmackeri*); Hainan Island (*E. t. mocquardi*); southern Thailand and West Malaysia (*E. t. ridleyi*); and Borneo and Sumatra (*E. t. grabowskyi*). The last two forms are the famous Cave Racers.

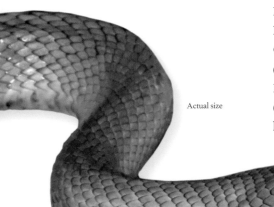

Actual size

FAMILY	Colubridae: Colubrinae
RISK FACTOR	Nonvenomous, constrictor
DISTRIBUTION	Southeast Asia: northeastern India, northern Myanmar and Vietnam, Tibet, China, and Taiwan
ELEVATION	1,480–9,840 ft (450–3,000 m) asl
HABITAT	Subtropical montane forest, vine forest, densely vegetated rocky outcrops, upland areas of mixed grass, scrub, bushes, and trees, and rice paddies
DIET	Small mammals
REPRODUCTION	Oviparous, with clutches of 2–10 eggs
CONSERVATION STATUS	IUCN Least Concern

ADULT LENGTH
3 ft 3 in–4 ft,
rarely 5 ft 7 in
(1.0–1.2 m,
rarely 1.7 m)

EUPREPIOPHIS MANDARINUS

MANDARIN RATSNAKE

(CANTOR, 1842)

173

The Mandarin Ratsnake is a poorly documented species that was infrequently seen in nature and for many years did not thrive in captivity. Its range includes southern, eastern, and central China, Taiwan, Tibet, northeastern India, and northern Myanmar and Vietnam. It is found in subtropical montane forest, vine forest, rocky habitats, and even rice paddies, and is rare at low elevations, being much more frequently encountered over 6,560 ft (2,000 m). Its prey in nature is also poorly documented, although one specimen did contain a shrew, while in captivity Mandarin Ratsnakes will feed on rodents. A secretive snake, it inhabits subterranean burrows, and as there is little space to coil and constrict, it will use its body to crush its prey against the burrow wall.

RELATED SPECIES

The genus *Euprepiophis* contains two other species, the Japanese Burrowing Ratsnake (*E. conspicillata*) and the even more infrequently encountered and Endangered Szechwan Ratsnake (*E. perlacea*). Both species also occur at higher elevations.

Actual size

The Mandarin Ratsnake is a moderately stout snake with a short tail and a rounded head, only slightly broader than the neck, with relatively small eyes and round pupils. The patterning of the Mandarin Ratsnake is unmistakable, being light gray, blue-gray, gray-brown, or dark gray with large, oval, black ocelli-like saddles with yellow edges and centers. The head bears two black chevrons, separated by yellow flashes, and a black snout tip.

FAMILY	Colubridae: Colubrinae
RISK FACTOR	Rear-fanged, very mildly venomous; harmless to humans
DISTRIBUTION	North America: southern USA and northern Mexico
ELEVATION	0–4,920 ft (0–1,500 m) asl
HABITAT	Arid thornbush, near watercourses
DIET	Spiders, centipedes, and other invertebrates
REPRODUCTION	Oviparous, with clutches of up to 3 eggs
CONSERVATION STATUS	IUCN Least Concern

ADULT LENGTH
5½–9 in,
rarely 19 in
(140–230 mm,
rarely 480 mm)

FICIMIA STRECKERI
TAMAULIPAN HOOKNOSE SNAKE
TAYLOR, 1931

The Tamaulipan Hooknose Snake inhabits the states of northeastern Mexico, as far south as Veracruz and Puebla, but only enters the United States along the lower Rio Grande, southeastern Texas. It is a small and relatively common, nocturnal, semi-fossorial snake found in arid thornbush habitats, and also near watercourses, such as irrigation ditches or ponds. It preys on spiders, although centipedes and other invertebrates are also taken. Hooknose snakes possess slightly enlarged, grooved rear fangs, but their oral secretions are pre-digestants and benumbing agents that only act on invertebrates. Its defensive tactic involves repeatedly forcing its cloacal lining in and out, making an audible popping sound. John Korn Strecker (1875–1933) was a Baylor University herpetologist.

RELATED SPECIES

Ficimia contains six other species of eastern hooknose snakes, occurring from Mexico to Central America. They are similar in appearance and ecology to the two western hooknose snakes (*Gyalopion*, page 177) of the southwestern USA and Mexico, and the Southwestern Hooknose Snake (*Pseudoficimia frontalis*, page 220) from Mexico.

The Tamaulipan Hooknose Snake is a short, stocky snake, with smooth scales, a short tail, a narrow head that terminates in a slightly upturned snout, and small eyes with round pupils. It is gray-brown to olive above with a series of irregular transverse dark cross-bands, a dark brown spot under the eye, and an immaculate white venter.

Actual size

FAMILY	Colubridae: Colubrinae
RISK FACTOR	Nonvenomous, constrictor
DISTRIBUTION	Southeast Asia: southern China, including Hainan Island, and northern Vietnam, including the Norway Islands (Vietnam)
ELEVATION	655–4,920 ft (200–1,500 m) asl
HABITAT	Primary rainforest
DIET	Probably small mammals and birds
REPRODUCTION	Oviparous, with clutches of up to 6 eggs
CONSERVATION STATUS	IUCN Least Concern

ADULT LENGTH
4 ft 3 in–5 ft 2 in,
rarely 6 ft 7 in
(1.3–1.6 m,
rarely 2.0 m)

GONYOSOMA BOULENGERI
RHINOCEROS RATSNAKE
(MOCQUARD, 1897)

175

The Rhinoceros Ratsnake, also known as the Vietnamese Horned Snake, is a curious species, with a long, fleshy projection of unknown purpose on its snout. In this respect it bears a strong resemblance to Baron's Bush Racer (*Philodryas baroni*, page 328) of South America. The Rhinoceros Ratsnake demonstrates a preference for primary rainforest, especially near water, where it is both diurnal and arboreal in behavior. The diet of the Rhinoceros Ratsnake in nature is poorly known but thought to comprise small mammals or birds. This species was named in honor of George Albert Boulenger (1858–1937), a Belgian-British zoologist at the British Museum (Natural History), and one of the most influential herpetologists of his generation.

RELATED SPECIES

Gonyosoma now contains six species of Asian snakes that were previously contained in different genera. *Gonyosoma boulengeri* was in genus *Rhynchophis*, as were the Rein Snake (*G. frenatum*) and the Green Bush Snake (*G. prasinum*). The Royal Treesnake (*G. margaritatum*, page 176) was in *Gonyophis*, while the Red-tailed Ratsnake (*G. oxycephalum*) and Celebes Black-tailed Ratsnake (*G. jansenii*) were already in *Gonyosoma*. Not all authors follow this arrangement.

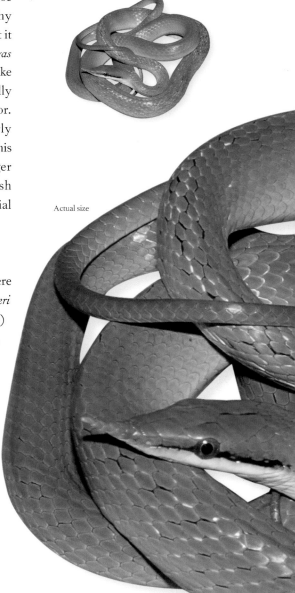

Actual size

The Rhinoceros Ratsnake is a laterally compressed snake with smooth scales, a long tail, and a narrow, elongate head that terminates in a long hornlike projection. It has relatively small eyes and round pupils. Coloration is generally green or bluish-green, paler on the venter, sometimes with a faint mottled pattern on the anterior body. A black line passes from the snout, through the eye, separating the green of the dorsum from the white, or yellow-green of the supralabials, chin, and throat.

FAMILY	Colubridae: Colubrinae
RISK FACTOR	Nonvenomous
DISTRIBUTION	Southeast Asia: Peninsular Malaysia, Singapore, and Borneo
ELEVATION	0–2,300 ft (0–700 m), possibly 6,560 ft (2,000 m) asl on Mt. Kinabalu
HABITAT	Lowland rainforest, possibly montane rainforest
DIET	Dietary preferences unknown
REPRODUCTION	Presumed oviparous, clutch size unknown
CONSERVATION STATUS	IUCN Least Concern

ADULT LENGTH
5 ft 7 in–6 ft 7 in
(1.7–2.0 m)

GONYOSOMA MARGARITATUM
ROYAL TREESNAKE
PETERS, 1871

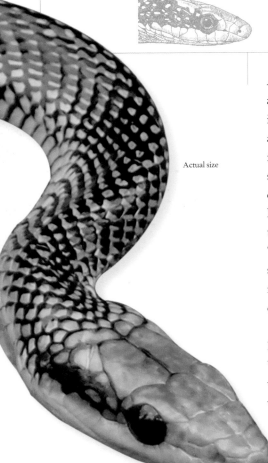

Actual size

Also known as the Rainbow Treesnake, the stunningly attractive Royal Treesnake is poorly known in nature. It inhabits lowland rainforest on the Malay Peninsula, Singapore, and Borneo, but there are reports that it may occur up to 6,560 ft (2,000 m) asl on Mt. Kinabalu in Sabah. The presence of this species is difficult to ascertain as it is a rainforest canopy-dweller that lives hundreds of feet above the forest floor. It has longitudinal keels on its ventral scales that enable it to climb the tall branchless trunks of rainforest emergents with ease. The only information regarding its diet is that a captive specimen refused everything except fish, an unlikely diet in nature for a diurnal canopy-dwelling snake. It is probably oviparous but even that is not known with certainty.

RELATED SPECIES

When the Royal Treesnake was contained in *Gonyophis* it was monotypic, but it now shares the genus *Gonyosoma* with five other species, including the Rhinoceros Ratsnake (*G. boulengeri*, page 175).

The Royal Treesnake is slender and laterally compressed with a whiplike tail, a long head, distinct from the neck, large eyes, and round pupils. The body is green anteriorly, becoming banded with yellow by the midbody, with green replaced by powder blue. Every scale is black-edged, the black becoming dominant on the tail, which is marked by widely separated yellow bands. The head is orange, with broad black postocular stripes, while the throat and undersides of the body are yellow with black scale edging. There are few colors absent from this species' livery.

FAMILY	Colubridae: Colubrinae
RISK FACTOR	Rear-fanged, very mildly venomous; harmless to humans
DISTRIBUTION	North America: southeastern USA and northwestern Mexico
ELEVATION	50–4,130 ft (15–1,260 m) asl
HABITAT	Grasslands, thorn scrub, creosote- and mesquite-covered foothills and canyons
DIET	Spiders, scorpions, centipedes
REPRODUCTION	Oviparous, but clutch size unknown
CONSERVATION STATUS	IUCN Least Concern

ADULT LENGTH
6½–9½ in,
rarely 14 in
(165–240 mm,
rarely 360 mm)

GYALOPION QUADRANGULARE
SONORAN HOOKNOSE SNAKE
(GÜNTHER, 1893)

177

Also known as the Thornscrub Hooknose Snake, the Sonoran Hooknose Snake occurs along the northwestern coast of Mexico, south to Nayarit, and just enters the United States in southern Arizona. It occurs in a variety of habitats, from desert thornbush on hills and in canyons, to creosote- or mesquite-covered hills, to grassy flatlands. It is semi-fossorial in habit and only seen on the surface after rain, and then only at night. Prey primarily consists of spiders, but scorpions, centipedes, and possibly insects and their larvae are also taken. Its defense involves cloacal popping, as described for the Tamaulipan Hooknose Snake (*Ficimia streckeri*, page 174). Hooknose snakes possess slightly enlarged, grooved rear fangs, but their oral secretions are pre-digestants and benumbing agents that specifically target invertebrates. They pose no threat to humans.

RELATED SPECIES

The closest relative of *Gyalopion quadrangulare* is the Chihuahuan Hooknose Snake (*G. canum*), which occurs in north-central Mexico and enters the United States in southeastern Arizona, southern New Mexico and southwestern Texas. Related genera include the eastern hooknose snakes (*Ficimia*, page 174) and the Southwestern Hooknose Snake (*Pseudoficimia frontalis*, page 220).

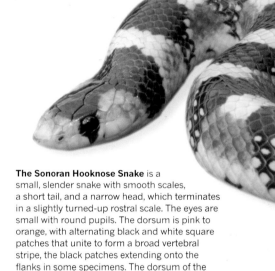

The Sonoran Hooknose Snake is a small, slender snake with smooth scales, a short tail, and a narrow head, which terminates in a slightly turned-up rostral scale. The eyes are small with round pupils. The dorsum is pink to orange, with alternating black and white square patches that unite to form a broad vertebral stripe, the black patches extending onto the flanks in some specimens. The dorsum of the head bears a black cap, which runs into the first of the black markings on the nape. The venter is cream, with or without black speckling.

Actual size

FAMILY	Colubridae: Colubrinae
RISK FACTOR	Nonvenomous
DISTRIBUTION	West and Central Africa: The Gambia to Uganda and Angola, also São Tomé
ELEVATION	0–7,280 ft (0–2,200 m) asl
HABITAT	Rainforest, gallery forest, savanna woodland, deciduous woodland, plantations, and villages
DIET	Lizards and frogs
REPRODUCTION	Oviparous, with clutches of 3–4 eggs
CONSERVATION STATUS	IUCN not listed

ADULT LENGTH
2 ft 4 in–4 ft
(0.7–1.2 m)

HAPSIDOPHRYS SMARAGDINUS
EMERALD SNAKE
(SCHLEGEL, 1837)

The Emerald Snake is a slender, laterally compressed snake with keeled scales, a very long tail, a long head, just distinct from the neck, large eyes, and round pupils. The dorsal color is dark or emerald green (*smaragdinus* means emerald green), the anterior body also being marked with several rows of pale blue and black spots, and the head by a dark stripe through the eye that separates the green dorsum from the yellow lips. The venter is pale green, while the throat is yellow.

The Emerald Snake occurs through West and Central Africa, from The Gambia to Uganda, and south to Angola. It is common in West Africa but less common in Central Africa. This species is also reported to occur on São Tomé in the Gulf of Guinea. It is a common arboreal and diurnal species that inhabits rainforests, gallery forest, deciduous woodland, savanna woodland mosaics, and plantations, and also occurs in close proximity to villages. It prefers habitats near water. A pair of longitudinal keels on the ventral scales enable it to scale straight tree trunks. It preys on small frogs such as hyperolid reed frogs, and a variety of lizards, from geckos to agamas. The long tail can be pseudautotomized, but unlike the autotomized tails of lizards it does not regenerate the lost tail.

RELATED SPECIES

Two other species are recognized, the Green-lined Green Snake (*Hapsidophrys lineatus*), which occurs in sympatry with *H. smaragdinus*, and the Príncipe Treesnake (*H. principis*), from Príncipe Island, the neighboring island to São Tomé, in the Gulf of Guinea. All three snakes were previously included in the genus *Gastropyxis*.

Actual size

FAMILY	Colubridae: Colubrinae
RISK FACTOR	Nonvenomous
DISTRIBUTION	Indian Ocean: Socotra Islands (Yemen)
ELEVATION	0–2,950 ft (0–900 m) asl
HABITAT	Rocky scrub near wadis, rivers, papyrus swamps, and in coastal areas
DIET	Lizards, small mammals, and possibly sea fish
REPRODUCTION	Oviparous, clutch size unknown
CONSERVATION STATUS	IUCN Near Threatened

ADULT LENGTH
Male
5 ft (1.5 m)

Female
3 ft 3 in (1.0 m)

HEMEROPHIS SOCOTRAE
SOCOTRAN RACER
(GÜNTHER, 1881)

179

The Socotran Racer is endemic to the Socotra Islands, a small Yemeni archipelago between Yemen and Somalia in the Arabian Sea. It is found on Socotra and the smaller islands of Darsa and Samha, known as The Brothers, but is not recorded from Abd al Kuri Island. It is terrestrial, crepuscular, and fast-moving, and it inhabits rocky scrub but is commonest in the vicinity of water, along rivers and around pools and wadis. It is also found in papyrus swamps and along the island coastline in sandy habitats. Prey consists of lizards and rodents, but one author reports that Socotran Racers feed on sea fish, an unlikely natural diet. The endemic Socotran Racer is threatened by habitat loss, road kills, and active persecution.

RELATED SPECIES

At one time the genus *Hemerophis* also contained the Kunene Racer (*Mopanveldophis ʒebrinus*, page 197) from Namibia. Nothing is known of the closest relatives of *Hemerophis socotrae*.

The Socotran Racer is an elongate snake with smooth scales, a long tail, and a narrow head with a squarish snout, large eyes, and round pupils. The head is black above, white on the sides. The anterior body is banded black and salmon-pink, the dark bands being 1.5 times wider than the pale interspaces, which become narrower and white, heavily flecked with black, on the posterior body, the tail being virtually entirely black. The venter is yellowish, reddish, or olive.

Actual size

FAMILY	Colubridae: Colubrinae
RISK FACTOR	Nonvenomous
DISTRIBUTION	Southwestern Europe and northwest Africa: Spain, Portugal, Italy, Morocco, Tunisia, and Algeria
ELEVATION	0–7,410 ft (0–2,260 m) asl
HABITAT	Rocky and scrubby hillsides, vineyards, olive groves, plantations, garbage dumps, cemeteries, dry stone walls, and abandoned settlements
DIET	Lizards, small mammals, snakes, birds and their eggs, reptile eggs, and insects
REPRODUCTION	Oviparous, with clutches of 5–11 eggs
CONSERVATION STATUS	IUCN Least Concern

ADULT LENGTH
3 ft 3 in–5 ft,
rarely 6 ft 7 in
(1.0–1.5 m,
rarely 2.0 m)

HEMORRHOIS HIPPOCREPIS
HORSESHOE WHIPSNAKE
(LINNAEUS, 1758)

180

The Horseshoe Whipsnake, which is sometimes called the Horseshoe Racer, is a fast-moving, diurnal snake found in southern Spain, Portugal, northern Morocco, Tunisia, and Algeria. It also inhabits the Italian Mediterranean islands of Sardinia and Pantelleria. The preferred habitat of the Horseshoe Whipsnake comprises open, scrub-covered rocky hillsides, but it also inhabits anthropogenic habitats such as plantations, olive groves, vineyards, abandoned buildings, and dry stone walls—anywhere with rocky ground, abundant prey, and escape routes. Its prey preferences are also wide, from the insects eaten by juveniles, to lizards, snakes, birds and their eggs, reptile eggs, and small mammals as adults. This is a highly alert predator that actively seeks and runs down its prey. It is also a nervous snake that avoids human confrontations, but will bite if cornered or handled. It is nonvenomous.

RELATED SPECIES

Hemorrhois hippocrepis has three close relatives: the Algerian Whipsnake (*H. algirus*) from North Africa, Asian Whipsnake (*H. nummifer*) from Greece to Egypt and east to Kazakhstan, and the Spotted Whipsnake (*H. ravergieri*), which occurs through the same region but farther east to Mongolia.

Actual size

The Horseshoe Whipsnake is a slender, smooth-scaled snake, with a long tail, a head just distinct from the neck, large eyes, and round pupils. The ground color varies from gray to yellow or olive with a dorsal pattern comprising a vertebral series of large, dark brown, coin-like or ocelli markings, often with light centers, and a smaller, less well-defined series on either flank. The common name comes from a bold horseshoe marking on the dorsum of the head in most specimens.

FAMILY	Colubridae: Colubrinae
RISK FACTOR	Nonvenomous
DISTRIBUTION	Southern Europe: northeast Spain, France, western Switzerland, Italy, Malta, Slovenia, Croatia, and Gyaros Island (Greece)
ELEVATION	0–6,890 ft (0–2,100 m) asl
HABITAT	Dry rocky slopes, forest edges and clearings, abandoned buildings, vineyards, railway embankments, and quarries
DIET	Lizards, small mammals, snakes, birds and their eggs, amphibians, and insects
REPRODUCTION	Oviparous, with clutches of up to 20 eggs
CONSERVATION STATUS	IUCN Least Concern

ADULT LENGTH
4 ft 7 in–5 ft 2 in
(1.4–1.6 m)

HIEROPHIS VIRIDIFLAVUS
WESTERN WHIPSNAKE
(LACÉPÈDE, 1789)

181

The Western Whipsnake occurs in southern Europe, from northeastern Spain, France, and Switzerland to Corsica, Sardinia, Italy, Sicily, Malta, Slovenia, and Croatia, in two distinct forms, a western patterned morphotype and an eastern melanistic morphotype. This is an alert, fast-moving, diurnal, and terrestrial inhabitant of rocky slopes and fields, forest edges and clearings, and anthropogenic habitats ranging from railway embankments to vineyards and quarries. It is a sun-loving species that actively hunts rodents, lizards, smaller snakes, and birds, with amphibians and insects taken by juveniles. Although a terrestrial species it is also an adept climber, which may be found basking or foraging in trees and bushes. Although nonvenomous, this is a nervous snake that will flee if the opportunity exists, but if captured it will bite freely and repeatedly.

The Western Whipsnake is a large snake with a long tail, smooth scales, a broad head that is distinct from the neck, large eyes, and round pupils. Juveniles may be pale brown with faint transverse markings and a dark-blotched head. Adults are dorsally black with fine yellow or yellow-green transverse markings on the anterior body, speckling on the posterior body, and fine stripes on the tail. The head bears numerous yellow-green spots while the lips, throat, and neck are yellow or yellow-green. The venter is yellow-gray with dark spotting.

RELATED SPECIES

Two subspecies are sometimes recognized, the nominate western form (*Hierophis viridiflavus viridiflavus*) and the usually melanistic eastern form (*H. v. carbonarius*). An isolated population on Gyaros Island, formerly known as *Dolichophis gyarosensis*, is now believed to have been introduced by man in historical times. The other members of the genus are the Andreas Racer (*H. andreanus*) from Iran, and the Balkan Racer (*H. gemonensis*), which occurs in the Balkans, Greece and Italy. The whipsnakes of genus *Dolichophis* (page 161) and the Slender Racer (*Orientocoluber spinalis*, page 204) are also related to *Hierophis*.

Actual size

FAMILY	Colubridae: Colubrinae
RISK FACTOR	Nonvenomous
DISTRIBUTION	North America: southern USA and northern Mexico
ELEVATION	1,480–5,970 ft (450–1,820 m) asl
HABITAT	Arid Chihuahuan Desert habitats, rocky canyons, with gravel soils and cacti, creosote, and mesquite, or rocky outcrops
DIET	Lizards, lizard eggs, small mammals, and frogs
REPRODUCTION	Oviparous, with clutches of 3–14 eggs
CONSERVATION STATUS	IUCN Least Concern

ADULT LENGTH
2 ft 4 in–3 ft 3 in,
rarely 5 ft
(0.7–1.0 m,
rarely 1.5 m)

182

LAMPROPELTIS ALTERNA
GRAY-BANDED KINGSNAKE
(BROWN, 1902)

The Gray-banded Kingsnake has smooth scales, a cylindrical body, and a distinct head, tapering to a pointed snout. The eyes are small and protruding, with round pupils. It is pale to dark gray dorsally, with a series of black-edged bands, narrow in *L. a. alterna* and broad with red or orange centers in *L.. a. blairi*.

The Gray-banded Kingsnake is found in southwestern Texas and southeastern New Mexico, and into Mexico as far south as Zacatecas state. It occurs in Chihuahuan Desert canyons, with gravel substrates vegetated with creosote bushes, mesquite and acacia trees, and cacti. Secretive by nature, it shelters during the day in rocky crevices, emerging as night falls to hunt its primary prey—sleeping diurnal lizards. It also takes lizard eggs, small mice, and frogs, but unlike many other kingsnakes and milksnakes it has not been recorded taking other snakes. Heavy rainfall may also trigger foraging activity in the Gray-banded Kingsnake. Prey is often captured in mammal burrows or rocky crevices because this rarely encountered kingsnake is reportedly an ineffectual constrictor that kills prey more easily by pressing it against a hard surface.

RELATED SPECIES

Lampropeltis alterna was previously a subspecies of the Mexican Kingsnake (*L. mexicana*), which occurs to its south. Some authors recognize two subspecies, the nominate subspecies (*L. a. alterna*) and Blair's Kingsnake (*L. a. blairi*). The related Mexican Kingsnake also contains two to three subspecies, the San Luis Potosi Kingsnake (*L. m. mexicana*), Thayer's Kingsnake (*L. m. thayeri*), and Durango Mountain Kingsnake (*L. m. greeri*), all from Mexico. Another close relative is Ruthven's Kingsnake or the Queretaro Mountain Kingsnake (*L. ruthveni*), found to the south of the Mexican Kingsnake's range.

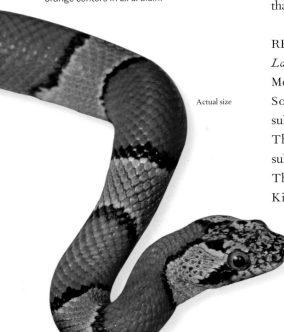

Actual size

FAMILY	Colubridae: Colubrinae
RISK FACTOR	Nonvenomous, constrictor
DISTRIBUTION	North America: southwestern USA and northwestern Mexico
ELEVATION	0–5,970 ft (0–1,820 m) asl
HABITAT	Desert, semidesert, canyons, open grasslands, prairie, agricultural habitats, and swamps
DIET	Insects, frogs, salamanders, lizards, other snakes, turtles and turtle eggs, birds and their eggs, small mammals
REPRODUCTION	Oviparous, with clutches of 2–24 eggs
CONSERVATION STATUS	IUCN not listed; *L. getula* Least Concern

ADULT LENGTH
2 ft 4 in–3 ft 3 in,
rarely 4 ft
(0.7–1.0 m,
rarely 1.2 m)

LAMPROPELTIS CALIFORNIAE
CALIFORNIAN KINGSNAKE
(BLAINVILLE, 1835)

183

The Californian Kingsnake is found from southern Oregon to California, Utah, Arizona, and the Baja California Peninsula and Sonora, Mexico. It inhabits coastal regions and also high desert locations, and may be encountered in almost any habitat, from desert and semidesert to agricultural environments and swamps, and it is often associated with water. A powerful constrictor, its prey preferences are as wide as its habitat preferences, with frogs, salamanders, lizards, turtles and their eggs, snakes (including venomous rattlesnakes), birds and their eggs, and small mammals all featuring in the diet. Juveniles may also eat large insects. The Californian Kingsnake is active by day or night. It is inoffensive when handled, and is a good pet species, certainly compared to the related Speckled Kingsnake (*Lampropeltis holbrooki*), which bites easily.

RELATED SPECIES

Lampropeltis californiae was formerly a subspecies of the Eastern Kingsnake (*L. getula*), but the latter is now split into five species, the other three being the Speckled Kingsnake (*L. holbrooki*), Desert Kingsnake (*L. splendida*), and Black Kingsnake (*L. nigra*). The Santa Catalina Island Kingsnake (*L. catalinensis*) was a former synonym of *L. californiae*.

The Californian Kingsnake is a cylindrical snake with smooth scales, a moderately long tail, a rounded head with moderately large eyes, and round pupils. Its patterning can be highly variable, there being a vertebrally striped "*californiae*" morphotype, and a banded "*boylii*" morphotype, both of which may occur in "coastal" livery—brown with cream or yellow bands/stripes, or "desert" livery—black with white bands/stripes. Californian Kingsnakes have yellowish heads with darker pigment on the posterior of the head, the snout, and on the sutures between the scutes.

Actual size

FAMILY	Colubridae: Colubrinae
RISK FACTOR	Nonvenomous, constrictor
DISTRIBUTION	North America: eastern USA
ELEVATION	0–3,000 ft (0–915 m) asl
HABITAT	Dry prairie, wet meadow, salt-grass savanna, rocky hillsides, riverine woodland, deciduous and mixed woodland
DIET	Small mammals and birds, amphibians, lizards, and snakes
REPRODUCTION	Oviparous, with clutches of 6–17 eggs
CONSERVATION STATUS	IUCN Least Concern

ADULT LENGTH
23¾–35½ in,
rarely 5 ft
(600–900 mm,
rarely 1.5 m)

184

LAMPROPELTIS CALLIGASTER
YELLOW-BELLIED KINGSNAKE
(HARLAN, 1827)

The Yellow-bellied Kingsnake is a slender species with smooth scales, a relatively short tail, a narrow, slightly pointed head, and moderately large eyes with round pupils. Patterning consists of a series of red to dark brown cross-bars, sometimes light-edged, on a ground color that varies from dark gray to gray-brown or pale gray. A corresponding series of lateral blotches is also present and dark rows of blotches are present on the pale venter. There is also a dark postocular stripe.

The Yellow-bellied Kingsnake is a secretive and rarely encountered species. The three races are found through eastern USA from eastern Texas, north to southern Iowa, and east to Maryland and the Carolinas. It is virtually absent from Florida, except for small populations in the Panhandle, and an isolated race around Lake Okeechobee in central Florida. Habitats range from arid prairie grassland to wet meadows and pastures, salt-grass savanna, and deciduous and mixed woodland. The western race feeds primarily on small mammals, while the eastern races are more inclined to take other reptiles. All three forms are generally docile when handled. During the day this species will shelter under flat rocks or in animal burrows, so it is only rarely encountered when moving abroad at night, especially following heavy rain.

RELATED SPECIES

Lampropeltis calligaster contains three subspecies: the Prairie Kingsnake (*L. c. calligaster*) from the west of the range, the Mole Kingsnake (*L. c. rhombomaculata*) from the east of the range, and the South Florida Mole Kingsnake (*L. c. occipitolineata*) from Lake Okeechobee environs. *Lampropeltis calligaster* may be the most basal (primitive) living member of the genus. Also near the base of *Lampropeltis* is the Short-tailed Snake (*L. extenuata*), from Florida, which was formerly in *Stilosoma*.

Actual size

FAMILY	Colubridae: Colubrinae
RISK FACTOR	Nonvenomous, constrictor
DISTRIBUTION	North America: southwestern USA and possibly northern Mexico
ELEVATION	2,790–8,860 ft (850–2,700 m) asl
HABITAT	Rocky, wooded canyons, scree and talus slopes, and montane deciduous and coniferous woodland, usually near water
DIET	Lizards, small mammals, birds, and other snakes
REPRODUCTION	Oviparous, with clutches of 1–9 eggs
CONSERVATION STATUS	IUCN Least Concern

ADULT LENGTH
27½–35½ in,
rarely 3 ft 3 in
(700–900 mm,
rarely 1.0 m)

LAMPROPELTIS PYROMELANA

ARIZONA MOUNTAIN KINGSNAKE

(COPE, 1867)

185

Originally known as the Sonoran Mountain Kingsnake, when it contained four subspecies and was more widely distributed, the now more localized Arizona Mountain Kingsnake is found in montane habitats from Utah and Nevada to Arizona and New Mexico in the USA, although it may cross the border into northern Sonora and Chihuahua, Mexico. It is found at high elevations, over 8,200 ft (2,500 m), but is much less frequently encountered below 4,920 ft (1,500 m) and does not occur in lowland habitats. Its montane populations are isolated and vulnerable to overcollecting, fire, or other threats. Preferred habitats include rocky wooded canyons with talus slopes, deciduous forest, and coniferous forest at the highest elevations. Prey consists of lizards and small mammals, with birds and snakes also taken.

RELATED SPECIES

Lampropeltis pyromelana contained four subspecies but one of these, the Chihuahuan Mountain Kingsnake (*L. knoblochi*), was elevated to specific status, with the Huachuca Mountain Kingsnake (*L. p. woodini*) in its synonymy. The Utah Mountain Kingsnake (*L. p. infralabialis*) has been synonymized with the Arizona Mountain Kingsnake (*L. p. pyromelana*). The recently described Webb's Kingsnake (*L. webbi*), from western Mexico, is also believed to be related to this group. *Lampropeltis pyromelana* also closely resembles the California Mountain Kingsnake (*L. ʒonata*, page 187) and venomous Sonoran Coralsnake (*Micruroides euryxanthus*, page 456).

Actual size

The Arizona Mountain Kingsnake is a cylindrical snake with smooth scales, a moderately long tail, and a narrow head, only lightly distinct from the neck, with bulbous eyes and round pupils. The snout tip is white or yellow, and there is a glossy black mask over the eyes and most of the posterior head, followed by another white/yellow band. Thereafter the body is patterned with distinctive, broad red bands, between narrow black bands, with narrow white/yellow interspaces.

FAMILY	Colubridae: Colubrinae
RISK FACTOR	Nonvenomous, constrictor
DISTRIBUTION	North America: eastern Canada and USA
ELEVATION	0–10,900 ft (0–3,330 m) asl
HABITAT	Rocky hillsides, deciduous and pine woodlands, scrub, marsh edges, riverine floodplains, and abandoned buildings
DIET	Small mammals, birds, eggs, lizards, small snakes, frogs, salamanders, and various invertebrates
REPRODUCTION	Oviparous, with clutches of 5–20 eggs
CONSERVATION STATUS	IUCN not listed

ADULT LENGTH
2 ft 4 in–3 ft 3 in,
rarely 4 ft 3 in
(0.7–1.0 m,
rarely 1.3 m)

186

LAMPROPELTIS TRIANGULUM
EASTERN MILKSNAKE
(LACÉPÈDE, 1789)

The Eastern Milksnake occurs from Ontario, Canada, south through the New England states of the USA to North Carolina, northern Alabama and Georgia, west to Arkansas and Kansas, and north to Wisconsin and Michigan, taking in the range of the former subspecies *syspila* and the northeastern Louisiana part of the former *amaura* range. Although tricolored, this species is not a banded coralsnake mimic, like the other milksnakes within *Lampropeltis triangulum* in its broadest definition. The Eastern Milksnake inhabits rocky hillsides, scree slopes, deciduous and coniferous woodlands, riverine floodplains, marsh edges, and abandoned buildings. Prey consists primarily of small rodents, but small snakes, lizards, birds, and both reptile and bird eggs are also eaten. When alarmed, Eastern Milksnakes vibrate their tails, which makes a rattling sound on dead leaves, like the venomous Copperhead (*Agkistrodon contortrix*, page 553).

The Eastern Milksnake has a narrow head, slightly distinct from the neck, and moderately large eyes with round pupils. Its normal pattern consists of red, orange, or brown dorsal saddles with black edges, on a gray or cream background. The dorsum of the head bears a brown or red V-shaped marking, which is contiguous with the first dark dorsal saddle. The other former *triangulum* species are classic tricolor red, black, and yellow banded, with the red and black pigment in contact.

Actual size

RELATED SPECIES

The Milksnake (*Lampropeltis triangulum*) was one of the most speciose and widely distributed American snake species, with 25 subspecies. Recent molecular advances demonstrate that at least seven species are involved. *Lampropeltis triangulum* is now confined to eastern North America. The other species are the Scarlet Milksnake (*L. elapsoides*) from southeastern USA, Western Milksnake (*L. gentilis*) from central USA, Northeast Mexican Milksnake (*L. annulata*), Mexican Milksnake (*L. polyzona*), Central American Milksnake (*L. abnorma*), and South American Milksnake (*L. micropholis*), which occurs as far south as Ecuador. No subspecies are currently recognized.

FAMILY	Colubridae: Colubrinae
RISK FACTOR	Nonvenomous, constrictor
DISTRIBUTION	North America: western USA and northwestern Mexico
ELEVATION	0–10,900 ft (0–3,330 m) asl
HABITAT	Deciduous woodland, coniferous forest, chaparral, riverine woodland, and rocky canyon slopes
DIET	Lizards, small birds and their eggs, and small snakes
REPRODUCTION	Oviparous, with clutches of 2–13 eggs
CONSERVATION STATUS	IUCN Least Concern

ADULT LENGTH
2 ft 7 in–3 ft 3 in,
rarely 4 ft
(0.8–1.0 m,
rarely 1.2 m)

LAMPROPELTIS ZONATA
CALIFORNIA MOUNTAIN KINGSNAKE
(LACÉPÈDE, 1789)

187

The range of the California Mountain Kingsnake is centered on California but extends north into Oregon, and there are isolated populations on the Oregon–Washington border, and also a population in Baja California, Mexico. It occurs in a variety of habitats from chaparral to deciduous woodland and coniferous forest at higher elevations, and riverine forest at lower elevations. South-facing slopes with rocks and vegetation are optimal habitats. The California Mountain Kingsnake appears to feed largely on reptiles, particularly lizards, which are chased and captured. Small snakes and birds and their eggs also feature in their diet. It has been suggested that the species' bright colors prompt attacks from the parent birds and that it may be guided to the nest based on the intensity of the birds' mobbing.

The California Mountain Kingsnake is a slender, smooth-scaled snake with a relatively short tail and a narrow head, barely distinct from the neck, with medium-sized eyes and round pupils. Its patterning is very similar to that of the Arizona and Chihuahuan Mountain Kingsnakes, with broad red bands between narrow black bands, and white or cream interspaces. The anterior of the head is black, although the snout is sometimes red, followed by a white band around the neck, and then the first black-red-black band.

RELATED SPECIES

Lampropeltis zonata has been listed with as many as seven subspecies, although currently none are recognized. The population from Todos Santos Island, in the Pacific off Baja California (*L. herrerae*), considered by some authors to be a subspecies of *L. zonata*, is listed as Critically Endangered by the IUCN. The closest relatives of *L. zonata* are the Arizona Mountain Kingsnake (*L. pyromelana*, page 185), Chihuahuan Mountain Kingsnake (*L. knoblochi*), and the recently described Webb's Kingsnake (*L. webbi*).

Actual size

FAMILY	Colubridae: Colubrinae
RISK FACTOR	Rear-fanged, mildly venomous; relatively harmless to humans
DISTRIBUTION	North, Central, and South America: southern Mexico to Argentina
ELEVATION	0–9,020 ft (0–2,750 m) asl
HABITAT	Lowland and low montane rainforest, and gallery forest in dry forest habitats
DIET	Frogs, lizards, snakes, birds and their eggs, and insects
REPRODUCTION	Oviparous, with clutches of 1–8 eggs
CONSERVATION STATUS	IUCN not listed

ADULT LENGTH
4–5 ft,
rarely 7 ft 7 in
(1.2–1.5 m,
rarely 2.3 m)

188

LEPTOPHIS AHAETULLA
GREEN PARROT SNAKE
(LINNAEUS, 1758)

The Green Parrot Snake is a large, slender snake with keeled scales arranged in oblique rows, a long tail and a broad head, distinct from the neck, with large eyes and round pupils. The overall color is green, lighter on the venter than the dorsum, with a black postocular stripe, pale yellow lips, and a black and golden iris.

The Green Parrot Snake, also known as the Green Frogger, occurs from Veracruz, southeastern Mexico, through Central and South America to Argentina and Uruguay. It is found in lowland and low montane rainforest, and also in riverine gallery forest in dry forest habitats. The preferred prey of the Green Parrot Snake comprises frogs, but lizards, snakes, birds and their eggs, and even large insects like grasshoppers are also taken. When it feels threatened the Green Parrot Snake gapes widely, exposing the fleshy interior of the mouth and its blue tongue. A large, rear-fanged, mildly venomous species, it has a long strike and bites easily. The effects of a bite from a large specimen include localized pain followed by numbness. It lays small clutches of eggs in epiphytic bromeliads.

RELATED SPECIES
This widely distributed species is represented by 10–11 subspecies occurring from southern Mexico (*Leptophis ahaetulla praestans*) to Paraguay, Uruguay, and Argentina (*L. a. marginatus*). The genus *Leptophis* includes a further ten species including several green species; the West Coast Parrot Snake (*L. diplotropis*), endemic to Mexico; Cope's Parrot Snake (*L. depressirostris*), from Honduras to Peru; and the Cloud Forest Parrot Snake (*L. modestus*), from Central America, which is considered Vulnerable by the IUCN.

Actual size

FAMILY	Colubridae: Colubrinae
RISK FACTOR	Nonvenomous
DISTRIBUTION	Southeast Asia: northeast India, Myanmar, Laos, and Vietnam
ELEVATION	2,000–6,000 ft (610–1,830 m) asl
HABITAT	Subtropical and montane forests, and bamboo thickets
DIET	Presumed frogs and tadpoles
REPRODUCTION	Oviparous, with clutches of 4–5 eggs
CONSERVATION STATUS	IUCN Least Concern

ADULT LENGTH
30 in
(760 mm)

LIOPELTIS FRENATUS
GÜNTHER'S
STRIPE-NECKED SNAKE
(GÜNTHER, 1858)

189

Günther's Stripe-necked Snake occurs in the northeast Indian states from Meghalaya and Assam to Arunachal Pradesh, and also in northern Myanmar, Laos, and Vietnam, where it inhabits subtropical and montane forests, especially those containing bamboo stands and thickets. This species is diurnal and believed to be terrestrial in habit. Specimens have been seen in close proximity to temporary pools containing frogs and their tadpoles and therefore these are its presumed prey, but its diet is otherwise undocumented. This is an oviparous species, which reportedly lays up to five eggs in the internodes of live bamboo stands. It is an inoffensive and completely harmless snake that does not bite when handled.

RELATED SPECIES

There are five other members of the genus *Liopeltis*, including Stoliczka's Stripe-necked Snake (*L. stoliczkae*), which occurs in sympatry with this species. The other species are the Philippine Four-striped Snake (*L. philippinus*), Sri Lankan Striped Snake (*L. calamaria*), Himalayan Stripe-necked Snake (*L. rappii*), and Malayan Ringneck Snake (*L. tricolor*).

Günther's Stripe-necked Snake is a small snake with smooth scales, a long, prehensile tail, and a head slightly distinct from the neck, with large eyes and round pupils. The dorsal color is mid-brown while the undersides are white. A broad, black postocular stripe exits the eye and narrows on the neck to form the upper of four fine, longitudinal lateral stripes, created by the black edging of the scales. The lips and throat are pale yellow.

Actual size

FAMILY	Colubridae: Colubrinae
RISK FACTOR	Nonvenomous
DISTRIBUTION	Southeast Asia: southeastern China to the Philippines, Indonesia, and Timor-Leste, and introduced widely
ELEVATION	0–2,300 ft (0–700 m) asl
HABITAT	Most habitats, including anthropogenic environments
DIET	Lizards, small mammals, and reptile eggs
REPRODUCTION	Oviparous, with clutches of 4–11 eggs
CONSERVATION STATUS	IUCN Least Concern

ADULT LENGTH
19¾–26½ in
(500–670 mm)

190

LYCODON CAPUCINUS
COMMON ISLAND WOLFSNAKE
(BOIE, 1827)

The Common Island Wolfsnake is a small snake with smooth scales, a moderately long tail, and a broad head that narrows abruptly before the snout, with bulbous eyes and vertically elliptical pupils. The general coloration is brown or gray above and immaculate white below, with the dorsal pigment also heavily flecked with white or pale yellow. The head is brown with a broad white or pale yellow collar, flecked with brown, and white lips with brown flecking.

The Common Island Wolfsnake occurs from southeastern China, though mainland Southeast Asia and the major archipelagos of the Philippines and Indonesia, to as far south as Timor-Leste. It is a good colonizer, which has been accidentally introduced to the Maldives, the Mascarenes, and Christmas Island (an Australian territory south of Java in the Indian Ocean), Micronesia, and New Guinea. It can occur almost anywhere, and it is especially common in anthropogenic habitats, entering inhabited and uninhabited buildings. Its primary prey consists of the geckos that also occur in human habitations, but skinks and rodents are also taken, and reptile eggs also feature in its diet. Although harmless, the Common Island Wolfsnake will vibrate its tail on leaves to warn potential predators to keep their distance, and it bites easily if handled.

RELATED SPECIES

The Asian genus *Lycodon* contains up to 50 species and is closely related to the Australo-Papuan genus *Stegonotus* (pages 235–236). *Lycodon capucinus* is very similar in appearance to, and was once a subspecies of, the Common Indian Wolfsnake (*L. aulicus*) from South Asia.

Actual size

FAMILY	Colubridae: Colubrinae
RISK FACTOR	Nonvenomous
DISTRIBUTION	East and Southeast Asia: Mongolia, eastern China and Russia, Hainan Island, Korea, Japanese Ryukyu Islands, Taiwan, Laos, and Vietnam
ELEVATION	1,310–3,610 ft (400–1,100 m) asl
HABITAT	Low montane habitats, rice paddies, and flood meadows
DIET	Frogs, lizards, fish, birds, and small mammals
REPRODUCTION	Oviparous, with clutches of 6–10 eggs
CONSERVATION STATUS	IUCN Least Concern, included in the Russian Red List

ADULT LENGTH
1 ft 8 in–3 ft 3 in,
occasionally 4 ft 3 in
(0.5–1.0 m,
occasionally 1.3 m)

LYCODON RUFOZONATUS
RED BANDED SNAKE
CANTOR, 1842

191

The Red Banded Snake inhabits Vietnam, Laos, China, Mongolia, Korea, and possibly also far eastern Russia. It also occurs on Hainan Island, Taiwan, and the Ryukyu Islands. It is generally a common, nocturnal snake that feeds primarily on frogs but also takes a wide variety of other vertebrates, including small fish, lizards, snakes, birds, and rodents. Its habitats include flood meadows and rice paddies at low montane elevations. When disturbed some specimens elevate the anterior body and further flatten the already flat head, biting readily if handled, while others are inoffensive and prefer thanatosis as a means of defense. The common name should not be hyphenated; this is a red snake with bands, not a red-banded snake.

The Red Banded Snake has a laterally compressed body, a broad, flattened head, and small eyes with vertical pupils. It is a red snake with a series of regular black dorsal bands that begin as a V-shaped marking on the neck and continue to the tail tip. The flanks also bear two rows of irregular black blotches. A black postocular stripe exits the rear of the eye. The venter is white.

RELATED SPECIES

This is one of eight species that spent the last century in the genus *Dinodon*, before being moved back into *Lycodon*. The main differences between the two genera relate to the arrangement of their teeth. *Lycodon rufozonatus* contains two subspecies, the mainland nominate subspecies (*L. r. rufozonatus*), which also occurs on the Tsushima Islands of Japan, and a broader-banded Okinawan subspecies (*L. r. walli*). Related species include the Rose Banded Snake (*L. rosozonatus*) and the Yellow Banded Snake (*L. flavozonatus*).

Actual size

FAMILY	Colubridae: Colubrinae
RISK FACTOR	Nonvenomous
DISTRIBUTION	North Africa, Arabia, and Western Asia: Mauritania to Egypt, Syria to Oman and Yemen, and western Iran
ELEVATION	0–7,550 ft (0–2,300 m) asl
HABITAT	Sandy deserts, especially where sand meets rock, gravel pans, and saltmarshes
DIET	Lizards and insects
REPRODUCTION	Oviparous, with clutches of 3–5 eggs
CONSERVATION STATUS	IUCN Least Concern

ADULT LENGTH
9¾–11¾ in,
rarely 17¼ in
(250–300 mm,
rarely 440 mm)

192

LYTORHYNCHUS DIADEMA
CROWNED
LEAFNOSE SNAKE
(DUMÉRIL, BIBRON & DUMÉRIL, 1854)

The Crowned Leafnose Snake is a slender snake with a long tail, smooth scales, a long, narrow head that terminates in a leaf-shaped rostral scale, and large eyes with vertically elliptical pupils. Across its range coloration is variable, from pinkish brown to tan or cream, with a series of irregular dorsal brown saddle markings and smaller lateral spots. The venter is white. A bold brown or red-brown crown marking is present on the head, with a band across the snout that continues backward through the eyes as a postocular stripe.

Also known as the Awl-headed Snake, the Crowned Leafnose Snake is found from Mauritania to Egypt in North Africa, through much of the Arabian Peninsula, and in western Iran, along the Iraq border and the coast of the Persian Gulf. It is a nocturnal inhabitant of sandy desert habitats, gravel pans, and saltmarshes. The Crowned Leafnose Snake has an enlarged leaf-shaped rostral scale to help it burrow in loose sand, but it also moves easily over the surface, leaving distinct trails. It feeds almost exclusively on lizards, especially sand dune-dwelling geckos. Although nonvenomous, the Crowned Leafnose Snake kills lizards by vigorous chewing, enabling its enlarged rear teeth to introduce its oral secretions into the prey, and it uses its coils for prey restraint until the weak secretions have taken effect.

RELATED SPECIES

Four subspecies are sometimes proposed, but only the widely distributed nominate form (*Lytorhynchus diadema diadema*) and a northeastern subspecies (*L. d. gaddi*), from Saudi Arabia, Iraq, and Iran, are widely recognized. Of the other six *Lytorhynchus* species, Gasperetti's Leafnose Snake (*L. gasperetti*) occurs in southwestern Saudi Arabia, while Kennedy's Leafnose Snake (*L. kennedyi*) is found in Jordan, Syria, and Iraq. Ridgeway's Leafnose Snake (*L. ridgewayi*) occurs in Iran, Afghanistan, and Pakistan, Leviton's Leafnose Snake (*L. levitoni*) is also in Iran, while Maynard's Leafnose Snake (*L. maynardi*) and the Sindh Leafnose Snake (*L. paradoxus*) are found in Pakistan.

Actual size

FAMILY	Colubridae: Colubrinae
RISK FACTOR	Rear-fanged, mildly venomous; harmless to humans
DISTRIBUTION	North Africa and Middle East: Western Sahara, Morocco, Algeria, Tunisia, Libya, Egypt, southwestern Israel; also Lampedusa Island, Italy
ELEVATION	0–8,200 ft (0–2,500 m) asl
HABITAT	Arid open habitats with sparse vegetation, rocky habitats, and around human habitations
DIET	Lizards, amphisbaenians, and small mammals
REPRODUCTION	Oviparous, with clutches of 5–7 eggs
CONSERVATION STATUS	IUCN Least Concern

ADULT LENGTH
15¾–19¾ in,
occasionally 21¼ in
(400–500 mm,
occasionally 550 mm)

MACROPROTODON CUCULLATUS
COMMON FALSE SMOOTH SNAKE
(GEOFFROY SAINT-HILAIRE, 1827)

193

Also known as the Cowled or Hooded Snake, the Common
False Smooth Snake resembles the true smooth snakes
(*Coronella*, page 155). It is a widely distributed North African
species, occurring from the Atlantic coast to the Middle East,
but only found in Europe on Lampedusa, an Italian island
between Tunisia and Sicily. It prefers open, arid rocky or sandy
habitats, with sparse vegetation, but this small species also
occurs around human habitations, using broken stone walls or
discarded trash as cover. A terrestrial species, it is usually
crepuscular, but it may also be active diurnally on overcast
days. Small lizards and amphisbaenians (worm-lizards) are its
main prey, with mice also taken. Prey is subdued, if not actually
killed, by weak venom injected via the snake's rear fangs.

RELATED SPECIES

The nominate subspecies (*Macroprotodon cucullatus cucullatus*)
occurs on the Mediterranean coast from Tunisia to Egypt, and
Israel, while another subspecies (*M. c. textilis*) inhabits the
Tunisian to Moroccan mountains, the Algerian Hoggar
region, and Lampedusa Island. Three other species
occur from Spain, Portugal, and Morocco
(*M. brevis*), the Morocco–Algeria border
region (*M. abubakeri*), and Algeria
to Tunisia, and the Balearic
Islands (*M. mauritanicus*).

The Common False Smooth Snake is a small,
slender snake with smooth scales, a relatively
short tail, a broad head, medium-sized eyes,
and round pupils. It is generally pale gray or
gray-brown with body patterning limited to faint
spotting of darker pigment. The dorsum of the
head is marked by a dark V-shape that runs into
a broad dark collar or nape band, the cowl or
mask, while a black stripe may also pass
through the eye.

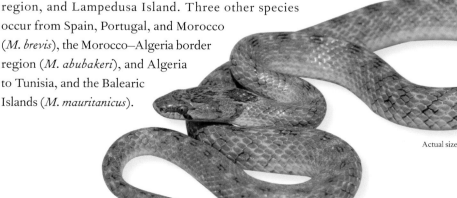

Actual size

FAMILY	Colubridae: Colubrinae
RISK FACTOR	Nonvenomous, constrictor
DISTRIBUTION	North America: southern USA to central Mexico
ELEVATION	8,200 ft (0–2,500 m) asl
HABITAT	Mesquite grassland, chaparral and prairie, farmland, thornbush and creosote scrub, deciduous and mixed woodland, and desert
DIET	Lizards, small mammals, birds, other snakes, reptile eggs, amphibians, and insects
REPRODUCTION	Oviparous, with clutches of 4–24 eggs
CONSERVATION STATUS	IUCN Least Concern

ADULT LENGTH
3 ft 3 in–5 ft,
occasionally 8 ft 6 in
(1.0–1.5 m,
occasionally 2.6 m)

194

MASTICOPHIS FLAGELLUM
COACHWHIP
(SHAW, 1802)

The Coachwhip is a slender snake with smooth scales, a long tail, a narrow head that tapers to a relatively pointed snout, and protruding supraocular scales over the large eyes, presenting a scowling expression. Coloration is variable, depending on subspecies, ranging from two-tone, with a black anterior body and head and a tan or red-brown posterior body and tail, to black, red-brown, yellow-brown, gray-brown, or pale pink throughout, with darker cross-bands.

The Coachwhip is a widely distributed species in the southern USA, found from Florida to California, almost coast to coast, although it is absent from the Mississippi Valley and coastal California. It also occurs farther south into central Mexico. This large, highly alert, fast-moving, diurnal snake is terrestrial but also hunts for prey in underground burrows. Prey mainly comprises lizards and small mammals, which are actively pursued or ambushed, but birds, amphibians, snakes including rattlesnakes, reptile eggs, and even insects are also eaten. The common name and the term *flagellum* both allude to an old wives' tale that this snake will coil its body around a person's legs and then lash them repeatedly with its whiplike tail. This is of course false. Coachwhips are nonvenomous, but bite quickly if handled.

RELATED SPECIES

This widely distributed species contains six subspecies: the Eastern Coachwhip (*Masticophis flagellum flagellum*), Western Coachwhip (*M. f. testaceus*), Lined Coachwhip (*M. f. lineatulus*), Red Coachwhip (*M. f. piceus*), Sonoran Coachwhip (*M. f. cingulum*), and San Joaquin Coachwhip (*M. f. ruddocki*). Related species include the Baja California Coachwhip (*M. fuliginosus*), a former subspecies, the Sonoran Whipsnake (*M. bilineatus*), and the Critically Endangered Clarion Island Whipsnake (*M. anthonyi*). All 11 species of *Masticophis* are included in the genus *Coluber* (page 152) by other authors.

Actual size

FAMILY	Colubridae: Colubrinae
RISK FACTOR	Nonvenomous
DISTRIBUTION	Northern South America: Colombia, Venezuela, the Guianas, Trinidad and Tobago, Brazil, Ecuador, Peru, and Bolivia
ELEVATION	0–7,280 ft (0–2,200 m) asl
HABITAT	Primary and secondary wet and dry tropical forests, forest edges, and plantations
DIET	Frogs, lizards, nestling birds, small mammals, and large insects
REPRODUCTION	Oviparous, with clutches of up to 5 eggs
CONSERVATION STATUS	IUCN not listed

ADULT LENGTH
3 ft 3 in–4 ft,
rarely 5 ft 2 in
(1.0–1.2 m,
rarely 1.6 m)

MASTIGODRYAS BODDAERTI
BODDAERT'S TROPICAL RACER
(SENTZEN, 1796)

195

Boddaert's Tropical Racer, also called the Tan Racer, is widely distributed across northern South America from Colombia to Trinidad and Tobago, where it is called "machete couesse," and south to Brazil and Bolivia. It is a closed-canopy species found in primary and secondary wet and dry forest, plantations, and forest-edge situations merging into savanna. It is also encountered close to semi-permanent ponds and other watercourses, or basking in sunny spots on trails or in treefall clearings. It can climb into low vegetation, but is more often seen on the forest floor. An alert, diurnal, terrestrial species, it hunts lizards, frogs, nestling birds, rodents, and large insects. Pieter Boddaert (1730–96) was a Dutch naturalist-physician and a friend of Carl Linnaeus.

Boddaert's Tropical Racer is a slender snake with smooth scales, a long tail, and an elongate head, just distinct from the neck, with large eyes and round pupils. Juveniles are brown above and white below and on the lips, with a dorsal pattern of alternative dark and light cross-bands and a faint dorsolateral stripe. Adults are red-brown or dark tan above with a much more distinctive yellow-brown stripe, no cross-bands, and a yellowish venter, throat, and lips.

RELATED SPECIES

The majority of the range is occupied by the nominate form (*Mastigodryas boddaerti boddaerti*), but a second subspecies may also occur in Colombia and Venezuela (*M. b. ruthveni*). The nominate subspecies occurs on Trinidad, but the Tobago and Little Tobago populations are treated as distinct (*M. b. dunni*). The genus *Mastigodryas* also contains a further 13 species distributed from southern Mexico to northern Argentina, and into the Lesser Antilles (*M. bruesi*). A number of species, including the Brazilian Tropical Racer (*M. bifossatus*), occur in sympatry with *M. boddaerti*.

Actual size

Actual size

FAMILY	Colubridae: Colubrinae
RISK FACTOR	Nonvenomous
DISTRIBUTION	Sub-Saharan Africa: Ethiopia and Somalia to Chad and Cameroon, to Swaziland and northeastern South Africa; possibly Yemen
ELEVATION	0–7,220 ft (0–2,200 m) asl
HABITAT	Desert, semidesert, coastal thickets, woodland, savanna, low montane grassland, and riverine and seasonally flooded habitats
DIET	Frogs and lizards
REPRODUCTION	Oviparous, with clutches of 2–3 eggs
CONSERVATION STATUS	IUCN not listed

ADULT LENGTH
Male
15¾–17¾ in
(400–450 mm)

Female
23¾–25½ in,
rarely 31½ in
(600–650 mm,
rarely 800 mm)

MEIZODON SEMIORNATUS
SEMIORNATE SMOOTH SNAKE
(PETERS, 1854)

The Semiornate Smooth Snake is a small, smooth-scaled, glossy snake with a flattened head, distinct from the neck, and large eyes with round pupils. Coloration is olive-brown, gray, or gray-brown with a pattern of irregular black cross-bands on the anterior body, those on the nape being broader and more distinct. The dorsum of the head is black and the undersides are white.

Actual size

The Semiornate Smooth Snake is a small, widely distributed species found from Ethiopia and Somalia to northeastern South Africa and Swaziland. It is also found farther west in South Sudan, Cameroon, and Chad. It inhabits arid and wet habitats, from desert, semidesert, dry woodland, and Sahel savannas, to montane grasslands, and around watercourses, both permanent and temporary. This species also inhabits coastal thickets and woodlands, which exhibit higher humidity than some of the inland habitats. The prey of the secretive, nonvenomous Semiornate Smooth Snake comprises frogs and lizards such as geckos. Although primarily terrestrial it can climb and it often shelters in hollow trees or behind tree bark, places where it may also encounter its preferred prey. It is inoffensive and does not bite when handled.

RELATED SPECIES

Two subspecies are recognized, the East and southern African populations belonging to the nominate subspecies (*Meizodon semiornatus semiornatus*), while the South Sudan, Cameroon, and Chad population is treated as a separate subspecies (*M. s. tchadensis*). There is also a possible record from Yemen. *Meizodon* contains four other species: the Western Crowned Smooth Snake (*M. coronatus*), Tana Delta Smooth Snake (*M. krameri*), Black-headed Smooth Snake (*M. plumiceps*), and Eastern Crowned Smooth Snake (*M. regularis*).

FAMILY	Colubridae: Colubrinae
RISK FACTOR	Nonvenomous
DISTRIBUTION	Southern Africa: northwest Namibia
ELEVATION	2,560 ft (780 m) asl
HABITAT	Rocky and scrubby riverine habitats, and savanna dominated by mopane trees, with scattered rocky outcrops
DIET	Lizards
REPRODUCTION	Oviparous, clutch size unknown
CONSERVATION STATUS	IUCN not listed

ADULT LENGTH
27½–31½ in
(700–800 mm)

MOPANVELDOPHIS ZEBRINUS
KUNENE RACER
(BROADLEY & SCHÄTTI, 2000)

197

The Kunene Racer is named for the Kunene River, which forms part of the border between Namibia and Angola, but is also known as the Zebra Racer due to its distinctive patterning. The holotype was accidentally run over by two herpetologists and the species is still only known from a few specimens. The habitat of the Kunune Racer comprises rocky, scrubby habitats and savanna woodland dominated by mopane trees, with scattered dolomite granite outcrops. It is diurnal and terrestrial and preys on geckos and skinks. It is also oviparous but the clutch size is unknown. The Kunene Racer's patterning may mimic that of the highly venomous Zebra Spitting Cobra (*Naja nigricincta*, page 474), with which it occurs in sympatry, the racer possibly being afforded some protection from predators by this subterfuge.

The Kunene Racer is a slender, smooth-scaled snake, with a long tail, a head just distinct from the neck, moderately large eyes, and round pupils. It is pale to olive-brown or gray, paler on the flanks and venter, with a dorsal pattern of broad, darker cross-bands and irregular spots on the flanks, the patterning fading on the tail. The head is gray-brown with yellowish labials.

RELATED SPECIES

This species was once contained in the genus *Hemerophis* with the endemic Socotran Racer (*H. socotrae*, page 179) from the Arabian Sea. *Hemerophis zebrinus* may also be related to the Flowered Racer (*Platyceps florulentus*) from north and northeastern Africa.

Actual size

FAMILY	Colubridae: Colubrinae
RISK FACTOR	Nonvenomous
DISTRIBUTION	South Asia: India, Nepal, Pakistan, Bangladesh, and Sri Lanka
ELEVATION	330–6,560 ft (100–2,000 m) asl
HABITAT	Wet and dry forests, parks, and gardens
DIET	Lizards, snakes, small mammals, reptile eggs, and insects
REPRODUCTION	Oviparous, with clutches of 3–9 eggs
CONSERVATION STATUS	IUCN not listed

ADULT LENGTH
13¾–27½ in
(350–700 mm)

OLIGODON ARNENSIS
BANDED KUKRI SNAKE
(SHAW, 1802)

The Banded Kukri Snake inhabits the Indian subcontinent, from Nepal to southern India, and the island of Sri Lanka. It occurs in a wide variety of habitats, both pristine and disturbed, from wet and dry forests to parks and gardens, and even human habitations. A nocturnal or crepuscular snake, it shelters during the day in tree holes, termite mounds, under logs, or in rocky crevices. At night it hunts lizards, small snakes, and mice in the leaf litter, but all kukri snakes also possess specialized sharp "kukri-blade" teeth to enable them to cut into the reptile eggs that form a major part of their diet. Juveniles also feed on insects and spiders. Although kukri snakes are generally inoffensive they can inflict deep slashing wounds with their bladelike teeth.

RELATED SPECIES

The genus *Oligodon* is the largest genus of the Colubridae, with 80 species. The species believed to be most closely related to *O. arnensis* is the Elegant Kukri Snake (*O. venustus*) from peninsular India. Other species occurring in sympatry include the White-barred Kukri Snake (*O. albocinctus*), Streaked Kukri Snake (*O. taeniolatus*), and Stripe-bellied Kukri Snake (*O. sublineatus*).

Actual size

The Banded Kukri Snake is a short, stocky snake with smooth scales, a short tail, and a rounded head with a blunt snout that terminates with a broad, shield-shaped rostral scale. The eyes are large with round pupils. The dorsum is pale brown with a series of light-edged, regularly spaced black cross-bands and a series of three inverted black chevrons on the head and neck. The undersides are white.

FAMILY	Colubridae: Colubrinae
RISK FACTOR	Nonvenomous
DISTRIBUTION	Southeast Asia: Malay Peninsula, Singapore, Borneo, Sumatra, and Java
ELEVATION	0–3,280 ft (0–1,000 m) asl.
HABITAT	Lowland and low montane dipterocarp forests, also disturbed habitats and gardens
DIET	Frogs, lizards, snakes, and reptile and birds' eggs
REPRODUCTION	Oviparous, with clutches of 4–5 eggs
CONSERVATION STATUS	IUCN Least Concern

ADULT LENGTH
23¾–27½ in
(600–700 mm)

OLIGODON OCTOLINEATUS
EIGHT-STRIPED KUKRI SNAKE
(SCHNEIDER, 1801)

199

The Eight-striped Kukri Snake occurs on the Malay Peninsula, on Singapore and the Greater Sunda Islands: Borneo, Sumatra, and Java. It inhabits pristine lowland and low montane dipterocarp forest but also occurs in secondary growth and disturbed or human-altered habitats such as gardens. It is nocturnal and terrestrial in habit, but can climb into bushes and hunt aloft. Prey is varied, from frogs and lizards to other snakes, and reptile and birds' eggs. The leathery shells of snake or lizard eggs are slit open by the Eight-striped Kukri Snake's remarkably sharp, specialized teeth. Although this is a generally inoffensive snake it will bite if handled and is capable of delivering a painful, bloody bite with the same teeth it uses to slice eggs.

The Eight-striped Kukri Snake is a slender, smooth-scaled snake with a short tail, a rounded head with an enlarged rostral scale, and eyes with round pupils. The dorsum is pale brown with three pairs of dark brown longitudinal stripes and an orange-brown vertebral stripe. Despite its name, the number of stripes may not necessarily number eight. Two dark brown inverted chevrons are present on the dorsum of the head, and the venter is pinkish white.

RELATED SPECIES
Oligodon octolineatus does not appear to be closely related to any other species, but other Southeast Asian kukri snakes occurring in sympatry include the Jewelled Kukri Snake (*Oligodon everetti*), Spotted Kukri Snake (*O. annulifer*), and Purple Kukri Snake (*O. purpurascens*, page 200).

Actual size

FAMILY	Colubridae: Colubrinae
RISK FACTOR	Nonvenomous
DISTRIBUTION	Southeast Asia: southern Thailand, Malay Peninsula, Singapore, Borneo, Sumatra, and Java
ELEVATION	0–6,040 ft (0–1,840 m) asl
HABITAT	Primary dipterocarp forest, secondary forest, and peat-swamp forests
DIET	Frogs, tadpoles, frogs' eggs, and lizard eggs
REPRODUCTION	Oviparous, with clutches of 8–13 eggs
CONSERVATION STATUS	IUCN Least Concern

ADULT LENGTH
31½–37½ in
(800–950 mm)

OLIGODON PURPURASCENS
PURPLE KUKRI SNAKE
(SCHLEGEL, 1837)

The Purple Kukri Snake is a medium-sized, smooth-scaled, stocky snake with a short tail, a squarish head that is slightly distinct from the neck, an enlarged rostral scale, and eyes with round pupils. The main coloration is more brown than purple, with a series of irregular, dorsal, darker brown, yellow-edged saddles or blotches. The venter is yellow or pinkish white, bright red in juveniles, while the throat and lips are yellow. Inverted chevrons are present on the head, but often obscured.

The Purple Kukri Snake is found from southern Thailand, through Peninsular Malaysia to Singapore, Sumatra, Java, and Borneo, where it has been found on the slopes of Mt. Kinabalu in Sabah. It is an inhabitant of lowland or low montane primary dipterocarp forest, secondary forest, and also peat swamp forest, where it lives a nocturnal, terrestrial existence. Unlike some other kukri snakes, which are distinctly marked with dark bands, the Purple Kukri Snake exhibits a subdued, cryptic pattern that helps it blend into the leaf litter of its forest habitats. It feeds on amphibian and reptile eggs but also takes frogs and tadpoles. In common with all kukri snakes, it is oviparous. Although generally inoffensive, like other kukri snakes it is capable of swiftly delivering a painful and bloody bite.

RELATED SPECIES

The patterning of *Oligodon purpurascens* is reminiscent of the venomous Mount Kinabalu Pitviper (*Garthius chaseni*, page 583), Javanese Flat-nosed Palm-pitviper (*Trimeresurus puniceus*, page 601), or Bornean Palm-pitviper (*T. borneensis*). *Oligodon purpurascens* occurs in sympatry with the Eight-striped Kukri Snake (*O. octolineatus*, page 199) and Rusty-banded Kukri Snake (*O. signatus*).

Actual size

FAMILY	Colubridae: Colubrinae
RISK FACTOR	Nonvenomous
DISTRIBUTION	East Asia: northeast China, eastern Russia, Korea, and Taiwan
ELEVATION	245–3,280 ft (75–1,000 m) asl
HABITAT	Swamps, rice paddies, ponds, streams, meadows, and gardens
DIET	Fish, frogs, lizards, small mammals, other snakes, and insects
REPRODUCTION	Viviparous, with litters of 8–25 neonates
CONSERVATION STATUS	IUCN Least Concern

ADULT LENGTH
19¾–27½ in,
rarely 35½ in
(500–700 mm,
rarely 900 mm)

OOCATOCHUS RUFODORSATUS
RED-BACKED RATSNAKE
(CANTOR, 1842)

201

Also known as the Frog-eating Ratsnake, the Red-backed
Ratsnake occurs from far-eastern Russia, south through
Korea, to eastern China and Taiwan. It occurs in low-lying
freshwater aquatic habitats such as swamps, marshes, rice
paddies, ponds, and streams. It is also encountered in meadows
and it enters gardens. It can be a very common species.
If threatened it flees to the water, where it remains submerged.
A terrestrial-aquatic species, it rarely climbs. Its prey is also
primarily aquatic, consisting of fish and frogs, but lizards,
mice, other snakes, and some insects are also taken. The Red-
backed Ratsnake is Asia's only known viviparous member of
the Colubrinae, all other species being oviparous.

RELATED SPECIES

Originally included in the ratsnake genus *Elaphe*, but now the
only species in the genus *Oocatochus*, *O. rufodorsatus* is related
to the equally viviparous Smooth Snake (*Coronella austriaca*,
page 155), rather than other ratsnakes.

The Red-backed Ratsnake is a small, slender
snake with smooth scales, a short tail, and a
narrow head, hardly broader than the neck,
with slightly protruding eyes and round pupils.
It is a boldly but variably marked species with
a gray or brown background color. Patterning
usually comprises a brown, orange, or yellow
vertebral stripe, which continues onto the tail,
with lateral patterning consisting of either
stripes or rows of blotches.

Actual size

FAMILY	Colubridae: Colubrinae
RISK FACTOR	Nonvenomous
DISTRIBUTION	North America: southeastern USA and northwest Mexico
ELEVATION	0–5,000 ft (0–1,525 m) asl
HABITAT	Marshland, swamps, lake edges, river and canal banks, wet grasslands and meadows, and wet woodland edges
DIET	Insects, spiders, millipedes, isopods, snails, and small frogs
REPRODUCTION	Oviparous, with clutches of 1–14 eggs
CONSERVATION STATUS	IUCN Least Concern

ADULT LENGTH
23¾–29½ in,
rarely 3 ft 10 in
(600–750 mm,
rarely 1.16 m)

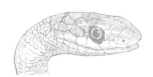

OPHEODRYS AESTIVUS
ROUGH GREENSNAKE
(LINNAEUS, 1766)

Distributed throughout the southeastern USA, from New Jersey to Texas, including the Florida Peninsula, and entering northwestern Mexico, the Rough Greensnake is primarily an arboreal species. It is found in a wide variety of wet habitats, from swamps to woodland, and from meadows to the banks of canals and rivers. It is an agile diurnal climber, very well camouflaged for arboreal life, being rendered almost invisible in the crowns of even sparsely foliated trees. It sleeps aloft and this is probably when it is most easily discovered. Rough Greensnakes feed on a wide variety of invertebrates, with insects and spiders their primary prey. They also take snails, millipedes, woodlice, and occasionally also small frogs. The use of pesticides may have an adverse effect on Rough Greensnake populations by reducing prey availability.

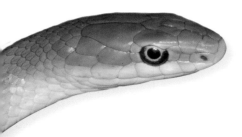

The Rough Greensnake is a slender snake with keeled scales, a long tail, and a head broader than the neck, with relatively large eyes and round pupils. It is dorsally green, darker on the dorsum than on the flanks, and the venter may be white or yellow, and with yellow-green lips and a yellowish throat.

RELATED SPECIES

Two subspecies are recognized, the Northern Rough Greensnake (*Opheodrys aestivus aestivus*) and the Florida Rough Greensnake (*O. a. carinatus*). The closest relative to *O. aestivus* is the Smooth Greensnake (*O. vernalis*), a species with smooth scales from northern and northeastern USA and southeastern Canada. Some authors place one or both species in the genus *Liochlorophis*.

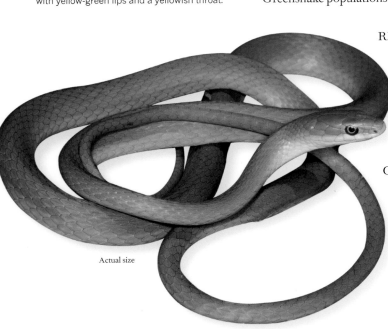

Actual size

FAMILY	Colubridae: Colubrinae
RISK FACTOR	Nonvenomous, constrictor
DISTRIBUTION	East and Southeast Asia: northeast India to eastern China and Taiwan, south to Singapore and Sumatra
ELEVATION	380–8,530 ft (115–2,600 m) asl
HABITAT	Low to medium montane forest, rainforest, forest edge situations, mossy forest floors, and bamboo thickets
DIET	Small mammals
REPRODUCTION	Oviparous, with clutches of 2–7 eggs
CONSERVATION STATUS	IUCN not listed

OREOCRYPTOPHIS PORPHYRACEUS
RED MOUNTAIN RATSNAKE
(CANTOR, 1839)

203

Also known as the Red Bamboo Trinket Snake, this distinctly patterned snake is found throughout East and Southeast Asia. It inhabits rainforest, montane, and hill forest to elevations up to 8,530 ft (2,600 m) asl. It is crepuscular and terrestrial in habit, and a rare and secretive species that burrows into deep mossy beds or hides on rocky slopes with stands of bamboo and tussock grass. Its prey consists of small mammals, mostly voles and shrews, which are killed by constriction. The Red Mountain Ratsnake is a slow-moving species and inoffensive, rarely biting even when handled. The generic name *Oreocryptophis* means "secretive mountain snake" (*Oreo* = mountain, *-crypto* = secretive, *-ophis* = snake).

The Red Mountain Ratsnake has a laterally compressed, smooth-scaled body, a narrow, elongate, squarish head, and small eyes and round pupils. Two color morphotypes are defined. One is dark red-brown with broad, pale-edged black bands, a black stripe through each eye, a black dorsal stripe on the head, and fine dorsolateral black stripes passing from band to band. The other is similar but the broad bands are red or orange, edged with black, and the background color is red or orange. Some specimens are striped rather than banded.

RELATED SPECIES

Oreocryptophis is a monotypic genus, previously included in *Elaphe*. Up to eight subspecies are recognized, the nominate form (*O. porphyraceus porphyraceus*) occurring in southwest China, northeast India, Nepal, northern Myanmar, and Thailand. Other subspecies are found in south and east China, Laos, Cambodia, and Vietnam (*O. p. vaillanti*); central China (*O. p. pulchra*); Taiwan (*O. p. kawakamii*); Hainan Island (*O. p. hainana*); northeastern Thailand (*O. p. coxi*); Peninsular Malaysia and Sumatra (*O. p. laticincta*), and Hong Kong, southern China, and possibly Laos and Vietnam (*O. p. nigrofasciata*).

Actual size

FAMILY	Colubridae: Colubrinae
RISK FACTOR	Nonvenomous
DISTRIBUTION	Southeast Asia: Russia, Mongolia, Kazakhstan, northern China, and Korean Peninsula
ELEVATION	4,080–6,230 ft (1,245–1,900 m) asl
HABITAT	Rocky or gravel semidesert, vegetated mountain slopes, streams, forests, and coastal thickets
DIET	Lizards
REPRODUCTION	Oviparous, with clutches of 4–9 eggs
CONSERVATION STATUS	IUCN not listed, listed in Kazakhstan Red Data Book

ADULT LENGTH
19¾–22½ in
(500–570 mm)

ORIENTOCOLUBER SPINALIS
SLENDER RACER
(PETERS, 1866)

The Slender Racer is a very poorly documented species. Its range largely lies within northern China and Mongolia but it also occurs in far-eastern Russia, on the Korean Peninsula, and in Kazakhstan, where it is listed in the Red Data Book of Endangered Species. In the interior of Asia it is found in arid habitats such as rocky or gravel semidesert, or rocky hillsides with shrubby cover, but it is also reported from forests and close to streams, and on the far-eastern coast it inhabits thickets. Its prey preferences are documented as comprising geckos and lacertid lizards. This species is nervous and fast-moving, but inoffensive if handled. It is oviparous with clutches of up to nine eggs.

RELATED SPECIES

Orientocoluber is a monotypic genus most closely allied to the Eurasian whipsnakes and racers of genera *Dolichophis* (page 161) and *Hierophis* (page 181), in which genus it was once listed, and the dwarf snakes (*Eirenis*, page 167).

Actual size

The Slender Racer is, as its name suggests, an extremely slender snake with weakly keeled scales, a long tail, and a head only slightly wider than the neck, with large eyes, round pupils, and shelved supraocular scales that present a scowling expression. It is gray to brown above, darker dorsally than laterally, and white below. A distinctive white vertebral stripe extends from the dorsum of the head to the tail, black flecks are present on the flanks, and there is black marbling on the head.

FAMILY	Colubridae: Colubrinae
RISK FACTOR	Rear-fanged, mildly venomous; harmless to humans
DISTRIBUTION	North, Central, and South America: southwest USA and Mexico, to eastern Bolivia, including Trinidad and Tobago, and Aruba Island
ELEVATION	0–6,280 ft (0–1,915 m) asl
HABITAT	Primary and secondary forest, gallery forest, woodland, disturbed habitats, and gardens
DIET	Lizards, small birds, frogs, small mammals, and insects
REPRODUCTION	Oviparous, with clutches of 3–8 eggs
CONSERVATION STATUS	IUCN not listed

ADULT LENGTH
3–5 ft 7 in
(0.9–1.7 m)

OXYBELIS AENEUS
BROWN VINESNAKE
(WAGLER, 1824)

205

The Brown Vinesnake is recorded from every territory from southwestern USA to eastern Bolivia, including many islands off the Caribbean and Pacific coasts. It inhabits both pristine and disturbed habitats, and is often found hunting small lizards, such as anoles, in low vegetation. It is also reported to take small birds, mice, frogs, and insects. When stalking prey the diurnal Brown Vinesnake elevates its head and anterior body and moves in a ponderous, punctuated way, with a swaying motion. This form of movement may assist it to locate camouflaged prey, and also blend in with the movements of the vegetation. Its venom is weak and designed for lizards, although one bite to a human caused localized swelling and blistering. Its main defense is to gape widely and expose its blue-black mouth lining.

The Brown Vinesnake is an extremely slender, laterally compressed snake with smooth, obliquely arranged scales, a long prehensile tail, and an elongate, pointed, narrow head, and eyes with round pupils. It is brown or gray-brown above and white below, with the supralabials and throat also white, the line between brown and white exhibiting a stark demarcation around the lower edge of the eye.

RELATED SPECIES

The American vinesnakes bear a strong resemblance to the Asian vinesnakes (*Ahaetulla*, pages 130–131) and the African twigsnakes (*Thelotornis*, pages 245–246), which are both genera of the Colubridae, although *Oxybelis* has round pupils while the other two genera possess horizontally elliptical pupils. In this respect *Oxybelis* more resembles the Caribbean treesnakes of genus *Uromacer* (page 344), although that genus is in the Dipsadidae.

Actual size

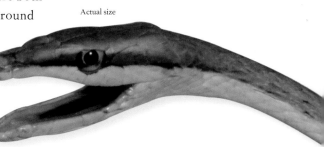

FAMILY	Colubridae: Colubrinae
RISK FACTOR	Rear-fanged, mildly venomous; harmless to humans
DISTRIBUTION	North, Central, and South America: southern Mexico to northeastern Bolivia
ELEVATION	0–5,250 ft (0–1,600 m) asl
HABITAT	Primary and secondary rainforest, montane forest, and riverine forest
DIET	Frogs, lizards, birds, and small mammals
REPRODUCTION	Oviparous, with clutches of 8–14 eggs
CONSERVATION STATUS	IUCN not listed

ADULT LENGTH
4 ft 3 in–6 ft 7 in,
rarely 7 ft 7 in
(1.3–2.0 m,
rarely 2.3 m)

206

OXYBELIS FULGIDUS
GREEN VINESNAKE
(DAUDIN, 1803)

The Green Vinesnake is slender with a long, moderately laterally compressed body, smooth scales in oblique rows, a long prehensile tail, and a long head, which terminates in a sharp, pointed snout. The small eyes have round pupils. The dorsum is emerald green, the lower flank and the lower half of the head are light green, and the undersides are green. A pale yellow-green stripe runs from the snout tip, under the eye to the angle of the jaw, and a white stripe runs along the lower flanks.

The Green Vinesnake occurs in sympatry with the Brown Vinesnake (*Oxybelis aeneus*, page 205) throughout much of its range, which extends from southern Mexico to northeastern Bolivia. A larger species, it is found in various forest habitats, from lowland to montane, primary to secondary, and wet to dry, but is also found in small trees and lower vegetation, and may be met with on the ground. It feeds on frogs, lizards, birds, and small mammals, killing them with venom injected through the enlarged rear fangs. Although like the Brown Vinesnake it is diurnal, it is less frequently encountered. Its venom is effective on lizards, frogs, and even small mice, but although bites to humans have caused localized pain and swelling there have been no more serious effects.

RELATED SPECIES

There are two other greenish *Oxybelis* vinesnakes in tropical America, the Short-headed Vinesnake (*O. brevirostris*), which occurs from Honduras to western Ecuador, and the endemic Isla de Roatán Vinesnake (*O. wilsoni*), which is more mustard yellow in color and which occurs on the largest of the Honduran Islas de la Bahía, while *O. fulgidus* is found on neighboring Isla de Utila.

Actual size

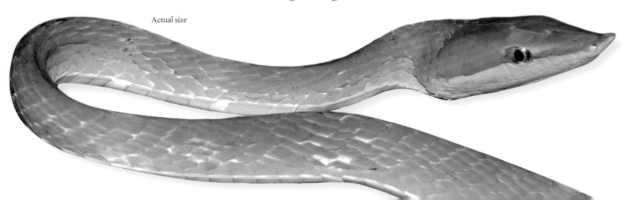

FAMILY	Colubridae: Colubrinae
RISK FACTOR	Nonvenomous, constrictor
DISTRIBUTION	North America: eastern USA
ELEVATION	0–1,970 ft (0–600 m) asl
HABITAT	Deciduous and mixed woodland, pinewoods, cypress stands, hardwood hammocks in sawgrass plains, mangrove thickets, farmland, and small islands
DIET	Mammals, birds, lizards, snakes, and frogs
REPRODUCTION	Oviparous, with clutches of 8–20, rarely 30 eggs
CONSERVATION STATUS	IUCN not listed

ADULT LENGTH
3 ft 3 in–6 ft,
occasionally 7 ft 7 in
(1.0–1.8 m,
occasionally 2.3 m)

PANTHEROPHIS ALLEGHANIENSIS
EASTERN RATSNAKE
(HOLBROOK, 1836)

The recently redefined Eastern Ratsnake occurs along the Atlantic seaboard, east of the Appalachians and Apalachicola River, from New England to the Florida Keys. A powerful constrictor, it is found in many habitats, including deciduous and mixed woodland, pinewoods, and hardwood hammocks on sawgrass plains. It also occurs in farmland, in coastal mangrove entanglements, and on small islands, such as the Keys. It is an adept climber, even of tall palms or pines, using its keeled ventral scales and powerful coils to scale the smoothest of trunks. Its prey preferences include most vertebrates small enough to overpower and swallow, especially mammals and birds. It sometimes takes domestic chickens and their eggs, hence the origin of the old name "Chicken Snake." Lizards, other snakes, and frogs are also included in the diet.

RELATED SPECIES

The American Ratsnake (*Elaphe obsoleta*) contained five to eight subspecies defined by color pattern, but modern molecular analysis led to the genus *Elaphe* being confined to Eurasian ratsnakes, with *Pantherophis* resurrected for American ratsnakes. It was determined that three separate species existed, *P. alleghaniensis* east of the Appalachians and Apalachicola River, the Midland Ratsnake (*P. spiloides*) between these barriers and the Mississippi, and the Western Ratsnake (*P. obsoletus*) west of the Mississippi. The former subspecies *E. o. obsoleta* had comprised the melanistic northern populations of all three taxa.

The Eastern Ratsnake is large snake, with a long head, large eyes, and round pupils. Its color pattern varies, northern populations being black with only faint stripes, central populations yellow with four black stripes, and southern populations yellow-gray with brown stripes (Gulf Hammock), orange with faint stripes (Everglades), or buff with brown stripes and blotches (Florida Keys). Juveniles are spotted, the origin of the generic name *Pantherophis*.

Actual size

FAMILY	Colubridae: Colubrinae
RISK FACTOR	Nonvenomous, constrictor
DISTRIBUTION	North America: southern USA and northern Mexico
ELEVATION	2,950–5,910 ft (900–1,800 m) asl
HABITAT	Semiarid woodland, rocky hillsides, wooded limestone canyons, and riparian habitats
DIET	Small mammals, birds and their eggs, and lizards
REPRODUCTION	Oviparous, with clutches of 4–15 eggs
CONSERVATION STATUS	IUCN not listed

ADULT LENGTH
4 ft–4 ft 7 in,
occasionally 5 ft 2 in
(1.2–1.4 m,
occasionally 1.6 m)

PANTHEROPHIS BAIRDI
BAIRD'S RATSNAKE
(YARROW, 1880)

Baird's Ratsnake is found in southwestern Texas, USA, and across the border into Coahuila, Nuevo León, and Tamaulipas in northeast Mexico. It is an inhabitant of arid upland habitats, including rocky hillsides and wooded limestone canyons, but is also found in riparian habitats and wetland areas on desert fringes. It also occurs around human habitations. Baird's Ratsnake is nocturnally active, especially following rainfall, and although primarily terrestrial it also climbs into low vegetation or onto rocky outcrops. It feeds on rodents, bats, birds and their eggs, and occasionally lizards, but given its secretive nocturnal existence it has been little studied in nature. It will vibrate its tail on dead leaves as a warning if confronted, and may bite if handled. Spencer Fullerton Baird (1823–87) was an American ornithologist and herpetologist.

Actual size

RELATED SPECIES
Pantherophis bairdi is related to the Cornsnake (*Pantherophis guttatus*, page 209), Western Ratsnake (*P. obsoletus*), and Eastern Foxsnake (*P. vulpinus*, page 210).

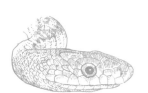

Baird's Ratsnake is a moderately large snake with smooth scales, a prehensile tail, a head that is distinct from the neck, and large eyes with round pupils. Its ground color varies from yellow to yellow-gray, gunmetal, or silver, with broad, pale gray longitudinal stripes. The yellow pigment predominates on the anterior body while the posterior body and tail are darker, but coloration varies across the geographical range. Juveniles are gray with darker gray bands rather than striped.

FAMILY	Colubridae: Colubrinae
RISK FACTOR	Nonvenomous, constrictor
DISTRIBUTION	North America: southeastern and eastern USA
ELEVATION	0–4,530 ft (0–1,380 m) asl
HABITAT	Hardwood forests, pine barrens, swamps, grassy plains, mangrove swamps, and around human habitations
DIET	Mammals, birds, lizards, snakes, frogs, and insects
REPRODUCTION	Oviparous, with clutches of 3–40 eggs
CONSERVATION STATUS	IUCN Least Concern

PANTHEROPHIS GUTTATUS
CORNSNAKE
(LINNAEUS, 1766)

ADULT LENGTH
2–5 ft,
rarely 6 ft
(0.6–1.5 m,
rarely 1.8 m)

209

The Cornsnake, or Red Ratsnake, is found throughout southeastern and eastern USA, from New Jersey to Florida, and west to Texas, in a wide variety of habitats including dry woodlands, pine barrens on sandy soil, freshwater swamps, hammocks in sawgrass plains, and mangrove swamps. Cornsnakes are also commonly encountered around farm buildings and even in suburban areas. This is a nocturnal predator, primarily of endothermic (warm-blooded) prey, rodents, and birds and their eggs, but it will also feed on frogs, lizards, sometimes other snakes, or insects. With its keeled ventral scales and muscular body, it is an excellent climber of both trees and buildings. Cornsnakes are inoffensive, rarely bite, and are popular in captivity. Many cultivars or color morphotypes are bred specifically for the vast pet trade.

Actual size

RELATED SPECIES

Formerly called *Elaphe guttata*, *Pantherophis guttatus* is related to the Great Plains Ratsnake (*P. emoryi*), a former subspecies, and Slowinski's Cornsnake (*P. slowinskii*), from Louisiana and Texas, named for the late Joseph Slowinski, killed by a Many-banded Krait (*Bungarus multicinctus*, page 447) in 2001. The former Rosy Ratsnake (*E. g. rosacea*), from the Florida Keys, is no longer recognized.

The Cornsnake varies in patterning across its range. The ground color may be gray, tan, or orange with a series of large, dark-edged, dark red rhomboid or oval saddles across the back, smaller markings on the flanks, and similarly colored postocular and dorsal stripes on the dorsum and sides of the head. The undersides are usually checkerboard black and white. Juveniles are gray with brown dorsal saddles. Adult amelanistic (lacking black) and anerythristic (lacking red) partial albinos are also encountered in nature.

FAMILY	Colubridae: Colubrinae
RISK FACTOR	Nonvenomous, constrictor
DISTRIBUTION	North America: northern USA and southeastern Canada
ELEVATION	500–1,500 ft (152–457 m) asl
HABITAT	Grasslands and pastures, open woodland, agricultural habitats, marshes, beaches, and around buildings
DIET	Mammals, birds and their eggs, lizards, frogs, and also snakes, salamanders, and earthworms
REPRODUCTION	Oviparous, with clutches of 7–20, rarely 29 eggs
CONSERVATION STATUS	IUCN Least Concern

ADULT LENGTH
3 ft 3 in–4 ft 7 in,
occasionally 6 ft
(1.0–1.4 m,
occasionally 1.8 m)

PANTHEROPHIS VULPINUS
EASTERN FOXSNAKE
(BAIRD & GIRARD, 1853)

210

The Eastern Foxsnake is gray to yellow, with patterning comprising a series of brown or black irregular dorsal blotches on the back and a small series on either flank. The head may bear a small dark marking in the center. The adult livery of some specimens is quite similar to that of the juveniles of other American ratsnakes, with dark blotches on a pale gray background.

The Eastern Foxsnake occurs around the Great Lakes, in Ontario, Canada, and Ohio, Michigan, Illinois, and Indiana, USA. Other American ratsnakes show a preference for forested or wooded habitats, but the Eastern Foxsnake prefers open habitats such as grasslands, swamps, and agricultural habitats. It also inhabits open woodlands, and occurs around buildings. Mammals form the bulk of the adult's diet, with birds also taken. Juveniles feed on lizards, frogs, and insects. Prey is killed by constriction. The Eastern Foxsnake is less studied in nature than its more commonly encountered relatives. It is an inoffensive species that rarely bites.

RELATED SPECIES
There were two subspecies of foxsnakes, the Western Foxsnake (*P. vulpinus*), from Michigan to Nebraska, and the Eastern Foxsnake (*P. gloydi*), from Michigan, northern Ohio, and Ontario. Molecular analysis has demonstrated that the division between the two populations is the Mississippi Valley, so *P. gloydi* was synonymized with *P. vulpinus* as the Eastern Foxsnake, and a new species was described as the Western Foxsnake (*P. ramspotti*).

Actual size

FAMILY	Colubridae: Colubrinae
RISK FACTOR	Nonvenomous
DISTRIBUTION	West, Central, and East Africa: Guinea-Bissau to Angola, and Kenya
ELEVATION	33–6,230 ft (10–1,900 m) asl
HABITAT	Rainforest, gallery forest, evergreen forest, deciduous woodland, and savanna woodland
DIET	Frogs
REPRODUCTION	Oviparous, with clutches of 1–4 eggs
CONSERVATION STATUS	IUCN not listed

ADULT LENGTH
17¾–27½ in,
rarely 35½ in
(450–700 mm,
rarely 900 mm)

PHILOTHAMNUS HETERODERMUS
FOREST GREENSNAKE
(HALLOWELL, 1857)

211

The Forest Greensnake is not always green, some specimens being brown. It inhabits deep forest, including rainforest, evergreen forest, gallery forest, and also deciduous woodland and savanna woodland, in a broad band through West, Central, and East Africa, from Guinea-Bissau to Kenya and south to Angola. It occurs in lowland and low montane forests to elevations of almost 6,560 ft (2,000 m). It is diurnal, and both arboreal and terrestrial in habit. The prey preferences of the Forest Greensnake comprise frogs, with toads reportedly refused. It is not known if lizards feature in the diet. This is generally a poorly known species.

RELATED SPECIES

The African genus *Philothamnus* contains 20 species of smooth-scaled treesnakes, which are closely related to the keel-scaled African treesnakes of genus *Hapsidophrys* (page 178). The closest relative of *P. heterodermus* is its former subspecies, the Thirteen-scaled Greensnake (*P. carinatus*), from Central Africa, which despite its name also has smooth dorsal scales. Harmless green treesnakes in Africa are often mistaken for green mambas (*Dendroaspis*, pages 450–451), and killled.

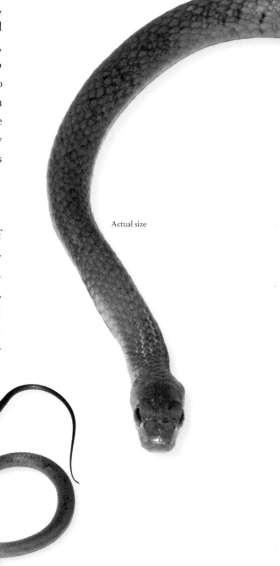

Actual size

The Forest Greensnake is a slender snake with obliquely arranged, smooth scales, a long tail, and a narrow, elongate head with large eyes and round pupils. Two color morphotypes exist, a green morphotype with a yellow-green venter, and a dark brown morphotype with a light brown venter. White spots are present on the anterior body scales but these remain concealed unless the snake inflates its throat defensively.

FAMILY	Colubridae: Colubrinae
RISK FACTOR	Nonvenomous
DISTRIBUTION	Sub-Saharan Africa: Ethiopia and Sudan to Guinea and South Africa
ELEVATION	0–6,560 ft (0–2,000 m) asl
HABITAT	Wet and dry forest and savanna, karoo scrub, coastal bush and forest, savanna woodland, semidesert, and riparian habitats
DIET	Lizards and frogs
REPRODUCTION	Oviparous, with clutches of 3–12 eggs
CONSERVATION STATUS	IUCN not listed

ADULT LENGTH
2¼–3½ ft,
occasionally 4 ft 3 in
(0.7–1.1 m,
occasionally 1.3 m)

212

PHILOTHAMNUS SEMIVARIEGATUS
SPOTTED BUSHSNAKE
(SMITH, 1840)

The Spotted Bushsnake is a slender snake with smooth, obliquely arranged scales, a long tail, and a head broader than the neck, with very large eyes and round pupils. Specimens may be bright green, blue-green, gray-green, or yellow-green, with a yellowish or greenish-white venter. The anterior of the body is heavily spotted or barred with black, and the head is green even in gray specimens.

The Spotted Bushsnake is an extremely attractive arboreal snake distributed very widely through Sub-Saharan Africa. It occurs in a range of wet and dry woodland, forest, and savanna habitats, from semidesert to riverine gallery forest and coastal bush. It is one of the most frequently encountered bushsnakes in Africa. It is alert, agile, fast-moving, and well suited to life aloft, being able to climb the trunks of trees swiftly using the keels on its ventral scales to obtain a purchase, and using its slender body to bridge caps between branches. Its prey consists of geckos, chameleons, and frogs. When threatened, the Spotted Bushsnake inflates its throat to expose the contrasting blue edges to its scales.

RELATED SPECIES

The closest relatives of the Spotted Bushsnake include the Western Green Snake (*Philothamnus angolensis*), which has a patchy distribution from KwaZulu-Natal to Cameroon, and the Elegant Greensnake (*P. nitidus*) from West, Central, and East Africa.

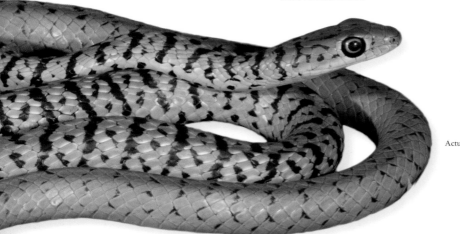

Actual size

FAMILY	Colubridae: Colubrinae
RISK FACTOR	Nonvenomous
DISTRIBUTION	North, Central, and South America: southern Mexico to Costa Rica
ELEVATION	0–4,660 ft (0–1,420 m) asl
HABITAT	Rainforest, lowland and low montane wet forest, and gallery forest in lowland dry forest
DIET	Birds and their eggs, small mammals, and lizards
REPRODUCTION	Oviparous, with clutches of 7–14 eggs
CONSERVATION STATUS	IUCN Least Concern

ADULT LENGTH
5–6 ft,
rarely 8 ft
(1.5–1.8 m,
rarely 2.4 m)

PHRYNONAX POECILONOTUS
PUFFING SNAKE
(GÜNTHER, 1858)

213

The Puffing Snake is a large, laterally compressed snake with smooth scales, a long tail, a broad head, and large eyes with round pupils. The dorsum is glossy olive, brown, or green with occasional scattered reddish or orange scales and obscure bands. The venter is yellowish, particularly on the throat and lips, while the lower flanks are pale orange. Juveniles may be banded, and almost melanistic specimens are known.

The Puffing Snake, also known as the Northern Birdsnake, is so named because of its defensive display, which consists of gaping widely, hissing loudly, and inflating the neck. It is found from southern Mexico to Honduras and Nicaragua, along the Caribbean versant, and onto the Pacific versant in Costa Rica. It primarily inhabits wet forest habitats from lowland rainforest to low montane moist forest, and where it occurs in dry forests it inhabits gallery forest along rivers. This species is diurnal, and both terrestrial and arboreal. It preys primarily on birds, but also takes their eggs, arboreal small mammals, bats, and lizards. Puffing Snakes are large nonvenomous snakes with a long reach, and they will bite if confronted but they are nonvenomous.

RELATED SPECIES

Three subspecies are recognized: the nominate form (*Phrynonax poecilonotus poecilonotus*) from southern Mexico to Honduras; a northern subspecies (*P. p. argus*) from the Yucatan Peninsula, Mexico; and a southern Central American form (*P. p. chrysobronchus*) from Nicaragua and Costa Rica. Specimens from Panama and northern South America are now attributed to the former subspecies *P. polylepis*. Close relatives are Shropshire's Puffing Snake (*P. shropshirei*) from Costa Rica, which may be a synonym of *P. poecilonotus*. These snakes were previously placed in the genus *Pseustes*.

Actual size

FAMILY	Colubridae: Colubrinae
RISK FACTOR	Nonvenomous
DISTRIBUTION	North America: southwestern USA and northwestern Mexico
ELEVATION	985–3,000 ft (300–915 m) asl
HABITAT	Rocky, sandy, or stony desert with mesquite, saltbush, thorn scrub, creosote, and saguaro cacti
DIET	Lizards and their eggs, insects
REPRODUCTION	Oviparous, with clutches of 2–6 eggs
CONSERVATION STATUS	IUCN Least Concern

ADULT LENGTH
9¾–15¾ in,
rarely 20 in
(250–400 mm,
rarely 510 mm)

PHYLLORHYNCHUS BROWNI
SADDLED LEAFNOSE SNAKE
STEJNEGER, 1890

214

The Saddled Leafnose Snake is a short, stocky little snake, with smooth scales, a short tail, and a short head, barely distinct from the neck, with an enlarged, protruding rostral scale and moderately large eyes with vertical pupils. Coloration is pinkish brown, white below, with regular large round dorsal saddles of mid- to dark brown, those on the anterior body being darkest and the others having dark edges, and a dark brown band across the anterior head and eyes.

The Saddled Leafnose Snake inhabits southern Arizona, Sonora, and Sinaloa. It is found in desert habitats with rocky, stony, or sandy substrates and with cover provided by mesquite, creosote bushes, saltbush, thorn scrub, or cacti such as saguaro. A small, nocturnal snake, the Saddled Leafnose Snake emerges from underground burrows at night, especially following rain. It hunts lizards, possibly geckos or side-blotched lizards, but also feeds on lizard eggs, excavated from the sand using the large, leaf-shaped rostral scale on its snout. Juveniles may feed on insects. It is rarely seen, with most specimens encountered crossing desert highways, being well camouflaged on the desert floor. Although small they will defend themselves by loud hissing, inflating their throats, and making mock but ineffectual strikes. Herbert Brown (1848–1913) was president of the Audubon Society of Arizona.

Actual size

RELATED SPECIES

The closest relative of *Phyllorhynchus browni* is the Spotted Leafnose Snake (*P. decurtatus*), with which it occurs in sympatry, although the Spotted Leafnose Snake also occurs in southern California and Baja California. Four subspecies are recognized, from Maricopa, Arizona (*P. b. lucidus*), southern Arizona and northern Sonora (*P. b. browni*), southern Sonora (*P. b. fortitus*), and Sinaloa (*P. b. klauberi*). The leafnose snakes are closely related to the lyresnakes (*Trimorphodon*, pages 248–249), also from southwest USA and northwest Mexico.

FAMILY	Colubridae: Colubrinae
RISK FACTOR	Nonvenomous, constrictor
DISTRIBUTION	North America: southwestern Canada, western and central USA, to central Mexico
ELEVATION	0–9,500 ft (0–2,895 m) asl
HABITAT	Desert and semidesert, prairies, deciduous woodlands, coniferous forests, agricultural land, and swamps
DIET	Mammals, birds and their eggs, and lizards
REPRODUCTION	Oviparous, with clutches of 2–24 eggs
CONSERVATION STATUS	IUCN Least Concern

ADULT LENGTH
4 ft–5 ft 2 in,
rarely 9 ft 2 in
(1.2–1.6 m,
rarely 2.8 m)

PITUOPHIS CATENIFER
GOPHERSNAKE
(BLAINVILLE, 1835)

215

The Gophersnake is a large, widely distributed snake, from southwestern Canada, through western and Midwestern USA, into northwestern and central Mexico. Among North American colubrids, probably only the indigo snakes (*Drymarchon*, pages 162–163) are larger than the Gophersnake, although island populations are often much smaller than mainland forms. It may be found in a wide variety of habitats, from desert to marsh, grassland to forest, and agricultural habitats, where it adopts a diurnal, crepuscular, or nocturnal habit depending on the season and climate. Gophersnakes are accomplished burrowers that dig to locate potential mammalian prey in subterranean burrows. Prey comprises mammals, from mice to rabbits, which are constricted on the surface or pinioned between coils and crushed against tunnel walls underground. Lizards, and birds and their eggs are also taken.

The Gophersnake is a powerfully muscular but relatively slender snake with keeled scales, a slightly pointed head, an enlarged rostral scale, and small eyes with round pupils. It is variably colored, being yellow, tan, brown, or gray, darker dorsally than laterally, with a pattern consisting of a bold series of dark dorsal blotches and a lateral pattern comprising several rows of irregular dark spots. The venter is usually immaculate yellow or tan.

RELATED SPECIES

There are up to ten subspecies, including *Pituophis catenifer sayi*, the loud hissing Bullsnake. Island endemics occur on Santa Cruz Island (*P. c. pumilus*) off California, and Coronado Island (*P. c. coronalis*), Cedros Island (*P. c. insulanus*), and San Martín Island (*P. c. fuliginatus*) in the Gulf of California. Other *Pituophis* include the Pinesnake (*P. melanoleucus*), of southeastern USA, Louisiana Pinesnake (*P. ruthveni*), Mexican Bullsnake (*P. deppei*), Cape Gophersnake (*P. vertebralis*), from Baja California, and Middle American Gophersnake (*P. lineaticollis*).

Actual size

FAMILY	Colubridae: Colubrinae
RISK FACTOR	Nonvenomous
DISTRIBUTION	Middle East: southern Israel, Palestine, southwestern Jordan, and northern Saudi Arabia
ELEVATION	0–5,410 ft (0–1,650 m) asl
HABITAT	Rocky and stony wadis and hillsides
DIET	Lizards, possibly small mammals
REPRODUCTION	Oviparous, clutch size unknown
CONSERVATION STATUS	IUCN Least Concern

ADULT LENGTH
19¾–23¾ in,
rarely 27½ in
(500–600 mm,
rarely 700 mm)

216

PLATYCEPS ELEGANTISSIMUS
ELEGANT RACER
(GÜNTHER, 1878)

The Elegant Racer is a small, slender snake with smooth scales, a long tail, a narrow pointed head with a projecting snout, large eyes, and round pupils. The patterning comprises a series of regular transverse black bands on an olive to cream background. The bands are broader dorsally than laterally, and become complete rings on the tail. The head bears two narrow black bands, followed by a broad nape band, and the venter is off-white or cream. Some specimens exhibit a narrow orange vertebral stripe, which may only be present in the pale interspaces.

The attractive Elegant Racer occurs in southern Israel and Palestine, southwestern Jordan, and northern and northwestern Saudi Arabia. It is an inhabitant of rocky or stony habitats such as dry wadis or rocky hillsides, and is only occasionally found in predominantly sandy habitats. It was believed that due to the high daytime temperatures the Elegant Racer was more nocturnal than diurnal, in stark contrast to its congeners, but recent research now suggests that it is diurnal, but cryptic in habits. It is an alert and fast-moving species that is rarely encountered in nature and therefore poorly known. Although a terrestrial species, one specimen was caught swimming in a wadi. The Elegant Racer feeds on small lizards, especially terrestrial geckos, but small rodents may also feature in its diet.

RELATED SPECIES

The genus *Platyceps* contains as many as 31 western Palearctic racers formerly included in the genus *Coluber*, which is now confined to the North American Racer (*C. constrictor*, page 152). Closely related species include the Sinai Racer (*P. sinai*), Thomas' Racer (*P. thomasi*), and Variable Racer (*P. variabilis*), from the Arabian Peninsula. The Sinai and Thomas' Racer are almost identical to vertebral-striped *P. elegantissimus*, while the Variable Racer is most like the unstriped forms.

Actual size

FAMILY	Colubridae: Colubrinae
RISK FACTOR	Nonvenomous
DISTRIBUTION	Southeastern Europe and southwestern Asia: the Balkans, Greece, Cyprus, Turkey, southwest Russia, Caucasus, Syria, Lebanon, Iraq, Turkmenistan, and northern Iran
ELEVATION	0–7,220 ft (0–2,200 m) asl
HABITAT	River valleys, rocky slopes, forest edges, scrubby hillsides, semidesert, and abandoned buildings
DIET	Lizards, rarely mice or insects
REPRODUCTION	Oviparous, with clutches of 3–16 eggs
CONSERVATION STATUS	IUCN Least Concern

ADULT LENGTH
31½–38½ in,
occasionally 3 ft 3 in
(800–980 mm,
occasionally 1.0 m)

PLATYCEPS NAJADUM
DAHL'S WHIPSNAKE
(EICHWALD, 1831)

217

Dahl's Whipsnake, also known as the Slender Racer, is a common and wide-ranging species, occurring from the Balkans, to Turkey, the Caucasus, and south to Iran. It is found in a variety of lowland and low montane habitats from semidesert to rocky hillsides, river valleys, abandoned buildings, forest edges, and scrubland. The highly alert and fast-moving Dahl's Whipsnake is active by day and hunts by pursuing and chasing down its lizard prey, which is either consumed alive or pressed against a rock by a body coil until it is dead. This species rarely employs constriction as a means of killing prey. Although lizards are the primary prey, nestling or juvenile mice, and large insects, may also be taken on occasion.

RELATED SPECIES

Up to six subspecies are recognized, from the Caucasus (*Platyceps najadum najadum*); southeastern Europe and Turkey (*P. n. dahlii*); Turkmenistan and Iran (*P. n. atayevi*); Zagros Mountains, Iran (*P. n. schmidtleri*); and southeast Azerbaijan (*P. n. albitemporalis*), while the status of the Kalymnos Island population (*P. n. kalymnensis*) and those on other Greek islands is open to question. The Glossy-bellied Racer (*P. ventromaculatus*), of western Asia and the Middle East, is a close relative.

Dahl's Whipsnake is extremely slender, and smooth-scaled, with a whiplike tail, a narrow, slightly pointed head, large eyes, and round pupils. The patterning, at least of western specimens, comprises a gray-brown head and a gray neck and anterior body, with a black, white-edged collar, and a series of black eyespots on the flanks that decrease in size posteriorly. Beyond the gray anterior section, the body and tail are uniform brown. The venter, throat, and lips are white, while white also extends onto the preoculars and postoculars. No other racer or whipsnake exhibits this pattern.

Actual size

FAMILY	Colubridae: Colubrinae
RISK FACTOR	Nonvenomous, but possesses mildly toxic saliva
DISTRIBUTION	Northeastern Africa, Middle East and southwest Asia: Eritrea, Ethiopia, Somalia, Yemen, Oman, Saudi Arabia, UAE, Israel, Jordan, Iraq, Iran, Afghanistan, Kazakhstan, Tajikistan, Turkmenistan, Uzbekistan, Pakistan, and northern India
ELEVATION	0–9,020 ft (0–2,750 m) asl
HABITAT	Rocky hillsides and wadis, stony coastal plains, and cultivated or wet habitats
DIET	Lizards, frogs, tadpoles, small mammals, birds, and snakes
REPRODUCTION	Oviparous, with clutches of 4–8 eggs
CONSERVATION STATUS	IUCN not listed

ADULT LENGTH
23¾–27½ in,
occasionally 4 ft 3 in
(600–700 mm,
occasionally 1.3 m)

218

PLATYCEPS RHODORACHIS
WADI RACER
(JAN, 1865)

The Wadi Racer is a slender snake, with smooth scales, a long tail, and a narrow, pointed head with large eyes and round pupils. It is generally tan, greenish gray, or olive-green in color, patterned with a series of dark bands or spots that may be present on the anterior body, the entire body and tail, or absent completely. The venter is whitish pink.

Also known as Jan's Cliff Racer, the Wadi Racer is a widely distributed species, found throughout the Middle East, from Israel to Yemen and Oman, and in western Asia to as far east as India. It also occurs in northeastern Africa, in Somalia, Ethiopia, and Eritrea. It inhabits arid habitats, rocky or stony plains, wadis, or hills, and although it is not reliant on water it is commonly encountered near watercourses. The Wadi Racer is diurnal, alert, and fast-moving, but crepuscular in hot weather. A skilled climber and swimmer, it preys on frogs, tadpoles, lizards, rodents, bats, birds, and snakes, including conspecifics, but it is said to avoid toads. Although not strictly venomous, the Wadi Racer chews to introduce a mildly neurotoxic saliva that subdues its prey. Bites to humans cause only localized itching.

RELATED SPECIES

Up to four subspecies are recognized, from Iran (*Platyceps rhodorachis rhodorachis*); Iran and Kazakhstan to Pakistan and India (*P. r. ladacensis*); Kashmir (*P. r. kashmirensis*); and northeast Africa (*P. r. subnigra*). The close relatives of *P. rhodorachis* include Rogers' Racer (*P. rogersi*) from North Africa and the Middle East. It also resembles the Spotted Desert Racer (*P. karelini*) from Iran and Turkmenistan.

Actual size

FAMILY	Colubridae: Colubrinae
RISK FACTOR	Nonvenomous, constrictor
DISTRIBUTION	North and Central America: southern Mexico, Guatemala, Belize, Honduras, and Nicaragua
ELEVATION	0–4,920 ft (0–1,500 m) asl
HABITAT	Tropical evergreen forest, semi-xeric thorn scrub, deciduous woodland, karst limestone escarpments, and lowland coastal swamps
DIET	Small mammals, birds, bats, and lizards
REPRODUCTION	Oviparous, with clutches of 4–9 eggs
CONSERVATION STATUS	IUCN Least Concern

ADULT LENGTH
4–5 ft,
rarely 5 ft 9 in
(1.2–1.5 m,
rarely 1.76 m)

PSEUDELAPHE FLAVIRUFA
TROPICAL RATSNAKE
(COPE, 1867)

219

Also known as the Mexican Night Snake, this ratsnake is distributed down the Caribbean coast of Mexico, Belize, and Honduras, excluding the Yucatán Peninsula but including the Honduran Islas de la Bahía, and Big Corn Island, Nicaragua. It also occurs in Guatemala and on the Pacific coast of Oaxaca and Chiapas, Mexico. It inhabits both mesic and xeric habitats, from swamps and evergreen forests to dry forest and thorn scrub. It is especially common in the coastal lowlands. The Tropical Ratsnake is a nocturnal predator of mice, rats, bats, birds, and sometimes lizards, which are killed by constriction. Although generally placid it will defend itself vigorously by flattening its head, raising its coils into an S-shape, vibrating its tail on dead leaves, and, if necessary, lunging with strikes and bites.

The Tropical Ratsnake is a moderately large snake with smooth scales, a long tail, a broad head, and large eyes with round pupils that become vertically elliptical in bright light. It is pale gray or brown with dorsal and lateral rows of irregular blotches, which may be red, brown, or black, often with black edges. Two similarly colored stripes run onto the back of the head. The venter is yellow-gray with small dark spots. The patterning of juveniles is lighter and more contrasting in its pigmentation.

RELATED SPECIES
Three subspecies are recognized: the Northern Tropical Ratsnake (*Pseudelaphe flavirufa flavirufa*); Matuda's Ratsnake (*P. f. matudai*) from southern Chiapas; and the Central American Tropical Ratsnake (*P. f. pardalina*). A fourth subspecies was elevated to specific status, the Yucatán Ratsnake (*P. phaescens*). The snakes most closely related to *P. flavirufa* are the glossy snakes (*Arizona elegans*, page 142) and long-nosed snakes (*Rhinocheilus*, page 225), but *P. flavirufa* is unlikely to be confused with the American Green Ratsnake (*Senticolis triaspis*), with which it occurs in sympatry.

Actual size

FAMILY	Colubridae: Colubrinae
RISK FACTOR	Rear-fanged, very mildly venomous; harmless to humans
DISTRIBUTION	North America: western Mexico
ELEVATION	0–3,610 ft (0–1,100 m) asl
HABITAT	Thornbush woodland, tropical semiarid and dry forest, and tropical deciduous forest
DIET	Spiders and insects
REPRODUCTION	Oviparous, with clutches of 3–30 eggs
CONSERVATION STATUS	IUCN Least Concern

ADULT LENGTH
15¾–19¾ in,
rarely 28 in
(400–500 mm,
rarely 710 mm)

220

PSEUDOFICIMIA FRONTALIS
SOUTHWESTERN HOOKNOSE SNAKE
(COPE, 1864)

The Southwestern Hooknose Snake is a Mexican endemic, found from southern Sonora and Sinaloa to Guerrero and Puebla, western Mexico, where it inhabits lowland and low montane thornbush woodland, tropical semiarid and dry forest, and deciduous forest. Its upturned snout is used for excavation of the substrate, and stomach contents reveal that this species preys on moth caterpillars and tarantula spiders. Female Southwestern Hooknose Snakes are oviparous and possess well-developed hemipenes like those of a male, a condition known as pseudohermaphroditism. This is a relatively rare species and poorly known in nature. Although it is a rear-fanged venomous snake, the Southwestern Hooknose Snake is not dangerous to humans.

The Southwestern Hooknose Snake is a moderately stout snake with a pointed head, upturned snout, and small eyes with round pupils. It is gray-brown with a pair of darker brown or red-brown stripes that run off the back of the head to initiate a series of similarly colored dorsal blotches connected by a broad yellow line, with smaller dark spots on the flanks, a dark stripe at an angle below the eye, and a dull orange venter.

RELATED SPECIES

This species is sometimes referred to as the "False Ficimia" because its generic name is *Pseudoficimia*, but this is a misinterpretation. The generic name indicates that while this species does not belong in *Ficimia* it is similar to snakes of that genus. Southwestern Hooknose Snake is a better name because it occurs to the south of the western hooknose snakes (*Gyalopion*, page 177), and to the west of the southern hooknoses (*Ficimia*, page 174). This is a monotypic genus.

Actual size

FAMILY	Colubridae: Colubrinae
RISK FACTOR	Nonvenomous
DISTRIBUTION	Asia: Iran and Turkmenistan to China and Taiwan, south to Sri Lanka and Peninsular Malaysia
ELEVATION	0–13,100 ft (0–4,000 m) asl
HABITAT	Wet and dry, lowland and montane forests, riverine forest, agricultural habitats, parks, and gardens
DIET	Small mammals, birds, lizards, frogs, and other snakes
REPRODUCTION	Oviparous, with clutches of 5–25 eggs
CONSERVATION STATUS	IUCN not listed

ADULT LENGTH
5 ft–6 ft 3 in,
rarely 12 ft 2 in
(1.5–1.9 m,
rarely 3.7 m)

PTYAS MUCOSA
DHARMAN RATSNAKE
(LINNAEUS, 1758)

221

The Dharman Ratsnake is a large, common, terrestrial and arboreal snake found widely across Asia, from Iran to China and south to Sri Lanka, Peninsular Malaysia, and Taiwan. It is found in many habitats, from lowland and montane forest, both wet and dry, to paddy fields and plantations, and even parks and gardens. It is a diurnal predator of vertebrates ranging from rodents to birds, frogs, lizards, and other snakes, with prey being pinioned by the coils, but it is itself the frequent prey of King Cobras (*Ophiophagus hannah*, page 480). Male Dharmans engage in combat during the mating period, entwining their bodies as they attempt to wrestle each other to the ground (see page 28). Females will actively guard their nests of eggs during incubation. Although nonvenomous, a large Dharman can deliver bloody bites.

The Dharman Ratsnake is a powerful, slightly laterally compressed snake, with smooth scales, a long tail, a slightly pointed head, large eyes, round pupils, and shelved supraocular scales. It is glossy brown, olive, or greenish with scattered black flecks on the dorsum, and white or yellow underneath. The supralabial scales and the scales of the throat are black-edged.

RELATED SPECIES

No subspecies are recognized across the vast range of *Ptyas mucosa*. The genus contains seven other species, with five centered on mainland Southeast Asia: the Malayan Keeled Ratsnake (*P. carinata*); Black-striped Ratsnake (*P. dhumnades*); White-bellied Ratsnake (*P. fusca*); Indo-Chinese Ratsnake (*P. korros*); and the stunning Asian Green Ratsnake (*P. nigromarginata*, page 222). The Luzon Mountain Ratsnake (*P. luzonensis*) is found in the north Philippines and the Sulawesi Black Ratsnake (*P. dipsas*) in Indonesia.

Actual size

FAMILY	Colubridae: Colubrinae
RISK FACTOR	Nonvenomous, constrictor
DISTRIBUTION	South and Southeast Asia: northeast India, Nepal, Bhutan, Bangladesh, western China, northern Myanmar, Laos, Vietnam, and Thailand
ELEVATION	1,640–7,710 ft (500–2,350 m) asl
HABITAT	Open woodland on plains and low hills; also disturbed areas
DIET	Small mammals, lizards, birds, and other snakes
REPRODUCTION	Oviparous, with clutches of 8–10 eggs
CONSERVATION STATUS	IUCN not listed

ADULT LENGTH
2 ft 4 in–3 ft 3 in,
rarely 8 ft 2 in
(0.7–1.0 m,
rarely 2.5 m)

PTYAS NIGROMARGINATA
ASIAN GREEN RATSNAKE
(BLYTH, 1854)

The Asian Green Ratsnake is a slender snake with smooth scales, a long tail, an elongate head, large eyes, and round pupils. The dorsum is bright green to olive-green with black scale edging. Juveniles bear four broad, black longitudinal stripes on the body and tail, but these are confined to the posterior body and tail in adults. The venter is yellow-green and the head is reddish brown with a white throat.

The stunning Asian Green Ratsnake is a diurnal, terrestrial, and arboreal inhabitant of open woodland at low to medium elevations, on plains and hills from Nepal and Bhutan to Sichuan and Yunnan in western China, and south to northern Myanmar, Thailand, Laos, and Vietnam. It also occurs in disturbed habitats. Its primary prey consists of rodents, which are killed by constriction or by being pinioned against solid objects by the ratsnake's coils, but it also takes lizards, birds, and other snakes. When cornered, its defenses may include lunging bites and exuding the foul-smelling contents of its cloacal glands, but it is nonvenomous and harmless to humans.

RELATED SPECIES

Ptyas nigromarginata is related to the Dharman Ratsnake (*P. mucosa*, page 221) and six other species of Indo-Chinese and Southeast Asian ratsnakes. Many of these species were previously included in the genus *Zaocys*.

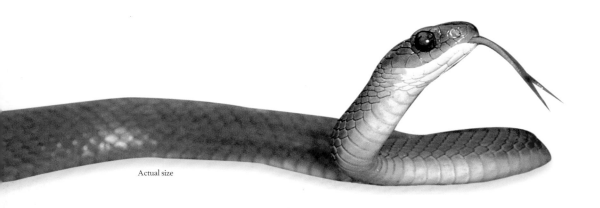

Actual size

FAMILY	Colubridae: Colubrinae
RISK FACTOR	Nonvenomous
DISTRIBUTION	West, Central, and East Africa: Guinea to Cameroon, south to Angola and west to Kenya, and on Bioko Island
ELEVATION	0–6,560 ft (0–2,000 m) asl
HABITAT	Rainforest, and other forest habitats
DIET	Frogs and toads
REPRODUCTION	Oviparous, with clutches of up to 17 eggs
CONSERVATION STATUS	IUCN not listed

ADULT LENGTH
3 ft–3 ft 7 in,
rarely 5 ft
(0.9–1.1 m,
rarely 1.5 m)

RHAMNOPHIS AETHIOPISSA
SPLENDID DAGGER-TOOTHED TREESNAKE
(GÜNTHER, 1862)

223

Also known as the Large-eyed Green Treesnake, this is a diurnal, arboreal inhabitant of rainforests and other tropical forest habitats, but not open habitats, from Guinea to Kenya, and south to Angola. It is also recorded from Bioko Island, formerly Fernando Pó, in the Gulf of Guinea. It is characterized by an enlarged vertebral scale row that assists the treesnake in bridging gaps. It also has enlarged rear maxillary teeth that are flanged anteriorly and posteriorly like those of the Dagger-toothed Vinesnake (*Xyelodontophis uluguruensis*), only slightly less so, hence its common name. It feeds on tree frogs and toads. The large eyes dominate the head and suggest that it is a very alert species with good eyesight, even in the low light conditions of the rainforest.

The Splendid Dagger-toothed Treesnake is a slender snake with obliquely arranged smooth scales, a long tail, and a head distinct from the neck, with especially large eyes and round pupils. Dorsally it is yellowish green with every scale black-edged, and ventrally it is green with or without black edges to the ventral and subcaudal scales.

RELATED SPECIES

Snakes of the genus *Rhamnophis* are closely related to the black treesnakes (*Thrasops*) and were originally included in that genus, but they also bear a strong resemblance to the dangerously venomous Boomslang (*Dispholidus typus*, page 160). *Rhamnophis aethiopissa* is represented by three subspecies. The nominate form (*R. a. aethiopissa*) occupies most of the range while two other subspecies occur in northern Angola and Zambia (*R. a. ituriensis*) and Uganda, Kenya, and Tanzania (*R. a. elgonensis*). The genus also contains Bates' Dagger-toothed Treesnake (*R. batesii*) from Central Africa.

Actual size

FAMILY	Colubridae: Colubrinae
RISK FACTOR	Rear-fanged, mildly venomous; harmless to humans
DISTRIBUTION	Northern South America: Colombia, Venezuela, the Guianas, Brazil, Peru, Bolivia, and Paraguay
ELEVATION	33–1,610 ft (10–490 m) asl
HABITAT	Tropical rainforest
DIET	Lizards
REPRODUCTION	Oviparous, with clutches of up to 3 eggs
CONSERVATION STATUS	IUCN not listed

ADULT LENGTH
2 ft 7 in–5 ft 2 in
(0.85–1.6 m)

RHINOBOTHRYUM LENTIGINOSUM
AMAZON BANDED SNAKE
(SCOPOLI, 1788)

224

The Amazon Banded Snake is an extremely slender snake with smooth scales, a long tail, a very narrow neck, and a broad, rounded head with large eyes and vertically elliptical pupils. It is a matte black snake with regularly spaced triads comprising a red band, tipped with black, between two white bands. The head is black with white scale suturing and scattered red spots.

The Amazon Banded Snake is an infrequently encountered species. It is found widely across northern South America, from Venezuela to Paraguay and the Guianas to Peru, but it only inhabits tropical rainforest and in this habitat it occurs primarily in the rainforest canopy far above the forest floor, where its presence may go unnoticed. It is nocturnal and arboreal and it hunts primarily arboreal lizards such as geckos, anoles, and tree runners. This is a rear-fanged venomous species, but it is docile when handled and its venom is designed to subdue lizards and is believed harmless to humans. It is oviparous but clutch sizes appear to be small, with three eggs being reported from one specimen. This is a very rare snake that requires much more study in nature.

RELATED SPECIES

A second species of *Rhinobothryum* is known, the Central American Banded Snake (*R. bovallii*). The genus *Rhinobothryum* is contained in a clade with the scorpion-eating snakes (*Stenorrhina*, page 237), shovelnose snakes (*Chionactis*, page 149), sandsnakes (*Chilomeniscus*, page 148), and groundsnakes (*Sonora*, page 232).

Actual size

FAMILY	Colubridae: Colubrinae
RISK FACTOR	Nonvenomous
DISTRIBUTION	North America: southwestern and southern USA, and northern Mexico
ELEVATION	0–6,230 ft (0–1,900 m) asl
HABITAT	Lowland desert, thornbush, acacia or mesquite semidesert, and dry prairie
DIET	Lizards, small mammals, birds, and large insects
REPRODUCTION	Oviparous, with clutches of 3–11 eggs
CONSERVATION STATUS	IUCN Least Concern

ADULT LENGTH
15¾–30 in,
occasionally 5 ft
(400–760 mm,
occasionally 1.5 m)

RHINOCHEILUS LECONTEI
LONG-NOSED SNAKE
BAIRD & GIRARD, 1853

225

The Long-nosed Snake is a common nocturnal snake of the southwestern deserts of the USA, and may often be encountered crossing roads. It also ranges south into northern Mexico. Its arid habitats range from thornbush or mesquite semidesert to dry prairies. It prefers sandy soils into which it may easily dig using its pointed snout and enlarged rostral scale. Prey consists of lizards and small mammals, but birds and grasshoppers are also eaten. Mice are constricted or pinioned against solid objects to kill them. Long-nosed Snakes rarely bite, even when handled, but they may twist their bodies and void the contents of their cloacal glands. Some specimens are said to defensively discharge blood from their nostrils. John Lawrence LeConte (1825–83) was a Civil War physician and naturalist.

The Long-nosed Snake is a slightly laterally compressed snake with smooth scales, a narrow, strongly pointed head, and small eyes with round pupils. Its patterning comprises alternating vertebral square blotches of black and red or orange, on a whitish-gray background, almost every pale scale being heavily infused with black, while the pale pigment also invades the black or red middorsal markings. The iris of the eye is also reddish.

RELATED SPECIES

Three subspecies are currently recognized, the Western Long-nosed Snake (*Rhinocheilus lecontei lecontei*), the Texas Long-nosed Snake (*R. l. tessellatus*), and the Pacific Coastal Long-nosed Snake (*R. l. antonii*). The Isla Cerralvo Long-nosed Snake (*R. etheridgei*), from the Gulf of California, was a former subspecies, but is now treated as a full species. The genus *Rhinocheilus* belongs to a clade that also includes the glossy snakes (*Arizona elegans*, page 142) and the Tropical Ratsnake (*Pseudelaphe flavirufa*, page 219), within a larger clade of ratsnakes and kingsnakes.

Actual size

FAMILY	Colubridae: Colubrinae
RISK FACTOR	Rear-fanged, mildly venomous; harmless to humans
DISTRIBUTION	Middle East: Syria, Israel, Palestine, Jordan, Egypt (Sinai), and Saudi Arabia
ELEVATION	165–5,910 ft (50–1,800 m) asl
HABITAT	Dry steppe, semidesert, wadis, rocky slopes, and gravel plains with sparse vegetation; also light oak forest, agricultural habitats, and abandoned buildings
DIET	Insects, crustaceans, centipedes, and small lizards
REPRODUCTION	Oviparous, clutch size unknown
CONSERVATION STATUS	IUCN not listed

ADULT LENGTH
15¾–19 in
(400–480 mm)

226

RHYNCHOCALAMUS MELANOCEPHALUS
PALESTINE
BLACK-HEADED SNAKE
(JAN, 1862)

The diminutive Palestine Black-headed Snake is a secretive fossorial species that shelters under stones during the day and becomes active at night, when it hunts a wide variety of insects, from ant larvae to locusts, and also feeds on woodlice, centipedes, and small geckos. It is distributed from Syria to the Sinai of Egypt and northern Saudi Arabia, generally in semidesert habitats such as wadis, rocky slopes, gravel pans, and arid steppes, but it may also be found in light oak woodland, agricultural areas, and abandoned buildings. Its natural history is poorly documented and although it is known to be oviparous its clutch size is not known.

The Palestine Black-headed Snake is a small, slender, cylindrical snake with smooth scales, a narrow head, just distinct from the neck, and moderately large eyes and round pupils. The body is uniform light brown above, and white below. The dorsum of the head and neck is glossy jet black, except for the contrasting white supralabials in some specimens.

RELATED SPECIES

Four other species of *Rhynchocalamus* are recognized, the Aden Black-headed Snake (*R. arabicus*) from Yemen and Oman, Baran's Black-headed Snake (*R. barani*) from southern Turkey, Dayan's Black-headed Snake (*R. dayanae*) from the Negev Mountains of Israel, and a former subspecies, Satunin's Black-headed Snake *(R. satunini)*, from eastern Turkey, the Caucasus, Iraq, and Iran. *Rhynchocalamus* is the sister taxon to the leafnose snake genus *Lytorhynchus* (page 192).

Actual size

FAMILY	Colubridae: Colubrinae
RISK FACTOR	Nonvenomous
DISTRIBUTION	North America: southwestern USA and northwestern Mexico
ELEVATION	785–7,280 ft (240–2,200 m) asl
HABITAT	Rocky desert arroyos, canyons, and hillsides, especially with cacti, thorn scrub, creosote, or saltbush vegetation; also chaparral
DIET	Lizards, small snakes, reptile eggs, small mammals, and birds
REPRODUCTION	Oviparous, with clutches of 3–12 eggs
CONSERVATION STATUS	IUCN Least Concern

ADULT LENGTH
26–35½ in,
occasionally 3 ft 10 in
(660–900 mm,
occasionally 1.17 m)

SALVADORA HEXALEPIS

WESTERN PATCHNOSE SNAKE
(COPE, 1867)

227

The Western Patchnose Snake is a common desert species found in both rocky and sandy habitats, from northern California and Nevada, USA, south to Baja California, Sonora, Chihuahua, and Sinaloa, Mexico. It may inhabit rocky slopes, canyons, or dry arroyos, especially those with desert vegetation. The enlarged rostral scale on the snout enables it to dig for prey, but it primarily hunts on the surface, during the day, being one of the few snakes active in the heat of midday. Its prey consists mostly of diurnal lizards, but small snakes, rodents, and birds are also taken. The Western Patchnose Snake is an alert snake with good eyesight, which largely avoids trouble by fleeing, but if cornered it will inflate its neck and make far-reaching strikes. It is nonvenomous and harmless to humans.

RELATED SPECIES

Four subspecies of *Salvadora hexalepis* are recognized: Desert Patchnose (*S. h. hexalepis*); Mohave Patchnose (*S. h. mojavensis*); Baja California Patchnose (*S. h. klauberi*); and Coastal Patchnose (*S. h. virgultea*). Two additional species, the Eastern or Mountain Patchnose (*S. grahamiae*) and Big Bend Patchnose (*S. deserticola*), also occur in the USA, while a further four species are found in Mexico: Mexican Patchnose (*S. mexicana*); Baird's Patchnose (*S. bairdi*); Oaxaca Patchnose (*S. intermedia*); and Pacific Patchnose (*S. lemniscatus*).

The Western Patchnose Snake is a muscular snake with smooth scales, a long tail, a head just distinct from the neck, large eyes and round pupils, and a large saddle-like rostral scale. It is generally patterned with pastel desert colors. The head is brown above, paler below, followed by a broad yellow to sandy stripe that runs the length of the body and tail. The flanks may be uniform brown or pale sand-colored with two brown longitudinal stripes, the uppermost being the broadest. The venter is white.

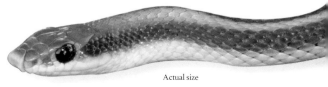

Actual size

FAMILY	Colubridae: Colubrinae
RISK FACTOR	Nonvenomous
DISTRIBUTION	West, Central, and East Africa: Sierra Leone and Ghana to Kenya, south to Zambia and Angola
ELEVATION	0–4,840 ft (0–1,475 m) asl
HABITAT	Coastal thicket, and wet and dry woodland and savanna
DIET	Small mammals
REPRODUCTION	Oviparous, with clutches of up to 48 eggs
CONSERVATION STATUS	IUCN not listed

ADULT LENGTH
3 ft 3 in–4 ft 3 in,
occasionally 5 ft 2 in
(1.0–1.3 m,
occasionally 1.6 m)

228

SCAPHIOPHIS ALBOPUNCTATUS
GRAY HOOKNOSED SNAKE
PETERS, 1870

The Gray Hooknosed Snake is a stocky-bodied snake with smooth scales, a short tail, and a narrow head, which is indistinct from the neck and terminates in a pronounced, pointed rostral scale. The eyes are small and the pupils are round. This species is gray or brown above with or without black spotting, and orange, cream, or white below.

The Gray Hooknosed Snake is distributed from Sierra Leone to Kenya and south to Zambia. It inhabits coastal thickets and woodland, but farther inland it occurs in both wet and dry woodland and savanna habitats. It is primarily fossorial, burrowing in loose soil using its enlarged, shovel-like rostral scale, the slit-like nostrils and tight-fitting lip scales preventing the ingress of soil during the burrowing process. It also uses animal burrows, hunting rodents below ground and killing them by pinioning them against the tunnel walls with its coils. When it feels threatened the Gray Hooknosed Snake will elevate its anterior body, gape widely to display the blue-black interior of its mouth, extend its tongue, and make mock strikes with such force that it may throw its body forward, but this is all bluff, it is nonvenomous and harmless to man.

RELATED SPECIES

A second species, the Ethiopian Hooknosed Snake (*Scaphiophis raffreyi*), occurs in Ethiopia, South Sudan, Eritrea, Uganda, and northwest Kenya. The African hooknosed snakes may be distantly related to the Asian kukri snakes (*Oligodon*, pages 198–200). They may also be may be confused with the beaked snakes (*Rhamphiophis*, page 376), although the hooknosed snakes' rostral scales are more pronounced, their bodies are more robust, and their eyes are smaller.

Actual size

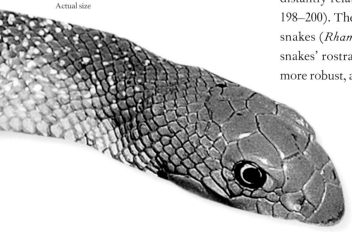

FAMILY	Colubridae: Colubrinae
RISK FACTOR	Rear-fanged, mildly venomous; harmless to humans
DISTRIBUTION	Central America: Guatemala, El Salvador, Honduras, Nicaragua, and Costa Rica
ELEVATION	0–5,020 ft (0–1,530 m) asl
HABITAT	Dry and wet, lowland and montane tropical forest
DIET	Centipedes
REPRODUCTION	Oviparous, with clutches of up to 7 eggs
CONSERVATION STATUS	IUCN Least Concern

ADULT LENGTH
17¾–19¼ in
(450–490 mm)

SCOLECOPHIS ATROCINCTUS
BLACK-BANDED CENTIPEDE SNAKE
(SCHLEGEL, 1837)

229

The Black-banded Centipede Snake occurs from Guatemala to northern Costa Rica along the Pacific versant of Central America, in both wet and dry tropical forests, in lowland and low montane locations. It is a terrestrial and semi-fossorial species that is also occasionally found climbing in low vegetation. Authorities differ as to whether this rare snake is diurnal, crepuscular, or nocturnal. It appears to prey exclusively on large, venomous, scolopendrid centipedes, which are usually eaten backward. It may also prey on insects or arachnids. Although a rear-fanged venomous snake, its mouth is too small to administer a bite to a human. The Black-banded Centipede Snake may be confused with the false coralsnakes (*Pliocercus*, page 285) and the true coralsnakes (*Micrurus*, pages 457–465), but most corals and coral mimics have fully encircling red bands.

The Black-banded Centipede Snake is a small, cylindrical snake with smooth scales, a short tail, and a rounded head, indistinct from the neck, with small eyes and round pupils. It is distinctively patterned with alternating black and white bands, the scales of the latter being black-tipped. A broad vivid red vertebral stripe runs the length of the body and tail between the black rings. The dorsum of the head is black-capped, with black rings through the nape, the eye, and the snout.

RELATED SPECIES
Scolecophis is a monotypic genus, and most closely related to the black-headed and centipede snakes of genus *Tantilla* (pages 240–241).

Actual size

FAMILY	Colubridae: Colubrinae
RISK FACTOR	Nonvenomous, constrictor
DISTRIBUTION	North and Central America: southwestern USA, Mexico, Belize, Guatemala, El Salvador, Honduras, Nicaragua, and Costa Rica
ELEVATION	0–7,960 ft (0–2,425 m) asl
HABITAT	Rocky canyons, semiarid chaparral, low montane mixed woodland, mesquite grassland, tropical dry forest, and thorn scrub
DIET	Small mammals, birds and their eggs, and lizards
REPRODUCTION	Oviparous, with clutches of 3–9 eggs
CONSERVATION STATUS	IUCN Least Concern, Endangered in New Mexico

ADULT LENGTH
Male
2 ft 4 in–3 ft 3 in
(0.7–1.0 m)

Female
3–4 ft,
rarely 6 ft
(0.9–1.2 m,
rarely 1.8 m)

230

SENTICOLIS TRIASPIS
AMERICAN GREEN RATSNAKE
(COPE, 1866)

The American Green Ratsnake is a muscular-bodied snake with a long tail and a long head, distinct from the neck, with moderately sized eyes and round pupils. Most of its scales are smooth but the median rows are weakly keeled. It is generally unicolor as an adult, usually green or olive and darker posteriorly than anteriorly, but some specimens are brown or red. Juveniles are gray with red or brown saddles.

The American Green Ratsnake is a medium-sized constrictor that just enters the United States in the Chihuahuan Desert of southern Arizona and New Mexico, where it is listed as Endangered. From the US border it is distributed throughout Mexico and Central America, as far south as Costa Rica. Preferred habitats are predominately arid, from rocky canyons to dry chaparral and thorn scrub, but it also occurs in moist habitats, such as low montane mixed forest. It is an alert and secretive terrestrial and arboreal species that avoids confrontation by fleeing, but if cornered it will defend itself vigorously. Adult American Green Ratsnakes prey on rodents and also birds and their eggs, shrews, and bats, while juveniles take lizards and small mice, prey being killed by constriction.

RELATED SPECIES

Three subspecies are recognized: the Western Green Ratsnake (*Senticolis triaspis intermedia*) from Arizona and Mexico, the nominate Yucatán Green Ratsnake (*S. t. triaspis*), and the Honduran Green Ratsnake (*S. t. mutabilis*) from Central America. *Senticolis triaspis* is unlikely to be confused with the Tropical Ratsnake (*Pseudelaphe flavirufa*, page 219), with which it occurs in sympatry.

Actual size

FAMILY	Colubridae: Colubrinae
RISK FACTOR	Nonvenomous
DISTRIBUTION	South America: Brazil, including Ilha de São Sebastião, and Paraguay
ELEVATION	590–3,490 ft (180–1,065 m) asl
HABITAT	Cerrado savanna, and Atlantic Forest habitats
DIET	Frogs, and possibly small mammals and lizards
REPRODUCTION	Oviparous, with clutches of 2–7 eggs
CONSERVATION STATUS	IUCN not listed

ADULT LENGTH
2 ft 7 in–3 ft 3 in
(0.8–1.0 m)

SIMOPHIS RHINOSTOMA

SÃO PAULO FALSE CORALSNAKE

(COPE, 1866)

231

The São Paulo False Coralsnake occurs in the southern Brazilian states of São Paulo, Minas Gerais, Goiás, and Mato Grosso do Sul, including Ilha de São Sebastião off the coast of São Paulo, as well as northeastern Brazil in Bahia, and west into Paraguay. It is associated with Atlantic coastal forest habitats but is also found in Cerrado savannas. It is a poorly known species with only limited natural history data available from nature, but it has a broad rostral scute of the kind associated with fossorial snakes. Although only frogs have been recorded as prey, some authors suggest it may also take small mammals or lizards. The São Paulo False Coralsnake mimics the aposematic patterning (warning colors) of the Amazonian Coralsnake (*Micrurus spixii*, page 464) and the Cerrado Coralsnake (*M. frontalis*).

The São Paulo False Coralsnake is a smooth-scaled, cylindrical snake with a relatively long tail and a long, pointed head, indistinct from the neck, which terminates in a large rostral scute. Its patterning consists of glossy triads comprising black-white-black-white-black between broad red interspaces. The scales of the white bands are black-tipped. The head is white with black suturing, followed by a curving black band and a narrow red band before the first triad.

RELATED SPECIES

Simophis is a monotypic genus. It bears a strong resemblance to several coralsnakes and false coralsnakes, but a suite of characters separate it from other colubrids, including other coral mimics.

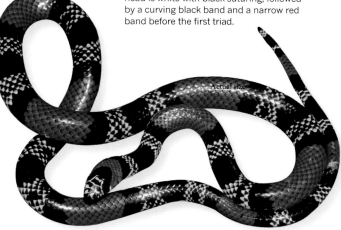

Actual size

FAMILY	Colubridae: Colubrinae
RISK FACTOR	Nonvenomous
DISTRIBUTION	North America: western USA and northern Mexico
ELEVATION	0–6,820 ft (0–2,080 m) asl
HABITAT	Desert and semidesert, dry grassland, rocky slopes, scrubland, and riparian habitats
DIET	Insects, spiders, scorpions, centipedes, earthworms, and occasional lizards
REPRODUCTION	Oviparous, with clutches of 3–6 eggs
CONSERVATION STATUS	IUCN Least Concern, protected in Oregon and Arkansas

ADULT LENGTH
8–12½ in,
occasionally 18 in
(180–320 mm,
occasionally 460 mm)

232

SONORA SEMIANNULATA
WESTERN GROUNDSNAKE
BAIRD & GIRARD, 1853

Widely distributed across southwestern USA, from Oregon and California to Texas and Missouri, and into Mexico as far south as Durango, the Western or Variable Groundsnake inhabits a range of arid habitats from desert and semidesert to dry grassland, rocky hillsides, and riparian habitats in arid areas. It is nocturnal and feeds almost exclusively on invertebrates, including not only beetles and grasshoppers, but also dangerous species such as black widow spiders, buthid scorpions, and scolopendrid centipedes. Small geckos also feature in the diet on occasion. As a small snake the Western Groundsnake has many predators and adopts a variety of defensive tactics ranging from thanatosis and spraying musk, to forming a loop by holding its own tail in its mouth, making itself hopefully harder to ingest.

RELATED SPECIES

Sonora semiannulata comprises two subspecies, the Western Groundsnake (*Sonora semiannulata semiannulata*) and the South Texas Groundsnake (*S. s. taylori*). *Sonora* also contains three endemic Mexican species: the Mexican Groundsnake (*S. mutabilis*); Michoacán Groundsnake (*S. michoacanensis*); and Filetail Groundsnake (*S. aemula*). *Sonora* belongs to a clade that includes the sandsnakes (*Chilomeniscus*, page 148) and the shovelnose snakes (*Chionactis*, page 149).

The Western Groundsnake is a small, smooth-scaled, cylindrical snake with a relatively short tail, a narrow, pointed head, and small eyes with round pupils. It is one of the most variably patterned snakes in North America, but patterning often comprises alternating black and white bands, the upper portions of the white bands being red. The snout may be white or red, followed by a black band.

Actual size

FAMILY	Colubridae: Colubrinae
RISK FACTOR	Nonvenomous, constrictor
DISTRIBUTION	North Africa, Middle East, and western Asia: Morocco to Egypt, Turkey to Oman, Turkmenistan to India
ELEVATION	0–7,960 ft (0–2,425 m) asl
HABITAT	Steppe, semidesert, wadis, rocky hillsides, sand dunes, and cultivated habitats
DIET	Small mammals, birds, and lizards
REPRODUCTION	Oviparous, with clutches of 3–16 eggs
CONSERVATION STATUS	IUCN not listed

ADULT LENGTH
2 ft 7 in–3 ft 3 in,
occasionally 4 ft 3 in
(0.8–1.0 m,
occasionally 1.3 m)

SPALEROSOPHIS DIADEMA

DIADEM SNAKE

(SCHLEGEL, 1837)

233

Also known as the Royal Snake, the Diadem Snake occurs across a vast swath of arid habitats from Morocco to Egypt, parts the Arabian Peninsula, and from Turkey to Kazakhstan and northwest India. The Diadem Snake is diurnal, but crepuscular or nocturnal in hot weather, sheltering in animal burrows when conditions are too hot. It is found in vegetated steppe and semidesert, on rocky, gravel, and sandy substrates. It is also common in cultivated habitats and around buildings where its primary prey, rats and mice, are most abundant. Birds and lizards also feature in its diet. Prey is killed by constriction, although toxic saliva is thought to play a part in subduing it. Due to its rodent-killing capacity, the Diadem Snake should be encouraged in agricultural communities.

The Diadem Snake has a muscular body, with keeled scales, a distinctive head, and large eyes with round pupils. Its body color is gray-brown, every scale spotted with dark brown. Patterning comprises a series of brown vertebral rhomboid markings and a second series of smaller brown spots on the flanks.

RELATED SPECIES

Three subspecies are recognized: the nominate form (*Spalerosophis diadema diadema*) occurs in Pakistan and northeast India; a western form (*S. d. cliffordi*) is found in North Africa, the Arabian Peninsula, Turkey, and Iraq; while Iran, Turkmenistan, and western Pakistan are inhabited by a third subspecies (*S. d. schirasianus*). The genus *Spalerosophis* also contains five other species across North Africa and western Asia. Its closest relatives include the Palearctic racers of genus *Platyceps* (pages 216–218).

Actual size

FAMILY	Colubridae: Colubrinae
RISK FACTOR	Nonvenomous, constrictor
DISTRIBUTION	North, Central, and South America: southern Mexico to Paraguay, including Trinidad, Tobago, and Isla Margarita
ELEVATION	0–6,560 ft (0–2,000 m) asl
HABITAT	Gallery forest, lowland and low montane wet forest, secondary growth
DIET	Mammals, birds and their eggs, and lizards
REPRODUCTION	Oviparous, with clutches of 5–12 eggs
CONSERVATION STATUS	IUCN not listed

ADULT LENGTH
3 ft 3 in–6 ft 7 in,
occasionally 8 ft 8 in
(1.0–2.0 m,
occasionally 2.65 m)

SPILOTES PULLATUS
TIGER RATSNAKE
LINNAEUS, 1758

234

Also known as the Thunder and Lightning Snake, or Chicken Snake, the Tiger Ratsnake is a common diurnal snake of lowland coastal or riverine habitats from southern Mexico, Central America, and South America, as far as the Guianas, Brazil, and Paraguay. It is terrestrial and arboreal and is frequently sighted sleeping in trees over lagoons or swimming across rivers. Primarily a snake of wet forest habitats, it inhabits dry forest in riparian situations. It is a powerful constrictor of mammals, from rats and bats to porcupines, but supplements this diet with birds, their eggs, and lizards. A large and bold snake, the Tiger Ratsnake will defend itself vigorously by inflating its throat and making far-reaching strikes. But it is nonvenomous, and because of its rodent-killing capabilities, it should be protected.

RELATED SPECIES

Genus *Spilotes* contains two other species, the Amazonian Puffing Snake (*S. sulphureus*), and a former subspecies elevated by some authors to specific status, the Ecuadorian Tiger Ratsnake (*S. megalolepis*). Although *S. pullatus* is often referred to as a "ratsnake" it is not closely related to any of the other American ratsnake genera, being closer to the puffing snakes of genus *Phrynonax* (page 213).

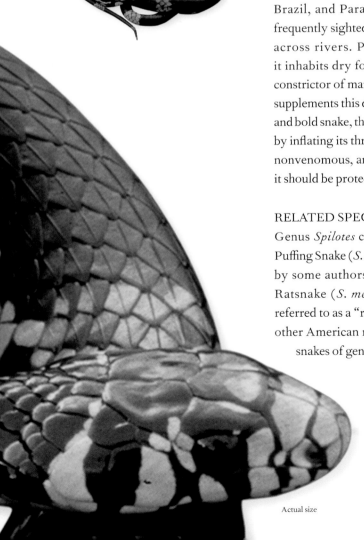

The Tiger Ratsnake is a powerful snake with smooth scales, a laterally compressed body, a long tail, a large head, and large eyes with round pupils. Its body, both dorsally and ventrally, is yellow with irregular black markings, the black markings being more dominant posteriorly, and the tail being all black. The head is yellow with black suturing and bands.

Actual size

FAMILY	Colubridae: Colubrinae
RISK FACTOR	Nonvenomous, possibly mildly toxic saliva
DISTRIBUTION	Southeast Asia: Indonesia
ELEVATION	0–2,130 ft (0–650 m) asl
HABITAT	Rainforest, plantations, and wetland areas
DIET	Frogs, lizards, small mammals, and reptile eggs
REPRODUCTION	Oviparous, clutch size unknown
CONSERVATION STATUS	IUCN not listed

ADULT LENGTH
2 ft 7 in–3 ft 3 in,
occasionally 5 ft
(0.8–1.0 m,
occasionally 1.55 m)

STEGONOTUS BATJANENSIS
NORTH MOLUCCAN GROUNDSNAKE
(GÜNTHER, 1865)

235

The largest member of the genus *Stegonotus*, the North Moluccan Groundsnake occurs in the Moluccas of eastern Indonesia. The holotype was collected on the island of Batjan, which is now known as Bacan. The North Moluccan Groundsnake inhabits rainforests, plantations, and low-lying wetland areas where it preys on frogs, lizards, and rodents. This species has enlarged rear teeth that are designed to shear through reptile eggs, another prey item, and these teeth may deliver bloody bites to humans picking them up. Although some discomfort may be felt, these snakes are technically nonvenomous.

RELATED SPECIES

The genus *Stegonotus* contains more than 20 other species, at least 12 from the New Guinea region, and species in Borneo, the Philippines, and Flores and Timor, in the Lesser Sunda Islands. The groundsnakes of genus *Stegonotus* are related to the wolfsnakes of genus *Lycodon* (pages 190–191).

The North Moluccan Groundsnake is a stout snake with smooth scales, a large, rounded head, and small eyes with vertically elliptical pupils. The head, body, and tail are brown, gray, or black above and white below, with a pattern comprising upward extensions of white, yellow, or orange pigment onto the flanks, these markings in some cases meeting at the midbody to form pale rings. There may be white or yellow bands around the neck and body and many of the scales on the sides of the head are pale-centered with dark edges.

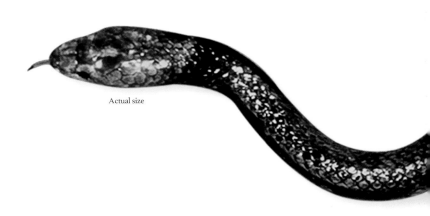

Actual size

FAMILY	Colubridae: Colubrinae
RISK FACTOR	Nonvenomous, possibly mildly toxic saliva
DISTRIBUTION	Australasia: Papua New Guinea, possibly northern Australia
ELEVATION	0–4,000 ft (0–1,220 m) asl
HABITAT	Rainforest, riverine forest, disturbed areas, wetland areas, plantations, and around human habitations
DIET	Frogs, lizards, small mammals, and reptile eggs
REPRODUCTION	Oviparous, with clutches of 6–12 eggs
CONSERVATION STATUS	IUCN not listed

ADULT LENGTH
2 ft 7 in–3 ft 3 in,
occasionally 4 ft 3 in
(0.8–1.0 m,
occasionally 1.3 m)

STEGONOTUS RETICULATUS
RETICULATED GROUNDSNAKE
BOULENGER, 1895

The Reticulated Groundsnake has been removed from the synonymy of the Slatey-gray Snake (*S. cucullatus*). It is distributed throughout the Papua New Guinea mainland and islands of Milne Bay, and is also thought to occur in northern Australia. One of the largest *Stegonotus* in New Guinea, achieving over 3 ft 3 in (1 m) in length, it inhabits low-lying wetland habitats, rainforest, and plantations, but also occurs in disturbed habitats and around human dwellings. Nocturnal and terrestrial, it can also climb, and it is often found crossing roads at night. Prey consists of frogs, lizards, mice, and reptile eggs. Groundsnakes bite vigorously, drawing blood but causing only local effects.

The Reticulated Groundsnake is a smooth-scaled snake with a head distinct from neck, and small, dark, protruding eyes. Every pale scale of the body is dark-edged, resulting in a "reticulate" pattern. The undersides are immaculate white. The head is usually dark above with pale lips, chin, and throat.

Actual size

RELATED SPECIES

Stegonotus cucullatus was a species-complex comprising several cryptic species, with *S. reticulatus* the most distinctive. The New Guinea mainland is home to numerous other *Stegonotus* species, with others in the D'Entrecasteaux and Trobriand archipelagos (*S. guentheri*), the Bismarck Archipelago (*S. heterurus*), Admiralty Islands (*S. admiraltiensis*), Raja Ampat Islands (*S. iridis*, *S. derooijae*), and Schouten Islands (*S. parvus*). Some species may be confused with the highly venomous New Guinea Small-eyed Snake (*Micropechis ikaheka*, page 518).

FAMILY	Colubridae: Colubrinae
RISK FACTOR	Rear-fanged, mildly venomous; harmless to humans
DISTRIBUTION	North and Central America: southern Mexico, Guatemala, Belize, El Salvador, Honduras, Nicaragua, Costa Rica, and Panama
ELEVATION	0–7,220 ft (0–2,200 m) asl
HABITAT	Lowland and low montane wet and dry forest, savanna woodland, and swampy areas
DIET	Arachnids and insects
REPRODUCTION	Oviparous, with clutches of 4–19 eggs
CONSERVATION STATUS	IUCN Least Concern

ADULT LENGTH
13¾–18½ in,
occasionally 33½ in
(350–470 mm,
occasionally 850 mm)

STENORRHINA FREMINVILLEI
BLOOD SNAKE
(DUMÉRIL, BIBRON & DUMÉRIL, 1854)

237

The Blood Snake is a member of the Central American scorpion-eating genus *Stenorrhina*. It inhabits lowland and low montane arid or semi-moist forest and also occurs in swampy areas where it adopts a secretive, diurnal, terrestrial, or semi-fossorial habit, living under logs and in leaf litter. Prey consists of scorpions and spiders, including tarantulas, but insects such as beetles and crickets are also eaten. Females are oviparous and may lay two clutches during the dry season. Although it is harmless, Costa Rican folklore holds that anybody bitten by this snake will die following massive hemorrhaging through the skin. This is hard to disprove as this inoffensive species cannot be induced to bite. Christophe-Paulin de La Poix Chevalier de Fréminville (1787–1848) was a French naval officer and naturalist.

RELATED SPECIES
Another species, Degenhardt's Scorpion-eating Snake (*Stenorrhina degenhardtii*), occurs as three subspecies, from southern Mexico to Venezuela and Peru. *Stenorrhina* differ from other snakes within their range due to their fused internasal and anterior nasal scales. *Stenorrhina* is most closely related to the sandsnakes (*Chilomeniscus*, page 148), shovelnose snakes (*Chironius*, page 150), and ground-snakes (*Sonora*, page 232).

The Blood Snake is a small, smooth-scaled snake with a narrow, squarish head that is indistinct from the neck, and small eyes with round pupils. Dorsally it is blood red, orange, or brown, either unicolor or longitudinally striped, with an unmarked venter.

Actual size

FAMILY	Colubridae: Colubrinae
RISK FACTOR	Nonvenomous
DISTRIBUTION	North America: Mexico
ELEVATION	655–3,280 ft (200–1,000 m) asl
HABITAT	Tropical deciduous forest, thorn forest, and pine–oak forest
DIET	Presumed insects and/or arachnids
REPRODUCTION	Presumed oviparous, clutch size unknown
CONSERVATION STATUS	IUCN Least Concern, Special Protection in Mexico

ADULT LENGTH
22¾–32 in
(580–810 mm)

238

SYMPHIMUS LEUCOSTOMUS
TEHUANTEPEC
WHITE-LIPPED SNAKE
COPE, 1869

The Tehuantepec White-lipped Snake is a slender snake with smooth scales, a long tail, and a broad, elongate head with large eyes and round pupils. Dorsally the snake is light brown with a broad, darker brown vertebral stripe. The head is brown above and yellow on the lips, with a yellow stripe running onto the neck, and the two colors separated by a dark stripe.

The Tehuantepec or Pacific White-lipped Snake inhabits tropical deciduous forest, thorn forest, and pine–oak forest, in the southwestern Mexican states of Jalisco, Michoacán, Guerrero, Oaxaca, and Chiapas. It is a rare snake, found at lowland or low montane elevations, and is probably known from fewer than 20 specimens in museum collections. Its natural history is also poorly documented. It is known to be diurnal, terrestrial, and semi-arboreal, and is also believed to feed on insects, although it may also take scorpions and spiders like its congener, the Mayan White-lipped Snake (*Symphimus mayae*). Like that species, it is probably oviparous. Although this species is listed as Least Concern by the IUCN it is considered Endangered in Chiapas by the Mexican SEMARNAT System, and afforded Special Protection.

RELATED SPECIES

The only other species of *Symphimus* is the Mayan White-lipped Snake (*S. mayae*), which occurs on the Yucatán Peninsula of Mexico and Belize. The genus *Symphimus* may be closely related to the American greensnakes (*Opheodrys*, page 202) and at one time the Mayan White-lipped Snake was included in that genus.

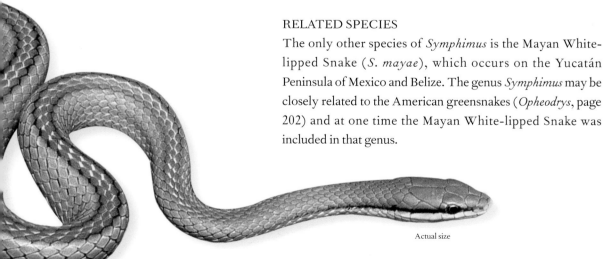

Actual size

FAMILY	Colubridae: Colubrinae
RISK FACTOR	Rear-fanged, mildly venomous; harmless to humans
DISTRIBUTION	North America: northwestern Mexico
ELEVATION	1,430–3,000 ft (435–915 m) asl
HABITAT	Tropical deciduous forest and thorn scrub
DIET	Presumed arthropods
REPRODUCTION	Presumed oviparous, clutch size unknown
CONSERVATION STATUS	IUCN not listed

ADULT LENGTH
17–21¼ in
(430–540 mm)

SYMPHOLIS LIPPIENS
MEXICAN SHORT-TAILED SNAKE
COPE, 1861

The Mexican Short-tailed Snake is distributed from southwestern Chihuahua to Sonora and Sinaloa, and down the northwestern Mexican coast to Jalisco. It is a rare snake that inhabits tropical deciduous forest and thorn scrub, where it is believed to lead a nocturnal and semi-fossorial existence. Given its close apparent relationship to Mexican scorpion- and spider-eating snakes it is likely that this species also feeds on arachnids, but no data regarding its prey preferences are available. It is an instantly recognizable snake, no other species within its range bearing the same distinctive pattern. The etymology of the specific name *lippiens*, which means "almost blind," is a reference to the very small eyes of this species.

The Mexican Short-tailed Snake is a cylindrical snake with smooth scales, a short tail, a head that is not distinct from the neck, and small eyes. Its patterning is distinctive, comprising regular, ragged-edged yellow or white rings on a black background, the interspaces being two to three times the width of the rings.

RELATED SPECIES
The genus *Sympholis* is monotypic and no subspecies are recognized. It is thought most closely related to the scorpion-eating snakes (*Stenorrhina*, page 237) and the hooknose snakes (*Ficimia*, page 174, *Gyalopion*, page 177, and *Pseudoficimia*, page 220).

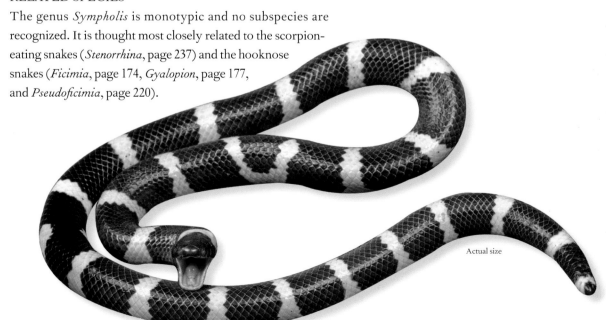

Actual size

FAMILY	Colubridae: Colubrinae
RISK FACTOR	Rear-fanged, mildly venomous; harmless to humans
DISTRIBUTION	South America: the Guianas to Argentina and Uruguay, Trinidad, Tobago, and Lesser Antilles
ELEVATION	0–10,100 ft (0–3,080 m) asl
HABITAT	Primary and secondary rainforest, Caatinga, Cerrado, savanna, and cultivated areas
DIET	Centipedes and insects
REPRODUCTION	Oviparous, with clutches of 1–3 eggs
CONSERVATION STATUS	IUCN not listed

ADULT LENGTH
11¾–15¼ in,
occasionally 17¾ in
(300–400 mm,
occasionally 450 mm)

TANTILLA MELANOCEPHALA
BLACK-HEADED
CENTIPEDE SNAKE
(LINNAEUS, 1758)

Actual size

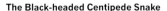

The Black-headed Centipede Snake
is a small, smooth-scaled snake with
a small head, indistinct from the neck,
and small eyes with round pupils. It is
dorsally brown with five longitudinal
black stripes, the vertebral stripe being
the boldest, and a yellow venter. The
head is black with white spots, and
a black nape band or collar is present.

The Black-headed Centipede Snake was originally thought to occur from Guatemala to northern Argentina, but several former populations have now been described as separate, but related, species and the mainland population of the Black-headed Centipede Snake is now confined to South America, although it is also reported from Trinidad, Tobago, and several islands of the Lesser Antilles. It is nocturnal, terrestrial, and semi-fossorial. It is found in rainforests and open habitats, at low and high elevations, sheltering under leaf litter, palm fronds, or in termite mounds. Where it occurs this is a very common species with several specimens being found under the same fallen palm fronds. Prey includes venomous scolopendrid centipedes and insects. In many parts of South America this is the only *Tantilla* species represented.

RELATED SPECIES

The American genus *Tantilla* contains 66 species. The *Tantilla melanocephala* species group contains at least eight other species: the Amulet Centipede Snake (*T. armillata*) and Red-headed Centipede Snake (*T. ruficeps*) in Central America; Honduran Centipede Snake (*T. lempira*); Andes Centipede Snake (*T. andinista*), Miyata's Centipede Snake (*T. miyatai*), and Mountain Centipede Snake (*T. insulamontana*) in Ecuador; and Masked Centipede Snake (*T. capistrata*) in Peru.

FAMILY	Colubridae: Colubrinae
RISK FACTOR	Rear-fanged, mildly venomous; harmless to humans
DISTRIBUTION	North America: southeastern USA
ELEVATION	0–33 ft (0–10 m) asl
HABITAT	Hardwood hammocks and pinewoods on oolitic limestone
DIET	Insects, spiders, scorpions, centipedes, snails, and small snakes
REPRODUCTION	Oviparous, clutch size unknown
CONSERVATION STATUS	IUCN Endangered, protected in Florida

ADULT LENGTH
6–9 in,
rarely 11½ in
(150–230 mm,
rarely 290 mm)

TANTILLA OOLITICA
RIM ROCK CROWNED SNAKE
TELFORD, 1966

241

The Rim Rock Crowned Snake is one of the smallest snakes in the United States, with one of the smallest ranges, inhabiting the hardwood hammocks and pinewoods on an oolitic (formed from rounded grains, or ooliths) limestone ridge running parallel to the coast, through Monroe and Miami-Dade counties, Florida, from Miami to Key Largo and continuing to Key West. Habitat loss due to development is the main threat to this tiny, inoffensive snake, one specimen even being found on a vacant lot in Miami. It is also one of the least studied of Florida's snakes. It is believed to prey on beetles, spiders, scorpions, centipedes, snails, and possibly smaller snakes. Probably its best chance of survival is the Key Largo wildlife refuge.

RELATED SPECIES

Eleven of the 64 species of *Tantilla* occur within the United States, but only three occur in the state of Florida: the Southeastern Crowned Snake (*T. coronata*) in the Panhandle, the Florida Crowned Snake (*T. relicta*) with three subspecies in the Peninsula, and *T. oolitica*.

Actual size

The Rim Rock Crowned Snake is a small, smooth-scaled snake with a cylindrical body, a head indistinct from the neck, and small eyes with round pupils. It is light tan above and pinkish brown or cream below, with a black dorsum to the head and a broad black band around the neck.

FAMILY	Colubridae: Colubrinae
RISK FACTOR	Rear-fanged, mildly venomous; harmless to humans
DISTRIBUTION	North and Central America: Mexico, northern Guatemala, and Belize
ELEVATION	0–985 ft (0–300 m) asl
HABITAT	Tropical evergreen forest and thorn forest
DIET	Presumed arthropods, especially centipedes
REPRODUCTION	Presumed oviparous, clutch size unknown
CONSERVATION STATUS	IUCN Least Concern

ADULT LENGTH
5¾–7 in
(145–180 mm)

242

TANTILLITA CANULA
YUCATÁN DWARF
SHORT-TAILED SNAKE
(COPE, 1875)

The Yucatán Dwarf Short-tailed Snake is found on the Yucatán Peninsula of southern Mexico, in the states of Campeche, Yucatán, and Quintana Roo, and in northern Guatemala and Belize, also on the Yucatán Peninsula. A lowland species, this tiny snake inhabits tropical evergreen forests and thorn forests below 985 ft (300 m). It is terrestrial or semi-fossorial, being found in leaf litter or under logs or rocks. Several live specimens have also been found by archeologists excavating Mayan temples, or by construction crews building new dwellings. Little is known of the natural history of this diminutive species, but it is believed to feed on arthropods, possibly centipedes, and probably also lays eggs.

Actual size

The Yucatán Dwarf Short-tailed Snake is small with smooth scales, a very short tail, a head indistinct from the neck, and small eyes with round pupils. It is brown above, slightly darker on the head, but it lacks the black cap and collar of many *Tantilla*, the only patterning being a faint pale vertebral stripe in some specimens. The venter is cream.

RELATED SPECIES

The genus *Tantillita* contains only two other species, the Speckled Dwarf Short-tailed Snake (*T. brevissima*) from southwestern Mexico and Guatemala, and Linton's Dwarf Short-tailed Snake (*T. lintoni*) from southern Mexico, Guatemala, and Belize. *Tantillita* is closely related to the centipede snakes (*Tantilla*, pages 240–241), and the Tehuantepec Striped Snake (*Geagras redimitus*).

FAMILY	Colubridae: Colubrinae
RISK FACTOR	Rear-fanged, mildly venomous; harmless to humans
DISTRIBUTION	Southwest Europe, Middle East, and western Asia: the Balkans, Albania, Greece, Turkey, Malta, Corfu, Cyprus, Rhodes, Israel, Syria, Lebanon, Sinai, Iraq, Iran, and Caucasus
ELEVATION	0–6,560 ft (0–2,000 m) asl
HABITAT	Rocky habitats, including perianthropic habitats such as railway embankments and dry stone walls
DIET	Lizards, small snakes, small mammals, and birds
REPRODUCTION	Oviparous, with clutches of 5–8 eggs
CONSERVATION STATUS	IUCN Least Concern

ADULT LENGTH
2 ft 7 in–3 ft 3 in,
rarely 4 ft
(0.8–1.0 m,
rarely 1.2 m)

TELESCOPUS FALLAX
EUROPEAN CATSNAKE
FLEISCHMANN, 1831

243

The European Catsnake is distributed through southeastern Europe, the Middle East, and western Asia, in arid rocky habitats, primarily in coastal regions and on islands. It is a nocturnal and terrestrial species that climbs rocks but not trees. It also inhabits man-made habitats, such as dry stone walls and railway embankments, which are home to large populations of geckos and lacertid lizards, its preferred prey. Small snakes, mice, and even nestling birds are also taken. The European Catsnake is a rear-fanged species that stalks its prey, strikes quickly, and chews to bring the fangs into play, using its coils to restrain the prey while the venom takes effect. Its defensive display involves flattening its already broad head, hissing, and striking, but its venom is not dangerous to humans.

The European Catsnake is a laterally compressed snake with smooth scales, a long tail, and a broad head with protruding eyes and vertically elliptical pupils. It is a gray to brown snake with a pattern comprising distinctive dark dorsal blotches and small lateral bars. The iris of the eye is yellow.

RELATED SPECIES

Seven subspecies are recognized, from mainland southeast Europe, Malta, Rhodes, and Turkey (*Telescopus fallax fallax*); Antikythera Island, Greece (*T. f. intermedius*); Crete (*T. f. pallidus*); Koufonisia Island, Greece (*T. f. multisquamatus*); Cyprus (*T. f. cyprianus*); southeast Turkey and the Middle East (*T. f. syriacus*); and eastern Turkey, northern Iran, and Transcaucasia (*T. f. iberus*). The genus *Telescopus* contains another 14 species in Africa, Arabia, and western Asia, including a former subspecies of *T. fallax* from Israel and Jordan (*T. hoogstraali*).

Actual size

FAMILY	Colubridae: Colubrinae
RISK FACTOR	Rear-fanged, mildly venomous; harmless to humans
DISTRIBUTION	Southern and East Africa: South Africa, Namibia, Swaziland, Botswana, Zambia, Mozambique, Malawi, Zimbabwe, DRC, Tanzania, and Kenya
ELEVATION	0–5,580 ft (0–1,700 m) asl
HABITAT	Rocky outcrops, sandveld, bushveld, lowveld, coastal thicket, savanna woodland
DIET	Lizards, small mammals, and birds
REPRODUCTION	Oviparous, with clutches of 6–20 eggs
CONSERVATION STATUS	IUCN not listed

ADULT LENGTH
Male
23¾–31½ in
(600–800 mm)

Female
31½–35½ in,
rarely 3 ft 9 in
(800–900 mm,
rarely 1.15 m)

244

TELESCOPUS SEMIANNULATUS
EASTERN TIGER SNAKE
SMITH, 1849

The Eastern Tiger Snake is a common nocturnal African snake, often seen crossing roads after rain. It occurs from South Africa and Namibia to Kenya and Tanzania, in rocky habitats and sandy savanna woodland, where it hunts geckos, sleeping chameleons, small birds, bats, and mice. Its venom is weak, requiring the Eastern Tiger Snake to also employ constriction to restrain its prey while it takes effect. Although primarily terrestrial, it scales rocks, buildings, or trees with ease. When encountered it may adopt a defensive posture, elevating its anterior body to form an S-shaped curve, flattening its already broad head, and making sudden and rapid strikes, with such force that the body is often thrown forward. Females can store the male's sperm and lay eggs every two months during the summer.

RELATED SPECIES

The nominate subspecies (*Telescopus semiannulata semiannulata*) occupies most of the range, with the Namibian population recognized as a separate subspecies (*T. s. polystictus*). A similar, second species, Beetz's Tiger Snake (*T. beetzi*), occurs in southern Namibia and into Namaqualand, South Africa. The Arabian Catsnake (*T. dhara*) occurs to the north in Kenya.

Actual size

The Eastern Tiger Snake is a slender, smooth-scaled snake with a broad head, distinct from the neck, and protruding eyes with vertically elliptical pupils. It is brown, orange, sandy yellow, or pinkish yellow, with a black nape band and a series of regularly spaced black blotches or bands on the dorsum. These bands are most numerous in the Namibian subspecies. The venter is pale yellow to orange.

FAMILY	Colubridae: Colubrinae
RISK FACTOR	Rear-fanged, highly venomous: procoagulants, and possibly anticoagulants or hemorrhagins
DISTRIBUTION	Southern Africa: southern Angola, Namibia, eastern Botswana, DRC, Malawi Mozambique, Zambia, Zimbabwe, Swaziland, and northeast South Africa
ELEVATION	0–6,000 ft (0–1,830 m) asl
HABITAT	Wet and dry savanna, and riparian woodland
DIET	Lizards, small birds, frogs, snakes, and bats
REPRODUCTION	Oviparous, with clutches of 4–18 eggs
CONSERVATION STATUS	IUCN Least Concern

ADULT LENGTH
2 ft 7 in–4 ft,
rarely 5 ft
(0.8–1.2 m,
rarely 1.5 m)

THELOTORNIS CAPENSIS
SOUTHEASTERN SAVANNA TWIGSNAKE
SMITH, 1849

245

The Southeastern Savanna Twigsnake, also known as the Savanna Vinesnake, inhabits open woodland and savanna habitats from Angola to KwaZulu-Natal, but it is absent from most of South Africa. A diurnal and highly arboreal snake, it is perfectly camouflaged for its lifestyle, hunting chameleons and small birds in the trees. The horizontal pupils provide it with amazing vision, enabling it to discern the most camouflaged of chameleons. Prey is killed with venom injected via the enlarged rear fangs. Its threat posture involves inflating its neck and flattening its head, but it is reluctant to bite. However, bites from this snake are extremely serious, the hemotoxins in the venom causing prolonged bleeding and renal failure. The eminent German herpetologist Robert Mertens (1894–1975) died several days after being bitten. There is no antivenom manufactured to treat twigsnake bites.

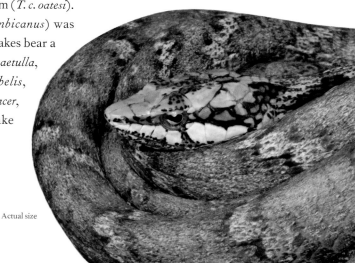

The Southeastern Savanna Twigsnake is an extremely slender snake with weakly keeled scales arranged in oblique rows, a long tail, an elongate, narrow head, and large eyes with horizontal pupils. It is pale gray to gray-brown in color, with patterning comprising diagonal pale bands and flecks of black and pink, altogether presenting a twig-like appearance. The dorsum of the head is brown, with a red-brown stripe through the eye.

RELATED SPECIES

Two subspecies are recognized, a nominate eastern form (*Thelotornis capensis capensis*), and a western form (*T. c. oatesi*). The Eastern Savanna Twigsnake (*T. mossambicanus*) was previously a subspecies of *T. capensis*. Twigsnakes bear a strong resemblance to the Asian vinesnakes (*Ahaetulla*, pages 130–131), American vinesnakes (*Oxybelis*, pages 205–206), Hispaniolan vinesnakes (*Uromacer*, page 344), and the Dagger-toothed Vinesnake (*Xyelodontophis uluguruensis*).

Actual size

FAMILY	Colubridae: Colubrinae
RISK FACTOR	Rear-fanged, highly venomous: probably procoagulants, possibly anticoagulants and hemorrhagins
DISTRIBUTION	West and Central Africa: Guinea to Ghana, Nigeria to DRC, Uganda and Tanzania, south to Angola and Zambia
ELEVATION	5,250–7,220 ft (1,600–2,200 m) asl
HABITAT	Rainforest, dense woodland, thickets, moist savanna reed beds
DIET	Lizards, birds, eggs, amphibians, and other snakes
REPRODUCTION	Oviparous, with clutches of 4–12 eggs
CONSERVATION STATUS	IUCN not listed

ADULT LENGTH
4 ft 3 in–5 ft 7 in
(1.3–1.7 m)

THELOTORNIS KIRTLANDI
FOREST TWIGSNAKE
(HALLOWELL, 1844)

246

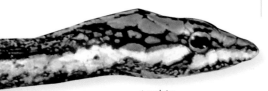

Actual size

The Forest Twigsnake is an extremely slender snake with weakly keeled scales arranged in oblique rows, a long tail, and an elongate, narrow head and large eyes with horizontal pupils. Its most striking characteristic is the bright green dorsum of the head, a character it shares with the Usumbara Twigsnake, which contrasts with the white of the lip scales. The body is mottled gray, green, and brown with black cross-bar markings.

The Forest Twigsnake, or Vinesnake, inhabits pockets of montane forest in East Africa, but occurs more widely in rainforest and dense woodland habitats farther west, and it also inhabits reed beds on moist savannas. It occurs as far west as Sierra Leone and Guinea and as far south as northern Angola and Zambia. Although sometimes called the "Birdsnake," it preys primarily on chameleons, geckos, and agamid lizards, with birds, their eggs, amphibians, and other snakes only occasional prey. Terrestrial lizards may be ambushed from above, in a tactic used by New Guinea Treeboas (*Candoia carinata*, page 115). Few bites are known for this species but it must be considered potentially dangerous given the fatalities attributed to the Southeastern Savanna Twigsnake (*Thelotornis capensis*, page 245), and the lack of antivenom. Jared Potter Kirtland (1793–1877) was an American naturalist.

RELATED SPECIES

The genus *Thelotornis* contains three other species, the Southeastern Savanna Twigsnake (*T. capensis*, page 245), the Eastern Savanna Twigsnake (*T. mossambicanus*), and the endangered Usumbara Twigsnake (*T. usumbaricus*) from East Africa.

FAMILY	Colubridae: Colubrinae
RISK FACTOR	Rear-fanged, venomous, potentially dangerous: postsynaptic neurotoxins
DISTRIBUTION	West and Central Africa: Senegal to Kenya, south to Angola and Zambia
ELEVATION	0–7,220 ft (0–2,200 m) asl
HABITAT	Rainforest, woodland, savanna woodland, gallery forest, and parks
DIET	Lizards, small mammals, birds and their eggs, and frogs
REPRODUCTION	Oviparous, with clutches of 7–14 eggs
CONSERVATION STATUS	IUCN not listed

ADULT LENGTH
4 ft 7 in–6 ft 7 in,
rarely 9 ft 2 in
(1.4–2.0 m,
rarely 2.8 m)

TOXICODRYAS BLANDINGII
BLANDING'S TREESNAKE
(HALLOWELL, 1844)

247

Blanding's Treesnake is Africa's largest nocturnal treesnake, occurring across West and Central Africa. It is found in tropical rainforest, riverine gallery forest, and savanna woodland. It prefers large trees, and rarely ventures to the ground. It hunts prey ranging from arboreal mammals to sleeping birds, their eggs, lizards, and frogs. Bats are often taken, the treesnake occupying trees near bat roosts or entering roof spaces in search of them, and it is believed able to smell sleeping birds in their nests. The venom of the rear-fanged Blanding's Treesnake contains powerful postsynaptic neurotoxins that interfere with neuromuscular transmissions. Bites have caused muscle pain and difficulty in breathing and there remains the potential for a serious snakebite. This species should be treated with caution. William Blanding (1772–1857) was an American naturalist.

Blanding's Treesnake is a large snake with smooth scales, a laterally compressed body, a long tail, and a very broad, flattened head, distinct from the neck, with large eyes and round pupils. The species is sexually dichromatic, males being glossy black above and yellowish below and on the lips, while females are matte brown with darker transverse bands.

RELATED SPECIES

There are two species in the African genus *Toxicodryas*, the other being the smaller Powdered Treesnake (*T. pulverulenta*). In the past both species were included in the Asian catsnake genus *Boiga* (pages 144–145). *Toxicodryas blandingii* is easily confused with the highly venomous tree cobras (*Pseudohaje*, page 481) or Forest Cobra (*Naja melanoleuca*, page 470).

Actual size

FAMILY	Colubridae: Colubrinae
RISK FACTOR	Rear-fanged, mildly venomous; harmless to humans
DISTRIBUTION	North America: southwestern USA and northern Mexico
ELEVATION	0–2,620 ft (0–800 m) asl
HABITAT	Rocky deserts, rocky escarpments, and mesquite, creosote, and saguaro habitats
DIET	Lizards, small mammals, and birds
REPRODUCTION	Oviparous, with clutches of 5–20 eggs
CONSERVATION STATUS	IUCN not listed

ADULT LENGTH
23¾–35½ in,
occasionally 4 ft
(600–900 mm,
occasionally 1.2 m)

248

TRIMORPHODON LAMBDA
SONORAN LYRESNAKE
COPE, 1886

The Sonoran Lyresnake is a laterally compressed snake with smooth scales and a broad head, distinct from the neck, with large, protruding eyes and vertically elliptical pupils. Its dorsal ground color is gray, brown, or pinkish, with a series of dark brown dorsal blotches, each with a transverse light brown center, the dark area extending almost to the venter on the flanks. The head bears two dark brown chevrons, the lyre-shaped marking, one through the eyes and the other to the angle of the jaw.

The Sonoran Lyresnake is found in Sonora, Mexico, and north into Arizona and Nevada, USA. It is found in rocky habitats, from escarpments to boulder-strewn desert floors with growth of creosote, mesquite, and saguaro cacti. Although primarily a terrestrial species, this snake can climb well if required and is at home scaling rocky outcrops and low desert vegetation. This is a relatively large snake that hunts primarily lizards at night, exploring crevices and fissures for prey, which is killed using venom injected via the rear fangs. Other prey is believed to include small rodents, bats, and birds. The warning display of the Sonoran Lyresnake involves elevating the anterior body in a threatening curve. The venom is believed weak and has no effect on humans, but caution is advised when handling large specimens.

RELATED SPECIES

No subspecies of *Trimorphodon lambda* are recognized, its former subspecies, the Sinaloan Lyresnake (*T. paucimaculatus*), being elevated to specific status. *Trimorphodon lambda* was once also a subspecies of the Western Lyresnake (*T. biscutatus*), along with the California Lyresnake (*T. lyrophanes*) and Central American Lyresnake (*T. quadruplex*). *Trimorphodon* is related to the leafnose snakes (*Phyllorhynchus*, page 214).

Actual size

FAMILY	Colubridae: Colubrinae
RISK FACTOR	Rear-fanged, mildly venomous; harmless to humans
DISTRIBUTION	North America: southern USA and northern Mexico
ELEVATION	2,950–6,070 ft (900–1,850 m) asl
HABITAT	Rocky outcrops and bluffs, and open rocky desert
DIET	Lizards, small mammals, and birds
REPRODUCTION	Oviparous, with clutches of 6–7 eggs
CONSERVATION STATUS	IUCN Least Concern, Threatened in Texas

ADULT LENGTH
23¾–35½ in,
rarely 3 ft 5 in
(600–900 mm,
rarely 1.04 m)

TRIMORPHODON VILKINSONII
TEXAS LYRESNAKE
COPE, 1886

249

The Texas Lyresnake is found in western Big Bend, Texas, southern New Mexico, and northern Chihuahua, Mexico. It is the easternmost member of the lyresnake genus *Trimorphodon*. It inhabits rocky bluffs but is also found in flat, open, rock-strewn desert and close to rivers. This is a secretive species that is rarely encountered. It is most active on nights with high humidity due to rainfall, when it searches the fissures and crevices for lizards, both active nocturnal species and sleeping diurnal species, but it also preys on mice, bats, and nestling birds. Prey is killed by the snake's weak venom, which is injected via its enlarged, grooved rear fangs. If handled the Texas Lyresnake will bite, but it is not dangerous to humans. Edward Wilkinson (1846–1918) was an amateur naturalist.

RELATED SPECIES

Trimorphodon vilkinsonii was once included as a subspecies of the Western Lyresnake complex (*T. biscutatus*, *T. lambda*, etc., page 248). Another species, the Mexican Lyresnake (*T. tau*), which has two subspecies, occurs farther south.

The Texas Lyresnake has a laterally compressed body, with smooth scales, a broad head, distinct from the neck, and large protruding eyes with vertically elliptical pupils. It is dorsally pale gray with a series of dark gray to black middorsal blotches, each edged with white or very pale gray, that extend down the flanks as narrow bands. A dark lyre-shaped marking is present on the dorsum of the head.

Actual size

FAMILY	Colubridae: Colubrinae
RISK FACTOR	Nonvenomous
DISTRIBUTION	Southeast Asia: Malay Peninsula, Borneo, and Sumatra
ELEVATION	490–3,610 ft (150–1,100 m) asl
HABITAT	Primary lowland and low montane rainforest, near water
DIET	Small mammals, lizards, frogs, and fish
REPRODUCTION	Oviparous, clutch size unknown
CONSERVATION STATUS	IUCN Least Concern

ADULT LENGTH
5 ft–7 ft 7 in,
occasionally 8 ft 2 in
(1.5–2.3 m,
occasionally 2.5 m)

XENELAPHIS ELLIPSIFER
ORNATE BROWNSNAKE
BOULENGER, 1900

The Ornate Brownsnake is a large but rare species known from only a few specimens or sightings from Borneo, Sumatra, and Peninsular Malaysia. It inhabits lowland and low montane rainforest and is most likely to be encountered near to watercourses. Brownsnakes of genus *Xenelaphis* are semi-aquatic, and they are reported to feed on fish. Adults also take small mammals to the size of squirrels, and juveniles are said to feed on lizards and frogs. This species can also climb into low bushes. It has only been sighted on rare occasions in nature, so natural history data are scarce. However, one specimen was repeatedly seen near the same stream, suggesting a small home range and site fidelity.

The Ornate Brownsnake is a large, robust snake with smooth scales, a long tail, a large head that is distinct from the neck, and large eyes with round pupils. It is dorsally pale cream with a series of large, red, dorsal ocelli markings with black edges, and a bright red to orange dorsum to the head. The supralabials and neck are yellow while the venter is white to cream.

RELATED SPECIES

A second species, the much commoner Malayan Brownsnake (*Xenelaphis hexagonotus*), occurs in southern Myanmar and Thailand, Vietnam, Malaysia, Sumatra, Java, and Borneo. This genus was originally contained in the Xenodermatinae (odd-scaled snakes), now Xenodermatidae (pages 548–551), but is now thought more closely related to Asian ratsnakes and racers of *Coelognathus* (page 151), *Gonyosoma* (pages 175–176), and *Ptyas* (pages 221–222).

Actual size

FAMILY	Colubridae: Colubrinae
RISK FACTOR	Nonvenomous, constrictor
DISTRIBUTION	Europe and western Asia: northeast Spain, France, Italy, Germany, Poland, Czech Republic, the Balkans, Greece, Bulgaria, Romania, Ukraine, Turkey, Russia, and Iran; introduced to UK
ELEVATION	0–5,580 ft (0–1,700 m) asl
HABITAT	Deciduous woodland, vineyards, orchards, riverbanks, water meadows, dry stone walls
DIET	Small mammals, birds, and lizards
REPRODUCTION	Oviparous, with clutches of 5–8 eggs
CONSERVATION STATUS	IUCN Least Concern

ADULT LENGTH
4 ft 7 in–5 ft 2 in,
rarely 7 ft 5 in
(1.4–1.6 m,
rarely 2.25 m)

ZAMENIS LONGISSIMUS
COMMON AESCULAPIAN SNAKE
(LAURENTI, 1768)

251

The Common Aesculapian Snake is linked with Asclepius, the Roman god of medicine, and featured on the Rod of Asclepius, the symbol of medicine. It is distributed from northeast Spain to Ukraine, and south to the Balkans, Greece, and Turkey. It was proposed that isolated populations in Germany and Austria were remnants of populations established by Roman legions, but they are more likely to represent refugia from a once wider distribution. Isolated populations are also known from eastern Turkey, Iran, and Russia. There is an introduced population in North Wales, which has thrived for decades, with another in London. This is a harmless predator of rodents and birds, with juveniles taking small mice and lizards. It is found in deciduous woodland, near rivers, and in perianthropic habitats.

RELATED SPECIES

No subspecies are currently recognized for *Zamenis longissimus*. Its nearest relative is the striped former subspecies, the Southern Italian Aesculapian Snake (*Z. lineatus*), which also occurs on Sicily. The Persian Ratsnake (*Z. persicus*), from Azerbaijan and Iran, was also a former subspecies.

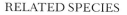

The Common Aesculapian Snake is a stout-bodied snake, with smooth scales, a long tail, a squarish head, just distinct from the neck, and large eyes with round pupils. Adults are fairly uniform olive-green or brown to gray with paler pigmentation on the undersides, throat, lips, and nape of the neck. Juveniles are more contrastingly marked, often with a checkerboard pattern of browns, a dark postocular stripe, and a black and yellow collar, which resembles that of the Western Grass Snake (*Natrix helvetica*, page 415).

Actual size

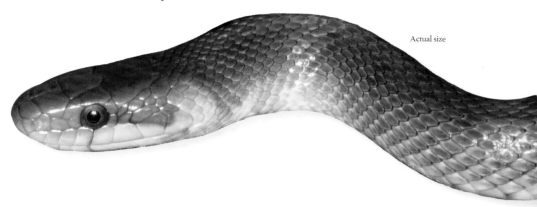

FAMILY	Colubridae: Colubrinae
RISK FACTOR	Nonvenomous, constrictor
DISTRIBUTION	Southwest Europe: Spain, Portugal, and southern France; also Minorca
ELEVATION	0–6,890 ft (0–2,100 m) asl
HABITAT	South-facing rocky hillsides, open woodland, olive groves, cork oak plantations, and vineyards
DIET	Small mammals, birds, and lizards
REPRODUCTION	Oviparous, with clutches of 6–12 eggs
CONSERVATION STATUS	IUCN Least Concern

ADULT LENGTH
4 ft 3 in–5 ft 2 in
(1.3–1.6 m)

252

ZAMENIS SCALARIS
LADDER SNAKE
(SCHINZ, 1822)

The Ladder Snake is a robust snake with smooth scales, a long tail, and a distinctive pointed head, with large eyes and round pupils. Adults vary from yellow to reddish brown or gray, with a bold pair of dark, longitudinal, dorsolateral stripes that are linked by regular transverse bars, which may be faint, presenting a ladder-like pattern. Juveniles exhibit the transverse bars across the midbody and a scattering of dark spots on the flanks.

The Ladder Snake is found throughout the Iberian Peninsula, except the extreme north. It also ranges into southern France and is recorded from Minorca in the Balearic Islands. It has a preference for open habitats, especially south-facing rocky slopes and dry, open woodland, but it also inhabits man-made habitats such as olive groves, vineyards, and cork oak plantations. It is generally diurnal and fond of basking, but it may become crepuscular or nocturnal in especially hot weather. The Ladder Snake is a fast-moving and alert predator of small mammals and birds, with juveniles taking mice and lizards. Prey may be pursued back to its own burrow, the Ladder Snake being equally at home underground. It climbs well and may be encountered in trees, hunting birds or their nestlings.

RELATED SPECIES

Zamenis scalaris was included in a separate genus, *Rhinechis*, due to its pointed snout, but molecular data places it within *Zamenis* with the Common Aesculapian Snake (*Z. longissimus*, page 251), Southern Italian Aesculapian Snake (*Z. lineatus*), Leopard Snake (*Z. situla*, page 253), Persian Ratsnake (*Z. persicus*), and Transcaucasian Ratsnake (*Z. hohenackeri*). Its closest relative is the Leopard Snake, from southeastern Europe.

Actual size

FAMILY	Colubridae: Colubrinae
RISK FACTOR	Nonvenomous, constrictor
DISTRIBUTION	Southeast Europe: southern Italy, Sicily, Malta, Croatia, Bosnia, Albania, Greece, Macedonia, Bulgaria, Turkey, and Crimea; also Corfu, Crete, and Rhodes
ELEVATION	0–5,250 ft (0–1,600 m) asl
HABITAT	Mediterranean habitats, from rocky slopes with dense vegetation, to open woodland, and human-mediated habitats such as plantations, groves, and dry stone walls
DIET	Small mammals, lizards, and birds
REPRODUCTION	Oviparous, with clutches of 2–7 eggs
CONSERVATION STATUS	IUCN Least Concern

ADULT LENGTH
27½–31½ in,
rarely 3 ft 3 in
(700–800 mm,
rarely 1.0 m)

ZAMENIS SITULA
LEOPARD SNAKE
(LINNAEUS, 1758)

253

The Leopard Snake has an extremely fragmented distribution from Sicily, Malta, and the heel of Italy, to the coastal strip from Croatia to Albania, Greece, and European Turkey, inland to Macedonia and Bulgaria, and across the Black Sea on the Crimean Peninsula. It is found on many islands, including Corfu, Crete, and Rhodes. This snake is an inhabitant of semiarid habitats such as well-vegetated rocky slopes, open woodland, and groves. Although it is diurnal it does not bask like the Ladder Snake (*Zamenis scalaris*, page 252), being more inclined to shelter close to vegetation. It preys on rodents, but lizards also feature in its diet. Leopard Snakes can climb, and birds or birds' eggs are also occasionally taken. Prey is killed by constriction although small animals may be eaten alive.

RELATED SPECIES

Zamenis situla is one of six species in the genus *Zamenis* but appears to be the sister species to the Ladder Snake (*Z. scalaris*, page 252) of the Iberian Peninsula, although it is also related to the Common Aesculapian Snake (*Z. longissimus*, page 251).

The Leopard Snake may be one of the most attractive snakes in Europe. The adults display a series of red, black-edged transverse dorsal bands on a gray background, with a red ocelli marking on the nape of the neck. The flanks exhibit a series of large black spots, and black bands are present on the head, between the eyes, from the eye to the lip, and from the eye to the angle of the jaw. In some specimens the red markings coalesce to form a pair of dorsal stripes.

Actual size

FAMILY	Colubridae: Grayiinae
RISK FACTOR	Nonvenomous
DISTRIBUTION	East, Central, and West Africa: Senegal to South Sudan, south to Angola
ELEVATION	0–4,540 ft (0–1,385 m) asl
HABITAT	Rivers and lakes in savanna habitats
DIET	Fish, frogs, and tadpoles
REPRODUCTION	Oviparous, with clutches of 9–20 eggs
CONSERVATION STATUS	IUCN not listed

ADULT LENGTH
3 ft 3 in–5 ft 7 in,
rarely 8 ft 2 in
(1.0–1.7 m,
rarely 2.5 m)

GRAYIA SMYTHII
SMITH'S AFRICAN WATERSNAKE
(LEACH, 1818)

Smith's African Watersnake is a stout-bodied snake with smooth scales, a long tail, and a robust head, with small eyes and round pupils. The dorsal coloration may be black or yellow-brown with faint, paler blotches, the remnants of the more distinctive bands of the juvenile livery. The head is brown with black-edged scales and a yellow throat and lips.

Smith's African Watersnake is a large snake that is found around the periphery of watercourses, such as lakes or rivers, usually in savanna regions, including Lakes Victoria, Albert, and Edward. It feeds on fish, with frogs and tadpoles also in its diet. It will hiss loudly and open its mouth in threat when confronted, but is reluctant to bite. However, its close resemblance to the highly venomous Banded Water Cobra (*Naja annulata*, page 466) counsels caution with wild specimens. Although the specific name is *smythii* it is believed this species was named for the Norwegian physician-naturalist Christen Smith (1785–1816), who died on the expedition that collected the holotype.

RELATED SPECIES

Grayia is the sole genus in Grayiinae, an endemic African subfamily of the Colubridae. It was named in honor of John Edward Gray (1800–1975), an eminent zoologist at The Natural History Museum in London. Apart from *G. smythii* the genus contains three other species: Caesar's African Water Snake (*G. caesar*) and the Ornate African Watersnake (*G. ornata*) in Central Africa, and Thollon's African Watersnake (*G. tholloni*) in East and Central Africa. This is one of the few Old World watersnake genera not contained in the Natricidae.

Actual size

FAMILY	Colubridae: *incertae sedis*
RISK FACTOR	Nonvenomous
DISTRIBUTION	South and Southeast Asia: northeast India, Myanmar, southwest China, and Tibet
ELEVATION	475–6,560 ft (145–2,000 m) asl
HABITAT	Wet evergreen forests
DIET	Earthworms and arthropods
REPRODUCTION	Oviparous, with clutches of up to 6 eggs
CONSERVATION STATUS	IUCN Data Deficient

ADULT LENGTH
20¼ in
(514 mm)

BLYTHIA RETICULATA

BLYTH'S IRIDESCENT SNAKE

(BLYTH, 1854)

255

Blyth's Iridescent Snake is an inhabitant of wet evergreen forests at low to medium elevations. Its natural history is poorly known due to its secretive nature. It is semi-fossorial to fossorial, being usually discovered under fallen logs, dense leaf litter, or piles of decomposing vegetation. The tail bears a short spine at its terminus. This spine is dug into the ground when the snake is burrowing, giving it an anchor point from which to apply force forward. Whether it is diurnal or nocturnal is not known. Prey consists primarily of earthworms but some authors suggest this species may also eat other soft-bodied soil invertebrates. It lays clutches of up to six eggs and is inoffensive when handled.

RELATED SPECIES

This species was described by the British naturalist Edward Blyth (1810–73), as *Calamaria reticulata*. When it was later transferred to a new monotypic genus, that genus was named *Blythia* in his honor. This genus is one of nine genera within the Colubridae that are treated as *incertae sedis*—"of uncertain placement" at the subfamily level.

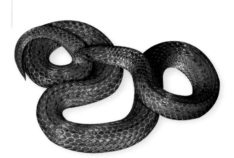

Blyth's Iridescent Snake is small with a cylindrical body, glossy, smooth scales, a short tail with a pointed tip, a narrow, pointed head that is indistinct from the neck, and small eyes. It is dorsally uniform black, blue-black, or even deep purple or olive, with a faint broken dark vertebral line sometimes present. Juveniles possess a yellow collar that is broken middorsally.

Actual size

FAMILY	Colubridae: *incertae sedis*
RISK FACTOR	Nonvenomous
DISTRIBUTION	Southeast Asia: Malaysia (Tioman Island)
ELEVATION	33 ft (10 m) asl
HABITAT	Lowland and submontane rainforest
DIET	Prey preferences not known, presumed arthropods and lizards
REPRODUCTION	Oviparous, clutch size unknown
CONSERVATION STATUS	IUCN Critically Endangered

ADULT LENGTH
17 in
(430 mm)

256

GONGYLOSOMA MUKUTENSE
PULAU TIOMAN GROUNDSNAKE
GRISMER, DAS & LEONG, 2003

The Pulau Tioman Groundsnake is a small, slender snake with a squarish, blunt head, slightly broader than the narrow neck, and moderately large eyes. The body is red-brown anteriorly, becoming brown-gray by midbody, with the undersides cream. The dorsum of the head and nape are darker brown with a white stripe passing along the lips, posterior to the eye and curving over the nape to form a chevron, this being the only species in the genus with such a marking.

The Pulau Tioman Groundsnake, or Mukut Smoothsnake, is endemic to Tioman Island, located off the east coast of the Malaysian Peninsula. Mukut is the small village where the holotype was collected, as it was being swallowed by a juvenile Malayan Keeled Ratsnake (*Ptyas carinatus*). Members of genus *Gongylosoma* inhabit lowland and submontane rainforest, where they are found in leaf-litter or beneath rocks or logs. The prey preferences of the Pulau Tioman Groundsnake are unknown but in common with its congenerics it probably feeds on spiders, insects and/or small lizards. The genus is oviparous. This is one of many small leaf-litter dwelling snakes in Southeast Asia, that are themselves the prey of larger ophiophagous snakes such as the coralsnakes (*Calliophis*) and kraits (*Bungarus*).

RELATED SPECIES

Gongylosoma is *incertae sedis* ("of uncertain placement") within the Colubridae. Four other species are included in the genus *Gongylosoma*: the Five-striped Groundsnake (*G. longicaudum*), which occurs in sympatry with this species, the Spotted Groundsnake (*G. baliodeirum*), Nicobar Groundsnake (*G. nicobariensis*), and the Indo-Chinese Groundsnake (*G. scriptum*) from Myanmar and Thailand.

Actual size

FAMILY	Colubridae: *incertae sedis*
RISK FACTOR	Nonvenomous
DISTRIBUTION	South Asia: southwestern India
ELEVATION	2,000–4,270 ft (610–1,300 m) asl
HABITAT	Hill forest, especially in riparian habitats
DIET	Slugs and earthworms
REPRODUCTION	Presumed oviparous, clutch size unknown
CONSERVATION STATUS	IUCN Least Concern

ADULT LENGTH
30¾ in
(780 mm)

RHABDOPS OLIVACEUS
INDIAN OLIVE FOREST SNAKE
(BEDDOME, 1863)

257

The Indian Olive Forest Snake is found in the ancient Western Ghats of southwestern India, an area of high rainfall and dense hill forests, with a high degree of endemicity in its fauna. The Indian Olive Forest Snake is found in hill forests, especially close to streams and creeks. At night it forages for earthworms and slugs, its primary prey, in moist conditions, in forest leaf litter or along stream beds. It uses its enlarged rostral scale to dig for its prey. It is terrestrial to semi-fossorial and may also be semi-aquatic, but it is rarely encountered and poorly known in nature. During the day it hides in crevices or under logs. Inoffensive, this snake will not bite, even when handled.

Actual size

RELATED SPECIES

The genus *Rhabdops* is *incertae sedis* within the Colubridae, "of uncertain placement," which means it cannot be pigeonholed into one of the four subfamilies. There are nine genera in this position, including *Blythia* (page 255). *Rhabdops* also contains a second species, the Trapezoid Forest Snake (*R. bicolor*), which occurs in northeast India, Myanmar, and southwest China.

The Indian Olive Forest Snake is a stocky snake with smooth, glossy scales, a head slightly broader than the neck, and small eyes with vertically elliptical pupils. It is olive-brown to yellow-green above, and yellow to brown below. The dorsum bears four lateral rows of black spots, and the head is as the body.

FAMILY	Sibynophiidae
RISK FACTOR	Nonvenomous
DISTRIBUTION	North and Central America: southern Mexico, Belize, Guatemala, Honduras, El Salvador, and Nicaragua
ELEVATION	0–5,090 ft (0–1,550 m) asl
HABITAT	Lowland and low montane wet and dry forest, and gallery forest
DIET	Lizards, especially skinks
REPRODUCTION	Oviparous, with clutches of 1–10 eggs
CONSERVATION STATUS	IUCN Least Concern

ADULT LENGTH
27½–36¼ in
(700–920 mm)

SCAPHIODONTOPHIS ANNULATUS
GUATEMALAN SPATULA-TOOTHED SNAKE
(DUMÉRIL, BIBRON & DUMÉRIL, 1854)

The Guatemalan Spatula-toothed Snake is a slender snake with a long tail, smooth scales, and a head only slightly distinct from the neck, with large eyes and round pupils. Most of the body and tail is brown above, either unicolor or with three fine longitudinal lines of small dark spots. The anterior body is brightly colored, being red with white-black-white or yellow-black-yellow bands, while the head may be brown, red, or black, with a yellow throat and lips.

The Guatemalan Spatula-toothed Snake, sometimes called the Neckband Snake or Half Coralsnake, is widely distributed from central Mexico to Nicaragua. A secretive, diurnal, terrestrial to semi-fossorial snake, it is found in leaf litter or under logs. It hunts lizards, especially smooth-scaled skinks, which possess tough osteoderms (bony plates) in the skin. The Guatemalan Spatula-toothed Snake has evolved specialized, hinged, spatula-like teeth to maintain purchase on a robust, writhing skink, which it then swallows in a few seconds. Many specimens have truncated tails, a possible sign that this species uses the tactic of caudal pseudautotomy to avoid predation, sacrificing part of its tail to save its life.

RELATED SPECIES

A second species of *Scaphiodontophis*, the Common Spatula-toothed Snake (*S. venustissimus*), occurs in southern Central America and Colombia. *Scaphiodontophis* and the Asian black-headed snakes, genus *Sibynophis* (page 259), comprise the small family Sibynophiidae, which some authors treat as a subfamily of the Colubridae.

Actual size

FAMILY	Sibynophiidae
RISK FACTOR	Nonvenomous
DISTRIBUTION	South and Southeast Asia: northern India, Nepal, southwestern China, Bhutan, Bangladesh, Myanmar, Thailand, Vietnam, Cambodia, and Peninsular Malaysia
ELEVATION	0–10,800 ft (0–3,280 m) asl
HABITAT	Lowland and montane forest, preferably with dense vegetation
DIET	Lizards, frogs, snakes, and insects
REPRODUCTION	Oviparous, with clutches of 4–6 eggs
CONSERVATION STATUS	IUCN Least Concern

ADULT LENGTH
30–33½ in
(760–850 mm)

SIBYNOPHIS COLLARIS
COLLARED BLACK-HEADED SNAKE
(GRAY, 1853)

259

The Collared Black-headed Snake is found from northwest India, through the Himalayan foothills to Myanmar and southwestern China, and south through Southeast Asia to Peninsular Malaysia. It occurs in lowland forests but also to over 9,840 ft (3,000 m) in montane forest, usually in dense vegetation. It may be diurnal or nocturnal, and is terrestrial to semi-fossorial in habit. Prey consists of lizards, especially skinks, for which it possesses specialized dentition, as well as frogs, other snakes, and, in the case of juveniles, insects. When specimens are handled they do not bite, but instead coil around the fingers and constrict. It will also exude a cloacal secretion, which is said to smell of tobacco.

RELATED SPECIES

The genus *Sibynophis* is the type genus of the Sibynophiidae. Eight other species of Asian black-headed snakes are described. Peninsular India is occupied by the Indian Black-headed Snake (*S. subpunctatus*), while the Arrow Black-headed Snake (*S. sagittarius*) also occurs in northeast India. The remaining species occur in Southeast Asia, China, Indonesia, and the Philippines. Some authors treat the Sibynophiidae as a subfamily of the Colubridae.

The Collared Black-headed Snake is a slender snake with smooth scales, a long tail, and a head no wider than the neck, with small eyes and round pupils. Dorsally it is brown or gray-brown, with a black vertebral line of spots. Ventrally it is yellow, while the head is gray above with a faint darker band posteriorly, the labials are white, and a broad black collar, edged posteriorly with white, is present posterior to the head.

Actual size

FAMILY	Dipsadidae: Carphophiinae
RISK FACTOR	Nonvenomous
DISTRIBUTION	North America: central USA
ELEVATION	195–2,000 ft (60–610 m) asl
HABITAT	Rocky wooded hillsides and wooded streamsides
DIET	Earthworms
REPRODUCTION	Oviparous, with clutches of 1–12 eggs
CONSERVATION STATUS	IUCN Least Concern

ADULT LENGTH
8–15½ in
(200–390 mm)

CARPHOPHIS VERMIS
WESTERN WORMSNAKE
(KENNICOTT, 1859)

260

The Western Wormsnake occurs in the Midwest USA from southern Iowa, western Wisconsin and Illinois, south to northwestern Texas and northern Louisiana. It is a secretive snake that inhabits rocky, wooded hillsides, and although not an open-country species, it may be found along forested streams that extend into prairie habitats. This is a semi-fossorial species that burrows into damp soil, the habitat of earthworms, its primary prey. It does not survive long when its forested habitat is cleared, being especially vulnerable to desiccation due to excessive water loss through its semipermeable skin. It is unable to burrow into already dry soil. Although earthworms form the majority of the diet, the Western Wormsnake also feeds on soft-bodied insect larvae and even small snakes, such as Ringneck Snakes (*Diadophis punctatus*, page 262).

RELATED SPECIES
A second species of wormsnake, the Eastern Wormsnake (*Carphophis amoenus*), occurs from New York to Illinois, and south to the Carolinas and Louisiana. The genus *Carphophis* is most closely related to the mudsnakes (*Farancia*, page 263). The Carphophiinae contains five North American genera, and one from Tibet.

The Western Wormsnake is a slender snake with smooth scales, and a head indistinct from the neck, with small, dark, round-pupilled eyes, and a short tail. It is glossy brown to dark gray dorsally, and yellowish, pinkish, or light brown on the venter and lower flanks.

Actual size

FAMILY	Dipsadidae: Carphophiinae
RISK FACTOR	Nonvenomous
DISTRIBUTION	North America: western USA and Canada
ELEVATION	0–6,890 ft (0–2,100 m) asl
HABITAT	Moist woodland, riparian habitats, open prairies, grasslands and meadows, rocky slopes, and suburban gardens
DIET	Slugs
REPRODUCTION	Oviparous, with clutches of 2–9 eggs
CONSERVATION STATUS	IUCN Least Concern

ADULT LENGTH
17¾–19 in
(450–483 mm)

CONTIA TENUIS
SHARPTAIL SNAKE
(BAIRD & GIRARD, 1852)

261

The Sharptail Snake is distributed from southwest British Columbia, including Vancouver Island, where it is one of only four snake species found, and south to central California. This species occurs in a wide variety of habitats, from moist coniferous and pine-oak woodlands to open grasslands and prairie, and cool montane meadows. It is also found on rocky slopes and in suburban gardens. It may be found under rocks or inside rotten logs where the conditions are damp, but it may venture deeper for hibernation. It uses its sharp caudal spine to force itself forward through the soil. This is a secretive and inoffensive species and its diet appears to consist almost entirely of slugs, or their eggs. Some authors report earthworms, insects, and salamanders in their diet, but this is disputed.

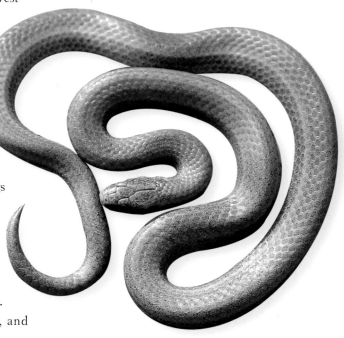

Actual size

The Sharptail Snake is a slender snake with smooth scales, a rounded head that is indistinct from the neck, small eyes with round pupils, and a short tail. It is brown above, gray on the flanks, and white on the undersides, every ventral scale being edged with black. A pale longitudinal line demarcates the dorsolateral margin. The snout tip and lips are also white.

RELATED SPECIES

A related species, the Forest Sharptail Snake (*Contia longicaudae*), also occurs in California and Oregon, and exhibits a longer tail and much more reduced black ventral markings than *C. tenuis*. The genus *Contia* is most closely related to the American hognose snakes (*Heterodon*, page 264).

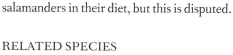

FAMILY	Dipsadidae: Carphophiinae
RISK FACTOR	Rear-fanged, mildly venomous; harmless to humans
DISTRIBUTION	North America and West Indies: southeast Canada, USA, and northern Mexico; introduced into the Cayman Islands
ELEVATION	0–7,870 ft (0–2,400 m) asl
HABITAT	Woodland, chaparral, rocky valleys, and railway lines
DIET	Salamanders, small frogs, small lizards, other snakes, earthworms, insect larvae, and slugs
REPRODUCTION	Oviparous, with clutches of 1–10 eggs, 18 in *D. p. regalis*
CONSERVATION STATUS	IUCN Least Concern

ADULT LENGTH
9¾–23¾ in,
rarely 34 in
(250–600 mm,
rarely 860 mm)

262

DIADOPHIS PUNCTATUS
RINGNECK SNAKE
(LINNAEUS, 1766)

The Ringneck Snake is a small snake with a rounded head, small eyes with round pupils, glossy, smooth scales, and a short tail. The dorsum is usually brown, gray, or blue-gray, but depending on subspecies, the venter may be yellow, orange, or red, while the throat is white, with a black-edged ring around the neck that is the same color as the venter. The undersides of some specimens are immaculate, others lightly spotted, while some subspecies exhibit a distinct line of midventral dark spots. Southern populations often lack the neck band.

The Ringneck Snake is one of the most widely distributed North American snakes, occurring from Nova Scotia in Canada to San Luis Potosí, eastern Mexico, through eastern and central USA, and on the western seaboard from Washington to Baja California. It has been introduced to Grand Cayman Island. Most Ringnecks are small (less than 23¾ in/600 mm), but the Regal Ringneck (*Diadophis punctatus regalis*), from Utah, may achieve 34 in (860 mm). Ringnecks inhabit woodland and rocky canyons, and sometimes occur in aggregations of over 100 individuals. They are also found in the scree between railway sleepers. Defensively they may display the tightly coiled, inverted, brightly colored tail, although the purpose of this display is not fully understood. This species is mildly venomous, and some people experience localized effects following bites.

RELATED SPECIES

The genus *Diadophis* is monotypic, but *D. punctatus* contains 7 to 14 geographical subspecies, some of which are difficult to tell apart. Some authors consider *D. punctatus* to be two species. The genus *Diadophis* is the sister clade to wormsnakes (*Carphophis*, page 260) and the Rainbow Snake and Mudsnake (*Farancia*, page 263).

Actual size

FAMILY	Dipsadidae: Carphophiinae
RISK FACTOR	Rear-fanged, mildly venomous; harmless to humans
DISTRIBUTION	North America: southeast USA
ELEVATION	0–490 ft (0–150 m) asl
HABITAT	Rivers, canals, lakes, cypress swamps, and coastal marshes
DIET	Fish (including eels), salamanders, sirens, frogs, and earthworms
REPRODUCTION	Oviparous, with clutches of 10–52 eggs
CONSERVATION STATUS	IUCN Least Concern, Endangered in Mississippi and Louisiana, possibly Extinct in Florida Peninsula

ADULT LENGTH
4 ft 7 in–5 ft 2 in
(1.4–1.6 m)

FARANCIA ERYTROGRAMMA
RAINBOW SNAKE
PALISOT DE BEAUVOIS, 1802

263

The Rainbow Snake occurs from southern Maryland to northern Florida and west to Louisiana, with a small isolated population in the Florida Peninsula. It is found in slow-moving freshwater habitats including rivers, lakes, cypress swamps, and canals, and it also occurs in brackish habitats, such as coastal marshes, mudflats, and tidal creeks. The female is oviparous, excavating a cavity in the soil for her eggs, and she may remain with them during incubation. This species is nocturnal in habit and feeds on aquatic vertebrates such as salamanders, sirens, frogs, and fish including eels, with earthworms also being taken. The enlarged rear teeth and Duvernoy's glands of the Rainbow Snake are used to subdue eels. If handled the Rainbow Snake does not bite, but may probe harmlessly with its caudal spine.

The Rainbow Snake is a slender, glossy, smooth-scaled snake with a rounded head, small eyes with round pupils, and a long tail with a spinous tip. Its color is iridescent blue-black to purple, with three longitudinal red stripes and red edging to the dorsal head scales. The lips and chin are yellow, and the venter is red to yellow with three rows of black spots.

RELATED SPECIES

Two subspecies of *Farancia erytrogramma* are known. The nominate Northern Rainbow Snake (*F. e. erytrogramma*) occupies the majority of the range, while the severely endangered South Florida Rainbow Snake (*F. e. seminola*) is only known from Fisheating Creek, Glades County, Florida. It has not been seen since 1952 and is the subject of a citizen science appeal for sightings. A second species of *Farancia* is also recognized, the Mudsnake (*F. abacura*), which has an eastern subspecies (*F. a. abacura*) and a western subspecies (*F. a. reinwardtii*), both in the United States.

Actual size

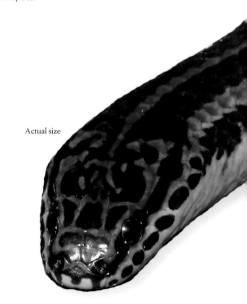

FAMILY	Dipsadidae: Carphophiinae
RISK FACTOR	Rear-fanged, mildly venomous
DISTRIBUTION	North America: northeastern Mexico and extreme southern Texas, New Mexico, and Arizona
ELEVATION	0–8,010 ft (0–2,440 m) asl
HABITAT	Woodland, flood plains, creosote desert, prairie, cultivated land
DIET	Toads, lizards, frogs, salamanders, turtles, reptile eggs, small birds, and small mammals
REPRODUCTION	Oviparous, with clutches of 4–25 eggs
CONSERVATION STATUS	IUCN not listed

ADULT LENGTH
15¾–23¾ in,
occasionally 30 in
(400–600 mm,
occasionally 760 mm)

HETERODON KENNERLEYI
MEXICAN HOGNOSE SNAKE
KENNICOTT, 1860

264

The Mexican Hognose Snake is a stocky snake with keeled scales, a short tail, and a short head that terminates in the upturned rostral scale of the snout. Its dorsal coloration is tan-brown with patterning comprising rows of black-edged dark brown blotches that become bands on the tail, and a series of dark chevron markings on the head. The venter is off-white, heavily speckled with black, and orange or yellow under the tail.

The Mexican Hognose Snake is found across northeastern Mexico and also enters the southern USA, in the border region from Texas to Arizona. It likes sandy or gravelly soils into which it can burrow, and it occurs in many habitats from farmland to woodland, floodplains, and creosote bush desert. It preys mostly on toads, with frogs, lizards, salamanders, reptile eggs, small mammals, birds, and even occasional turtles, also taken. The turned-up snout is used for excavating prey from the soil. Toads that inflate themselves, to avoid predation, are deflated by the enlarged rear teeth and subdued by the snake's Duvernoy's gland secretions. Hognose snakes rely on bluff and rarely bite, but some people do experience localized reactions to their secretions, so bites should be avoided. They also employ thanatosis, rolling upside down with their mouths agape, tongue protruding loosely, as if dead.

RELATED SPECIES

Four other North American hognose snakes are known: the Eastern Hognose Snake (*Heterodon platirhinos*), from Ontario to Florida and Texas; the Western Hognose Snake (*H. nasicus*), from Alberta to Manitoba, and south to Texas; the Dusky Hognose Snake (*H. gloydi*) in Texas and Oklahoma; and the Southern Hognose Snake (*H. simus*) from Lousiana to the Carolinas and Florida. The Mexican Hognose Snake is a former subspecies of the Western Hognose Snake.

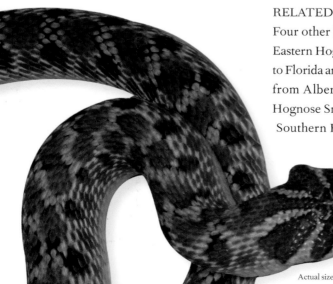

Actual size

FAMILY	Dipsadidae: Carphophiinae
RISK FACTOR	Rear-fanged, mildly venomous
DISTRIBUTION	Central Asia: Tibet
ELEVATION	9,840–16,100 ft (3,000–4,900 m) asl
HABITAT	Hot-spring rivulets, pools, meadows, and marshes
DIET	Small frogs and fish
REPRODUCTION	Believed to be oviparous, with clutches of up to 6 eggs; possibly viviparous
CONSERVATION STATUS	IUCN Near Threatened

ADULT LENGTH
17¾–24 in
(450–610 mm)

THERMOPHIS BAILEYI
XIZANG HOT-SPRING SNAKE
(WALL, 1907)

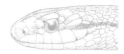

265

The Xizang Hot-spring Snake, or Tibetan Hot-spring Snake, is a high-elevation species occurring in the Lhasa region. It inhabits the streams, ponds, meadows, and marshes surrounding the hot springs of the Tibetan Plateau. Poorly known, it is believed to feed on the small alpine frogs and fish that also inhabit this barren landscape. Although it is believed to be oviparous it may be viviparous, a safer strategy for snakes living at high elevations. According to the IUCN, this species is threatened by the development of geothermal hydroelectric stations and habitat disturbance. This species was named in honor of Lieutenant Colonel Frederick Bailey (1882–1967), a British officer and spy.

The Xizang Hot-spring Snake is a slender snake with a long tail and keeled scales. The head is slightly pointed, and the eyes relatively small with round pupils. Its dorsal coloration is olive-green to brown with transverse rows of dark brown spots, the middorsal series forming a punctuated longitudinal stripe. The undersides and lips are pale yellow.

RELATED SPECIES

The genus *Thermophis* also contains the Shangri-La Hot-spring Snake (*T. shangrila*), from Yunnan, China, and the Endangered Sichuan Hot-spring Snake (*T. zhaoermii*). These snakes are the only known non-American members of the Dipsadidae and may illustrate the radiation of snake families across the Bering land bridge between Siberia and Alaska during the Quaternary glacial period. Their closest relative is the Ringneck Snake (*Diadophis punctatus*, page 262).

Actual size

FAMILY	Dipsadidae: Dipsadinae
RISK FACTOR	Nonvenomous
DISTRIBUTION	North America: southeast Mexico
ELEVATION	0–6,230 ft (0–1,900 m) asl
HABITAT	Lowland and low montane wet tropical forest, upland pine forest, coffee and sugar plantations
DIET	Earthworms and soft-bodied invertebrates
REPRODUCTION	Oviparous, with clutches of 3–5 eggs
CONSERVATION STATUS	IUCN Least Concern

ADULT LENGTH
11–15¾ in,
rarely 20½ in
(280–400 mm,
rarely 520 mm)

ADELPHICOS QUADRIVIGATUM
VERACRUZ EARTHSNAKE
JAN, 1862

The small Veracruz Earthsnake occurs from the Mexican state of Tamaulipas to Veracruz. It is an inhabitant of lowland and low montane wet forest and upland pine forest, where it shelters and hunts under rotten logs or in decaying leaf litter. It is also found in coffee plantations, in the dense leaf litter at the bases of trees, and in sugar plantations under logs. It is not found in plantations that engage in regular burns of the ground cover. The Veracruz Earthsnake feeds on earthworms and soft-bodied invertebrates, and is, itself, predated by ophiophagous snakes, such as coralsnakes (*Micrurus*, pages 457–465). Eggs are frequently laid in termite mounds, including the arboreal nests of tree-dwelling termites.

The Veracruz Earthsnake is small with glossy, smooth scales; a small head that is indistinct from the neck; small, dark eyes, and a short tail. The dorsal coloration is olive-green or red-brown with a series of dark brown longitudinal stripes, the lateral stripes being bolder than the vertebral ones, and all scales are speckled with dark spots. The undersides are yellow, sometimes pink, and also heavily infused with dark spots.

RELATED SPECIES

There are three subspecies recognized, although some authors treat them as species. The genus *Adelphicos* also contains another six species from Oaxaca, Mexico (*A. latifasciatum* and *A. visoninum*), Chiapas, Mexico (*A. nigrilatum*), Guatemala (*A. daryi* and *A ibarrorum*), and both Mexico and Guatemala (*A. sargii* and *A. veraepacis*). Although *A. quadrivigatum* is not considered threatened, several of the more localized species are listed as Vulnerable or Endangered by the IUCN.

Actual size

FAMILY	Dipsadidae: Dipsadinae
RISK FACTOR	Nonvenomous
DISTRIBUTION	North and Central America: Mexico, Belize, Guatemala, and Honduras
ELEVATION	0–5,250 ft (0–1,600 m) asl
HABITAT	Lowland and low montane wet tropical forest, and coffee plantations
DIET	Frogs, small lizards, and centipedes
REPRODUCTION	Probably oviparous, clutch size unknown
CONSERVATION STATUS	IUCN Least Concern

ADULT LENGTH
13¾–15¾ in,
rarely 28¾ in
(350–400 mm,
rarely 730 mm)

AMASTRIDIUM SAPPERI
NORTHERN RUSTYHEAD SNAKE
(WERNER, 1903)

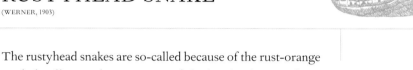

267

The rustyhead snakes are so-called because of the rust-orange nuchal collar on the posterior of the head. The Northern Rustyhead Snake occurs from Tamaulipas to Oaxaca in Mexico, and in Belize, Guatemala, and Honduras, but not the Yucatán Peninsula. This small diurnal snake has been collected in wet tropical forest, on streamsides, and in coffee plantations, where it takes advantage of any ground cover to hide or hunt. Specimens have contained small frogs, lizards, and centipedes, but little is known of their natural behavior. This species' reproductive strategy is unknown but its congener, the Southern Rustyhead Snake (*Amastridium veliferum*), is oviparous. Karl Theodor Sapper (1866–1945) was a German volcanologist, linguist, and explorer who traveled widely in Mexico and Central America during the late nineteenth century.

RELATED SPECIES

The Southern Rustyhead Snake (*Amastridium veliferum*) occurs from Costa Rica and Panama to Colombia. Some authors consider *Amastridium* to be monotypic and reduce *A. sapperi* to a subspecies of *A. veliferum*.

The Northern Rustyhead Snake is a small snake with smooth scales, a squarish head, small eyes with round pupils, and a long tail. It is black or dark brown in color, although the dorsum of the head may be lighter brown or gray. Faint longitudinal stripes may be visible in some specimens. Its most obvious feature is the rust-orange nuchal collar that contrasts strongly with the dorsal color and provides the species with their common names.

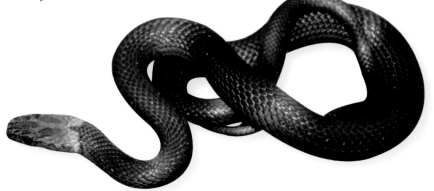

Actual size

FAMILY	Dipsadidae: Dipsadinae
RISK FACTOR	Nonvenomous
DISTRIBUTION	South America: Ecuador and Peru
ELEVATION	1,970–7,550 ft (600–2,300 m) asl
HABITAT	Humid tropical forest, cloud forest, secondary growth, and coffee plantations
DIET	Earthworms and possibly small mammals
REPRODUCTION	Oviparous, with clutches of up to 12 eggs
CONSERVATION STATUS	IUCN not listed

ADULT LENGTH
10–41 in,
occasionally 3 ft 7 in
(255–1,040 mm,
occasionally 1.1 m)

268

ATRACTUS GIGAS
GIANT ARROW EARTHSNAKE
MYERS & SCHARGEL, 2006

The Giant Arrow Earthsnake is a robust species with smooth scales, a rounded head that is hardly distinct from the neck, small eyes with round pupils, and a short tail. Adults may be uniform reddish brown or exhibit a reticulate pattern, every scale being spotted with brown on a pale background. Juveniles are dark with a series of irregular transverse reddish cross-bands, and yellowish infusions on the head scales. This species may possess a checkerboard or heavily spotted venter. Some specimens are more melanistic, a possible adaptation for life at higher elevations.

Actual size

The Giant Arrow Earthsnake is the largest known species of the largest snake genus, and inhabits the Ecuadorian and Peruvian Andes. Described from a single 3 ft 5 in (1.04 m) female specimen as recently as 2006, it is now known from over a dozen specimens. The largest male found so far was a 10 in (255 mm) subadult. The Giant Arrow Earthsnake inhabits montane humid forest, cloud forest, and secondary growth, but is also found in plantations. It is semi-fossorial, in leaf litter or under fallen logs, but may also be seen crossing trails in the early morning or late afternoon. The diet of this moderately large, docile snake is poorly known. It is reported to feed on earthworms but the feces of one specimen contained small mammal remains.

RELATED SPECIES

The South American genus *Atractus* is the most specious snake genus with 143 species currently described. The name *Atractus* means "arrow." The genus exhibits considerable diversity and complexity. *Atractus* just enters Central America in eastern Panama, being replaced by the closely related genus *Geophis* (pages 277–278) in Central America.

FAMILY	Dipsadidae: Dipsadinae
RISK FACTOR	Nonvenomous
DISTRIBUTION	South America: Colombia, Venezuela, Suriname, French Guiana, Brazil, Peru, and Bolivia
ELEVATION	295–1,640 ft (90–500 m) asl
HABITAT	Primary lowland rainforest, riverine forest, and recently deforested areas
DIET	Earthworms and insects
REPRODUCTION	Oviparous, with clutches of 3–6 eggs
CONSERVATION STATUS	IUCN not listed

ADULT LENGTH
20½–24 in
(520–610 mm)

ATRACTUS LATIFRONS
BROADHEAD ARROW EARTHSNAKE
(GÜNTHER, 1868)

The Broadhead Arrow Earthsnake, or the Wedgetail Earthsnake, is an Amazonian member of the largest snake genus, *Atractus*. It has been recorded from every country in northern South America except Ecuador and Guyana. Its preferred habitat comprises pristine lowland primary rainforest or riverine gallery forest, but it is found more easily in recently cleared forest plots. It is semi-fossorial, living in the forest-floor leaf litter, where it feeds on earthworms and insects. It is a harmless but polymorphic species and caution is advised with specimens whose patterning resembles the highly venomous coralsnakes (*Micrurus*, pages 457–465).

RELATED SPECIES

Tricolor *Atractus latifrons* and some of its congeners, including the Black Arrow Earthsnake (*A. elaps*) and the Guianan Arrow Earthsnake (*A. badius*), resemble other Amazonian coral-mimics such as the Aesculapian False Coralsnake (*Erythrolamprus aesculapii*, page 311) and the true coralsnakes.

The Broadhead Arrow Earthsnake is a slender species with smooth scales, a slightly pointed head, small eyes, round pupils, and a relatively short tail. It is polymorphic, occurring in several color morphotypes, ranging from bicolor (black with narrow yellow rings), to tricolor (red with black-yellow-black or black-yellow-black-yellow-black rings), to virtually unicolor with rings only present on the head and tail. Juveniles are more brightly marked, while in old adults the markings may be obscured by dark pigment.

Actual size

FAMILY	Dipsadidae: Dipsadinae
RISK FACTOR	Nonvenomous
DISTRIBUTION	North America: Central Mexico
ELEVATION	5,410–6,040 ft (1,650–1,840 m) asl
HABITAT	Cloud forest and pine–oak forest on karst limestone
DIET	Ants and beetle larvae
REPRODUCTION	Oviparous, clutch size unknown
CONSERVATION STATUS	IUCN Endangered

ADULT LENGTH
5–13¾ in
(130–350 mm)

CHERSODROMUS RUBRIVENTRIS
RED-BELLIED EARTHRUNNER
(TAYLOR, 1949)

The Red-bellied Earthrunner is confined to the Sierra Madre Occidental, in the Mexican states of Querétaro, San Luis Potosí, and Hidalgo, where it inhabits cloud forest and pine–oak forest, on rocky karst slopes at relatively high elevations. It is considered Endangered by the IUCN due to habitat alteration and fragmentation caused by agricultural practices. This is a rare and secretive snake that is infrequently encountered, but it is most likely to be found under rotting logs, boulders, and in leaf litter during the day. Stomach contents of specimens have revealed a diet of beetle larvae and adult ants, which are probably foraged at night under ground cover, although it is possible the Red-bellied Earthrunner may also forage on the surface.

The Red-bellied Earthrunner is small with smooth, shiny scales, a squarish head, small eyes, and a short tail. It is black above and red on the ventral surfaces, with a broad yellow collar around the neck, which continues onto the throat. The chin is black.

RELATED SPECIES

A second species is recognized, Leibmann's Earthrunner (*Chersodromus liebmanni*), a more abundant species from Oaxaca and Veracruz, Mexico. The nearest relatives to *Chersodromus* may be the coffee snakes (*Ninia*, page 283) of Central America, which both look similar and appear to exhibit a parallel natural history.

Actual size

FAMILY	Dipsadidae: Dipsadinae
RISK FACTOR	Rear-fanged, mildly venomous; harmless to humans
DISTRIBUTION	North and Central America: Mexico, Guatemala, El Salvador, Honduras, Nicaragua, and Costa Rica
ELEVATION	0–4,270 ft (0–1,300 m) asl
HABITAT	Dry tropical and wet tropical deciduous and evergreen forests
DIET	Frogs, lizards, small snakes, and reptile eggs
REPRODUCTION	Oviparous, with clutches of 1–6 eggs
CONSERVATION STATUS	IUCN Least Concern

ADULT LENGTH
19¾–22½ in
(500–570 mm)

CONIOPHANES PICEIVITTIS
MESOAMERICAN STRIPED SNAKE
COPE, 1870

271

The Mesoamerican Striped Snake, also called Cope's Striped Snake, is found from northern Mexico to Costa Rica and is recorded from every country in between except Belize, being replaced on the Yucatán Peninsula by the related and similarly patterned Schmidt's Striped Snake (*Coniophanes schmidti*). Preferred habitats include lowland and low montane wet and dry tropical forest, both deciduous and evergreen. This is a terrestrial species that hunts frogs, lizards, and smaller snakes, such as blindsnakes, both diurnally and nocturnally, but it is not commonly encountered, being most likely to be found under rocks or rotten logs. It is also documented to forage for reptile eggs. Although members of the genus *Coniophanes* are rear-fanged, their venom is mild and they are not considered dangerous to humans.

The Mesoamerican Striped Snake is a small, smooth-scaled snake with a pointed head, distinct from the neck, small eyes, and a long tail. It is black above with a pair of yellow dorsolateral stripes running from the snout, over the eyes, and along the body onto the tail. The undersides are pale brown, this coloration extending onto the lower flanks.

RELATED SPECIES
Two subspecies are recognized, the nominate form (*Coniophanes piceivittis piceivittis*) from Mexico to Costa Rica, and a northern form (*C. p. frangivirgatus*) from Veracruz and Tamaulipas, Mexico. A further 16 species of *Coniophanes* are found in Mexico, Central America, and on inshore islands in the western Caribbean. *Coniophanes piceivittis* occurs in sympatry with the Yellow-bellied Snake (*C. fissidens*), the Spot-bellied Snake (*C. bipunctatus*), and the Regal Black-striped Snake (*C. imperialis*). *Coniophanes* is closely related to the brownsnakes (*Rhadinaea*, page 287).

Actual size

FAMILY	Dipsadidae: Dipsadinae
RISK FACTOR	Nonvenomous
DISTRIBUTION	Central and South America: Panama, Colombia, and Ecuador
ELEVATION	330–4,990 ft (100–1,520 m) asl
HABITAT	Lowland and low montane evergreen forests and clearings
DIET	Undetermined, possibly frogs or lizards
REPRODUCTION	Oviparous, clutch size unknown
CONSERVATION STATUS	IUCN Data Deficient, locally endangered

ADULT LENGTH
19¾–27 in
(500–690 mm)

272

DIAPHOROLEPIS WAGNERI
HUMPBACK TERRIER SNAKE
JAN, 1863

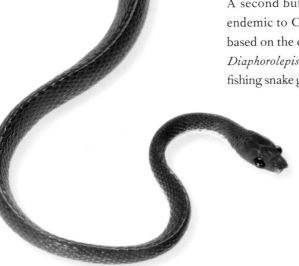

Actual size

The curious common name of this species is a translation from the Spanish "Culebra terrier jorobada." The only other name available, "Ecuadorian Frog-eating Snake," may be misleading as authors do not agree on whether this species preys on frogs or lizards. The most distinctive characteristic in this extremely rare snake is its enlarged, bicarinate vertebral scale row. It occurs in the Pacific versant of the Ecuadorian and Colombian Andes, and also in the Darién region of eastern Panama. It is a highly agile inhabitant of lowland and low montane evergreen forests, especially near water, and is reportedly most often encountered after rain, but authors differ as to whether it is diurnal or nocturnal. This species is named after the German naturalist and traveler Moritz Wagner (1813–87), who collected the holotype.

RELATED SPECIES

A second but even rarer species (*Diaphorolepis laevis*) is endemic to Colombia and may have smooth dorsal scales, based on the epithet *laevis* (Latin for "smooth"). The genus *Diaphorolepis* is most closely related to the South American fishing snake genus *Synophis* (page 293).

The Humpback Terrier Snake is a slender snake with strongly keeled scales, which are bicarinate on the vertebral row. Its square-nosed head is distinct from the narrow neck, the eyes are large with round pupils, and it has a very long tail. It is brown on the dorsum, and bright yellow on the venter, throat, and supralabials, with a sharp demarcation between the two colors on the outer ventral scales, and brown pigmentation midventrally, the underside of the tail being completely brown.

FAMILY	Dipsadidae: Dipsadinae
RISK FACTOR	Nonvenomous
DISTRIBUTION	North and Central America: Mexico and Belize
ELEVATION	0–985 ft (0–300 m) asl
HABITAT	Dry tropical thorn forest and seasonally wet forest
DIET	Snails and slugs
REPRODUCTION	Oviparous, with clutches of 2–5 eggs
CONSERVATION STATUS	IUCN Least Concern

ADULT LENGTH
14½–20¾ in
(370–528 mm)

DIPSAS BREVIFACIES
YUCATÁN THIRST SNAKE
(COPE, 1866)

273

Central and South America contain 36 species of "thirst snakes" or "snail-suckers." The Yucatán or Short-faced Thirst Snake is the most northerly species, ocurring on the Yucatán Peninsula of Mexico and Belize. Arboreal and nocturnal, it is common in dry thorn forests. Thirst snakes feed exclusively on slugs and snails. Slugs are swallowed tail first, but snails must be extracted from their shells. This is accomplished by holding the shell firmly in the coils, inserting the lower jaw and hooking the snail with the teeth. Then, applying pressure to the shell, the lower jaw is withdrawn with the prize, a process assisted by infralabial gland secretions that subdue the snail.

RELATED SPECIES

No other *Dipsas* occur within the range of this species, but it may be confused with Sartorius' Terrestrial Snail-sucker (*Tropidodipsas sartorii*, page 296), with which it occurs in sympatry. That species has weakly keeled scales and a mental groove, but *Dipsas* has smooth scales and lacks a mental groove.

The Yucatán Thirst Snake is an extremely slender, laterally compressed, smooth-scaled snake, with a broad, rounded head, large, bulbous eyes, vertically elliptical pupils, and a long tail. It is shiny black dorsally and ventrally with regularly spaced broad salmon-pink to red bands, half to one-third the width of the black interspaces, the first red band forming a broad nuchal collar.

Actual size

FAMILY	Dipsadidae: Dipsadinae
RISK FACTOR	Nonvenomous
DISTRIBUTION	South America: Colombia, Venezuela, the Guianas, Brazil, Ecuador, Peru, Bolivia, Paraguay, and northeastern Argentina
ELEVATION	0–3,280 ft (0–1,000 m) asl
HABITAT	Amazonian rainforest
DIET	Snails and slugs
REPRODUCTION	Oviparous, with clutches of 2–6 eggs
CONSERVATION STATUS	IUCN not listed

ADULT LENGTH
23¾–31½ in,
rarely 35¾ in
(600–800 mm,
rarely 910 mm)

274

DIPSAS INDICA
SOUTH AMERICAN THIRST SNAKE
LAURENTI, 1768

The South American Thirst Snake is a slender snake with smooth scales, a laterally compressed body, and a long, prehensile tail. The head is rounded, with large eyes and vertically elliptical pupils. This species is dorsally light brown with a series of irregular dark brown saddle markings and a scattering of white spots. Some specimens are almost unicolor. The undersides are immaculate white.

The South American Thirst Snake may be the mostly widely distributed member of the genus *Dipsas*. It may also be one of the largest, with specimens approaching 3 ft 3 in (1 m) known. It is found throughout northern South America, east of the Andes, from the Caribbean coast to northeastern Argentina, but it is not commonly encountered. The infrequency of encounters may be as a result of this species being highly arboreal, often inhabiting the canopy, and nocturnal in the heavily forested Amazon and Guianan forests. As with other thirst snakes, it takes both slugs and snails, but given its larger size it can predate larger snails than its congeners. The method by which snails are extracted from their shells is explained for the Yucatán Thirst Snake (*D. brevifacies*, page 273).

RELATED SPECIES

The nominate subspecies (*Dipsas indica indica*) occupies most of the range, while the Ecuadorian population is treated as a separate subspecies (*D. i. ecuadoriensis*). Peter's Snail-eater (*D. petersi*), from southeastern Brazil, is also a former subspecies. Thirst snakes, also known as snail-eaters or snail-suckers, are members of a global guild of snakes referred to by herpetologists as "goo-eaters."

Actual size

FAMILY	Dipsadidae: Dipsadinae
RISK FACTOR	Rear-fanged, mildly venomous; harmless to humans
DISTRIBUTION	Central and South America: Honduras, Nicaragua, Costa Rica, Panama, and Colombia
ELEVATION	0–4,220 ft (0–1,285 m) asl
HABITAT	Lowland moist forest, evergreen forest, gallery forest, dry forest, and savanna
DIET	Frogs, lizards, small snakes, and reptile eggs
REPRODUCTION	Oviparous, with clutches of 2–5 eggs
CONSERVATION STATUS	IUCN not listed

ADULT LENGTH
19¾–21¼ in
(500–550 mm)

ENULIOPHIS SCLATERI
COLOMBIAN LONGTAIL SNAKE
(BOULENGER, 1894)

275

The Colombian Longtail Snake occurs in Honduras, Nicaragua, Costa Rica, Panama, and Colombia, in lowland wet forests, evergreen forest, and the gallery forests along rivers. It is also found in dry forest and arid savannas. It is active by both day and night but is a secretive species that inhabits forest floor leaf litter and rotten logs. The prey of the Colombian Longtail Snake comprises reptiles and amphibians, and also reptile eggs, which are punctured by the enlarged rear fangs. The long tail of this species is extremely fragile and is often broken. Females are oviparous and lay between two and five eggs. Although mildly venomous and rear-fanged, this species is harmless to humans. Philip Lutley Sclater (1829–1913) was a British ornithologist and biogeographer.

The Colombian Longtail Snake is long and slender with smooth scales and a rounded head, which is broader than the neck, and a long tail. It is dark gray to black with a distinctive white head interrupted by black pigment around the eyes and on the snout tip.

RELATED SPECIES

Enuliophis sclateri was previously included in the genus *Enulius* (page 276), and some authors still recognize that taxonomic arrangement.

Actual size

FAMILY	Dipsadidae: Dipsadinae
RISK FACTOR	Rear-fanged, mildly venomous, harmless to human
DISTRIBUTION	Central America: southwestern Mexico to Colombia
ELEVATION	0–5,910 ft (0–1,800 m) asl
HABITAT	Lowland and midmontane wet and dry tropical or deciduous forest, savanna, thornscrub
DIET	Termites, ants, and reptile eggs
REPRODUCTION	Oviparous, with clutches of 1–2 eggs
CONSERVATION STATUS	IUCN Least Concern

ADULT LENGTH
5¾–19¾ in
(400–500 mm)

ENULIUS FLAVITORQUES
PACIFIC LONGTAIL SNAKE
(COPE, 1869)

276

The Pacific Longtail Snake occurs from Jalisco, on the
southwestern coast of Mexico, through Central America to
Colombia, being found up to elevations of 5,910 ft (1,800 m)
asl, inhabiting both wet and dry forest. It also inhabits savannas
or thornscrub and is active both diurnally and nocturnally,
although it is a secretive, semi-fossorial species that inhabits
rotten logs and hides under rocks or in crevices in the soil.
It feeds on termites, ants, and their larvae, and also reptile eggs.
The scientific name suggests this species has a yellow color but
unicolor specimens lacking such a marking are common. The
tails of longtail snakes are thick but fragile and they are often
truncated. This is an oviparous snake that lays up to two eggs.

RELATED SPECIES

Three other species of *Enulius* are recognized: the Endangered
Roatán Longtail Snake (*E. ruatanensis*) and Guanaja Longtail
Snake (*E. bifoveatus*), both from the Islas de la Bahia, and
the Mexico Longtail Snake (*E. oligostichus*) from southern
Mexico. The related Colombian Longtail Snake (*Enuliophis
sclateri*, page 275) was once also included in
the genus *Enulius*. Three subspecies are
recognized, two in Mexico (*E. f. sumichrasti*
and *E. f. unicolor*), while the nominate
subspecies occurs in Central America.

The Pacific Longtail Snake is a small, extremely
slender snake with smooth glossy scales, a
rounded head slightly broader than the neck,
and a long tail. It is gray in color, darker above
than below, sometimes with a broad off-white or
yellow collar, but frequently lacking this marking.

Actual size

FAMILY	Dipsadidae: Dipsadinae
RISK FACTOR	Rear-fanged, mildly venomous; harmless to humans
DISTRIBUTION	Central America: Costa Rica and Panama
ELEVATION	50–6,940 ft (15–2,115 m) asl
HABITAT	Low montane and lowland rainforest, and rough grassland or cleared areas
DIET	Earthworms
REPRODUCTION	Oviparous, with clutches of 3–6 eggs
CONSERVATION STATUS	IUCN Least Concern

ADULT LENGTH
11¾–18 in
(300–460 mm)

GEOPHIS BRACHYCEPHALUS
SHORT-HEADED EARTHSNAKE
COPE, 1871

277

The Short-headed Earthsnake is found in central Costa Rica and western Panama, with a small disjunct population in eastern Panama, just east of the Canal, possibly attributed to this species. It is an inhabitant of low montane and lowland rainforest, although it is much more common at higher elevations than in the lowlands. It is also found in disturbed areas and low montane pastures. It may be found under rocks, logs, fallen branches, and other forest-floor debris. As with all other *Geophis* earthsnakes for which dietary preferences are known, this species feeds exclusively on earthworms. Although a rear-fanged, mildly venomous species, it is completely harmless to humans, does not bite, and could do no damage if it did. Females are egg-layers.

RELATED SPECIES

The genus *Geophis* takes over from the South American genus *Atractus* (pages 268–269) in Central America, with few of the 50 known species entering eastern Panama or the Chaco of Colombia. Colombian specimens of *G. brachycephalus* are regarded as representing a different species, the Colombian Earthsnake (*G. nigroalbus*).

The Short-headed Earthsnake is a small, keel-scaled snake with a blunt head, small eyes, and a short tail. Its coloration and patterning consist of a brown, gray, or black dorsum, with a fine white nuchal collar in juveniles, and either a pair of red-orange dorsolateral longitudinal stripes or orange-red blotches on the flanks that may form crossbars. The venter is immaculate white. This species has a shorter, more rounded head than some of its congeners.

Actual size

FAMILY	Dipsadidae: Dipsadinae
RISK FACTOR	Rear-fanged, mildly venomous; harmless to humans
DISTRIBUTION	North America: Mexico
ELEVATION	2,620–8,530 ft (800–2,600 m) asl
HABITAT	Low montane rainforest
DIET	Earthworms, possibly soft-bodied invertebrates
REPRODUCTION	Oviparous, with clutches of up to 4 eggs
CONSERVATION STATUS	IUCN Data Deficient

ADULT LENGTH
11¾–16½ in
(300–420 mm)

278

GEOPHIS LATIFRONTALIS
SAN LUIS POTOSÍ EARTHSNAKE
GARMAN, 1883

The San Luis Potosí Earthsnake is one of the northernmost members of the North and Central American earthsnake genus *Geophis*. It is found only in the Mexican states of San Luis Potosí, Hidalgo, and Querétaro. It inhabits low montane rainforest where it exists as a semi-fossorial burrower living under rotten logs, rocks, and forest-floor debris. It may also be found under trash discarded by humans. In common with its congeners it is an oviparous species with clutches of up to four eggs. It feeds on earthworms, and possibly other soft-bodied invertebrates. The San Luis Potosí Earthsnake defends itself by smearing noxious-smelling cloacal secretions on its perceived enemy. A mildly venomous rear-fanged species, it is of no threat to humans.

RELATED SPECIES

Two subspecies of *Geophis latifrontalis* are recognized, the unicolor nominate San Luis Potosí Earthsnake (*G. l. latifrontalis*) and the Red-banded Mountain Earthsnake (*G. l. semiannulatus*). A further 30 species of *Geophis* earthsnakes occur in Mexico but none enter the United States.

The San Luis Potosí Earthsnake is a small, smooth-scaled snake with a rounded head, small eyes, and a short tail. It may be uniform brown or gray above, with some specimens exhibiting body patterning comprising a series of orange or red-orange rings or broken bands. The venter is off-white but may be flecked with grayish brown.

Actual size

FAMILY	Dipsadidae: Dipsadinae
RISK FACTOR	Nonvenomous
DISTRIBUTION	Central America: Guatemala, Honduras, Nicaragua, Costa Rica, and Panama
ELEVATION	0–4,920 ft (0–1,500 m) asl
HABITAT	Streams and ponds in lowland and low montane rainforest
DIET	Possibly freshwater shrimps, fish, or frogs
REPRODUCTION	Oviparous, with clutches of 4–7 eggs
CONSERVATION STATUS	IUCN Least Concern

ADULT LENGTH
28¾–33½ in
(730–850 mm)

HYDROMORPHUS CONCOLOR
CENTRAL AMERICAN WATERSNAKE
PETERS, 1859

279

The widely distributed, but infrequently encountered, Central American Watersnake is found from northeastern Guatemala to Panama. It is an aquatic or semi-aquatic inhabitant of lowland and low montane rainforest where it occurs along streams, and in or near ponds. It shelters under piles of decaying vegetation or on the muddy bottom and, being aquatic, it has relatively dorsally positioned nostrils to enable it to breathe at the surface. It is most active at night and it may even be found crossing roads after heavy rain. It is believed to feed on freshwater shrimps, fish, or frogs but there are few definitive reports regarding the diet of this species. Females are oviparous, laying clutches of four to seven eggs. The Central American Watersnake is harmless to humans.

The Central American Watersnake is a shiny, smooth-scaled snake with a cylindrical body, a narrow head that is barely distinct from the neck, small eyes, and a short tail. It is uniform light or dark brown above, although the dorsum may also bear scattered dark brown spots, while the venter is yellowish or pale brown.

RELATED SPECIES

A second species of *Hydromorphus* is recognized, the Panamanian Watersnake (*H. dunni*), which is known only from a single low montane locality in western Panama. The genus *Hydromorphus* is most closely related to the Central American swamp snake genus *Tretanorhinus* (page 294).

Actual size

FAMILY	Dipsadidae: Dipsadinae
RISK FACTOR	Mildly venomous; harmless to humans
DISTRIBUTION	North America: Canada, USA, and Mexico
ELEVATION	0–5,070 ft (0–1,545 m) asl
HABITAT	Desert and semidesert
DIET	Lizards; also toads, snakes, invertebrates, and small mammals
REPRODUCTION	Oviparous, with clutches of 2–9 eggs
CONSERVATION STATUS	IUCN not listed

ADULT LENGTH
8–11¾ in,
rarely 23¾ in
(200–300 mm,
rarely 600 mm)

280

DESERT NIGHTSNAKE
COPE, 1860

The most northerly member of the genus *Hypsiglena*, the Desert Nightsnake is found from southern British Columbia in Canada, throughout western USA, to the northwest Mexican states of Chihuahua and Sonora. As its common name suggests it is a nocturnally active denizen of desert and semidesert. It shelters during the day, either burrowed into the sand or in the burrows of small mammals, reptiles, or spiders, emerging at night to hunt its primary prey of small lizards. Desert Nightsnakes also occasionally take small snakes, mice, toads, and large invertebrates. Although its toxic oral secretions will rapidly paralyze small prey, the small size of this snake, its reluctance to bite, and its lack of large rear fangs mean it is considered harmless to humans. Desert Nightsnakes may be found on desert roads at night.

RELATED SPECIES

Hypsiglena chlorophaea was formerly treated as a subspecies of the Collared Nightsnake (*H. torquata*), which is now confined to Mexico. As a separate species it contains four subspecies: the nominate Sonoran Nightsnake (*H. c. chlorophaea*), the Northern Desert Nightsnake (*H. c. deserticola*), the Mesa Verde Nightsnake (*H. c. loreala*), and the Tiburon Island Nightsnake (*H. c. tiburonensis*). A former subspecies, the Santa Catalina Island Nightsnake (*H. catalinae*) has been elevated to specific status. *Hypsiglena* contains a further six species in western USA and Mexico.

The Desert Nightsnake is a small snake with smooth scales, an angular, pointed head that is distinct from the neck, and eyes with vertically elliptical pupils. Coloration is predominantly pastel shades, like its habitat, with pale grays or browns, rows of darker blotches of the same color, and large blotches over the neck forming a broken collar.

Actual size

FAMILY	Dipsadidae: Dipsadinae
RISK FACTOR	Rear-fanged, mildly venomous; harmless to humans
DISTRIBUTION	North, Central, and South America: southern Mexico to northern Argentina
ELEVATION	0–6,560 ft (0–2,000 m) asl
HABITAT	Low montane and lowland rainforest and dry forest, plantations, and secondary growth
DIET	Lizards and frogs
REPRODUCTION	Oviparous, with clutches of 1–3, rarely 8, eggs
CONSERVATION STATUS	IUCN not listed

ADULT LENGTH
3 ft 3 in–4 ft 3 in
(1.0–1.3 m)

IMANTODES CENCHOA
COMMON BLUNT-HEADED TREESNAKE
LINNAEUS, 1758

281

Although the Common Blunt-headed Treesnake is a very common and widely distributed species, it often evades discovery due to its nocturnal and highly arboreal habits, its cryptic coloration, and liana-like body shape, combined with its ability to remain stationary for long periods. This species is primarily an inhabitant of pristine wet or dry forest, especially near water, but it is also found in plantations and secondary growth. It is believed to sleep during the day, inside bromeliads or under bark, but even then it is difficult to find. Sometimes specimens are seen on the ground, crossing trails. It preys on frogs and lizards, particularly anoles, which are captured while they sleep, and it uses its weak venom and constriction to subdue its prey.

RELATED SPECIES

The genus *Imantodes* contains seven other species. Three widely distributed species, with which *I. cenchoa* occurs in sympatry, are the Central American Blunt-headed Treesnake (*I. gemmistratus*), the Western Blunt-headed Treesnake (*I. inornatus*), and the Amazon Blunt-headed Treesnake (*I. lentiferus*). Other species are more localized, in the Yucatán, Mexico (*I. tenuissimus*), Darién Gap, Panama (*I. phantasma*), Chocó, Colombia (*I. chocoensis*), and Santander, Colombia (*I. guane*). These snakes also resemble the slug- and snail-eating snakes (*Dipsas*, pages 273–274 and *Sibon*, pages 290–291).

Actual size

The Common Blunt-headed Treesnake is an extremely slender and elongate, laterally compressed snake with smooth scales, a long tail, and a bulbous head with large protruding eyes and vertically elliptical pupils. It is orange to tan dorsally with a series of bold, black-edged, dark orange to red vertebral saddles, and a yellow iris to the eye.

FAMILY	Dipsadidae: Dipsadinae
RISK FACTOR	Rear-fanged, mildly venomous; harmless to humans
DISTRIBUTION	South America: Panama to northern Argentina, also Trinidad, Tobago, and Isla Margarita
ELEVATION	0–7,550 ft (0–2,300 m) asl
HABITAT	Lowland and upland secondary forest, gallery forest, and edge situations
DIET	Frogs and their eggs, and lizards
REPRODUCTION	Oviparous, with clutches of 3–6 eggs
CONSERVATION STATUS	IUCN not listed

ADULT LENGTH
15¾–27½ in,
rarely 31½ in
(400–700 mm,
rarely 800 mm)

282

LEPTODEIRA ANNULATA
BANDED CAT-EYED SNAKE
(LINNAEUS, 1758)

The Banded Cat-eyed Snake is a relatively slender, laterally compressed snake with smooth scales, a long tail, and a broad head with large, protruding eyes and vertically elliptical pupils. It is brown to orange with a dorsal pattern of black or dark brown diamonds or blotches that may coalesce to form an irregular zigzag pattern.

The Banded or Common Cat-eyed Snake is so-called because it has the vertically elliptical pupils of a nocturnal predator. It inhabits secondary growth, disturbed habitats, savanna–forest edge situations, and the vegetation around or along watercourses, where it hunts tree frogs and their eggs, with lizards also featuring in its diet. It is also frequently encountered around buildings and as a semi-arboreal species it may be found moving over the ground at night. Although a rear-fanged venomous species, the commonly encountered Banded Cat-eyed Snake is generally inoffensive and its venom is too weak to harm humans. It is widely distributed through the northern half of South America, east of the Andes, and it also occurs on several offshore islands, including Trinidad, Tobago, and Isla Margarita, off Venezuela.

RELATED SPECIES

Three subspecies are recognized, the nominate form (*Leptodeira annulata annulata*) in the Amazon Basin, a northern form (*L. a. ashmeadi*), and southern form (*L. a. pulchriceps*). The genus *Leptodeira* also contains a further 11 species. The Northern Cat-eyed Snake (*L. septentrionalis*) occurs as far north as Texas, while Baker's Cat-eyed Snake (*L. bakeri*) is found in Venezuela and on Aruba Island. *Leptodeira* is related to the treesnakes of genus *Imantodes* (page 281).

Actual size

FAMILY	Dipsadidae: Dipsadinae
RISK FACTOR	Rear-fanged, mildly venomous; harmless to humans
DISTRIBUTION	North and Central America: southern Mexico, Belize, Guatemala, El Salvador, Honduras, Nicaragua, Costa Rica, and Panama
ELEVATION	0–6,560 ft (0–2,000 m) asl
HABITAT	Lowland and low montane rainforest, dry forest, savanna, pasture, and coffee plantations
DIET	Earthworms, slugs, and snails
REPRODUCTION	Oviparous, with clutches of 1–4 eggs
CONSERVATION STATUS	IUCN Least Concern

ADULT LENGTH
6–15¼ in
(150–386 mm)

NINIA SEBAE

RED COFFEE SNAKE

(DUMÉRIL, BIBRON & DUMÉRIL, 1854)

283

The tiny, forest floor-dwelling Red Coffee Snake is a mimic of several Central American coralsnake species (*Micrurus*, pages 457–465), but is itself harmless. It is found in a wide variety of closed canopy and open habitats from Mexico to Panama, hiding under logs or in leaf litter during the day. At night it is active, hunting earthworms, slugs, and snails, as part of a guild of snakes referred to by ecologists as "goo-eaters." Snakes in this guild possess special oral secretions to enable them to feed on slimy prey without becoming "glued-up" with their defensive secretions. Snakes of the genus *Ninia* are referred to as "coffee snakes" because they are often found in coffee plantations. The Red Coffee Snake is named for Albertus Seba (1665–1736), a Dutch collector and author of *Cabinet of Natural Curiosities*.

The Red Coffee Snake is a diminutive snake with smooth scales, a head distinct from the neck, small eyes with vertically elliptical pupils, and a moderately long tail. It is red to pink above with a black cap over the dorsum of the head, and a yellow and black collar. It is immaculate white below.

RELATED SPECIES

Four subspecies are recognized, from the Yucatán, Mexico (*Ninia s. morleyi*), Guatemala (*N. s. punctulata*), and southeast Nicaragua to Panama (*N. s. immaculata*), with the nominate form occupying the remainder of the range. Ten other species of coffee snake occur from Mexico to Trinidad. Espinal's Coffee Snake (*N. espinali*), from Honduras and El Salvador, is considered to be Near Threatened by the IUCN, while a recently described species(*N. franciscoi*) is known only from its Trinidadian holotype.

Actual size

FAMILY	Dipsadidae: Dipsadinae
RISK FACTOR	Rear-fanged, mildly venomous; harmless to humans
DISTRIBUTION	Central and South America: Honduras, Nicaragua, Costa Rica, Panama, Colombia, and Ecuador
ELEVATION	0–3,280 ft (0–1,000 m) asl
HABITAT	Lowland to low montane rainforest
DIET	Frogs, salamanders, possibly lizards
REPRODUCTION	Oviparous, with clutches of up to 3 eggs
CONSERVATION STATUS	IUCN Least Concern

ADULT LENGTH
8–15 in,
rarely 17 in
(200–380 mm,
rarely 430 mm)

284

NOTHOPSIS RUGOSUS
ROUGH COFFEE SNAKE
COPE, 1871

The Rough Coffee Snake is unlike any other snake within its range. It has been linked with the xenodermatid snakes of Southeast Asia, such as the Dragon Snake (*Xenodermus javanicus*, page 551), due to its unusual appearance. Unique features of the Rough Coffee Snake include the fragmentation of most of its dorsal head scutes into numerous granular scales, and the keeling of the head and body scales, a morphology usually associated with boas or pitvipers. Distributed from Honduras to Ecuador, it is a rare snake known from only a few specimens. It is a secretive inhabitant of tropical rainforest floors, hiding under logs or in leaf litter. Prey includes salamanders and small frogs, and possibly also small lizards. It has enlarged rear teeth and possibly Duvernoy's glands, but is harmless to humans.

Actual size

The Rough Coffee Snake is a slender snake that is characterized by its strongly keeled body scales, and rugose, granular head scales. It has a relatively long tail, a head distinct from the neck, and small eyes with vertical pupils. It may be dark brown or light yellowish brown with darker brown reticulations, wavy stripes, and blotches.

RELATED SPECIES

Despite appearances, the monotypic *Nothopsis rugosus* is related to other Central American snakes such as the blunt-headed treesnakes (*Imantodes*, page 281) and the cat-eyed snakes (*Leptodeira*, page 282).

FAMILY	Dipsadidae: Dipsadinae
RISK FACTOR	Rear-fanged, venomous
DISTRIBUTION	North and Central America: southern Mexico, Guatemala, Belize, Honduras, and El Salvador
ELEVATION	0–6,500 ft (0–1,980 m) asl
HABITAT	Lowland and low montane rainforest and dry forest, also coffee plantations and cultivated areas
DIET	Frogs, salamanders, and amphibian eggs
REPRODUCTION	Oviparous, with clutches of 4–8 eggs
CONSERVATION STATUS	IUCN Least Concern

ADULT LENGTH
11¾–19¾ in,
rarely 30¾ in
(300–500 mm,
rarely 780 mm)

PLIOCERCUS ELAPOIDES
VARIEGATED FALSE CORALSNAKE
COPE, 1860

285

As its common name suggests, this is an extremely variably patterned snake. It occurs across a wide area of Mexico and northern Central America and may be difficult to identify to species. The Variegated False Coralsnake is an excellent mimic of several highly venomous coralsnakes (*Micrurus*, pages 457–465), its pattern seemingly varying to match that of the local coralsnakes across its range. Although not lethal, bites should be avoided as they have caused localized pain and swelling, and even severe pain and tissue discoloration that may take weeks to recover. A wet or dry forest species, this snake is also found in coffee plantations and other cultivated areas, where it is active by day or night on the forest floor. It feeds on frogs, salamanders, and amphibian eggs.

The Variegated False Coralsnake
is a moderately slender snake with a rounded head, distinct from the neck, and round pupils to the eyes. It has smooth scales and the tail is often truncated. Many populations exhibit a banded coralsnake-mimic pattern of red-yellow-black-yellow-red, but some specimens (*P. e. wilmarai*) lack the black bands.

RELATED SPECIES

Five subspecies are recognized, from Puebla to Chiapas (*Pliocercus elapoides elapoides*), Veracruz to Yucatán and Honduras (*P. e. aequalis*), Oaxaca to Guatemala (*P. e. diastema*), Pacific coastal Oaxaca (*P. e. occidentalis*), and Caribbean coastal Veracruz (*P. e. wilmarai*). The genus *Pliocercus* also contains three other species: Andrew's False Coralsnake (*P. andrewsi*) on the northern Yucatán Peninsula, the Northern False Coralsnake (*P. bicolor*) from Tamaulipas to Veracuz, and the Southern False Coralsnake (*P. euryzonus*) from Guatemala to Peru. *Pliocercus* is closely related to the thick-tailed snakes of genus *Urotheca* (page 297).

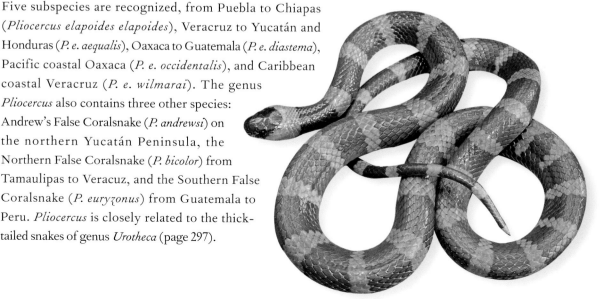

Actual size

FAMILY	Dipsadidae: Dipsadinae
RISK FACTOR	Rear-fanged, mildly venomous; harmless to humans
DISTRIBUTION	North America: southern Mexico
ELEVATION	330–5,910 ft (100–1,800 m) asl
HABITAT	Tropical semi-deciduous forest
DIET	Lizards and frogs
REPRODUCTION	Oviparous, clutch size unknown
CONSERVATION STATUS	IUCN Least Concern

ADULT LENGTH
15¾–19¾ in,
rarely 27½ in
(400–500 mm,
rarely 700 mm)

286

PSEUDOLEPTODEIRA LATIFASCIATA
FALSE CAT-EYED SNAKE
(GÜNTHER, 1894)

The boldly patterned False Cat-eyed Snake occurs in the southwest Mexican states of Colima, Oaxaca, Guerrero, Michoacán, Morelos, Puebla, and Jalisco, where it inhabits tropical semi-deciduous forest and coastal lowland forest. Terrestrial and nocturnal, it is not a commonly encountered species. Although it resembles the cat-eyed snakes of genus *Leptodeira*, it is not closely related to them and it possesses ungrooved, rather than strongly grooved, rear fangs, among other defining characteristics. This species is known to enter buildings in search of geckos, but it also eats frogs. The False Cat-eyed Snake is a poorly known species that is believed to be oviparous, but its clutch size is unknown.

The False Cat-eyed Snake is a laterally compressed snake with smooth scales, a relatively long tail, and a distinct head with moderate eyes and vertically elliptical pupils. The head is white with a heavy infusion of black on every scale and a red dorsum and neck, followed by a black nape band. The body is white with broad black or brown saddles that almost form complete rings, the white interspaces being unmarked and relatively narrow.

RELATED SPECIES

Although *Pseudoleptodeira latifasciata* does resemble members of the genus *Leptodeira* it is more closely related to the nightsnakes of genus *Hypsiglena* (page 280), and it was at one time included in that genus. *Pseudoleptodeira* is now a monotypic genus, Uribe's Cat-eyed Snake (*L. uribei*) being transferred to *Leptodeira*.

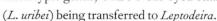

Actual size

FAMILY	Dipsadidae: Dipsadinae
RISK FACTOR	Rear-fanged, mildly venomous; harmless to humans
DISTRIBUTION	North, Central, and South America: southeastern Mexico, Belize, Guatemala, Honduras, Nicaragua, Costa Rica, Panama, Colombia, and Ecuador
ELEVATION	0–5,740 ft (0–1,750 m) asl
HABITAT	Lowland and low montane rainforest, secondary growth, and coffee plantations
DIET	Frogs, salamanders, amphibian eggs, lizards, and earthworms
REPRODUCTION	Oviparous, with clutches of 1–4 eggs
CONSERVATION STATUS	IUCN not listed

ADULT LENGTH
13¾–15¼ in,
rarely 18½ in
(350–400 mm,
rarely 470 mm)

RHADINAEA DECORATA
CENTRAL AMERICAN GRACEFUL BROWNSNAKE
(GÜNTHER, 1858)

287

Also known as the Adorned Graceful Brownsnake, this species is found from southeastern Mexico, through Central America, to Colombia and Ecuador in northwestern South America. It is a diurnal leaf-litter inhabitant of lowland and low montane rainforest, and also inhabits secondary growth areas and coffee plantations, though it is less common in disturbed habitats than pristine rainforest. It preys on terrestrial frogs and their eggs, and also salamanders, lizards, and earthworms. It appears to be most active on overcast days or following rain. As a small snake, the female Central American Graceful Brownsnake produces clutches comprising up to four eggs.

RELATED SPECIES

The 21 snakes of genus *Rhadinaea* are also known as forest snakes or littersnakes. Eleven species belong to the *R. decorata* group, but only *R. decorata* itself occurs outside of Mexico. *Rhadinaea* is related to the dwarf brownsnakes (*Rhadinella*, page 289) and the leaf-litter snakes (*Taeniophallus*, page 340).

The Central American Graceful Brownsnake is a slender, smooth-scaled snake, with a long, often truncated tail, a narrow but distinct and pointed head, and moderately large eyes with round pupils. It is dark brown with a pair of pale brown longitudinal stripes, which begin on the neck as yellow stripes and broaden as they continue onto the body. A white stripe passes along the lips, while broken yellow or white lines exit the rear of the eye.

Actual size

FAMILY	Dipsadidae: Dipsadinae
RISK FACTOR	Rear-fanged, mildly venomous; harmless to humans
DISTRIBUTION	North America: southeastern USA
ELEVATION	0–625 ft (0–190 m) asl
HABITAT	Lowland pinewoods, hardwood hammocks, and coastal marshes
DIET	Frogs, salamanders, and lizards
REPRODUCTION	Oviparous, with clutches of 1–4 eggs
CONSERVATION STATUS	IUCN Least Concern

ADULT LENGTH
9¾–11¾ in,
rarely 15¾ in
(250–300 mm,
rarely 400 mm)

RHADINAEA FLAVILATA
PINEWOODS SNAKE
(COPE, 1871)

The only member of the genus *Rhadinaea* in the United States, the Pinewoods Snake is found from the Carolinas to Florida and west to southern Louisiana and Mississippi. It inhabits pinewood flatlands and adjacent hardwood hammocks in Florida, and coastal marshes and islands in the Carolinas, and adopts a semi-fossorial existence, living under logs or bark or buried in the sandy soil. It is said to estivate in crayfish burrows during dry weather. Prey consists of small frogs, salamanders, and lizards, which are subdued by the snake's weak venom. Like many small snakes it has many enemies, such as ophiophagous snakes, birds of prey, and carnivorous mammals, and it is also vulnerable to habitat destruction, fragmentation, or alteration. Pinewoods Snakes do not bite, but expel a noxious secretion from the cloacal glands.

Actual size

The Pinewoods Snake is a small, smooth-scaled snake with a short tail that ends in a terminal spine, a head slightly broader than the neck, and moderately large eyes with round pupils. It is red-brown in color, paler below, with yellow or white lips and a red iris.

RELATED SPECIES

Rhadinaea flavilata is the only member of its genus in the United States. A close relative may be the Crowned Graceful Brownsnake (*R. laureata*) from central Mexico.

FAMILY	Dipsadidae: Dipsadinae
RISK FACTOR	Rear-fanged, mildly venomous; harmless to humans
DISTRIBUTION	Central America: Costa Rica
ELEVATION	3,810–7,220 ft (1,160–2,200 m) asl
HABITAT	Low montane wet forest and rainforest
DIET	Reptile eggs
REPRODUCTION	Oviparous, with clutches of up to 6 eggs
CONSERVATION STATUS	IUCN Least Concern

ADULT LENGTH
14–17½ in
(360–445 mm)

RHADINELLA SERPERASTER
COSTA RICAN GRACEFUL BROWNSNAKE
(COPE, 1871)

289

The Costa Rican Graceful Brownsnake, also known as the Striped Littersnake, is confined to low to medium elevations in the Cordilleras de Tilarán and Talamanca in north and central Costa Rica. This small snake is terrestrial or semi-fossorial, being located under logs on the rainforest floor but sometimes found as deep as 20 in (50 cm) below the surface in soft, humid rainforest soil. It is believed to feed on reptile eggs, especially those of earthsnakes of the genus *Geophis* (pages 277–278). Although it is technically venomous it is inoffensive and harmless to humans. It is oviparous, with clutches of up to six eggs being laid.

RELATED SPECIES

The genus *Rhadinella* contains 17 species, most of which previously constituted the *Rhadinaea godmani* group. Most species are extremely localized in their distribution, the only other species occurring in Costa Rica being the widely distributed Godman's Graceful Brownsnake (*R. godmani*).

The Costa Rican Graceful Brownsnake is a diminutive snake with smooth scales, a relatively short tail, and a slightly pointed head with small eyes and round pupils. The dorsum of the head is dark brown, the lips are white, and the body is brown with a series of tan and dark brown longitudinal stripes.

Actual size

FAMILY	Dipsadidae: Dipsadinae
RISK FACTOR	Nonvenomous
DISTRIBUTION	Central America: Honduras, Nicaragua, Costa Rica, and Panama
ELEVATION	0–2,620 ft (0–800 m) asl
HABITAT	Low montane wet forest and rainforest
DIET	Slugs and snails, and possibly earthworms
REPRODUCTION	Oviparous, clutch size unknown
CONSERVATION STATUS	IUCN Least Concern

ADULT LENGTH
19¾–27½ in
(500–700 mm)

SIBON LONGIFRENIS
LICHEN SNAIL-EATER
(STEJNEGER, 1909)

The Lichen Snail-eater occurs in Honduras, Nicaragua, Costa Rica, and Panama where it inhabits lowland wet forest and rainforest, especially near water. Nocturnal and arboreal, its cryptic camouflage means it is rarely encountered, and less well known than some of its congeners. It patterning resembles that of lichen-patterned Eyelash Palm-pitvipers (*Bothriechis schlegelii*, page 558), but whether this is coincidental or mimicry is not known. The Lichen Snail-eater feeds on slugs and snails, withdrawing the latter from their shells by hooking them with the teeth of the lower jaw. It also feeds on earthworms. Oviparous, like its congeners, its clutch size is unknown.

RELATED SPECIES
Sibon longifrenis appears similar to the recently described Costa Rican Snail-eater (*Sibon lamari*). Nine other snail-eaters occur in Costa Rica, five in *Sibon*, three in *Dipsas*, and Sartorius' Terrestrial Snail-sucker (*Tropidodipsas sartorii*, page 296).

The Lichen Snail-eater has a laterally compressed body, a long tail, a broad head with protruding eyes, and catlike, vertically elliptical pupils. Its patterning consists of a complex green, brown, and gray lichen pattern with white spots on the lower flanks.

Actual size

FAMILY	Dipsadidae: Dipsadinae
RISK FACTOR	Nonvenomous
DISTRIBUTION	North, Central, and South America: southeast Mexico to Brazil and Ecuador, the Guianas, Trinidad and Tobago, and Isla Margarita
ELEVATION	0–8,630 ft (0–2,630 m) asl
HABITAT	Low montane wet forest and rainforest, dry forest, gallery forest, and secondary growth
DIET	Slugs and snails, and possibly tree frogs' eggs
REPRODUCTION	Oviparous, with clutches of 3–9 eggs
CONSERVATION STATUS	IUCN not listed

ADULT LENGTH
27½–32¾ in
(700–830 mm)

SIBON NEBULATUS
CLOUDY SNAIL-EATER
(LINNAEUS, 1758)

The Cloudy Snail-eater is the most widely distributed member of genus *Sibon*, occurring from southeastern Mexico to Ecuador, west of the Andes, and the Guianas east of the Andes. It occurs in a variety of low- to medium-elevation rainforest and wet forest habitats, but also inhabits dry forest, gallery forest along rivers, and secondary growth. In common with other *Sibon* snakes, this species is nocturnal and arboreal, and feeds on slugs and snails. Numerous snail-eaters have been found on trees hosting large arboreal snail populations, but they were found to be absent from trees lacking snails. This species is also thought to feed on tree frogs' eggs. This is a commonly encountered snake that is also known to enter roof spaces in houses.

The Cloudy Snail-eater is a slender snake with a laterally compressed body, a long tail, a broad head, and protruding eyes with vertically elliptical pupils. It is gray in color with irregular light-edged, dark gray bands that continue under the venter, and dark speckling on the interspaces. Specimens may vary considerably.

RELATED SPECIES

The genus *Sibon* contains 16 species of snail-eaters or snail-suckers, but none resemble *Sibon nebulatus*, the most widely distributed member of the genus. Its closest relatives are believed to be the Tegucigalpa Snail-eater (*S. carri*) from Honduras and El Salvador, and the Imbabura Snail-eater (*S. dunni*) from Ecuador, which together form the *S. nebulatus* group.

Actual size

FAMILY	Dipsadidae: Dipsadinae
RISK FACTOR	Nonvenomous
DISTRIBUTION	South America: southeastern Brazil
ELEVATION	0–2,100 ft (0–640 m) asl
HABITAT	Low montane wet forest and dry forest, also disturbed and suburban areas
DIET	Slugs and snails
REPRODUCTION	Oviparous, with clutch size unknown
CONSERVATION STATUS	IUCN not listed

ADULT LENGTH
19¾–27½ in
(500–700 mm)

SIBYNOMORPHUS NEUWIEDI
EASTERN SLUG-EATER
(IHERING, 1911)

292

The Eastern Slug-eater is a relatively stocky, laterally compressed, smooth-scaled snake with a broad head and small, protruding eyes with vertical pupils. It is pale brown with a pattern of broad, dark brown transverse bars that may or may not meet middorsally, but patterning may be variable.

The Eastern Slug-eater is found in southeastern Brazil from Bahia to Rio Grande do Sul, and also on the Ilhas São Sebastião and São Vicente. It inhabits the Atlantic coastal forests but is also found in disturbed and suburban areas. Unlike the genera *Sibon* and *Dipsas*, snakes in *Sibynomorphus* are terrestrial, or semi-arboreal, and feed on terrestrial slugs and snails. Slug- and snail-eaters are generally known as "goo-eaters." They have specialized oral glands that secrete toxins to combat their prey's secretions, which would otherwise gum up their mouths, but these oral secretions are not dangerous to humans and the snakes are inoffensive. Slug- and snail-eaters are known as "dormideiras," or sleeping snakes, in Brazil. Maximilian Alexander Philipp, Prinz zu Wied-Neuwied (1782–1867) was a Prussian naturalist who explored Amazonia.

RELATED SPECIES
The genus *Sibynomorphus* contains 11 species of South American slug- or snail-eating snakes. Other Brazilian species include the White-collared Slug-eater (*S. mikanii*), the Southern Slug-eater (*S. ventrimaculatus*), and the Bolivian Slug-eater (*S. turgidus*).

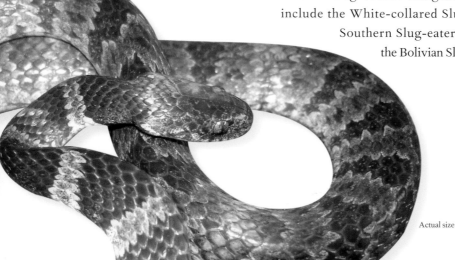

Actual size

FAMILY	Dipsadidae: Dipsadinae
RISK FACTOR	Nonvenomous
DISTRIBUTION	South America: Colombia
ELEVATION	655–5,580 ft (200–1,700 m) asl
HABITAT	Lowland rainforest and Andean cloud forest
DIET	Unknown but probably lizards
REPRODUCTION	Oviparous, with clutches of 2–8 eggs
CONSERVATION STATUS	IUCN not listed

ADULT LENGTH
8–15¾ in,
rarely 31½ in
(200–400 mm,
rarely 800 mm)

SYNOPHIS BICOLOR
BICOLORED FISHING SNAKE
PERACCA, 1896

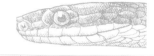

293

The common name, "fishing snakes," applied to snakes of the genus *Synophis*, is a misnomer, as none of the nine species described actually feed on fish. The diets of these snakes are poorly documented but seem to feature small lizards. The type species of the genus, the Bicolored Fishing Snake, occurs at low to medium elevations in Colombia, but authorities now believe that this species may actually constitute three similar but distinct species from the Ecuadorian Chocó, Ecuadorian Andes, and Colombian Andes. The first of these three forms inhabits lowland rainforest while the other two occur in Andean cloud forest. Specimens of all three have been found in low vegetation or leaf litter, and they are reportedly active by day and night.

RELATED SPECIES

The genus *Synophis* comprises nine species, one of which, the Valle de Cauca Fishing Snake (*S. plectovertebralis*), which was only described in 2001 from Colombia, is listed as Critically Endangered by the IUCN. Closely related genera include the terrier snakes (*Emmochliophis*) and the frog-eating snakes (*Diaphorolepis*, page 272).

The Bicolored Fishing Snake is a slender species with a laterally compressed body, keeled scales, a relatively broad head with bulbous eyes, round pupils, and a long tail. As its common name suggests it is two-tone, being brown above and bright yellow below.

Actual size

FAMILY	Dipsadidae: Dipsadinae
RISK FACTOR	Nonvenomous
DISTRIBUTION	North, Central, and South America: southeast Mexico, Guatemala, Belize, Honduras, Nicaragua, Costa Rica, Panama, and Colombia
ELEVATION	0–4,130 ft (0–1,260 m) asl
HABITAT	Lowland wet forest, tree swamps, freshwater streams, and mangrove forest
DIET	Fish, frogs, and tadpoles
REPRODUCTION	Oviparous, with clutches of 6–9 eggs
CONSERVATION STATUS	IUCN Least Concern

ADULT LENGTH
Male
8–25¼ in
(200–640 mm)

Female
22¾–35¾ in
(575–900 mm)

294

TRETANORHINUS NIGROLUTEUS
ORANGE-BELLIED SWAMP SNAKE
COPE, 1861

The Orange-bellied Swamp Snake is a glossy, smooth-scaled snake, with a long tail and a narrow but distinct head with small eyes. It is brown above and bright orange below, with the demarcation line along the lower flanks.

The Orange-bellied Swamp Snake occurs at low and medium elevations from southeastern Mexico, through Central America, including the Honduran Islas de la Bahía and the Nicaraguan Corn Islands, and into Colombia. It can be very common in parts of its range but is reportedly a rare snake in Costa Rica. It is found in aquatic habitats, from freshwater streams and tree swamps to brackish mangrove forests, where it is nocturnally active. Prey consists of small fish, frogs, and their tadpoles, which may be actively hunted or ambushed by the snake lying in wait on the bottom of the watercourse. When resting the snake keeps its snout at the surface, ready to dive if danger threatens. Small specimens are taken by waders and other waterbirds.

RELATED SPECIES

Four subspecies are recognized by some authors, but they are poorly defined and are generally not accepted. *Tretanorhinus* also contains three other species, Mocquard's Swamp Snake (*T. mocquardi*) in Panama and Ecuador, the Striped Swamp Snake (*T. taeniatus*) in Colombia and Ecuador, and the Caribbean Swamp Snake (*T. variabilis*) on Cuba and the Cayman Islands.

Actual size

FAMILY	Dipsadidae: Dipsadinae
RISK FACTOR	Nonvenomous
DISTRIBUTION	Central America: Panama
ELEVATION	0–3,610 ft (0–1,100 m) asl
HABITAT	Lowland rainforest, secondary regrowth, and gardens and agricultural patches
DIET	Probably salamanders and lizards
REPRODUCTION	Presumed oviparous, with clutches of 1–2 eggs
CONSERVATION STATUS	IUCN Vulnerable

ADULT LENGTH
8–11¾ in
(200–300 mm)

TRIMETOPON BARBOURI
CANAL ZONE GROUNDSNAKE
DUNN, 1930

The Canal Zone Groundsnake is endemic to Panama, including Barro Colorado Island. It was once believed confined to the Canal Zone, but is now known to be more widely distributed. The IUCN list this species as Vulnerable due to habitat loss. It is a semi-fossorial species that inhabits leaf litter in rainforest and secondary growth areas, but it is also found in gardens and agricultural plots. It is crepuscular in habit, but its prey preferences are unknown, although other members of the genus feed on salamanders and lizards so it may be expected to exhibit a similar diet. Likewise, it is thought to be oviparous with small clutch sizes, based on the data from congeners. Thomas Barbour (1884–1946) was an American zoologist with a specialist interest in Central America.

Actual size

The Canal Zone Groundsnake is a small snake with smooth scales, a slender body, a head indistinct from the neck, and moderate-sized eyes with round pupils. It is brown above with a pair of dark-edged, pale brown longitudinal stripes. The lower flanks are gray and the venter is white. A pair of pale spots are present on the neck in some specimens.

RELATED SPECIES
Genus *Trimetopon* contains five other leaf litter-dwelling species in Costa Rica and western Panama: the Cartago Groundsnake (*T. gracile*), San José Groundsnake (*T. pliolepis*), Reventazon Groundsnake (*T. simile*), Chiriquí Groundsnake (*T. slevini*), and Siquirres Groundsnake (*T. viquezi*).

FAMILY	Dipsadidae: Dipsadinae
RISK FACTOR	Rear-fanged, mildly venomous; harmless to humans
DISTRIBUTION	North and Central America: Mexico, Guatemala, Belize, Honduras, El Salvador, Nicaragua, and Costa Rica
ELEVATION	0–8,010 ft (0–2,440 m) asl
HABITAT	Lowland dry and gallery forest, lowland and premontane wet forest, and secondary growth
DIET	Slugs and snails
REPRODUCTION	Oviparous, with clutches of 3–5 eggs
CONSERVATION STATUS	IUCN Least Concern

ADULT LENGTH
Male
15¾–28¼ in
(400–720 mm)

Female
15¾–34 in
(400–860 mm)

296

TROPIDODIPSAS SARTORII

SARTORIUS' TERRESTRIAL SNAIL-SUCKER

COPE, 1863

Sartorius' Terrestrial Snail-sucker is a glossy, smooth-scaled snake with a long tail and a bulbous head with small, protruding eyes. It is black with well-spaced yellow bands, the first being across the rear of the head.

Most snail-eating snakes are arboreal (*Dipsas*, pages 273–274, and *Sibon*, pages 290–291), but those in genus *Tropidodipsas* are primarily terrestrial. Sartorius' Terrestrial Snail-sucker is the most widely distributed member of the genus, being found from southern Mexico to Costa Rica. It inhabits both wet and dry lowland tropical forest, gallery forest, wet premontane forest, and secondary growth, and is active nocturnally. It is also found in areas of secondary growth and is most abundant in limestone areas, entering caves and crevices in search of prey or shelter. It feeds exclusively on slugs and snails, which are subdued by the oral toxins of the snail-sucker, these secretions also combating the mucus produced by the mollusks. Christian Carl Wilhelm Sartorius (1796–1872) was a German naturalist who collected in Mexico.

RELATED SPECIES

Two subspecies of *Tropidodipsas sartorii* are recognized, a western or Mexico form (*T. s. macdougalli*) and an eastern or Central American form (*T. s. sartorii*). Six other Latin American species are known: the Western Snail-eater (*T. annulifera*), Banded Snail-eater (*T. fasciata*), Fischer's Snail-eater (*T. fischeri*), Philippi's Snail-eater (*T. philippii*), Sonoran Snail-eater (*T. repleta*), and Zweifel's Snail-eater (*T. zweifeli*).

Actual size

FAMILY	Dipsadidae: Dipsadinae
RISK FACTOR	Rear-fanged, mildly venomous; harmless to humans
DISTRIBUTION	Central America: Honduras, Nicaragua, Costa Rica, and Panama
ELEVATION	0–6,890 ft (0–2,100 m) asl
HABITAT	Lowland and low montane wet and dry forest
DIET	Frogs, salamanders, and lizards
REPRODUCTION	Oviparous, with clutches of 3–5 eggs
CONSERVATION STATUS	IUCN Least Concern

ADULT LENGTH
11¾–19¼ in,
rarely 26½ in
(300–490 mm,
rarely 670 mm)

UROTHECA GUENTHERI
GÜNTHER'S BROWNSNAKE
(DUNN, 1938)

297

Günther's Brownsnake is found from northeastern Honduras to central Panama, at low and medium elevations in rainforest habitats, especially over limestone, where it feeds on frogs, salamanders, and lizards. This diurnal snake is secretive, hiding inside rotten logs or in leaf litter when it is not active. Members of the genus *Urotheca* have long tails, but many specimens are found with truncated tails due to their practice of caudal pseudautotomy as a defense against predation. This is a defense similar to the caudal autotomy practiced by lizards such as skinks and geckos. Unlike the lizard, snakes that practice pseudautotomy do not regenerate their lost tails. Albert Günther (1830–1914) was a German-born zoologist who worked in the British Museum of Natural History.

Günther's Brownsnake is a small, smooth-scaled snake with a very long, but often truncated, tail, a head only slightly distinct from the neck, and medium-sized eyes with round pupils. It is brown dorsally and orange ventrally, with a pair of fine yellow longitudinal stripes on either flank, the upper ones beginning as a pair of yellow nape spots.

RELATED SPECIES

The genus *Urotheca* contains seven other species distributed from Honduras to Peru. *Urotheca* is related to graceful brownsnakes (*Rhadinaea*, pages 287–288) and the false coralsnakes (*Pliocercus*, page 285).

Actual size

FAMILY	Dipsadidae: Xenodontinae
RISK FACTOR	Rear-fanged, mildly venomous; harmless to humans
DISTRIBUTION	West Indies: Great Bird Island (Antigua)
ELEVATION	0–98 ft (0–30 m) asl
HABITAT	Open, dry, rocky scrubland
DIET	Lizards and small mammals
REPRODUCTION	Oviparous, with clutches up to 11 eggs (based on two females)
CONSERVATION STATUS	IUCN Critically Endangered, as *A. antiguae*

ADULT LENGTH
23¾–30½ in,
rarely 3 ft 3 in
(600–776 mm,
rarely 1.0 m)

ALSOPHIS SAJDAKI
GREAT BIRD ISLAND RACER
HENDERSON, 1990

298

The Great Bird Island Racer is a former subspecies of the Antiguan Racer (*Alsophis antiguae*). It is one of the most endangered snakes in the world, the Antiguan population going extinct at the end of the nineteenth century following the accidental introduction of rats, and deliberate introduction of mongooses to control the rats. In 1995 the Great Bird Island population comprised 50–70 individuals. Conservation management measures, including rat eradication, captive breeding, and reintroduction to York, Green, and Rabbit Islands, have raised the wild racer population to over 1,000 individuals. Harmless to humans, these diurnally active racers inhabit low-lying rocky scrubland and prey on lizards and mice, ambushed from cover in the leaf litter. This species is still Critically Endangered according to the IUCN. Richard A. Sajdak is an American herpetologist.

The Great Bird Island Racer is a smooth-scaled snake with a slightly elongate, pointed head, and small eyes with round pupils. This species is sexually dichromatic, males being dark brown with cream markings while females are silver-gray with pale brown markings. Specimens may be blotched, striped, or spotted. Females also possess larger heads than males and achieve slightly larger overall sizes.

RELATED SPECIES

In addition to *Alsophis antiguae* and *A. sajdaki*, the genus *Alsophis* contains a further seven species of Antillean racers. These include the Guadeloupe Racer (*A. antillensis*), Dominican Racer (*A. sibonius*), Montserrat Racer (*A. manselli*), Terre-de-Bas Racer (*A. danforthi*), Terre-de-Haut Racer (*A. sanctonum*), Anguilla or Leeward Island Racer (*A. rijgersmaei*), and Orange-bellied Saba Racer (*A. rufiventris*). The last three of these are also listed as Endangered by the IUCN.

Actual size

FAMILY	Dipsadidae: Xenodontinae
RISK FACTOR	Rear-fanged, venomous: possibly hemorrhagins
DISTRIBUTION	South America: central and northeastern Brazil
ELEVATION	755–2,820 ft (230–860 m) asl
HABITAT	Open Cerrado on sandy soils
DIET	Amphisbaenians
REPRODUCTION	Oviparous, with clutches of up to 3 eggs
CONSERVATION STATUS	IUCN Data Deficient

ADULT LENGTH
17¾–25 in
(450–634 mm)

APOSTOLEPIS AMMODITES
TWIN-COLLARED CERRADO SANDSNAKE

FERRAREZZI, ERRITTO BARBO & ESPAÑA ALBUQUERQUE, 2005

No common English name exists for snakes of the genus *Apostolepis*, so Twin-collared Cerrado Sandsnake is coined to combine its most distinctive characteristic, two white rings around the neck, with its distribution and habitat preferences. It is found in the arid sandy habitat known as Cerrado, which occurs across a wide swath of central and northeastern Brazil, where it adopts a primarily fossorial existence in the sandy soil. The only prey recorded from the guts of museum specimens have been amphisbaenians, or worm-lizards, a group of similarly fossorial reptiles considered distinct from other lizards. Twin-collared Cerrado Sandsnakes may be rear-fanged and venomous, but they have extremely small mouths. Even so, there are snakebites on record from *A. dimidiata*, which has a hemorrhagic venom, and caution is recommended with any unusual snake species.

The Twin-collared Cerrado Sandsnake is an extremely slender, elongate snake with smooth scales, a narrow head that is indistinct from the neck, small eyes, and a shovel-shaped snout for burrowing. Like many of its congeners it is red in color with a black tail tip and a series of black and white rings around the head and neck, but it differs from its relatives in possessing two broad white rings, rather than a single white collar.

RELATED SPECIES

The genus *Apostolepis* contains 33 species. *Apostolepis ammodites* belongs to the *A. assimilis* group, which may contain five other species, *A. assimilis* from southeastern and central Brazil, Paraguay and Argentina, *A. cearensis* from Ceará, Brazil, *A. arenaria* and *A. gaboi* from Bahia, Brazil, and *A. quirogai* from Rio Grande do Sul, Brazil and Misiones, Argentina.

Actual size

FAMILY	Dipsadidae: Xenodontinae
RISK FACTOR	Probably rear-fanged, mildly venomous
DISTRIBUTION	West Indies: Cuba, including Isla de la Juventud
ELEVATION	0–330 ft (0–100 m) asl
HABITAT	Open grasslands and rocky pasture
DIET	Amphisbaenians and small snakes
REPRODUCTION	Presumed oviparous, but unknown
CONSERVATION STATUS	IUCN Least Concern

ADULT LENGTH
15¾–18 in
(400–460 mm)

ARRHYTON TAENIATUM
BROAD-STRIPED RACERLET
GÜNTHER, 1858

The Broad-striped Racerlet is a small, smooth-scaled snake with a slightly pointed and compressed head that is just distinct from the neck, small eyes with vertically elliptical pupils, a pattern comprising two pale longitudinal dorsolateral stripes on a mahogany background, and a yellow venter.

Secretive, semi-fossorial to fossorial, and probably nocturnal based on its vertical pupils, this small snake is rarely encountered unless it is excavated from the earth by a plow or found sheltering under a flat stone. It is believed more fossorial in its habits than some of its long-tailed congeners, which are more terrestrial and less adapted for burrowing. It inhabits open grassland and pasture, especially areas with large stones or rocks, and is recorded from eastern, western, and southern Cuba, and Isla de la Juventud. The prey of the Broad-striped Racerlet comprises other fossorial reptiles, such as amphisbaenians (commonly known as worm-lizards), blindsnakes (Typhlopidae), and threadsnakes (Leptotyphlopidae). The Cuban racerlets are poorly known, with very few ecological observations documented.

RELATED SPECIES

The genus *Arrhyton* contains eight species of Cuban racerlets, but only the equally widely distributed Cuban Short-tailed Racerlet (*A. vittatum*) also occurs on the Isla de la Juventud. These species can be distinguished on the presence (*A. vittatum*) or absence (*A. taeniatum*) of a loreal scale, between the preocular and nasal scales, and by the longer tail of *A. vittatum*. The remaining species are much more localized in their distribution on the Cuban mainland.

Actual size

FAMILY	Dipsadidae: Xenodontinae
RISK FACTOR	Rear-fanged, venomous: possibly cytotoxins, composition unknown
DISTRIBUTION	South America: southeast Brazil, east Bolivia, Paraguay, Uruguay, and Argentina
ELEVATION	98–2,890 ft (30–880 m) asl
HABITAT	Wet Atlantic forests, dry Chaco woodland, and arid desert
DIET	Snakes, lizards, and small mammals
REPRODUCTION	Oviparous, with clutches of 4–15 eggs
CONSERVATION STATUS	IUCN not yet assessed

ADULT LENGTH
5–6 ft
(1.5–1.8 m)

BOIRUNA MACULATA
BLACK-TAILED MUSSURANA
(BOULENGER, 1896)

301

The Black-tailed Mussurana inhabits a wide range of different environments from southeastern Brazil to Paraguay, Bolivia, Uruguay, and Argentina. It is found in wet coastal forests and dry Chaco woodlands and even occurs in arid elevated desert habitats. This is a species that goes through a considerable ontogenetic color change from juvenile to adult. The generic name *Boiruna* is derived from a Tupi-Guarani term, "Mboi-r-ú," which means "eats snakes." Mussuranas are ophiophagous species that prey primarily on other species, including highly venomous lanceheads of genus *Bothrops* (pages 560–569). They are immune to the pitvipers' bites. Lizards and small mammals are also taken. Bites to humans have caused localized pain and swelling similar to bites from lanceheads, although much less serious. However, bites from large specimens to children may be life-threatening.

The Black-tailed Mussurana is a large, smooth-scaled snake with a relatively large head and moderately large eyes. Juveniles are black on the dorsum of the head with a white, yellow, or red nape band behind and a broad black vertebral stripe. The remainder of the dorsal scales are red with black tips, the undersides and lips being white. Adults are much darker, the black pigment overwhelming all other dorsal patterns.

RELATED SPECIES

The genus *Boiruna* contains a second species, the Sertão Mussurana (*B. sertaneja*), from the open xeric habitats of northeastern Brazil. Several other genera of mussuranas also occur in South America, including *Clelia* (page 304), *Mussurana* (page 323), and *Paraphimophis* (page 326). Adult *B. maculata* can be distinguished from most other mussuranas within their range by the black pigment on the underside of the posterior body and tail. Closely related genera include the Brazilian Birdsnake (*Rhachidelus brazili*) and the scarlet snakes (*Pseudoboa*, page 333).

Actual size

FAMILY	Dipsadidae: Xenodontinae
RISK FACTOR	Rear-fanged, venomous: hemotoxins and hemorrhagins
DISTRIBUTION	West Indies: Puerto Rico, and the US and British Virgin islands
ELEVATION	0–1,480 ft (0–450 m) asl
HABITAT	Rainforest, open pasture, rocky hillsides, coconut plantations and groves, gardens, mangrove swamps, and beaches
DIET	Lizards, frogs, and snakes
REPRODUCTION	Oviparous, with clutches of 4–10 eggs
CONSERVATION STATUS	IUCN not yet assessed

ADULT LENGTH
2 ft 7 in–3 ft 3 in
(0.8–1.0 m)

BORIKENOPHIS PORTORICENSIS
PUERTO RICAN RACER
(REINHARDT & LÜTKEN, 1862)

The Puerto Rican Racer is found in Puerto Rico and the US and British Virgin islands. It is diurnally active in habitats ranging from rainforest to pastures and plantations. It may be found on beaches or under human trash, but is an early morning snake that is rarely seen after 10 a.m. Although terrestrial it is also highly agile and has been found up to 245 ft (75 m) from the ground in rainforest trees. It preys on lizards, frogs, and smaller snakes, which are killed by its hemolytic and hemorrhagic venom (see pages 34–35), although prey often has to be restrained with the coils while the venom takes effect. Bites to humans have caused severe but not life-threatening results. There are no highly venomous snakes on Puerto Rico or the Virgin Islands.

RELATED SPECIES

There are six recognized subspecies of *Borikenophis portoricensis*. The nominate form and one other subspecies (*B. p. prymnus*) occur on Puerto Rico, while other subspecies are found on Isla de Vieques (*B. p. aphantus*), Isla Culebra and St. Thomas (*B. p. richardi*), Buck Island, off St. Thomas (*B. p. nicholsi*), and Anegada and the British Virgin Islands (*B. p. anegadae*). The genus *Borikenophis* contained two other species, the Isla Mona Racer (*B. variegatus*), and the St. Croix Racer (*B. sanctaecrucis*), from the US Virgin Islands, but now listed as Extinct by the IUCN.

Actual size

The Puerto Rican Racer is a slender, long-tailed snake with a long, pointed head, slightly distinct from the neck, and large eyes with round pupils. It may be variably patterned, depending on which island it originated from, but commonly it is brown or gray above with fine black longitudinal stripes and pale gray to white below and on the lips. A black stripe may pass from the snout, through the eye.

FAMILY	Dipsadidae: Xenodontinae
RISK FACTOR	Nonvenomous
DISTRIBUTION	West Indies: Cuba, including Isla de la Juventud
ELEVATION	0–330 ft (0–100 m) asl
HABITAT	Palm forest, coastal scrub, dry scrub, wooded areas, brackish lagoons, pastures, fields, and houses
DIET	Frogs and lizards
REPRODUCTION	Oviparous, with clutches of up to 3 eggs
CONSERVATION STATUS	IUCN Least Concern

ADULT LENGTH
15¾–19¾ in
(400–500 mm)

CARAIBA ANDREAE
CUBAN BLACK AND WHITE RACER
(REINHARDT & LÜTKEN, 1862)

303

Also known as the Cuban Lesser Racer, this is one of the most widely distributed Cuban snakes. The Cuban Black and White Racer is found throughout mainland Cuba, the Isla de la Juventud, and the smaller cays off the north and south coasts. This snake is an inhabitant of palm forest, coastal scrub, and brackish lagoons, but it is also found in agricultural areas and will enter houses. It is terrestrial, diurnally active, and highly alert, hunting frogs and lizards, and using a burst of speed to escape predators. It is also sometimes fossorial as specimens have been turned up by plows. Captain F. Andréa was the Danish sea captain who collected the original type series for the Copenhagen Museum.

RELATED SPECIES

Although *Caraiba andreae* is a monotypic species, it does contain six subspecies, three from mainland Cuba and three from the Isla de la Juventud, the Cayo Cantiles, and the Cayos Santa María and Guajaba. Its closest relative is the Hispaniolan Racer (*Haitiophis anomalus*), which some authors also place in *Caraiba*. Both these species are related to the Greater Antillean racers (*Cubophis*, page 307).

The Cuban Black and White Racer is a small, smooth-scaled snake with a distinctive head, large eyes with round pupils, and a long tail. Although variable, it is often black above with occasional white spots, while the black scales of the flanks are heavily marked with white and the undersides and lips are white with black suturing.

Actual size

FAMILY	Dipsadidae: Xenodontinae
RISK FACTOR	Rear-fanged, venomous, also constrictor
DISTRIBUTION	Central and South America: Belize and Guatemala, to Trinidad and Lesser Antilles, and south to northwest Argentina
ELEVATION	0–8,200 ft (0–2,500 m) asl
HABITAT	Lowland forests, secondary growth, cultivated areas, and along ditches and roads
DIET	Snakes, lizards, and small mammals
REPRODUCTION	Oviparous, with clutches of 10–22 eggs
CONSERVATION STATUS	IUCN not yet assessed, CITES Appendix II

ADULT LENGTH
Male
3 ft 3 in–6 ft
(1.0–1.8 m)

Female
5 ft–8 ft 6 in
(1.5–2.6 m)

304

CLELIA CLELIA
COMMON MUSSURANA
(DAUDIN, 1803)

Actual size

Also known as the Ratonera, this is a widely distributed mussurana, found from Belize and Guatemala to Trinidad, and south to northeastern Argentina. It occurs in lowland habitats, from pristine rainforest to secondary growth, and is encountered in human-mediated habitats, along ditches or roads, or around buildings, where it hunts its primary prey, sympatric pitvipers (*Atropoides*, page 556; *Bothrops*, pages 560–569; *Porthidium*, page 593) and rattlesnakes (*Crotalus*, pages 572–581). Mussuranas are immune to the bites of their quarry species. They also prey on lizards and small mammals. Following an initial venomous bite to the neck, the Common Mussurana will also use constriction to subdue its prey. It is generally inoffensive, and does not bite if handled, but caution is advised, as it is a large snake. Females are larger than males.

RELATED SPECIES

The Common Mussurana is represented by two subspecies, the nominate form inhabiting most of the range, with an island endemic (*C. c. groomei*) on Grenada in the Lesser Antilles. The genus *Clelia* also contains six other species of mussuranas. Reports of *Clelia clelia* from southern Mexico refer to the Mexican Mussurana (*C. scytalina*), while those from St. Lucia refer to the endemic St. Lucia Mussurana (*C. errabunda*). Other related genera of South American mussuranas include *Boiruna* (page 301), *Mussurana* (page 323), and *Paraphimophis* (page 326).

The Common Mussurana is a large, powerful snake with a stout body, glossy, smooth scales, and a large head with small eyes and round pupils. Adults are gunmetal gray above, without markings, and white below, while juveniles are red above and white below, with a black head and a broad yellow or cream neck band, followed by a dorsal black patch. The juvenile resembles juveniles of the scarlet snakes (*Pseudoboa*, page 333) and adult coffee snakes (*Ninia*, page 283).

FAMILY	Dipsadidae: Xenodontinae
RISK FACTOR	Rear-fanged, mildly venomous: possibly anticoagulants; potentially dangerous
DISTRIBUTION	North and Central America: Mexico, Guatemala, El Salvador, Honduras, Nicaragua, and Costa Rica
ELEVATION	0–4,920 ft (0–1,500 m) asl
HABITAT	Dry savannas, beaches, dry tropical forest, roads, and open areas in wet tropical forest
DIET	Lizards, snakes, frogs, and small mammals or birds
REPRODUCTION	Oviparous, with clutches of 5–10 eggs
CONSERVATION STATUS	IUCN Least Concern

ADULT LENGTH
3 ft–3 ft 10 in
(0.7–1.16 m)

CONOPHIS LINEATUS
CENTRAL AMERICAN ROAD GUARDER
(DUMÉRIL, BIBRON & DUMÉRIL, 1854)

305

The Central American Road Guarder is the most widely distributed member of the genus *Conophis*, occurring from central Mexico to Costa Rica. It inhabits savanna grassland, dry tropical forest, beaches, and wet tropical forest, and is both terrestrial and arboreal. As a "road guarder," a snake that predates other snakes, it often preys on highly venomous sympatric coralsnakes (*Micrurus*, pages 457–465), but its prey consists primarily of lizards, hence the alternative name "lizard killer." Small mammals, birds, and frogs are also taken. The long rear fangs inject fairly toxic venom. This is an irascible species that will bite readily if handled. Given that snakebite victims report an intense painful burning sensation and swelling, widespread tingling, and prolonged bleeding from the bite site, road guarders should be considered potentially dangerous and treated with caution.

Actual size

RELATED SPECIES

The genus *Conophis* also contains four localized species, treated as synonyms of subspecies by some authors: the Yucatan Road Guarder (*C. concolor*), a former unicolor subspecies of *C. lineatus*; the Tuxtlas Road Guarder (*C. morai*); the Southwest Mexican Road Guarder (*C. vittatus*); and the Beautiful Road Guarder (*C. pulcher*) from Chiapas, Guatemala, and El Salvador. Nevermann's Road Guarder (*Crisantophis nevermanni*, page 306), from Guatemala to Costa Rica, was also once included in *Conophis*.

The Central American Road Guarder is a smooth-scaled snake with a narrow head, a pointed snout, moderately large eyes, round pupils, and a long tail. It may be brown or olive and it is patterned dorsally with a series of light and dark longitudinal stripes, although the color, width, and number of these stripes varies. The throat and belly are immaculate white.

FAMILY	Dipsadidae: Xenodontinae
RISK FACTOR	Rear-fanged, mildly venomous
DISTRIBUTION	Central America: Guatemala, El Salvador, Honduras, Nicaragua, and Costa Rica
ELEVATION	0–4,580 ft (0–1,395 m) asl
HABITAT	Lowland dry forest and submontane wet forest, especially along streams, lakes, and rice paddies
DIET	Frogs, toads, and small snakes, including cannibalism
REPRODUCTION	Oviparous, with clutches of up to 10 eggs
CONSERVATION STATUS	IUCN Least Concern

ADULT LENGTH
23¾–32½ in
(600–828 mm)

306

CRISANTOPHIS NEVERMANNI
NEVERMANN'S ROAD GUARDER
(DUNN, 1937)

Nevermann's Road Guarder is a smooth-scaled, relatively slender snake with an elongate head, moderately large eyes, and a long tail. It is black above with a series of four fine yellow longitudinal stripes, the lower stripes being twice the width of the upper stripes, and a yellow to cream venter.

Nevermann's Road Guarder is found along the Pacific versant of Central America from Guatemala to Costa Rica, and on the Atlantic versant in Nicaragua. It inhabits dry lowland forest but has also been found in submontane wet forest. Although it is called a "road guarder," a snake that eats other snakes, it primarily takes frogs and lizards, but snakes are also eaten, including smaller specimens of its own species. It uses a combination of venom and constriction to subdue its prey. The effects of a bite to a human are unknown and, as with other "unknown" rear-fanged snakes, caution is recommended. Often called Dunn's Road Guarder, this species was described by Dunn in honor of his German coleopterist friend Wilhelm Nevermann (1881–1938).

RELATED SPECIES

The monotypic genus *Crisantophis* is possibly most closely related to the Central American Road Guarder (*Conophis lineatus*, page 305), but it is considered *incertae sedis* (of unknown status) within the Dipsadinae by some authors.

Actual size

FAMILY	Dipsadidae: Xenodontinae
RISK FACTOR	Rear-fanged, mildly venomous
DISTRIBUTION	West Indies: Cuba, including Isla de la Juventud
ELEVATION	0–985 ft (0–300 m) asl
HABITAT	Hillsides, scrubland, beaches, pasture, wooded areas, mangrove swamps, and around human habitations
DIET	Frogs, lizards, snakes, birds, and small mammals
REPRODUCTION	Oviparous, with clutches of 10–24 eggs
CONSERVATION STATUS	IUCN not yet assessed

ADULT LENGTH
3 ft–4 ft 3 in
(0.95–1.3 m)

CUBOPHIS CANTHERIGERUS
CUBAN RACER
(BIBRON, 1840)

A widespread species, the Cuban Racer is found throughout Cuba in many different habitats, from beaches and mangrove swamps to fields and scrubby hillsides. It is also found close to human habitations and may be found sheltering under human trash. It is an active predator of frogs, lizards, snakes, birds, bats, and rodents, but its venom is weak so it also relies on constriction to subdue its prey. There are no accounts of bites to humans, but caution is recommended when handling any rear-fanged species of a reasonable size. Although primarily a terrestrial, diurnal snake, which may be seen dashing about after prey, it has also been observed to be active at night.

RELATED SPECIES

There are four subspecies of *Cubophis cantherigerus* distributed throughout mainland Cuba, on Isla de la Juventud, and on small islands and cays (low banks) around the coast. The Greater Antillean racer genus *Cubophis* also contains five other species: the Swan Island Racer (*C. brooksi*), Grand Cayman Racer (*C. caymanus*), Cayman Brac Racer (*C. fuscicauda*), and Little Cayman Racer (*C. ruttyi*), all former subspecies of *C. cantherigerus*, and the Bahamian Racer (*C. vudii*). Closely related taxa are the Cuban Black and White Racer (*Caraiba andreae*, page 303) and the Hispaniolan Racer (*Haitiophis anomalus*).

The Cuban Racer is a smooth-scaled snake with an elongate head, large eyes with round pupils, and a long tail. It is variably patterned, but many specimens are gray to light brown with dark suturing around every scale, presenting a reticulate effect. A black cap is often present on the posterior of the head, and dark postocular stripes are also present.

Actual size

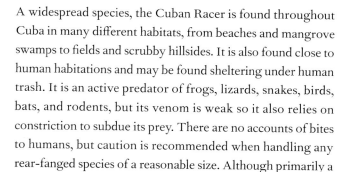

FAMILY	Dipsadidae: Xenodontinae
RISK FACTOR	Rear-fanged, mildly venomous
DISTRIBUTION	South America: Colombia, Ecuador, Brazil, Peru, Bolivia, and the Guianas
ELEVATION	165–1,640 ft (50–500 m) asl
HABITAT	Rainforest clearings or trails
DIET	Lizard eggs
REPRODUCTION	Oviparous, with clutches of 2–3 eggs
CONSERVATION STATUS	IUCN not yet assessed

ADULT LENGTH
15¾–21¼ in
(400–540 mm)

DREPANOIDES ANOMALUS
LIZARD EGG-EATING SNAKE
(JAN, 1863)

The Lizard Egg-eating Snake is a small, smooth-scaled snake with a head slightly distinct from the neck, small eyes, and a long tail that is often truncated. It is bright red with black scale tips, which provides a reticulate effect, and a glossy black head, with a white cream band across the black nape collar, which extends forward onto the lips.

The Lizard Egg-eating Snake is a western Amazonian species found in southern Colombia, western Brazil, eastern Ecuador, Peru, and northern Bolivia, but it is also reported from the Guianas in the northeast. It is a nocturnal, forest-floor species that is usually found in clearings or along trails, where it forages for lizard eggs, its only documented diet. It has also been reported to be both semi-fossorial and semi-arboreal. This is a small snake, and females only lay two or three eggs. This species may easily be mistaken for a juvenile mussurana (*Clelia*, page 304) or scarlet snake (*Pseudoboa*, page 333), with which it occurs in sympatry, and from which it can be distinguished by the lack of a loreal scale. Although rear-fanged and venomous it is not thought to be dangerous to humans.

RELATED SPECIES

The genus *Drepanoides* is monotypic, and closely related to the snake-eating mussuranas (*Boiruna*, page 301; *Clelia*, page 304; *Mussurana*, page 323; and *Paraphimophis*, page 326).

Actual size

FAMILY	Dipsadidae: Xenodontinae
RISK FACTOR	Nonvenomous
DISTRIBUTION	South America: southeastern and southern Brazil, and northern Argentina
ELEVATION	1,670–3,360 ft (510–1,025 m) asl
HABITAT	Atlantic coastal forests
DIET	Frogs, and possibly lizards
REPRODUCTION	Oviparous, clutch size unknown
CONSERVATION STATUS	IUCN not yet assessed

ADULT LENGTH
23¾–31½ in
(600–800 mm)

ECHINANTHERA CYANOPLEURA
YELLOW-BELLIED FOREST SNAKE
(COPE, 1885)

First described from a holotype collected at São Joao de Montenegro in Rio Grande do Sul, southeastern Brazil, this species is now known to occur as far north as Rio de Janeiro and as far south as Misiones, Argentina. It is primarily an inhabitant of the threatened Atlantic coastal forests, but has also been found farther inland. Its prey consists of frogs and possibly small lizards. It is a diurnal species that inhabits leaf litter on the forest floor, and it is rarely encountered at night. Females are oviparous but the clutch size is unknown. If disturbed this nonvenomous snake defends itself by body-flattening and expelling the foul-smelling contents of its cloacal glands.

The Yellow-bellied Forest Snake is a small, smooth-scaled snake with a long tail and a head slightly distinct from the neck, with moderately large eyes and round pupils. It is brown above, darker on the flanks than the dorsum, with pale punctuated longitudinal stripes. The undersides are yellow, as is a broken collar around the neck, but the lips are white.

RELATED SPECIES

There are five other species in the genus *Echinanthera*, distributed from Bahia to Rio Grande do Sul in eastern Brazil and into Uruguay and northeast Argentina, although the Undulated Forest Snake (*E. undulata*) is found in southeastern Colombia. The most similar species is the Line-headed Forest Snake (*E. cephalostriata*). The closest related genus is another leaf-litter genus, *Taeniophallus* (page 340).

Actual size

FAMILY	Dipsadidae: Xenodontinae
RISK FACTOR	Rear-fanged, venomous; potentially dangerous
DISTRIBUTION	South America: southeastern Brazil
ELEVATION	0–1,150 ft (0–350 m) asl
HABITAT	Lowland and coastal forests
DIET	Earthworms, amphisbaenians, and small snakes
REPRODUCTION	Oviparous, clutch size unknown
CONSERVATION STATUS	IUCN not yet assessed

ADULT LENGTH
1 ft 8 in–3 ft 3 in
(0.5–1.0 m)

ELAPOMORPHUS QUINQUELINEATUS
FIVE-LINED BURROWING SNAKE
(RADDI, 1820)

310

The Five-lined Burrowing Snake is a rare leaf-litter inhabitant of southeastern Brazilian lowland rainforest, from southern Bahia to Rio Grande do Sul. Adults prey on amphisbaenians, (also known as worm-lizards) and small snakes, while juveniles eat earthworms. The commonly used name, Raddi's Lizard-eating Snake, seems inappropriate given this diet. Although known to be oviparous, its clutch size is unknown. This is a diurnal, terrestrial, or fossorial species that shelters under logs or in litter-filled tree buttresses. If handled it will thrash about violently, expel its cloacal gland contents, and may bite. The venom has not been studied, but despite the relatively small size of its head this snake should be treated with caution because *Elapomorphus* is closely related to the genus *Phalotris* (page 327), which has caused a life-threatening snakebite.

RELATED SPECIES

The genus *Elapomorphus* contains a second species, Wucherer's Burrowing Snake (*E. wuchereri*) from Bahia and Espírito Santo, eastern Brazil. Genera related to *Elapomorphus* include *Phalotris* (page 327) and *Apostolepis* (page 299).

The Five-lined Burrowing Snake is a smooth-scaled snake with a short tail, a narrow head that is indistinct from the neck, and small eyes with round pupils. It is brown above and yellow on the flanks and undersides, with a yellowish collar around the neck and five dark brown longitudinal stripes.

Actual size

FAMILY	Dipsadidae: Xenodontinae
RISK FACTOR	Rear-fanged, mildly venomous
DISTRIBUTION	South America: Colombia to Trinidad and northeast Argentina
ELEVATION	0–7,550 ft (0–2,300 m) asl
HABITAT	Rainforest
DIET	Other snakes, amphisbaenians, and lizards
REPRODUCTION	Oviparous, with clutches of up to 5 eggs
CONSERVATION STATUS	IUCN not yet assessed

ADULT LENGTH
25½–31½ in
(650–800 mm)

ERYTHROLAMPRUS AESCULAPII
AESCULAPIAN FALSE CORALSNAKE
(LINNAEUS, 1758)

311

The Aesculapian False Coralsnake is a common Amazonian coralsnake mimic found throughout most of northern South America, east of the Andes, including Trinidad. It occurs as far south as northeastern Argentina. Across its extensive range it demonstrates a variation of tricolor banded patterns to match the coralsnakes (*Micrurus*, pages 457–465) within its range, but it differs from true coralsnakes by possessing a loreal scale. The Aesculapian False Coralsnake is a forest-floor species that inhabits leaf litter in primary rainforest habitats, often being found along trails or in freshly cleared plots. It preys on other snakes, amphisbaenians (worm-lizards), and lizards, which it kills with a venomous bite, but it is not believed to possess venom toxic enough to affect humans. Caution is recommended with any coralsnake mimic, in case it isn't a mimic.

The Aesculapian False Coralsnake is a smooth-scaled snake with a narrow head, only slightly distinct from the neck, a rounded, white-tipped snout, and moderately large eyes with round pupils. A tricolor snake, it is banded black-white-black, including over the head, with broad red interspaces, although older animals are more melanistic. White and red scales are black-tipped. There are geographical variations in this pattern to match the local coralsnakes.

Actual size

RELATED SPECIES

Erythrolamprus aesculapii is represented by four subspecies, the nominate form in the Amazon, and subspecies in the Brazilian Atlantic forests (*E. a. monozona*), Bolivia (*E. a. tetrazona*), and southeastern Brazil to Argentina (*E. a. venustissimus*). Until relatively recently the genus *Erythrolamprus* contained only five other species, but it has been greatly expanded, with the inclusion of many species formerly in *Liophis*. There are now 50 species in the genus *Erythrolamprus*.

FAMILY	Dipsadidae: Xenodontinae
RISK FACTOR	Rear-fanged, mildly venomous; may contain an anticoagulant
DISTRIBUTION	South America: Colombia to the Guianas and Trinidad, and south to northern Argentina
ELEVATION	0–9,840 ft (0–3,000 m) asl
HABITAT	Rainforest, secondary growth, and cultivated areas, usually near water
DIET	Frogs and lizards
REPRODUCTION	Oviparous, with clutches of 1–6 eggs
CONSERVATION STATUS	IUCN not yet assessed

ADULT LENGTH
13¾–21¾ in,
rarely 27½ in
(350–550 mm,
rarely 700 mm)

312

ERYTHROLAMPRUS REGINAE
ROYAL GROUNDSNAKE
(LINNAEUS, 1758)

The Royal Groundsnake is a smooth-scaled snake with a rounded head, large eyes, and round pupils. It is brown or olive-green above with a speckling of yellow and black. The undersides are checkerboarded with black and either yellow or red.

Also sometimes referred to as the Common Swamp Snake, this is a widely distributed and frequently encountered American snake that occurs from Colombia to northern Argentina. It inhabits both primary and secondary rainforest and also cultivated gardens and cultivated plots, but is usually associated with wetlands. Its preference is for shallow watercourses with abundant littoral or floating vegetation. Prey consists of frogs, geckos, and the microteiid lizards that also occupy aquatic habitats and low vegetation into which the snake may climb. Some relatives of the Royal Groundsnake are known to possess toxic saliva containing an anticoagulant, but whether it would have any effect on humans is unknown. Its defense involves neck-flattening to expose the contrasting orange and blue interstitial skin, but this snake is generally inoffensive. The name *reginae* means "royal."

Actual size

RELATED SPECIES

Erythrolamprus reginae was formerly in the genus *Liophis*, along with 42 other species now synonymized within the genus *Erythrolamprus*. Three subspecies of *E. reginae* are recognized, the nominate northern South American form (*E. r. reginae*), an Amazonian and Atlantic coastal form (*E. r. semilineatus*), and a southern Brazilian, Paraguayan, and northern Argentinian form (*E. r. macrosoma*). The Military Groundsnake (*E. miliaris*) is a close relative of *E. reginae*.

FAMILY	Dipsadidae: Xenodontinae
RISK FACTOR	Rear-fanged, mildly venomous
DISTRIBUTION	South America: southeastern Brazil
ELEVATION	1,820 ft (555 m) asl
HABITAT	Gallery forest and watercourses over muddy substrates
DIET	Earthworms
REPRODUCTION	Viviparous, litter size unknown
CONSERVATION STATUS	IUCN not yet assessed

ADULT LENGTH
11¾–19¾ in
(300–500 mm)

GOMESOPHIS BRASILIENSIS
BRAZILIAN BALLSNAKE
(GOMES, 1918)

313

The Brazilian Ballsnake is confined to the southeastern states of Brazil, from Minas Gerais to Rio Grande do Sul. It is terrestrial to semi-aquatic, inhabiting shallow watercourses and gallery forest, and is most often found burrowing in muddy substrates in these habitats. This species is believed to feed exclusively on earthworms and is a member of a guild of snakes known as "goo-eaters"— feeders on earthworms, slugs, or snails. Although a rear-fanged venomous species, the Brazilian Ballsnake is so named because its primary defense is to roll into a defensive ball in the same manner as the Ball Python (*Python regius*, page 99), Northern Rubber Boa (*Charina bottae*, page 120), and Asian pipesnakes (*Cylindrophis*, pages 73–75). It is unusual within the Xenodontinae in being live-bearing, which is probably an aquatic adaptation.

Actual size

RELATED SPECIES

This species was originally described as *Tachymenis brasiliensis* by the Brazilian herpetologist João Florêncio Gomes (1886–1919), before being transferred to the newly created monotypic genus *Gomesophis* in 1959. *Gomesophis* is closely related to the scrub snakes (*Tachymenis*, page 339), the pampas snakes (*Tomodon*, page 342), and the mock vipers (*Thamnodynastes*, page 341).

The Brazilian Ballsnake is a glossy, smooth-scaled snake with a narrow, pointed head that is indistinct from the neck, small eyes, and a short tail. It is light brown above, often with either broad or fine brown longitudinal stripes, and dark brown on the flanks.

FAMILY	Dipsadidae: Xenodontinae
RISK FACTOR	Rear-fanged, mildly venomous
DISTRIBUTION	South America: Colombia to Trinidad, south to Brazil and Bolivia
ELEVATION	0–7,910 ft (0–2,410 m) asl
HABITAT	Slow rivers, grassy marshes, swamps
DIET	Fish, frogs, lizards, and earthworms
REPRODUCTION	Oviparous, with clutches of 4–20 eggs
CONSERVATION STATUS	IUCN not yet assessed

ADULT LENGTH
23¾–31½ in,
rarely 3 ft 4 in
(600–800 mm,
rarely 1.02 m)

314

HELICOPS ANGULATUS
BANDED KEELED WATERSNAKE
(LINNAEUS, 1758)

The Banded Keeled Watersnake is a small snake with strongly keeled scales, a short tail, and a moderately broad head, with small, dorsolaterally positioned eyes and round pupils. Patterning comprises irregular dark or reddish bands on a light brown background. The bands may continue onto the cream, gray, orange, or red belly. The subcaudal scales are also keeled.

The Banded Keeled Watersnake is an inhabitant of slow-moving rivers and heavily vegetated marshes and swamps throughout northern South America, from Colombia, east to Trinidad, and south to Bolivia, and Goiás in central Brazil. It is nocturnal and semi-aquatic and preys on fish, frogs, and occasionally the lizards associated with aquatic habitats. Giant earthworms are also included in its diet when other prey is less abundant. Although generally inoffensive, this species can be truculent if disturbed, flattening its body in defense and voiding the contents of its cloacal glands if handled. It may also bite easily, with bites causing localized inflammation, swelling, and pain, but it is not considered dangerous to humans. Large females may lay up to 20 eggs in a single clutch.

RELATED SPECIES

The genus *Helicops* is the equivalent of the North American watersnake or keelback genus *Nerodia* (pages 417–420), albeit in a different family. The genus contains 17 other species, with the Spotted Keeled Watersnake (*H. leopardinus*) occurring in sympatry with *H. angulatus* through much of its range. The São Paulo Keeled Watersnake (*H. gomesi*) is a close relative of *H. angulatus*.

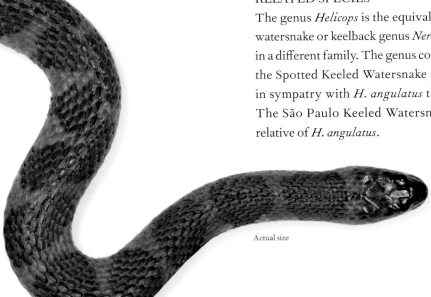

Actual size

FAMILY	Dipsadidae: Xenodontinae
RISK FACTOR	Rear-fanged, mildly venomous
DISTRIBUTION	South America: the Guianas and Brazil to northern Argentina
ELEVATION	0–7,910 ft (0–2,410 m) asl
HABITAT	Slow rivers, small ponds, and grassy marshes and swamps
DIET	Frogs and fish
REPRODUCTION	Viviparous, with litters of 7–31 neonates
CONSERVATION STATUS	IUCN not listed

ADULT LENGTH
19¾–31½ in
(500–800 mm)

HELICOPS LEOPARDINUS
SPOTTED WATERSNAKE
(SCHLEGEL, 1837)

The Spotted Watersnake is an inhabitant of slow-moving rivers, small quiet ponds, and heavily vegetated marshes and swamps, especially those with floating littoral vegetation, throughout northern South America, from the Guianas and Brazil, where it is common and widely distributed, as far south as the Rio Parana region of Argentina. It is nocturnal and semi-aquatic in habit, preying on frogs and fish. Although generally inoffensive all members of this genus will bite if handled, with bites causing localized inflammation, swelling and pain, but these watersnakes are not considered dangerous to humans. The Spotted Watersnake is viviparous, with litters of 7–31 neonates, in contrast to its congener, *Helicops angulatus*, which lays eggs.

RELATED SPECIES

The genus *Helicops* contains 18 species. *Helicops leopardinus* occurs in sympatry with the South American Banded Keeled Watersnake (*H. angulatus*, page 314) across much of its range, especially in Brazil, although the two species differ in their reproductive strategy.

Actual size

The Spotted Watersnake is a small snake with strongly keeled scales, a short tail, and a moderately broad head, distinct from the neck, with small, dorsolaterally positioned eyes and round pupils. The dorsal patterning consists of four rows of dark spots on a gray background, some of the spots being occasionally connected by narrow bands, while the undersides exhibit a checkerboard of red and black posteriorly and white and black anteriorly.

FAMILY	Dipsadidae: Xenodontinae
RISK FACTOR	Rear-fanged, venomous
DISTRIBUTION	South America: the Guianas to northern Argentina
ELEVATION	0–330 ft (0–100 m) asl
HABITAT	Slow rivers, marshes, and lakes
DIET	Fish, frogs, small mammals, birds, and other reptiles
REPRODUCTION	Oviparous, with clutches of 20–30 eggs
CONSERVATION STATUS	IUCN not yet assessed

ADULT LENGTH
6 ft 7 in–9 ft,
possibly up to 10 ft
(2.0–2.7 m,
possibly up to 3.0 m)

HYDRODYNASTES GIGAS
GIANT FALSE WATER COBRA
(DUMÉRIL, BIBRON & DUMÉRIL, 1854)

The Giant False Water Cobra is a large, smooth-scaled snake with a moderately broad, pointed head and eyes with round pupils. It is mottled brown above with a series of darker-centered and black-edged rhomboid blotches on the back and a broad, black postocular stripe that runs onto the neck. It is pale yellow to grayish brown below. The defensive hood is very impressive.

Called a "false cobra" because of its ability to spread a narrow, cobra-like hood when threatened, albeit horizontally and not vertically, this is a popular species in collections. It is a large, showy, diurnal snake, but not one without problems as people respond in different ways to its bites, some experiencing no symptoms but others suffering gross swelling and localized pain. As a large, rear-fanged venomous snake it demands respect. It inhabits lowland aquatic habitats from the Guianas, through eastern Amazonia, to northern Argentina, with a preference for slow-moving rivers and marshes, though it is also encountered away from water. This is one of South America's largest snakes, after the boas. It preys on a wide variety of aquatic and terrestrial vertebrates and is an adept climber.

RELATED SPECIES

Two additional species are included in genus *Hydrodynastes*, which was once known as *Cyclagras*: Hermann's Watersnake (*H. bicinctus*), also from northern South America, and the recently described Black False Water Cobra (*H. melanogigas*) from Tocantins state, central Brazil. This genus appears to be most closely related to the monotypic Amaral's Groundsnake (*Caaeteboia amarali*) and the flat-headed snakes (*Xenopholis*, page 347). *Hydrodynastes gigas* is known locality as Boipevaçu.

Actual size

FAMILY	Dipsadidae: Xenodontinae
RISK FACTOR	Rear-fanged, mildly venomous
DISTRIBUTION	South America: Amazonian Brazil, Venezuela, Colombia, Ecuador, and Peru
ELEVATION	0–820 ft (0–250 m) asl
HABITAT	Slow and shallow rivers, marshes, and lakes
DIET	Freshwater eels and possibly caecilians
REPRODUCTION	Oviparous or viviparous, with clutches or litters of up to 7
CONSERVATION STATUS	IUCN Least Concern

ADULT LENGTH
2 ft 7 in–3 ft 3 in
(0.8–1.0 m)

HYDROPS MARTII
AMAZONIAN SMOOTH-SCALED WATERSNAKE
(WAGLER, 1824)

317

Also known as the Coral Mudsnake, this brightly colored Amazonian species is recorded from Pará in the mouth of the Brazilian Amazon, to Venezuela, Colombia, Peru, and Ecuador, in the western Amazon. It may be both diurnal and nocturnal and shows a preference for shallow, slow-moving waterways, including oxbow lakes, swamps, and the side channels of larger rivers. Its prey is fairly specialized and consists of synbranchid eels, and possibly also caecilians (legless amphibians), which can be swallowed easily by narrow-mouthed snakes. The reproductive strategy of this species appears to be bimodal, with both egg-laying and live-bearing reported. Nothing is known of the toxicity of its saliva or venom so caution is recommended. This species is named for Carl Friedrich Philipp von Martius (1794–1868), a German botanist who visited Brazil.

The Amazonian Smooth-scaled Watersnake is a small snake with a narrow head and small eyes. It is stunningly patterned, with white-edged black bands, or rings, separated by red interspaces, these becoming pale yellow on the venter. The head bears a black-edged white band across the snout, and a white and black collar.

RELATED SPECIES

The genus *Hydrops* contains two other species, the widely distributed Triangle Watersnake (*H. triangularis*), which has six subspecies across northern and central South America, and the recently described Pantanal Watersnake (*H. caesurus*), from Paraguay, southern Brazil, and northern Argentina. The most closely related genera are the smooth-scaled glossy pondsnakes (*Pseudoeryx*, page 334), and the South American keeled watersnakes (*Helicops*, pages 314–315), which have keeled scales.

Actual size

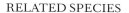

FAMILY	Dipsadidae: Xenodontinae
RISK FACTOR	Nonvenomous
DISTRIBUTION	West Indies: Hispaniola (Haiti and Dominican Republic)
ELEVATION	0–5,580 ft (0–1,700 m) asl
HABITAT	Dry coastal forest and lowland cactus scrub
DIET	Lizards
REPRODUCTION	Oviparous, with clutches of 3–15 eggs
CONSERVATION STATUS	IUCN not yet assessed

ADULT LENGTH
27½–31½ in
(700–800 mm)

318

HYPSIRHYNCHUS FEROX
HISPANIOLAN CAT-EYED SNAKE
GÜNTHER, 1858

The Hispaniolan Cat-eyed Snake occurs in both Haiti and the Dominican Republic, and on Île de la Gonâve (Dominican Republic) and Isla Saona (Haiti). It is a common snake of arid lowland and submontane habitats such as mesic woodland and xeric cactus scrub, where it may be either terrestrial or arboreal. Despite its vertically elliptical, catlike pupils, which suggest nocturnal activity, this snake is often seen during the height of the day. It is an active forager or ambush predator of both terrestrial and arboreal lizards. Juveniles show a preference for anoles. Larger, more terrestrial lizards such as ameivas, galliwasps, and curlytail lizards feature more in the adults' diet. Females are oviparous.

The Hispaniolan Cat-eyed Snake is a small snake with smooth scales, a long tail, a narrow, pointed head, a slightly upturned snout, and small eyes with vertically elliptical pupils. Specimens are dorsally brown or gray, usually with a narrow, dark, viper-like vertebral zigzag pattern. The shape and patterning of this species resembles that of the unrelated bark snakes (*Hemirhagerrhis*, page 369) of Africa.

RELATED SPECIES

Three subspecies are recognized, from the mainland (*Hypsirhynchus ferox ferox*), Île de la Gonâve (*H. f. exedrus*), and Isla Saona (*H. f. paracrousis*). This species' closest relative is the Tiburon Cat-eyed snake (*H. scalaris*), a former Haitian subspecies. The genus *Hypsirhynchus* contains between two and eight species, because some authors place the Hispaniolan and Jamaican species into different genera (*Antillophis*, *Ocyophis*, and *Schwartzophis*). It is feared the Jamaican Racer (*H. ater*) and the La Vega Racer (*H. melanichnus*), from Hispaniola, may be extinct.

Actual size

FAMILY	Dipsadidae: Xenodontinae
RISK FACTOR	Rear-fanged, mildly venomous
DISTRIBUTION	West Indies: Hispaniola (Haiti and Dominican Republic)
ELEVATION	0–3,280 ft (0–1,000 m) asl
HABITAT	Dry coastal forest, especially on limestone karst; also coffee plantations
DIET	Frogs, lizards, and small mammals
REPRODUCTION	Oviparous, with clutches of up to 12 eggs
CONSERVATION STATUS	IUCN not yet assessed

ADULT LENGTH
3 ft–3 ft 3 in
(0.9–1.0 m)

IALTRIS DORSALIS

HISPANIOLAN FANGED RACER

(GÜNTHER, 1858)

319

The Hispaniolan Fanged Racer is a diurnally active, terrestrial predator of frogs, lizards, and small mammals. It occurs throughout mainland Hispaniola, and on the Haitian islands of Île-à-Vache, Île de la Gonâve, and Île de la Tortue. It frequents dry lowland habitats such as mesic woodland, especially woodland on limestone karst outcrops, but is also found in coffee plantations. Although a rear-fanged venomous species, the Hispaniolan Fanged Racer is not believed to be harmful to humans, and specimens kept in captivity have not attempted to bite. There are no dangerous snakes on Hispaniola or its satellite islands, but caution is advised when handling any large specimens. Females are oviparous.

The Hispaniolan Fanged Racer is a smooth-scaled snake with a long tail, a moderately broad head with a pointed snout, and small eyes with round pupils. It is usually gray in color, with darker gray flecking on the dorsum of the body, a dark gray nape band, a fine dark gray "W" on the back of the head with the bars forming the postocular stripes, and a fine vertebral stripe. The undersides are brown or gray.

RELATED SPECIES

There are three other species in the genus *Ialtris*, all from Hispaniola. The Barreras Fanged Racer (*I. agyrtes*) occurs in southwestern Dominican Republic, the Haitian Fanged Racer (*I. haetianus*) occurs in both Haiti and the Dominican Republic, and Parish's Fanged Racer (*I. parishi*) is endemic to Île de la Tortue. Some authors place the Haitian Fanged Racer in genus *Darlingtonia*. *Ialtris* (and *Darlingtonia*) are closest to the Hispaniolan and Jamaica racers of genus *Hypsirhynchus* (page 318).

Actual size

FAMILY	Dipsadidae: Xenodontinae
RISK FACTOR	Nonvenomous
DISTRIBUTION	Central and South America: Panama, Colombia, Venezuela, the Guianas, and northern Brazil
ELEVATION	0–2,950 ft (0–900 m) asl
HABITAT	Savanna, wetlands, gallery forest, dry deciduous forest, and thorn scrub
DIET	Frogs
REPRODUCTION	Oviparous, with clutches of 5–7 eggs
CONSERVATION STATUS	IUCN not yet assessed

ADULT LENGTH
13¾–19¾ in,
rarely 23¾ in
(350–500 mm,
rarely 600 mm)

320

LYGOPHIS LINEATUS
NORTHERN LINED
GROUNDSNAKE
(LINNAEUS, 1758)

The genus *Lygophis* contains eight species, with the Northern Lined Groundsnake being the northernmost, occurring from Panama, through Colombia, Venezuela and the Guianas, to Brazil and the mouth of the Amazon. The Northern Lined Groundsnake is a diurnal and terrestrial species that occupies a variety of habitats across its range, from seasonally flooded savanna to dry deciduous woodland, and from wetlands and moist gallery forest to dry thorn scrub. It preys primarily, if not exclusively, on amphibians, mostly frogs. This is a nonvenomous species of no danger to humans. It is generally inoffensive and does not bite. Females produce clutches of five to seven eggs.

The Northern Lined Groundsnake is a smooth-scaled snake with a long tail, a narrow, pointed head, and moderately large eyes with round pupils. It is light brown above and white to gray below, with three broad, dark brown stripes that begin as a middorsal stripe and two lateral stripes on the head and continue longitudinally down the body. The lateral stripes may become less evident posteriorly.

RELATED SPECIES

The other species of *Lygophis* are found through southeastern and central Brazil, Bolivia, Paraguay, Uruguay, and Argentina. The Southern Lined Groundsnake (*L. meridionalis*) from southeastern Brazil, Bolivia, Paraguay, and Argentina is a former subspecies. Until recently, *Lygophis* was synonymized within *Liophis*, itself now a synonym of *Erythrolamprus* (pages 311–312). It is now considered the sister taxon to a clade comprising *Erythrolamprus* and *Xenodon* (page 345).

Actual size

FAMILY	Dipsadidae: Xenodontinae
RISK FACTOR	Nonvenomous
DISTRIBUTION	West Indies: Puerto Rico
ELEVATION	0–795 ft (0–243 m) asl
HABITAT	Rainforest, coastal forest, dry woods, cactus thicket, open pasture, coconut groves, and gardens
DIET	Frogs, their eggs, tadpoles, and lizards
REPRODUCTION	Oviparous, with clutches of 6–18 eggs
CONSERVATION STATUS	IUCN Least Concern

ADULT LENGTH
5¾–19¾ in
(400–500 mm)

MAGLIOPHIS STAHLI
PUERTO RICAN RACERLET
(STEJNEGER, 1904)

321

The Puerto Rican Racerlet inhabits the majority of Puerto Rico, only being absent from the southern coastal region where a different species occurs. It is generally a secretive, diurnal inhabitant of leaf litter and piles of dead vegetation. It may be found in wet habitats such as rainforest or coastal forest, or in dry woodland, cactus thicket, and also in open pasture. Being relatively small and easily overlooked it can also adapt to living in man-made monocultures such as coconut groves and plantations, and it may inhabit gardens. It is a harmless snake that preys on frogs and small lizards such as anoles or geckos, and it also eats tadpoles and frogs' eggs when they are available. Females are oviparous, laying up to 18 eggs. Agustín Stahl (1842–1917) was a Puerto Rican medical doctor and naturalist.

The Puerto Rican Racerlet is a small, smooth-scaled species with a narrow, slightly pointed head, moderately large eyes with round pupils and a relatively long tail. It is dark gray-brown on the flanks, chestnut brown on the dorsum, separated by a black stripe, which begins on the snout and runs through the eye. The lips are off-white to light gray and the undersides may be yellow-tan, greenish-tan, or orange-brown with brown speckling and a large spot on the outer ventral scales.

RELATED SPECIES

The Puerto Rican Racerlet was previously a subspecies of the Virgin Island Racerlet (*Magliophis exiguus*). It shares Puerto Rico with another sub-species (*M. e. subspadix*), which inhabits the south, while the nominate sub-species occurs on the British and US Virgin Islands. The genus *Magliophis* is closely related to the Greater Antillean racers i.e. *Cubophis* (page 307), *Borikenophis* (page 302), *Haitiophis*, *Caraiba* (page 303), and *Alsophis* (page 298).

Actual size

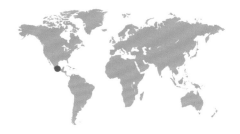

FAMILY	Dipsadidae: Xenodontinae
RISK FACTOR	Nonvenomous, constrictor
DISTRIBUTION	North America: western Mexico
ELEVATION	0–6,230 ft (0–1,900 m) asl
HABITAT	Tropical deciduous woodland, semi-deciduous woodland, and pine-oak forest
DIET	Lizards
REPRODUCTION	Oviparous, with clutches of up to 10 eggs
CONSERVATION STATUS	IUCN Least Concern

ADULT LENGTH
Male
15¾–22½ in
(400–570 mm)

Female
19¾–28¼ in
(500–720 mm)

322

MANOLEPIS PUTNAMI
MEXICAN THIN-SCALED SNAKE
(JAN, 1863)

The Mexican Thin-scaled Snake is a small snake, with smooth to slightly carinate scales, a slender body, a relatively long tail, and a narrow, pointed, angular head with large eyes and round pupils. It is brown to yellow dorsally, with a broad, dark vertebral stripe, which is light-centered in females but fully dark in males. The head is light or dark gray, flecked with black and with black marks on the dorsum and under the eyes. The undersides of the head and body are dark in females, and much lighter in males.

Also known as the Ridgehead Snake, the Mexican Thin-scaled Snake is distributed along the Pacific coast of Mexico from Nayarit to Chiapas. It inhabits tropical deciduous and semi-deciduous woodland and pine–oak forest and is a diurnal predator of terrestrial lizards such as whiptails and spiny lizards, which are killed by constriction. The Mexican Thin-scaled Snake is reported to be most active in the wet season and is also unusual in that it exhibits a degree of sexual dimorphism, females being much larger, and dichromatism, with females generally being much darker than males, though with a pale center to the distinctive vertebral stripe. In common with most xenodontines, this species is oviparous. Frederic Ward Putnam (1839–1915) was an American anthropologist, zoologist, and museum curator.

RELATED SPECIES

The monotypic genus *Manolepis* is most closely related to the South American smooth watersnakes (*Hydrops*, page 317), South American keeled watersnakes (*Helicops*, pages 314–315), and glossy pondsnakes (*Pseudoeryx*, page 334).

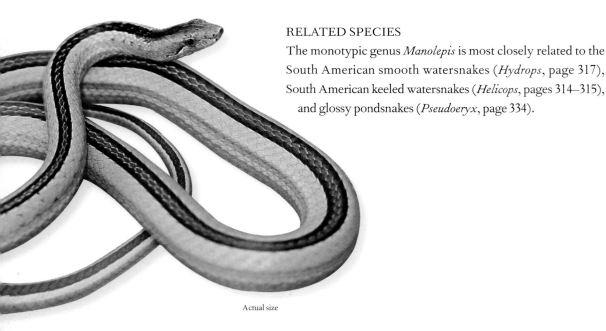

Actual size

FAMILY	Dipsadidae: Xenodontinae
RISK FACTOR	Rear-fanged, mildly venomous; also constrictor
DISTRIBUTION	South America: central and southern Brazil, Paraguay, and northern Argentina
ELEVATION	410–1,530 ft (125–465 m) asl
HABITAT	Chaco dry forest and seasonally flooded Pantanal wetlands, to the transition with wet coastal forest
DIET	Snakes, lizards, small mammals, and frogs
REPRODUCTION	Oviparous, clutch size uncertain, probably 7–10 eggs
CONSERVATION STATUS	IUCN Least Concern

ADULT LENGTH
Male
19¾–25½ in
(500–650 mm)

Female
19¾–29½ in,
rarely 39 in
(500–750 mm,
rarely 990 mm)

323

MUSSURANA BICOLOR
BICOLORED MUSSURANA
(PERACCA, 1904)

The Bicolored Mussurana is the smallest species of mussurana in South America, and the only species that fails to achieve 3 ft 3 in (1 m) in total length. Terrestrial and nocturnal, it occurs in central and southern Brazil, Paraguay, and northern Argentina, where it inhabits dry Chaco forest and seasonally flooded Pantanal wetlands. It is also present in the transition zone to the wet Atlantic coastal forests, but is nowhere a commonly encountered species. It is therefore relatively little studied. It preys on snakes, including South American keeled watersnakes (*Helicops*, pages 314–315), but is also a generalist vertebrate predator that takes frogs, lizards, and small mammals. Prey is dispatched with a combination of venom and constriction. This species is generally inoffensive and rarely bites, but as with most rear-fanged species care is advised.

The Bicolored Mussurana is a small snake, with smooth, glossy scales, a rounded, flattened head, and moderately small eyes with round pupils. It exhibits an ontogenetic color change, from a juvenile and immature adult with red flanks, a dark red to brown dorsum, and a pale venter, to an adult with a totally black dorsum, contrastingly white lips, and pale undersides and sometimes white bands on the dorsum.

RELATED SPECIES
The genus *Mussurana* contains two other species, the Montane Mussurana (*M. montana*) and Quim's Mussurana (*M. quimi*), named in honor of the herpetologist Joaquim (Quim) Cavalheiro. Both species are found in southeastern Brazil. This genus is closely related to the other mussurana genera, *Boiruna* (page 301), *Clelia* (page 304), and *Paraphimophis* (page 326), and the Lizard Egg-eating Snake (*Drepanoides*, page 308).

Actual size

FAMILY	Dipsadidae: Xenodontinae
RISK FACTOR	Rear-fanged, mildly venomous
DISTRIBUTION	South America: Colombia, Venezuela, northern Brazil, Ecuador, Peru, and Bolivia
ELEVATION	655–3,280 ft (200–1,000 m) asl
HABITAT	Primary and secondary rainforest, and cultivated gardens
DIET	Lizards, and possibly other vertebrates
REPRODUCTION	Oviparous, with clutches of up to 17 eggs
CONSERVATION STATUS	IUCN not yet assessed

ADULT LENGTH
31½–35¾ in
(800–910 mm)

OXYRHOPUS FORMOSUS
BEAUTIFUL CALICO SNAKE
(WIED-NEUWIED, 1820)

324

Actual size

The Beautiful Calico Snake is a slender snake, with smooth scales, a slightly pointed head, small eyes with round pupils, and a long tail. It may be banded dark gray and red like the Broad-banded Calico Snake (*O. petolarius*, page 325) or it may be orange to red throughout, with each scale tipped with black, a black cap over the posterior of the head being present or absent, and a yellow snout and lips. The iris is often bright red, contrasting strongly with the head markings.

Also known as the Yellow-headed Calico Snake, the Beautiful Calico Snake is found across northern South America from Colombia to Peru and Brazil. It usually inhabits pristine rainforest habitats but is also found in secondary growth areas and cultivated plots, and it may be nocturnal or crepuscular, and terrestrial to semi-arboreal in habit. The only prey documented for this species was a microteiid lizard, but with the knowledge that other calico snakes also prey on frogs, snakes, and rodents it is reasonable to assume that the diet of the Beautiful Calico Snake is more catholic than just lizards. The prey is restrained in the snake's constricting coils while the slow-acting venom takes effect. Calico snakes are usual inoffensive and not considered dangerous to humans.

RELATED SPECIES

Generally called calico snakes, the 14 brightly colored species of the genus *Oxyrhopus* are distributed from Mexico to Argentina. Several are false coralsnakes, mimicking highly venomous true coralsnakes (*Micrurus*, pages 457–465) within their ranges. The calico snakes form a sister clade to a group comprising the mussuranas (*Boiruna*, page 301; *Clelia*, page 304; *Mussurana*, page 323; and *Paraphimophis*, page 326) and their kin. *Oxyrhopus formosus* is most closely related to the Brazilian Calico Snake (*O. trigeminus*), a distinctive false coralsnake.

FAMILY	Dipsadidae: Xenodontinae
RISK FACTOR	Rear-fanged, mildly venomous; also constrictor
DISTRIBUTION	North, Central, and South America: southeastern Mexico to Panama, Colombia to the Guianas and Trinidad, and south to Argentina
ELEVATION	0–9,020 ft (0–2,750 m) asl
HABITAT	Lowland and low montane wet forests and savannas
DIET	Lizards, small mammals, frogs, snakes, and birds
REPRODUCTION	Oviparous, with clutches of 5–15 eggs
CONSERVATION STATUS	IUCN not yet assessed

ADULT LENGTH
Male
36¼–44½ in
(920–1,130 mm)

Female
38½–47¼ in
(980–1,200 mm)

325

OXYRHOPUS PETOLARIUS
BROAD-BANDED CALICO SNAKE
(LINNAEUS, 1758)

The Broad-banded Calico Snake is the most widely distributed member of its genus, being found from Mexico to Argentina, in lowland and low montane wet forests and savannas. It is a nocturnal species that spends most of its time on the ground, especially after heavy rainfall, but also climbs well and is encountered during the day. It preys on small lizards, with frogs and rodents also featuring in its diet. Birds and snakes are also sometimes taken, including rustyhead snakes (*Amastridium*, page 267). Prey is killed with a combination of weak venom and constriction. The Broad-banded Calico Snake is another false coralsnake that may gain a degree of protection by mimicking more dangerous true coralsnakes (*Micrurus*, pages 457–465). Inoffensive and disinclined to bite, it is not believed dangerous to humans.

The Broad-banded Calico Snake is a medium-sized snake with a slender body, smooth scales, a long tail, an elongate head that is slightly broader than the neck, and small eyes with round pupils. It is usually light or dark gray with a distinctive series of red or orange bands over the back, these bands often being lighter, even white, anteriorly, and beginning with a broad band over the posterior of the head behind a black-capped snout. The undersides are pale gray.

RELATED SPECIES
Three subspecies are recognized, the nominate form (*Oxyrhopus petolarius petolarius*) occurring across northern South America from Colombia to Trinidad and Tobago, with an Amazonian subspecies (*O. p. digitalis*) throughout the Amazon Basin to Argentina, and a northern form (*O. p. sebae*) in Mexico and Central America.

Actual size

FAMILY	Dipsadidae: Xenodontinae
RISK FACTOR	Rear-fanged, mildly venomous; also constrictor
DISTRIBUTION	South America: southern Brazil, Uruguay, and Argentina
ELEVATION	No elevations available
HABITAT	Open habitats, grassland, wetlands, sandy areas with coarse grass, dunes, suburban areas
DIET	Snakes, lizards, birds, and small mammals
REPRODUCTION	Oviparous, with clutches of up 7–8 eggs
CONSERVATION STATUS	IUCN not yet assessed

ADULT LENGTH
31½–35½ in,
rarely 4 ft 3 in
(800–900 mm,
rarely 1.3 m)

PARAPHIMOPHIS RUSTICUS
BROWN MUSSURANA
(COPE, 1878)

Actual size

The Brown Mussurana was formerly included in the genus *Clelia*, the mussuranas or snake-eaters. A southern species, from southern Brazil, Uruguay, and northern and central Argentina, it is not a large snake, rarely exceeding 3 ft 3 in (1 m) in total length. It is terrestrial and nocturnal, but unlike many other mussurana species it inhabits semiarid to wet open habitats such as grasslands, wetlands, and sand dune areas, and is also found on the outskirts of towns. This is an inoffensive snake that does not bite when handled. Its prey consists of small snakes, such as small swamp snakes (*Erythrolamprus reginae*, page 312), as well as lizards, small birds, and rodents, which are killed by constriction and venom injected via the rear fangs. Females are oviparous, producing clutches of seven to eight eggs between February and March.

RELATED SPECIES
The genus *Paraphimophis* was created to include this single species, which has also been placed in *Clelia* (page 304), *Oxyrhopus* (pages 324–325), and *Pseudoboa* (page 333). Within the mussurana clade, its closest relatives are the pampas snakes (*Phimophis*, page 331).

The Brown Mussurana is a small, muscular snake with smooth, glossy scales, a narrow, pointed head, large eyes, and round pupils. It is generally uniform brown, though more yellow-brown on the lower flanks and venter than on the dorsum.

FAMILY	Dipsadidae: Xenodontinae
RISK FACTOR	Rear-fanged, venomous: possibly hemorrhagins
DISTRIBUTION	South America: southeastern Brazil, Bolivia, Paraguay, Uruguay, and northern Argentina
ELEVATION	0–1,640 ft (0–500 m) asl
HABITAT	Meadows and rocky hillsides, sandbanks, and mammal burrows
DIET	Amphisbaenians, lizards, snakes, slugs, insects, and earthworms
REPRODUCTION	Oviparous, with clutches of 1–8 eggs
CONSERVATION STATUS	IUCN Least Concern

ADULT LENGTH
11¾–27½ ft
(300–700 mm)

PHALOTRIS LEMNISCATUS
ARGENTINE BLACK-HEADED SNAKE
(DUMÉRIL, BIBRON & DUMÉRIL, 1854)

327

Sometimes called Duméril's Diadem Snake, the Argentine Black-headed Snake is a terrestrial or semi-fossorial species found in meadows, on rocky hillsides, and on sandy banks in southeastern Brazil, Bolivia, Paraguay, Uruguay, and northern Argentina. It is nocturnal and secretive, often sheltering in mammal burrows underground during the day. Due to its small mouth gape, its prey is usually elongate in shape. It takes amphisbaenians (worm-lizards), slender lizards, small snakes, earthworms, slugs, and some insects. Although a small snake, with a small mouth and rear fangs, it possesses remarkably toxic venom and one herpetologist, who was bitten between the fingers, suffered an extremely serious envenoming with renal failure and bleeding, and almost died. Therefore this species, and all its generally inoffensive relatives, must be treated with extreme caution.

Actual size

RELATED SPECIES

Four subspecies are recognized, the nominate form (*Phalotris lemniscatus lemniscatus*) from northeastern Argentina, and three subspecies in the rest of the range (*P. l. trilineatus*, *P. l. divittatus*, and *P. l. iheringi*). Some authors treat these taxa as full species. *Phalotris* also contains 15 other species, mostly distributed though southern South America. These snakes were previously contained within the genus *Elapomorphus* (page 310), which is still recognized for two species.

The Argentine Black-headed Snake is a very slender, smooth-scaled snake with a short tail, a narrow head that is indistinct from the neck, small eyes, and round pupils. It is brown above and yellowish below, with a pair of orange paravertebral stripes along the back and a yellow collar around the neck, posterior to the black head.

FAMILY	Dipsadidae: Xenodontinae
RISK FACTOR	Rear-fanged, venomous: possibly hemorrhagins
DISTRIBUTION	South America: Bolivia, Paraguay, and northern Argentina
ELEVATION	490–985 ft (150–300 m) asl
HABITAT	Open grassland, rocky hillsides, and salt pans
DIET	Frogs, lizards, and birds; possibly small mammals
REPRODUCTION	Oviparous, with clutches of 4–11 eggs
CONSERVATION STATUS	IUCN not yet assessed

ADULT LENGTH
3 ft 3 in–5 ft,
occasionally 6 ft
(1.0–1.5 m,
occasionally 1.8 m)

PHILODRYAS BARONI
BARÓN'S BUSH RACER
BERG, 1895

Probably the most instantly recognizable member of its genus, Barón's Bush Racer is a South American species from Bolivia, northwestern Paraguay, and northern Argentina. It is an open country snake, found in grasslands, on rocky hillsides, and even on salt pans. Although primarily terrestrial, it easily adopts an arboreal lifestyle when required. It feeds on frogs, lizards, and birds, and may also take small mammals. As with all members of its genus, Barón's Bush Racer is irascible, biting at any opportunity and chewing vigorously to bring its rear fangs into play. Given the proven toxicity of its congener, Lichtenstein's Green Racer (*Philodryas olfersii*, page 330), snakebites from this species are to be avoided, especially as this is a popular species in captivity with owners often unaware of its potential. Manuel Barón Morlat collected the holotype.

RELATED SPECIES

The genus *Philodryas* contains 23 species. *Philodryas baroni* can be distinguished from its two most similar, sympatric species, the Patagonian Bush Racer (*P. patagoniensis*) and the Argentine Bush Racer (*P. trilineata*), by its distinctive fleshy snout protuberance. Barón's Bush Racer is also closely related to Natterer's Bush Racer (*P. nattereri*) from eastern and southern Brazil, which also lacks a snout protuberance.

Barón's Bush Racer is a large, smooth-scaled snake with a moderately long tail and a narrow head that terminates in a characteristically upturned fleshy snout protuberance. Its eyes are large with round pupils. Two color morphotypes are known, an all-green morphotype that is paler green below and on the lips, and a brown morphotype, which has a dark brown vertebral stripe, a paler brown lateral stripe that begins on the snout, and white lips and undersides that may have a pinkish tinge.

Actual size

FAMILY	Dipsadidae: Xenodontinae
RISK FACTOR	Rear-fanged, highly venomous: contains an anticoagulant and proteolytic; also a constrictor
DISTRIBUTION	South America: central Chile
ELEVATION	0–6,230 ft (0–1,900 m) asl
HABITAT	Dry hillsides, grassland, agricultural habitats, and dry stone walls
DIET	Small mammals, frogs, birds, and lizards
REPRODUCTION	Oviparous, with clutches of 6–8 eggs
CONSERVATION STATUS	IUCN Least Concern

ADULT LENGTH
3 ft 3 in–5 ft
(1.0–1.5 m)

PHILODRYAS CHAMISSONIS
CHILEAN LONG-TAILED BUSH RACER
(WIEGMANN, 1835)

329

The Chilean Long-tailed Bush Racer is endemic to Chile, being found only in the area between the Atacama Desert and Bío-Bío. It is an inhabitant of relatively dry habitats, including rocky hillsides, open grasslands, and dry stone walls around cultivated areas. It preys on lizards, frogs, nestling birds, and small mammals, including young rabbits. Prey is killed with a combination of venom and constriction. This species can be truculent and may bite freely if handled. Other members of the genus *Philodryas* have caused painful and potentially serious snakebites, and, similarly, extensive edema has been observed in cases of severe envenoming following bites by this species. Recovery may take four to six days, so caution is advised with all Chilean rear-fanged snakes.

The Chilean Long-tailed Bush Racer is a moderately stout snake with smooth scales, a long tail, and a head slightly broader than the neck, with large eyes and round pupils. It is generally brown in color with a series of four yellow longitudinal stripes, edged with black, white lips, and an off-white underbelly.

RELATED SPECIES
Chile is home to only six snake species, four of which are in the genus *Philodryas*, but were previously included in *Dromicus*: *P. chamissonis*, the Elegant Bush Racer (*P. elegans*), the Peruvian Bush Racer (*P. tachymenoides*), and Simons' Bush Racer (*P. simonsii*). There are also two species of scrub snakes (*Tachymenis*, page 339). Chile is the only mainland American country to lack any front-fanged pitvipers or coralsnakes, but its rear-fanged species warrant caution.

Actual size

FAMILY	Dipsadidae: Xenodontinae
RISK FACTOR	Rear-fanged, venomous: contains hemorrhagins, procoagulants, myotoxins, and postsynaptic neurotoxins
DISTRIBUTION	South America: Colombia, Venezuela, the Guianas, Brazil, Peru, Bolivia, Paraguay, Uruguay, and northern Argentina
ELEVATION	0–1,640 ft (0–500 m) asl
HABITAT	Savanna, degraded forest, scrubland, and cultivated areas
DIET	Frogs, lizards, birds, snakes, and small mammals
REPRODUCTION	Oviparous, with clutches of 7–8 eggs
CONSERVATION STATUS	IUCN not yet assessed

ADULT LENGTH
29½–37½ in,
rarely 3 ft 7 in
(750–950 mm,
rarely 1.1 m)

330

PHILODRYAS OLFERSII
LICHTENSTEIN'S GREEN RACER
(LICHTENSTEIN, 1823)

Lichtenstein's Green Racer is a relatively slender snake with smooth scales, a long tail, a head slightly wider than the neck, and large eyes with round pupils. It is grass- or yellow-green above, often with blue on the throat, and paler green below. A fine black line runs through the eye above the pale green lips.

Usually called Lichtenstein's Green Racer, this species should really be called Von Olfers' Green Racer, because the German zoologist Martin Lichtenstein (1780–1857) named it in honor of his compatriot, Ignaz Franz Werner Maria von Olfers (1793–1871). Lichtenstein's Green Racer is widely distributed through most of northern and central South America, east of the Andes. It is diurnal, terrestrial and semi-arboreal, and inhabits degraded forest, grassland, scrubland, and cultivated areas. It feeds on frogs, lizards, birds, other snakes, and rodents, and possesses a particularly toxic venom. Numerous snakebites are on record, with painful swelling and discoloration common, and there is also the possible fatality of a small child. This species, and its conspecifics, should therefore be treated with respect and caution.

RELATED SPECIES

Three subspecies are recognized, the nominate eastern subspecies (*Philodryas olfersii olfersii*), a northern subspecies (*P. o. herbeus*), and a western and southern subspecies (*P. o. latirostris*). The species most similar in appearance to *P. olfersii* is the Common Green Racer (*P. viridissima*), with which it occurs in sympatry, but the two can be separated by their ventral scale counts, which come to fewer than 205 in *P. olfersii*, and over 205 in *P. viridissima*.

Actual size

FAMILY	Dipsadidae: Xenodontinae
RISK FACTOR	Rear-fanged, mildly venomous; also constrictor
DISTRIBUTION	Central and South America: Panama, Colombia, Venezuela, Guyana, Suriname, and French Guiana; possibly northern Brazil
ELEVATION	0–2,950 ft (0–900 m) asl
HABITAT	Coastal savanna, riverine grasslands, dry woodland, thornbush woodland, gallery forest, and deciduous forest
DIET	Lizards and frogs
REPRODUCTION	Oviparous, with clutches of 4–7 eggs
CONSERVATION STATUS	IUCN not yet assessed

ADULT LENGTH
19¾–25½ in
(500–650 mm)

PHIMOPHIS GUIANENSIS
GUIANAN SHOVEL-NOSED PAMPAS SNAKE
(TROSCHEL, 1848)

331

The Guianan Shovel-nosed Pampas Snake is a distinctive but rarely encountered species. Found in Panama and northern South America, it is an inhabitant of relatively arid habitats, from savanna to thornbush and dry deciduous forest. In the Guianas it is largely confined to the narrow coastal savanna strip, but it is more widely distributed in Venezuela, where it occurs in the Llanos region, and in the savannas and gallery forest to the north and south of the Orinoco. It is nocturnal, and both terrestrial and semi-fossorial in habit. Prey consists primarily of lizards, and sometimes frogs. When handled these small snakes are reluctant to bite, but will void the contents of their cloacal glands or defecate defensively. Although technically rear-fanged and venomous, they are not generally considered dangerous to humans.

The Guianan Shovel-nosed Pampas Snake is a smooth-scaled snake with a relatively short tail, a broad head that terminates in an upturned shovel-snout, and moderately large eyes with round pupils. Dorsal coloration is fairly variable, being yellow or white with brown, yellow, or red mottling on every scale, while the venter and lower flanks are immaculate white or pale yellow. The dorsum of the head and neck are distinctively dark.

RELATED SPECIES

Two other species are recognized within the genus *Phimophis*: the Eastern Shovel-nosed Pampas Snake (*P. guerini*), from the Guianas, eastern and southern Brazil, Paraguay, and northern Argentina, and the Striped Shovel-nosed Pampas Snake (*P. vittatus*) from southern Bolivia, Paraguay, and northern Argentina. The genus *Phimophis*, and its close relative, the Brown Mussurana (*Paraphimophis rusticus*, page 326), form the sister clade to all the other mussuranas (*Boiruna*, page 301; *Clelia*, page 304; *Mussurana*, page 323; and *Paraphimophis*, page 326, and the scarlet snakes (*Pseudoboa*, page 333).

Actual size

FAMILY	Dipsadidae: Xenodontinae
RISK FACTOR	Nonvenomous, constrictor
DISTRIBUTION	Pacific Islands: Galapagos Islands (Ecuador)
ELEVATION	0–2,620 ft (0–800 m) asl
HABITAT	Coastal and arid habitats
DIET	Lizards, small mammals, birds, and insects
REPRODUCTION	Oviparous, clutch size unknown
CONSERVATION STATUS	IUCN not yet assessed

ADULT LENGTH
Male
19¾–24 in
(500–610 mm)

Female
19¾–21¼ in
(500–540 mm)

332

PSEUDALSOPHIS STEINDACHNERI
STRIPED GALAPAGOS RACER
(VAN DENBURGH, 1912)

The Striped Galapagos Racer is a small, smooth-scaled snake with a long tail, a narrow, elongate head, large eyes, and round pupils. It is dorsally dark brown with a pair of dorsolateral yellow to buff stripes, which begin behind the eyes, and a white throat, lips, and underbelly. The Española Racer (*Pseudalsophis hoodensis*) is similarly patterned.

The Striped Galapagos Racer is found on the central Galapagos Islands of Baltra, Rábida, Santa Cruz, and Santiago. It inhabits coastal and inland arid areas and hunts geckos, lava lizards, rodents, nestling birds, and possibly large insects. Vertebrate prey is killed by constriction. Even given the popularity of the Galapagos Islands, its snakes are poorly studied, herpetologists possibly concentrating more on the famous tortoises and iguanas. Yet before Charles Darwin famously visited the Galapagos in 1835, the presence of the racers was noted by the English buccaneer and naturalist William Dampier (1651–1715). Although the racers are believed to be oviparous, there are no data for clutch sizes available. Franz Steindachner (1834–1919) was an Austrian zoologist.

RELATED SPECIES

The genus *Pseudalsophis* contains seven species. Only the Western Elegant Racer (*P. elegans*) occurs on the mainland, in Ecuador, Peru, and northern Chile. *Pseudalsophis steindachneri* occurs in sympatry with the Central Galapagos Racer (*P. dorsalis*). The Eastern Galapagos Racer (*P. biserialis*) inhabits San Cristóbal, Floreana, and neighboring islands, the Española Racer (*P. hoodensis*) occurs on Española, and the Western Galapagos Racer (*P. occidentalis*) and Banded Galapagos Racer (*P. slevini*) occur on Isabela, Fernandina, and neighboring islands. Several of these species contain island endemic subspecies.

Actual size

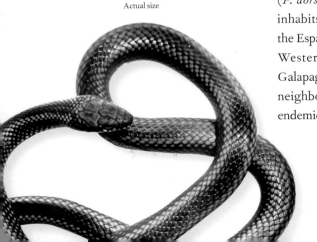

FAMILY	Dipsadidae: Xenodontinae
RISK FACTOR	Rear-fanged, mildly venomous
DISTRIBUTION	South America: northern Brazil
ELEVATION	No elevation available
HABITAT	Primary rainforest and streamsides
DIET	Snakes
REPRODUCTION	Oviparous, clutch size unknown
CONSERVATION STATUS	IUCN not yet assessed

ADULT LENGTH
3 ft 3 in–3 ft 7 in
(1.0–1.09 m)

PSEUDOBOA MARTINSI
BLACK-STRIPED SCARLET SNAKE
ZAHER, OLIVEIRA & FRANCO, 2008

333

Snakes in the genus *Pseudoboa* are often called "false boas," a literal translation of the generic name, although they do not resemble boas. Scarlet snake or false mussurana are more apt. The Black-striped Scarlet Snake is a stunning species, only described in 2008, from the northern Brazilian states of Pará, Amazonas, Roraima, and Rondônia. It may yet be discovered in neighboring countries. It inhabits primary rainforest and is associated with watercourses. One specimen contained the remnants of a snake, the common prey of mussuranas. Nothing else is known of its natural history, other than it occurs in leaf litter or under logs, and is oviparous. It does not bite or defend itself in any way. The species is named for the Brazilian herpetologist Marcio Martins from Universidade do São Paulo.

The Black-striped Scarlet Snake has smooth, glossy scales, a long tail, and a head only slightly distinct from the neck, with small eyes and round pupils. It is bright red above with a black snout, a broad red posterior head and nape band, and a wide black vertebral stripe that continues the length of the body and tail. The undersides are immaculate white. Juveniles are similarly patterned, but with a broad white collar, a common feature of juveniles of all species in the genus.

RELATED SPECIES
The genus *Pseudoboa* contains five other species. The Northern Scarlet Snake (*P. neuwiedii*) and the Amazonian Scarlet Snake (*P. coronata*), bright red species with black heads, occur in northern South America. The Serrana False Mussurana (*P. serrana*), of southeastern Brazil, and Haas' False Mussurana (*P. haasi*), of southern Brazil and Argentina, are red with black stripes and black heads. The Black False Mussurana (*P. nigra*), of eastern Brazil and Paraguay, is black with white markings. Closely related species include the Black-tailed Mussurana (*Boiruna maculata*, page 301) and the Brazilian Birdsnake (*Rhachidelus brazili*).

Actual size

FAMILY	Dipsadidae: Xenodontinae
RISK FACTOR	Nonvenomous
DISTRIBUTION	South America: Colombia, Venezuela, the Guianas, Brazil, Peru, Bolivia, Paraguay, and northern Argentina
ELEVATION	0–1,350 ft (0–410 m) asl
HABITAT	Slow-moving watercourses, oxbow lakes, marshes, canals, and ponds
DIET	Fish (including eels), and tadpoles
REPRODUCTION	Oviparous, clutch size up to 49 eggs; possibly also viviparous
CONSERVATION STATUS	IUCN Least Concern

ADULT LENGTH
3 ft 3 in–3 ft 7 in,
occasionally 5 ft
(1.0–1.1 m,
occasionally 1.5 m)

334

PSEUDOERYX PLICATILIS
SOUTH AMERICAN GLOSSY PONDSNAKE
(LINNAEUS, 1758)

The South American Glossy Pondsnake is a fairly large snake with glossy, smooth scales, a short tail, and a short head with large, slightly forward-facing eyes and round pupils. It is medium to dark brown above with a pair of dark brown lateral stripes that run off the head and down the body, broken by a pale spot on the neck, with a pair of yellow-brown stripes above and yellow on the lips below. The undersides are cream, or red in juveniles, with rows of small black spots.

The South American Glossy Pondsnake is found through northern and central South America, east of the Andes, from Colombia to northern Argentina. This species is nocturnal or diurnal and inhabits slow-moving rivers, oxbow lakes, ponds, canals, and marshes where the water flow is gentle. It feeds almost exclusively on fish, including eels, but also takes frog tadpoles. It is generally believed to be oviparous, with large clutches of eggs, but some authors suggest that certain populations may be viviparous. The usual defense of the South American Glossy Pondsnake is to hide its head under its coils and flatten its body, and although nonvenomous, it bites easily if handled. This species goes through an ontogenetic color change from light, bright juvenile to drab, dark adult livery.

RELATED SPECIES

There are two subspecies recognized, the nominate form (*Pseudoeryx plicatilis plicatilis*), which occupies the bulk of the range, and a southwestern form (*P. p. mimeticus*), from Amazonian Bolivia and neighboring Brazil. Another species has recently been described, the Lake Maracaibo Pondsnake (*P. relictualis*), from northwestern Venezuela. *Pseudoeryx* is most closely related to the South American smooth watersnakes (*Hydrops*, page 317) and keeled watersnakes (*Helicops*, pages 314–315). In habit it appears similar to the Rainbow Snake and Mudsnake (*Farancia*, page 263) of North America.

Actual size

FAMILY	Dipsadidae: Xenodontinae
RISK FACTOR	Rear-fanged, mildly venomous
DISTRIBUTION	South America: eastern Brazil
ELEVATION	0–410 ft (0–125 m) asl
HABITAT	Cerrado savanna woodland, campo grasslands, and Caatinga thorn forest and scrubland
DIET	Lizards
REPRODUCTION	Oviparous, with clutches of up to 4 eggs
CONSERVATION STATUS	IUCN not yet assessed

ADULT LENGTH
14½ in
(370 mm)

PSOMOPHIS JOBERTI

JOBERT'S GROUNDSNAKE

(SAUVAGE, 1884)

335

Jobert's Groundsnake is a diurnal and terrestrial inhabitant of lowland semiarid Cerrado savanna woodland, campo grasslands, and Caatinga thorn forest and arid scrubland. It is found in eastern Brazil, from Pará to Rio Grande do Norte, and south to Minas Gerais and São Paulo, although its collection localities are scattered over a wide area, some being separated by areas of wet forest, in which it does not occur, so its distribution is unlikely to be continuous. The Ilha de Marajó is one such isolated savanna area. Jobert's Groundsnake preys on lizards, and is an oviparous species, with a clutch of four eggs documented. The holotype was collected on the Ilha de Marajó, in the mouth of the Amazon, by Clément-Léger-Nicolas Jobert (1840–1910), a French zoologist and botanist who documented the Amerindian use of curare (arrow poison).

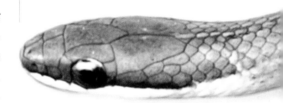

Jobert's Groundsnake is a slender snake with smooth scales, a long tail, and a head slightly wider than the neck, with large eyes and round pupils. It is brown above with a dark brown vertebral stripe the length of the body and tail, and fainter brown lateral stripes. A dark stripe runs from the snout, through the eye, to the angle of the jaw, above the white lips. The undersides are pale brown to cream.

RELATED SPECIES

Two additional species are recognized in the genus, the Spirit Groundsnake (*Psomophis genimaculatus*) from southwestern Brazil, Bolivia, Paraguay, and northern Argentina, and the Wide Groundsnake (*P. obtusus*) from southern Brazil, Paraguay, Uruguay, and northern Argentina, both of which can be distinguished from *P. joberti* by a downward-pointing dark patch beyond the last supralabial. The most closely related genera are the Central American road guarders (*Conophis*, page 305 and *Crisantophis*, page 306).

Actual size

FAMILY	Dipsadidae: Xenodontinae
RISK FACTOR	Rear-fanged, mildly venomous
DISTRIBUTION	South America: northeastern Brazil
ELEVATION	985–1,970 ft (300–600 m) asl
HABITAT	Caatinga woodland and savanna
DIET	Lizards
REPRODUCTION	Oviparous, with clutches of up to 4 eggs
CONSERVATION STATUS	IUCN not yet assessed

ADULT LENGTH
19¼–19¾ in
(490–500 mm)

RODRIGUESOPHIS IGLESIASI
IGLESIAS' LONG-NOSED PAMPAS SNAKE
(GOMES, 1915)

Iglesias' Long-nosed Pampas Snake is a small, smooth-scaled snake with a short tail, a remarkably long, pointed head, and small eyes with vertically elliptical pupils. It is generally pinkish or orange above and yellow or white below, with a pinkish dorsum to the head, followed by a broad black patch in the nape region.

Iglesias' Long-nosed Pampas Snake is the most widely distributed member of its genus, occurring in the Brazilian states of Bahia, Minas Gerais, Piauí, and Tocantins. It forms part of the highly specialized psammophilous (sandy habitat) herpetofauna associated with the São Francisco River. This species inhabits arid Caatinga woodland and savanna grasslands, where it exhibits a nocturnal and fossorial lifestyle in the sandy substrate. Its prey comprises equally fossorial microteiid lizards. Fossorial snakes are, by definition, difficult to study in nature. This species is known to be oviparous, but there are no clutch size data available. Miguel Trefaut Urbano Rodrigues is a Portuguese herpetologist, and an expert on the fossorial herpetofauna of the São Francisco River. Francisco Iglesias was a Brazilian zoologist. The holotype was lost in the Instituto Butantan fire of May 2010.

RELATED SPECIES

Two other long-nosed pampas snakes are recognized, the Santo Inacio Long-nosed Pampas Snake (*Rodriguesophis chui*), and the Ibiraba Long-nosed Pampas Snake (*R. scriptorcibatus*). The name of the second species, *scriptorcibatus*, means "the one that eats writers," which would be incomprehensible without the knowledge that it preys on microteiid lizards (*Calyptommatus*), which are locally known as "escrivoes" (clerks or writers), due to the tracks they leave in the sand.

Actual size

FAMILY	Dipsadidae: Xenodontinae
RISK FACTOR	Rear-fanged, mildly venomous
DISTRIBUTION	South America: Andean Colombia and Ecuador
ELEVATION	985 or 3,280–6,200 ft (300 or 1,000–1,890 m) asl
HABITAT	Moist montane habitats
DIET	Prey preferences unknown
REPRODUCTION	Oviparous, clutch size unknown
CONSERVATION STATUS	IUCN not yet assessed

ADULT LENGTH
19¼–23¼ in
(490–590 mm)

SAPHENOPHIS BOURSIERI
BOURCIER'S ANDEAN SNAKE
(JAN, 1867)

Bourcier's Andean Snake is the most widely distributed species in the genus, but even this species is poorly represented in museum collections. It is found at elevations between 3,280 and 6,200 ft (1,000 and 1,890 m) in the southern Colombian and Ecuadorian Andes, but there are also records from 985 ft (300 m) in the Amazonian lowlands of Ecuador. All five members of genus *Saphenophis* are poorly known. They occur in humid habitats and are likely diurnal due to the cool temperatures that exist at higher elevations at night, and they are oviparous, though clutch size data are lacking. Nothing is known of their prey preferences. They are rear-fanged and mildly venomous, but unlikely to be dangerous to humans. Jules Bourcier (1797–1873) was the French Consul to Ecuador and a hummingbird collector.

Bourcier's Andean Snake is a slender snake with smooth scales, a long tail, a head just distinct from the neck, large eyes, and round pupils. It is light brown above, with a pair of darker brown, yellow-edged, longitudinal stripes, although these may be faint on the anterior body. The lips, throat, and underbelly are white or yellow, the former flecked with black, and the latter with scattered black checkerboard markings.

RELATED SPECIES

The five species in genus *Saphenophis* are found at elevations of up to 10,500 ft (3,200 m) in the Ecuadorian and Colombian Andes. Four are only known from their type localities, the Antioquia Snake (*S. antioquiensis*) at 8,400 ft (2,560 m) in Colombia, Atahuallpa Snake (*S. atahuallpae*) at 8,200 ft (2,500 m) in Ecuador, Von Sneidern's Andean Snake (*S. sneiderni*) from El Tambo, at 5,730 ft (1,745 m), in Colombia, and the Cauca Snake (*S. tristriatus*), at up to 10,500 ft (3,200 m) in the Cauca Valley, Colombia.

Actual size

FAMILY	Dipsadidae: Xenodontinae
RISK FACTOR	Rear-fanged, mildly venomous
DISTRIBUTION	Central and South America: Costa Rica, Panama, Colombia, Venezuela, the Guianas, Trinidad, Brazil, Ecuador, Peru, Bolivia, and Paraguay
ELEVATION	0–625 ft (0–190 m) asl
HABITAT	Lowland rainforest, humid evergreen forest, and edge situations
DIET	Lizards and large insects
REPRODUCTION	Oviparous, with clutches of up to 6 eggs
CONSERVATION STATUS	IUCN Least Concern

ADULT LENGTH
3 ft 3 in–4 ft
(1.0–1.2 m)

SIPHLOPHIS COMPRESSUS
RED-HEADED LIANA SNAKE
(DAUDIN, 1803)

The Red-headed Liana Snake is an extremely elongate snake with a slender, laterally compressed body (hence the name *compressus*), which earns it the alternative name of Tropical Flatsnake. It has smooth scales, a long, prehensile tail, and a moderately broad head with large eyes and vertically elliptical pupils. The body is red to pink with equally spaced black bands, while the head is usually red with an orange or yellowish collar, followed by a broad black dorsal patch. The iris of the eye is also red.

Also known as the Tropical Flatsnake, the Red-headed Liana Snake is a widely distributed species, found from Costa Rica, to Trinidad and Paraguay. It is a highly arboreal, sometimes terrestrial, nocturnal or crepuscular species that inhabits lowland rainforest, tropical evergreen forest, and edge situations, where tropical forests meet savanna. It is believed to feed primarily on lizards, which are killed by a combination of venom and constriction. Large arthropods are reportedly also taken. The Red-headed Liana Snake is oviparous, with clutches of up to six eggs being found in parasol and leaf-cutter ant nests, presumably for protection from oophagous (egg-eating) animals. More than one clutch of eggs may be found in the same ants' nest. Nothing is known about the effects of the venom on humans.

RELATED SPECIES

Some authors retain the monotypic genus *Tripanurgos* for this species. From Panama to Bolivia it occurs in sympatry with the Common Liana Snake (*Siphlophis cervinus*), which is banded black and white on the body, with a vertebral rows of large red spots. The remaining five species in the genus are confined to northwestern, eastern, and southeastern Brazil. The liana snakes form a clade which is the sister clade to the mussuranas (*Boiruna*, page 301; *Clelia*, page 304; *Mussurana*, page 323; and *Paraphimophis*, page 326), and the calico snakes (*Oxyrhopus*, pages 324–325).

Actual size

FAMILY	Dipsadidae: Xenodontinae
RISK FACTOR	Rear-fanged, mildly venomous: proteolytic and hemolytic; potentially dangerous
DISTRIBUTION	South America: Chile and Argentina
ELEVATION	0–6,560 ft (0–2,000 m) asl
HABITAT	Pastures, meadows, wet forests, and coastal scrub
DIET	Frogs, toads, and lizards
REPRODUCTION	Viviparous, with litters of 6–12 neonates
CONSERVATION STATUS	IUCN Least Concern

ADULT LENGTH
23¾–27½ in
(600–700 mm)

TACHYMENIS CHILENSIS
CHILEAN SHORT-TAILED SCRUB SNAKE
(SCHLEGEL, 1837)

339

Also known as the Chilean Slender Snake, the Chilean Short-tailed Scrub Snake is the southernmost snake in Chile, occurring as far south as Puerto Montt and the Chiloé Archipelago. It is also found in the Andes to 6,560 ft (2,000 m) asl, and in Argentina, at similar elevations. Preferred habitats include pastures and meadows, humid forests, and coastal scrub. This species is found in relatively cool areas and is diurnal in activity, despite its vertically elliptical pupils, and live-bearing in reproductive strategy. Its prey consists of frogs, toads, and lizards, which are killed by the snake's venom. Snakebites to humans have resulted in severe symptoms, including pain and swelling, so *Tachymenis* should be treated with respect. Chile is the only American mainland country to lack any front-fanged pitvipers or coralsnakes.

The Chilean Short-tailed Scrub Snake is a slender snake with smooth scales, a short tail, a head just distinct from the neck, relatively small eyes, and vertically elliptical pupils. It is brown above with a series of light brown longitudinal stripes, edged with black, which run the length of the body and tail. The dorsum of the head is brown with dark flecking, and the sides and lips lighter, but interrupted by a series of black stripes which radiate from the eye.

RELATED SPECIES
Chile is home to only two genera of snakes, *Tachymenis* (two species) and *Philodryas* (four species, pages 328–330). Two subspecies of *T. chilensis* are recognized, a northern form (*T. c. coronellina*) and the nominate, southern form (*T. c. chilensis*). The second species of *Tachymenis* occurring in Chile is the Peruvian Scrub Snake (*T. peruviana*), which just enters northern Chile. Four other species inhabit Peru and Bolivia. *Tachymenis* is related to the mock vipers (*Thamnodynastes*, page 341) and Argentine Mock Lancehead (*Pseudotomodon trigonatus*).

Actual size

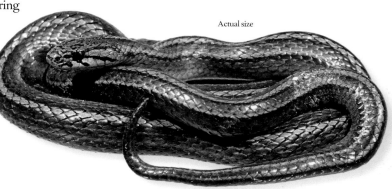

FAMILY	Dipsadidae: Xenodontinae
RISK FACTOR	Nonvenomous
DISTRIBUTION	South America: Colombia, Venezuela, the Guianas, northern Brazil, Ecuador, Peru, and Bolivia
ELEVATION	98–6,560 ft (30–2,000 m) asl
HABITAT	Secondary and primary rainforest, forest clearings, and agricultural habitats
DIET	Lizards
REPRODUCTION	Oviparous, with clutches of 2–3 eggs
CONSERVATION STATUS	IUCN not yet assessed

ADULT LENGTH
11¾–19¾ in
(300–500 mm)

340

TAENIOPHALLUS BREVIROSTRIS
SHORT-NOSED
LEAF-LITTER SNAKE
(PETERS, 1863)

The Short-nosed Leaf-litter Snake is a small species with smooth scales, a slender body, a long tail, and a narrow head with moderately large eyes and round pupils. It is gray-brown above, with five dark brown longitudinal stripes. A dark stripe though the eye separates the dorsal coloration from the pale lips, which are heavily flecked with black. The undersides are pale brown to yellow-green, while the chin may be reddish.

The Short-nosed Leaf-litter Snake is one of many small, secretive, inoffensive snakes to be found on the forest floor in northern South America. Diurnal in activity, it is found in accumulated leaf litter in pristine rainforest tree buttresses and in disturbed rainforest habitats, under cut logs in clearings, and even in areas turned over to agriculture. As with many of these smaller snake species, its natural history is poorly documented in nature. It is known to be oviparous, producing small clutches of eggs, and it feeds on small lizards such as microteiids. Due to its small mouth gape it is also likely to feed on earthworms and other soft-bodied invertebrates, but it is possible that it takes more dangerous prey, such as centipedes taken by the similar-sized genus *Tantilla* (pages 240–241).

RELATED SPECIES

There are nine species of *Taeniophallus* snakes in South America and there are many other similar-sized genera with which they can be confused. Several congeners also occur in northern South America, the Venezuelan Leaf-litter Snake (*T. nebularis*), Guianan Leaf-litter Snake (*T. nicagus*), and Bahia Leaf-litter Snake (*T. occipitalis*). The forest snakes (*Echinanthera*, page 309) are most closely related to *Taeniophallus*.

Actual size

FAMILY	Dipsadidae: Xenodontinae
RISK FACTOR	Rear-fanged, mildly venomous
DISTRIBUTION	South America: Colombia, Venezuela, the Guianas, northern Brazil, Ecuador, Peru, Bolivia, Argentina, and Uruguay
ELEVATION	0–1,640 ft (0–500 m) asl
HABITAT	Savanna, primary rainforest, wetland areas, coastal forest, and cultivated habitats
DIET	Frogs, small lizards, and arthropods
REPRODUCTION	Viviparous, litter size unknown
CONSERVATION STATUS	IUCN Least Concern

ADULT LENGTH
13¾–17¾ in,
occasionally 23¾ in
(350–450 mm,
occasionally 600 mm)

THAMNODYNASTES PALLIDUS
PALE MOCK VIPER
(LINNAEUS, 1758)

341

The Pale Mock Viper is a widely distributed member of the genus, while some of the other species are much more localized. It is a small snake that inhabits a variety of habitats from savanna to rainforest, wetland habitats, and cultivated areas, from Colombia to Bolivia, across Amazonia, and south to Argentina. It is terrestrial, semi-aquatic, and occasionally semi-arboreal in habit, and feeds on frogs, small lizards, and arthropods, including beetle larvae. Other species of *Thamnodynastes* also take small mammals or fish, and this prey may also potentially feature in the diet of the Pale Mock Viper. The vertically elliptical pupils are indicative of its primarily nocturnal activity cycle, but this species is also active by day. Mock vipers are live-bearers. They are also mildly venomous, but they are not considered a snakebite risk.

The Pale Mock Viper is a small, slender snake with keeled scales, an elongate head that is broader than the neck, and large eyes with vertically elliptical pupils. It is pale brown or light orange with a pair of very faint longitudinal stripes and light and dark flecking on the scales.

Actual size

RELATED SPECIES
The genus *Thamnodynastes* contains a further 18 species which are fairly similar in appearance, with at least six other species occurring within the range of *T. pallidus*. In the Amazon this is reputedly the only species with 17 scale rows at midbody. The *pallidus* group includes the Long-tailed Mock Viper (*T. longicaudus*) from northeastern Brazil, and the recently described Sertão Mock Viper (*T. sertanejo*). *Thamnodynastes* resemble the Asian mock vipers (*Psammodynastes*, page 397), but are not related. Their closest relatives are the pampas snakes (*Tomodon*, page 342) and the Brazilian Keeled Watersnake (*Ptychophis flavovirgatus*).

FAMILY	Dipsadidae: Xenodontinae
RISK FACTOR	Rear-fanged, mildly venomous
DISTRIBUTION	South America: southern and southeastern Brazil, northern Argentina
ELEVATION	0–625 ft (0–190 m) asl
HABITAT	Grasslands and forests
DIET	Slugs
REPRODUCTION	Viviparous, with litters of 4–26 neonates.
CONSERVATION STATUS	IUCN not yet assessed

ADULT LENGTH
15¾–18 in
(400–460 mm)

TOMODON DORSATUS
SLUG-EATING MOCK VIPER
(DUMÉRIL, BIBRON & DUMÉRIL, 1854)

342

Found from São Paulo and the Atlantic forests of southern Brazil, to Misiones in extreme northeastern Argentina, the Slug-eating Mock Viper is a specialist slug-eating snake that inhabits grasslands and forests. Live-bearing, females produce up to 26 neonates. The Slug-eating Mock Viper is generally inoffensive, with a threat posture that involves body-flattening, which, when combined with its distinctive patterning, mimics the highly venomous lanceheads (*Bothrops*, pages 560–569). It is known locality as "falsa yarará" (false jararaca or lancehead). It is capable of delivering a bite that may result in pain and swelling, so caution is recommended around this species.

The Slug-eating Mock Viper is a relatively stout snake with smooth scales, a short tail, and a broad head with moderately large eyes and vertically elliptical pupils. It is dorsally pale gray or brown with darker blotches or chevrons on the dorsum and a row of pale spots along the midline. A fine dark stripe runs from the rear of the eye to the angle of the jaw.

RELATED SPECIES
Snakes in the genus *Tomodon* are also known as pampas snakes and *T. dorsatus* occurs in sympatry with the Eyed Pampas Snake (*T. ocellatus*). A third species, the Bolivian Pampas Snake (*T. orestes*), occurs in Bolivia and Argentina. These three snakes are closely related to another mock viper genus, *Thamnodynastes* (page 341).

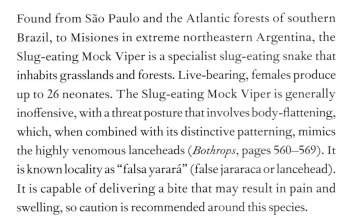

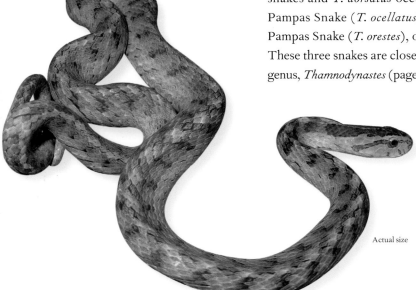

Actual size

FAMILY	Dipsadidae: Xenodontinae
RISK FACTOR	Rear-fanged, mildly venomous
DISTRIBUTION	South America: southeastern Brazil
ELEVATION	0–330 ft (0–100 m) asl
HABITAT	Atlantic coastal forest and secondary growth
DIET	Lizards, frogs, and small mammals
REPRODUCTION	Oviparous, clutch size unknown
CONSERVATION STATUS	IUCN Least Concern

ADULT LENGTH
35½ in
(900 mm)

TROPIDODRYAS SERRA
SERRA SNAKE
(SCHLEGEL, 1837)

343

The Serra Snake is found in southeastern Brazil, from Bahia to Santa Catarina, and also on Santo Amaro Island. It is endemic to the highly endangered lowland Atlantic coastal forest, and associated secondary regrowth. It is a diurnally active snake, and both terrestrial and arboreal in habit. The Serra Snake preys on small lizards, frogs, and small mammals. Juveniles wiggle their white tail tips, which resemble a worm, to lure prey within strike range, in the same manner as the unrelated death adders (*Acanthophis*, pages 484–485) and cantils (*Agkistrodon*, pages 553–555). The Serra Snake is oviparous but its clutch size is unrecorded.

The Serra Snake is an elongate snake with an exceptionally long tail and an arrow-shaped head, with small eyes and round pupils. It is light gray or brown in color with a vertebral pattern consisting of white-edged black squares, separated by broad brown interspaces, brown marbling on the head, a brown stripe through the eye, and a white tail tip in juveniles.

RELATED SPECIES

A second species also occurs in southeastern Brazil, the Jiboinha (*Tropidodryas striaticeps*), which is much darker brown than *T. serra*. This species does not appear to be closely related to any other xenodontines, but comprises the sister clade to the flat-headed snakes (*Xenopholis*, page 347), false water cobras (*Hydrodynastes*, page 316), and Amaral's Groundsnake (*Caaeteboia amarali*).

Actual size

FAMILY	Dipsadidae: Xenodontinae
RISK FACTOR	Rear-fanged, mildly venomous
DISTRIBUTION	West Indies: Hispaniola (Haiti and Dominican Republic)
ELEVATION	0–5,000 ft (0–1,525 m) asl
HABITAT	Dry forest to tropical rainforest
DIET	Lizards and frogs
REPRODUCTION	Oviparous, with clutches of up to 5 eggs
CONSERVATION STATUS	IUCN not yet assessed

ADULT LENGTH
2 ft 7 in–4 ft 7 in
(0.8–1.4 m)

344

UROMACER CATESBYI
HISPANIOLAN
BLUNT-NOSED VINESNAKE
(SCHLEGEL, 1837)

The Hispaniolan Blunt-nosed Vinesnake
is a slender-bodied snake with smooth
scales, a long, prehensile tail, and a narrow,
pointed head, though not as pronounced as
its congeners, with large eyes and round
pupils. Adults are green, with or without
a white, light green, or blue ventrolateral
stripe, while juveniles may be brown or
gray with green heads.

The Hispaniolan Blunt-nosed Vinesnake is a highly arboreal
and sometimes terrestrial species, found in a wide variety of
habitats throughout Hispaniola (Haiti and the Dominican
Republic) and on its satellite islands. It may be found in tropical
rainforest but also in dry woodland, and is usually found aloft.
It is an active predator of small lizards and frogs, stalking
them by day in the canopy, or along branches. The *Uromacer*
vinesnakes are the West Indian equivalent of Asia's *Dryophis*,
Africa's *Thelotornis*, and mainland America's *Oxybelis*; that is,
elongate, camouflaged, diurnal, rear-fanged treesnakes. These
genera may represent examples of convergent evolution—
unrelated species that look and act similarly in different
locations. Although venomous, these snakes are not considered
a snakebite risk. Mark Catesby (1683–1749) was an English
naturalist who traveled widely in the Americas.

RELATED SPECIES

There are eight subspecies of *Uromacer catesbyi* in Haiti and the
Dominican Republic, with endemic island populations on Île de
la Tortue (*U. c. scandax*), Île de la Gonâve (*U. c. frondicolor*),
Îles Cayemites (*U. c. cereolineatus*), and Île-à-Vache
(*U. c. insulaevaccarum*), all of which are Haitian islands, and
Isla Saona, Dominican Republic (*U. c. inchausteguii*).
There are also two other species of Hispaniolan
vinesnakes, both with much longer snouts than
U. catesbyi: the Southeast Hispaniolan Vinesnake
(*U. frenatus*) and Hispaniolan Long-nosed Vinesnake
(*U. oxyrhynchus*). Both may be other colors apart from green.

Actual size

FAMILY	Dipsadidae: Xenodontinae
RISK FACTOR	Rear-fanged, mildly venomous
DISTRIBUTION	South America: southern Brazil, Paraguay, Uruguay, and Argentina
ELEVATION	No elevations available
HABITAT	Open grassland, rocky hillsides, sandy habitats, and pinelands
DIET	Lizards, frogs, tadpoles, fish, and insect larvae
REPRODUCTION	Oviparous, with clutches of 3–15 eggs
CONSERVATION STATUS	IUCN not yet assessed

ADULT LENGTH
Male
15¾–19¾ in
(400–500 mm)
Female
15¾–23¾ in,
occasionally 37½ in
(400–600 mm,
occasionally 950 mm)

345

XENODON DORBIGNYI

D'ORBIGNY'S HOGNOSE SNAKE

(BIBRON, 1854)

D'Orbigny's Hognose Snake is a generally inoffensive snake that actively mimics highly venomous lanceheads, such as the Urutu (*Bothrops alternatus*, page 561), for protection, and it is often called "falsa yarará" (false jararaca or lancehead). It also buries its head under its coils and exposes the red and black underside of its tail, to intimidate potential predators with coralsnake colors. D'Orbigny's Hogsnake Snake is found in southern Brazil, Paraguay, Uruguay, and Argentina as far south as Río Negro Province. It usually inhabits open sandy grassland, rocky hillsides, or beach-type habitats where it can quickly burrow under logs or rocks. It preys on lizards, frogs, tadpoles, fish, and beetle larvae, and although mildly venomous, is not considered a snakebite risk. Alcide d'Orbigny (1802–57) was a French naturalist and traveler who lived in Argentina.

D'Orbigny's Hognose Snake is a relatively stout snake with smooth scales, a short tail, and a short, broad head with a distinctive upturned snout and large eyes with round pupils. It is dorsally pale gray with several dark gray chevrons on the head that continue onto the body as three series of large ocelli markings. The undersides are pale with a black checkerboard pattern and a heavy infusion of red pigment.

RELATED SPECIES

The six species of South American hognose snakes, formerly in genus *Lystrophis*, are now placed in *Xenodon*, although some authors retain the original generic name. They resemble the North American hognose snakes (*Heterodon*, page 264), but are more closely related to the false coralsnakes and groundsnakes of genus *Erythrolamprus* (pages 311–312). The Mato Grosso Hognose (*X. matogrossensis*), Beautiful Hognose (*X. pulcher*), Half-banded Hognose (*X. semicinctus*), and Jan's Hognose (*X. histricus*) are dorsally banded or patterned red, black, and white, as coralsnakes mimics, while Natterer's Hognose (*X. nattereri*) is blotched with light and dark brown.

Actual size

FAMILY	Dipsadidae: Xenodontinae
RISK FACTOR	Rear-fanged, mildly venomous
DISTRIBUTION	North, Central, and South America: southern Mexico, to the Guianas, Brazil, and Bolivia
ELEVATION	0–6,230 ft (0–1,900 m) asl
HABITAT	Rainforest, dry deciduous forest, evergreen forest, and low montane forest, especially near water
DIET	Toads and frogs
REPRODUCTION	Oviparous, with clutches of 5–15 eggs
CONSERVATION STATUS	IUCN not yet assessed

ADULT LENGTH
Male
23¾–27½ in
(600–700 mm)

Female
23¾–38¼ in
(600–1,000 mm)

XENODON RABDOCEPHALUS
NORTHERN FALSE LANCEHEAD
(WIED-NEUWIED, 1824)

The Northern False Lancehead is a stout-bodied, viperine snake with smooth scales, a short tail, and a broad head with small eyes and round pupils. Its patterning comprises alternating irregular, white-edged light and dark brown bands, with a dark brown arrow marking on the dorsum of the head. The undersides are cream or gray with extensive dark mottling.

The Northern False Lancehead is the only *Xenodon* in Mexico and Central America. It also occurs across northern South America. It inhabits a wide range of habitats, from tropical rainforest to dry forest, at low and mid-montane elevations, usually near water. Although it superficially resembles the numerous pitvipers, with which it occurs in sympatry, it has round, rather than vertical, pupils and it lacks the pitvipers' heat-sensitive pits. This species is a toad specialist, even feeding on the large, poisonous cane toad. Toads habitually inflate themselves to avoid predation but the Northern False Lancehead easily deflates its prey by puncturing the body cavity and lungs with its long rear fangs. Frogs or tadpoles are taken less often. Bites to humans result in swelling, localized pain, and extensive bleeding.

RELATED SPECIES

Aside from the South American hognose snakes, formerly included in *Lystrophis*, there are also six species of false lanceheads in the genus *Xenodon*, several of which were previously included in the genus *Waglerophis*. The northern South American species include the Amazonian False Lancehead (*X. severus*), Merrem's False Lancehead (*X. merremi*), and Werner's False Lancehead (*X. werneri*), while Günther's False Lancehead (*X. guentheri*) and Neuwied's False Lancehead (*X. neuwiedii*) are found farther south.

Actual size

FAMILY	Dipsadidae: Xenodontinae
RISK FACTOR	Nonvenomous
DISTRIBUTION	South America: Colombia, the Guianas, Brazil, Ecuador, Peru, and Bolivia
ELEVATION	0–4,920 ft (0–1,500 m) asl
HABITAT	Primary and secondary rainforest
DIET	Juvenile frogs
REPRODUCTION	Oviparous, with clutches of 4 eggs
CONSERVATION STATUS	IUCN Least Concern

ADULT LENGTH
13¾–15¾ in
(350–400 mm)

XENOPHOLIS SCALARIS
LADDER
FLAT-HEADED SNAKE
(WUCHERER, 1861)

347

The flat-headed snakes are terrestrial or semi-fossorial species that inhabit primary and secondary rainforest, especially in association with small rainforest pools. The Ladder Flat-headed Snake may be either diurnal or nocturnal in habit. It is found across northern Amazonia, from Colombia to the Guianas and Brazil, and south to Peru and Bolivia. Many small leaf-litter-dwelling snakes are poorly documented in nature, due to the infrequency of encounters and their secretive habits, and this species is no exception. They are nonvenomous predators of small frogs. Females are oviparous, with clutches of four eggs on record. The defense of the Ladder Flat-headed Snake involves body-flattening to the thinness of a ribbon, a tactic intended to make the snake appear larger than it is in reality, and hopefully deter predators.

The Ladder Flat-headed Snake is a smooth-scaled species with a short tail and a flattened head, just distinct from the neck, with small eyes and vertically elliptical pupils. It is dorsally light red-brown with a black vertebral stripe and a series of black blotches along either side of the middorsal line, which may fuse to form short cross-bands, hence "ladder." The lips and lower flanks are yellow, while the venter is white.

RELATED SPECIES

Two other species are recognized, the Undulated Flat-headed Snake (*Xenopholis undulatus*), from eastern and southeastern Brazil, and Paraguay, and Werdings' Flat-headed Snake (*X. werdingorum*), from eastern Bolivia and possibly southwestern Brazil. The flat-headed snakes are related to the Jiboinha and the Serra Snake (*Tropidodryas*, page 343) and Amaral's Groundsnake (*Caaeteboia amarali*).

Actual size

FAMILY	Lamprophiidae: Aparallactinae
RISK FACTOR	Rear-fanged, mildly venomous: venom composition unknown
DISTRIBUTION	Southern and East Africa: South Africa, Swaziland, Mozambique, Malawi, Zimbabwe, Zambia, Namibia, Angola, southern DRC, Kenya, Tanzania, and Somalia
ELEVATION	0–4,920 ft (0–1,500 m) asl
HABITAT	Moist savanna grasslands and lowland forests
DIET	Fossorial snakes, lizards, and amphisbaenians
REPRODUCTION	Oviparous, with clutches of 6 eggs
CONSERVATION STATUS	IUCN not listed

ADULT LENGTH
Male
19¾–21¾ in
(500–550 mm)

Female
2 ft 4 in–3 ft 7 in
(0.7–1.1 m)

AMBLYODIPSAS POLYLEPIS
COMMON PURPLE-GLOSSED SNAKE
(BOCAGE, 1873)

The Common Purple-glossed Snake is, as its common name suggests, a highly glossy snake with a uniform purple-brown to black coloration. It has a relatively stout body, a blunt tail, and a dorsally compressed head with a slightly pointed snout and set-back lower jaw, for burrowing. The eyes are very small.

The Common Purple-glossed Snake has a wide distribution, from Somalia to KwaZulu-Natal, South Africa, and westward to Angola and Namibia. It shows a preference for wet grassland and lowland forests, but is often only seen after rain. It feeds on fossorial reptiles, including blindsnakes, legless skinks, and amphisbaenians (worm-lizards). Although this species is venomous, prey may be constricted before it is swallowed, suggesting the venom is not especially toxic. Although purple-glossed snakes are not considered dangerous to humans, their venom composition, yield, and toxicity have not been studied, and the lack of snakebites may be due to their placid nature. They prefer to bury their heads under their coils rather than bite, and when threatened by a predator will elevate the tail as a mock head, to draw attention away from the real head.

RELATED SPECIES

Most of the range of *Amblyodipsas polylepis* is occupied by the nominate subspecies, while a second subspecies (*A. p. hildebrandtii*) is found from Tanzania to Somalia. The genus *Amblyodipsas* contains a further eight species, from the Western Purple-glossed Snake (*A. unicolor*) of West and Central Africa, to the Natal Purple-glossed Snake (*A. concolor*) in South Africa. Purple-glossed snakes may be confused with the Natal Blacksnake (*Macrelaps microlepidotus*, page 351) and the venomous side-stabbing or stiletto snakes (*Atractaspis*, pages 355–358).

Actual size

FAMILY	Lamprophiidae: Aparallactinae
RISK FACTOR	Rear-fanged, mildly venomous: venom composition unknown
DISTRIBUTION	Southern and East Africa: South Africa, Lesotho, Swaziland, central Mozambique, Zimbabwe, Zambia, eastern Botswana, Namibia (Caprivi Strip), and coastal Tanzania
ELEVATION	0–5,580 ft (0–1,700 m) asl
HABITAT	Wet savanna grassland, lowland woodland, and termitaria
DIET	Centipedes
REPRODUCTION	Oviparous, with clutches of 2–4 eggs
CONSERVATION STATUS	IUCN Least Concern

ADULT LENGTH
8–11¾ in
(200–300 mm)

APARALLACTUS CAPENSIS
CAPE CENTIPEDE-EATER
SMITH, 1849

349

The Cape Centipede-eater is a widely distributed snake, occurring down the eastern side of East and southern Africa, from Zambia to the Eastern Cape of South Africa, and west to the Caprivi Strip of Namibia, but it is absent from southern Mozambique. It inhabits wet savanna and lowland woodland, and is especially common around termitaria. Its prey consists entirely of centipedes, which are bitten and eventually killed by multiple bites. If the centipede turns and bites the snake, the snake will release its prey, but start again shortly afterward and eventually swallow the centipede headfirst. Centipede-eaters will struggle and try to bite if handled but their mouths are small and their fangs short, and no ill effects have been documented. Their venom is unstudied but not considered highly toxic to humans.

Actual size

The Cape Centipede-eater is a slender snake with shiny, smooth scales and a head no wider than the body. It is yellow or red-brown above, paler on the flanks, becoming off-white below. Its most defining characteristic is the black cap that covers the head and extends onto the neck, where it forms a band, contrasting with the white of the posterior labial scales.

RELATED SPECIES

There are 11 species of *Aparallactus* in sub-Saharan Africa, from the Western Black Centipede-eater (*A. niger*) of West Africa, to the Malindi Centipede-eater (*A. turneri*) of coastal Kenya and the Usambara Centipede-eater (*A. werneri*) of Tanzania. *Aparallactus capensis* occurs in sympatry with the Plumbeous Centipede-eater (*A. lunulatus*) and Eastern Black Centipede-eater (*A. guentheri*). Through its range four subspecies are recognized.

FAMILY	Lamprophiidae: Aparallactinae
RISK FACTOR	Rear-fanged, mildly venomous: venom composition unknown
DISTRIBUTION	East Africa: Tanzania, Zambia, Zimbabwe, and southern DRC
ELEVATION	1,920–4,270 ft (585–1,300 m) asl
HABITAT	Sandy savanna grasslands, and fields
DIET	Fossorial lizards and amphisbaenians, and small snakes
REPRODUCTION	Oviparous, with clutches of 6 eggs
CONSERVATION STATUS	IUCN not listed

ADULT LENGTH
15¾–19¾ in
(400–500 mm)

350

CHILORHINOPHIS GERARDI
GERARD'S
TWO-HEADED SNAKE
(BOULENGER, 1913)

Gerard's Two-headed Snake is so-called because its tail closely resembles its head, and when it is threatened it will hide its head and elevate and writhe its tail to distract its enemy, giving the impression of a head at either end. This is different from true "two-headed" or dicephalic snakes, which are birth deformities with two heads at the same end of a single body (see page 31). Gerard's Two-headed Snake is secretive, fossorial, and nocturnal, and only seen on the surface after rain. It feeds on slender fossorial skinks, amphisbaenians (worm-lizards), and small snakes, including conspecifics. Its preferred habitat is savanna grasslands on sandy substrates, but it may also be excavated in agricultural fields. Nothing is known of its venom, but its mouth is small and probably incapable of delivering a snakebite.

Gerard's Two-headed Snake is an elongate snake with a short tail and a small head, which is indistinct from the neck. The body is golden yellow with three bold black longitudinal stripes, the head is black with yellow spots and a black nape band around the neck, and the tail is black above and light blue or white on the flanks. The venter is orange, while the throat is white.

RELATED SPECIES

Two subspecies are recognized, the nominate form (*Chilorhinophis gerardi gerardi*) from Zimbabwe and eastern Zambia and a northern subspecies (*C. g. tanganyikae*) from northern Zambia, southern Democratic Republic of the Congo, and Tanzania. The genus *Chilorhinophis* also contains a second species, Butler's Two-headed Snake (*C. butleri*), from South Sudan, Tanzania, and also Mozambique, where the population was formerly treated as a separate species (*C. carpenteri*).

Actual size

FAMILY	Lamprophiidae: Aparallactinae
RISK FACTOR	Rear-fanged, mildly venomous: sarafotoxins
DISTRIBUTION	Southern Africa: South Africa
ELEVATION	0–1,720 ft (0–525 m) asl
HABITAT	Coastal bush, lowland forest, creeksides, and gardens
DIET	Burrowing frogs, legless lizards, snakes, and small mammals
REPRODUCTION	Oviparous, with clutches of 3–10 eggs
CONSERVATION STATUS	IUCN not listed, Near Threatened (South Africa)

ADULT LENGTH
27½–35½ in,
rarely 3 ft 7 in
(700–900 mm,
rarely 1.1 m)

MACRELAPS MICROLEPIDOTUS
NATAL BLACKSNAKE
(GÜNTHER, 1860)

351

The Natal Blacksnake is confined to the coastal strip from East London, in the Eastern Cape, South Africa, to the Mozambique border, with an isolated population at Stutterheim, Eastern Cape. It is only seen abroad on nights when the air is warm and humid. During the day it shelters in moist leaf litter or damp soil. It feeds on a wider range of small vertebrates than other aparallactines (of the subfamily Aparallactinae), from burrowing rain frogs to legless skinks, and small snakes. Small mammals are also taken. Prey is held by constricting coils and killed by a venomous bite. While this species is generally placid, there are two accounts of snakebites leading to collapse within 30 minutes. The venom is little studied but contains a potentially dangerous cardiotoxin known as a sarafotoxin, and the rear fangs of this species are large. There is no antivenom.

The Natal Blacksnake is a stout-bodied snake, with smooth, shiny scales and a short tail. The head is relatively broad, dorsally compressed, and slightly pointed, with small eyes.

RELATED SPECIES
Macrelaps is a monotypic genus, but *M. microlepidotus* may easily be confused with the venomous Bibron's Stiletto Snake (*Atractaspis bibronii*, page 355), the innocuous Common Purple-glossed Snake (*Amblyodipsas polylepis*, page 348), or the Natal Purple-glossed Snake (*A. concolor*), with which it occurs in sympatry.

Actual size

FAMILY	Lamprophiidae: Aparallactinae
RISK FACTOR	Rear-fanged, mildly venomous: venom composition unknown
DISTRIBUTION	Middle East: northern Israel, northwestern Jordan, Lebanon, and western Syria
ELEVATION	985 ft (300 m) bsl to 5,910 ft (1,800 m) asl
HABITAT	Rocky hillsides, with scrubby or bushy cover
DIET	Lizards and snakes
REPRODUCTION	Believed viviparous, litter size unknown
CONSERVATION STATUS	IUCN Least Concern

ADULT LENGTH
11¾–14½ in
(300–370 mm)

MICRELAPS MUELLERI
MÜLLER'S SNAKE
BOETTGER, 1880

Müller's Snake is a slender, smooth-scaled snake with a variable patterning that ranges from equal-width black and pale yellow or pinkish rings to broad pale bands with occasional irregular black rings, or even pale above and black on the lower flanks, but in all specimens the head is black. The head is depressed with extremely small eyes.

Müller's Snake inhabits vegetated rocky slopes, in Lebanon, western Syria, Israel, and northwestern Jordan. It is found to elevations of 5,910 ft (1,800 m), but also below sea-level in the Jordan Valley. It is a fossorial snake that is most likely to be encountered sheltering under flat stones during droughts. Its prey consists of lizards, mostly fossorial cylindrical skinks or rock-dwelling snake-eyed skinks, and snakes. Prey is subdued with a venomous bite, but due to its small size and the limited gape of its mouth it is not considered dangerous to humans, although its venom has not been studied. East African *Micrelaps* lay eggs, but Müller's Snake is thought to be viviparous.

RELATED SPECIES

The genus *Micrelaps* contains five species, split between the Middle East and East Africa. *M. muelleri* and Tchernov's Chainling (*M. tchernovi*) occur in Israel and Jordan, the latter exhibiting black saddles rather than rings like *M. muelleri*. In East Africa, the Kenyan Bicolored Snake (*M. bicoloratus*) and Desert Black-headed Snake (*M. boettgeri*) are found in Kenya and Tanzania, while the Somali Black-headed Snake (*M. vaillanti*) inhabits Somalia and Ethiopia.

Actual size

FAMILY	Lamprophiidae: Aparallactinae
RISK FACTOR	Rear-fanged, mildly venomous: venom composition unknown
DISTRIBUTION	East Africa: Kenya, Tanzania, Zambia, and DRC
ELEVATION	0–5,580 ft (0–1,700 m) asl
HABITAT	Wooded savanna, gallery forest, thick forest, and woodland
DIET	Snakes
REPRODUCTION	Presumed oviparous, clutch size unknown
CONSERVATION STATUS	IUCN not listed

ADULT LENGTH
17¾–27½ in,
rarely 33 in
(450–700 mm,
rarely 840 mm)

POLEMON CHRISTYI

CHRISTY'S SNAKE-EATER

(BOULENGER, 1903)

353

Christy's Snake-eater inhabits heavily forested or wooded habitats, including savanna woodland, from Kenya to Zambia and west into the Democratic Republic of the Congo. It is a secretive snake, only seen on the surface at night, after rain. It lives in the leaf litter or in animal holes and burrows, and preys entirely on snakes, from blindsnakes and threadsnakes, to herald snakes (*Crotaphopeltis*, page 156), and even its conspecifics. It has been documented killing and eating snakes of close to its own size. It is otherwise generally inoffensive and reluctant to bite, but its venom composition, yield, and toxicity are unknown. Other *Polemon* lay eggs, so it is presumed Christy's Snake-eater is also oviparous. This species is named in honor of the Edinburgh physician, army doctor, and zoologist Cuthbert Christy (1863–1932).

Christy's Snake-eater is a short snake with a stumpy tail and a head slightly broader than the neck. It is usually iridescent gunmetal gray or blue-black, while the dark gray ventral scales may be edged with white. Juveniles are often brown with white undersides.

RELATED SPECIES

The genus *Polemon* contains 13 species, mostly in West and Central Africa, from Reinhardt's Snake-eater (*P. acanthias*) and the Guinea Snake-eater (*P. barthii*) in West Africa, to the Democratic Republic of the Congo, home to nine species. *Polemon* is marginally represented in East Africa and absent from South Africa. *Polemon christyi* may easily be confused with the highly venomous side-stabbing or stiletto snakes (*Atractaspis*, pages 355–358).

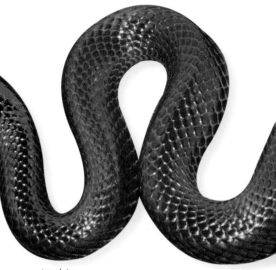

Actual size

FAMILY	Lamprophiidae: Aparallactinae
RISK FACTOR	Rear-fanged, mildly venomous: venom composition unknown
DISTRIBUTION	Southern Africa: Botswana, Zimbabwe, Namibia, southern Angola, and South Africa
ELEVATION	98–5,580 ft (30–1,700 m) asl
HABITAT	Savanna grasslands, deep alluvial sand, and termitaria
DIET	Amphisbaenians and fossorial skinks
REPRODUCTION	Oviparous, with clutches of 3–4 eggs
CONSERVATION STATUS	IUCN not listed

ADULT LENGTH
19¾–25½ in,
rarely 28¼ in
(500–650 mm,
rarely 720 mm)

354

XENOCALAMUS BICOLOR
SLENDER QUILL-SNOUTED SNAKE
(GÜNTHER, 1868)

Quill-snouted snakes are so named because they resemble a nineteenth-century pen or a porcupine's quill, due to their enlarged rostral scales, a feature that reaches its most extreme in the Slender Quill-snouted Snake, in which the rostral scale protrudes forward to a sharp point. Quill-snouted snakes are fossorial, the quill-snout acting as a burrowing aid. They inhabit savanna grassland, often living at depth in loose, alluvial sands, though they may be encountered on the surface after rain, under logs, or inside termitaria. They prey on elongate reptiles, primarily amphisbaenians, and also take fossorial skinks. The Slender Quill-snouted Snake occurs in Zimbabwe, Botswana (especially the Kalahari), northern Namibia, and Limpopo Province, South Africa. It is inoffensive and does not bite when handled. Its venom composition is unknown.

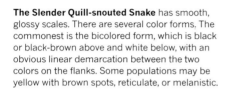

Actual size

The Slender Quill-snouted Snake has smooth, glossy scales. There are several color forms. The commonest is the bicolored form, which is black or black-brown above and white below, with an obvious linear demarcation between the two colors on the flanks. Some populations may be yellow with brown spots, reticulate, or melanistic.

RELATED SPECIES

Six subspecies of *Xenocalamus bicolor* are recognized, some with very limited ranges, such as the Waterberg Quill-snouted Snake (*X. b. australis*) of Limpopo Province. The nominate Bicolored Quill-snouted Snake (*X. b. bicolor*) occurs in Namibia, Botswana, and Zimbabwe, while the Striped Quill-snouted Snake (*X. lineatus*) of Zimbabwe, Mozambique, and Limpopo was also a former subspecies. Five other species of *Xenocalamus* are found from Congo and the Democratic Republic of the Congo to Mozambique.

FAMILY	Lamprophiidae: Atractaspidinae
RISK FACTOR	Horizontal front-fanged, venomous: sarafotoxins, and possibly cytotoxins
DISTRIBUTION	East and southern Africa: Kenya to South Africa (KwaZulu-Natal), west to Namibia and Angola
ELEVATION	0–5,910 ft (0–1,800 m) asl
HABITAT	Lowland forest, wet savanna, grassland, fynbos, karoo scrub, semidesert, and desert
DIET	Lizards, snakes, frogs, and small mammals
REPRODUCTION	Oviparous, with clutches 3–7 eggs
CONSERVATION STATUS	IUCN not listed

ADULT LENGTH
11¾–19¾ in,
rarely 27½ in
(300–500 mm,
rarely 700 mm)

ATRACTASPIS BIBRONII
BIBRON'S STILETTO SNAKE
SMITH, 1849

355

Also known as the Southern Burrowing Asp, Bibron's Stiletto Snake has a wide range, from Kenya to South Africa, and westward to Namibia and Angola. It is found in a wide variety of habitats, from wet coastal forests in KwaZulu-Natal, South Africa, to the Namib Desert. Fossorial, it only moves onto the surface on wet nights. Prey consists of lizards, frogs, small snakes, shrews, and mice, which are killed by a venomous bite. Bibron's Stiletto Snakes are also responsible for numerous defensive snakebites to humans. All members of this genus possess large, moveable, horizontal front fangs, allowing them to strike sideways without opening their mouths. Bites are exceedingly painful and cause prolonged swelling, but no deaths are on record for this species. Antivenom is ineffective. Gabriel Bibron (1806–48) was an influential French zoologist.

RELATED SPECIES

The genus *Atractaspis* contains 22 species of burrowing asps or side-stabbing snakes, although the term "mole vipers" fell into disuse when the genus was removed from the Viperidae. Across its range *A. bibronii* occurs in sympatry with several other species; for example, in northern South Africa it occurs with the Beaked Stiletto Snake (*A. duerdeni*), which has a more pointed snout and larger eyes. In KwaZulu-Natal it may also be confused with the Natal Blacksnake (*Macrelaps microlepidotus*, page 351).

Bibron's Stiletto Snake is a glossy black, black-brown, or purple-brown snake with pinkish, cream, or white undersides. It has small eyes and a pointed snout and adopts a defensive posture which involves arching the neck, as here, and pointing the snout to the ground—although these snakes can arch much more than this. From this position they strike sideways rapidly.

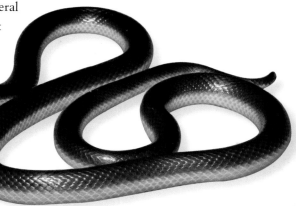

Actual size

FAMILY	Lamprophiidae: Atractaspidinae
RISK FACTOR	Horizontal front-fanged, venomous: sarafotoxins, and possibly cytotoxins
DISTRIBUTION	West and Central Africa: Liberia to Congo and DRC
ELEVATION	33–3,170 ft (10–965 m) asl
HABITAT	Rainforest and oil-palm plantations
DIET	Presumed reptiles and small mammals
REPRODUCTION	Presumed oviparous, clutch size unknown
CONSERVATION STATUS	IUCN not listed

ADULT LENGTH
11¾–19¾ in,
rarely 26¾ in
(300–500 mm,
rarely 680 mm)

356

ATRACTASPIS CORPULENTA
CORPULENT
BURROWING ASP
(HALLOWELL, 1854)

The Corpulent Burrowing Asp is a stocky-bodied or corpulent snake with a slightly pointed head and smooth scales. It is generally uniform black, dark gray, gunmetal, or dark brown in color, although specimens belonging to the West African subspecies may have white tail tips, hence the name *leucura*.

The Corpulent Burrowing Asp has a punctuated distribution across West and Central Africa, from Liberia to Congo and the Democratic Republic of the Congo. It is primarily a rainforest inhabitant, but is also encountered in oil-palm plantations. Like other burrowing asps, this is a fossorial snake that is only encountered moving around on the surface at night, and then usually only after rain. Although the natural history of this species is poorly known, it may be expected to prey upon lizards, other snakes, and possibly small mammals, and to lay eggs, in common with other burrowing asps. Snakebites to humans are rare but have been recorded, with effects ranging from symptomless to localized pain and swelling. However, fatalities from other *Atractaspis* are known, so care is cautioned with all members of the genus.

RELATED SPECIES

Three geographically separated subspecies of *Atractaspis corpulenta* are recognized: the nominate form (*A. c. corpulenta*) from Nigeria to the Congo and Democratic Republic of the Congo, a southern form (*A. c. kivuensis*) from the Democratic Republic of the Congo, and a western form (*A. c. leucura*) in Ghana, Côte d'Ivoire, and Liberia. *Atractaspis corpulenta* may be confused with other stout-bodied species within its range, such as the Central African Burrowing Asp (*A. boulengeri*), Benin Burrowing Asp (*A. dahomeyensis*), and Variable Burrowing Asp (*A. irregularis*).

Actual size

FAMILY	Lamprophiidae: Atractaspidinae
RISK FACTOR	Horizontal-fanged, venomous: sarafotoxins, and possibly also cytotoxins, procoagulants, or hemorrhagins
DISTRIBUTION	Middle East: Israel, Palestine, Jordan, Sinai (Egypt), and western Saudi Arabia
ELEVATION	1,000 ft (305 m) bsl to 7,000 ft (2,135 m) asl
HABITAT	Arid and semiarid rocky hillsides and wadis with coarse scrub, inhabiting animal burrows
DIET	Amphibians, lizards, snakes, and small mammals
REPRODUCTION	Oviparous, clutch size unknown
CONSERVATION STATUS	IUCN Least Concern, Vulnerable in Egypt

ADULT LENGTH
19¾–23¾ in,
occasionally 35 in
(500–600 mm,
occasionally 890 mm)

ATRACTASPIS ENGADDENSIS
ISRAELI BURROWING ASP
HAAS, 1950

357

Also known as the Ein Gedi Burrowing Asp, or Palestinian Burrowing Asp, the Israeli Burrowing Asp is probably the most dangerous member in the genus *Atractaspis*. It is found in Israel, Palestine, Jordan, Egypt's Sinai Peninsula, and on the Red Sea coast of Saudi Arabia, with an isolated population in central Saudi Arabia. Although it is recorded up to 7,000 ft (2,135 m) asl, the type locality at Ein Gedi, on the Dead Sea's western shore, is located at 1,000 ft (305 m) bsl. This species inhabits arid rocky hills and wadis, where it explores animal burrows for frogs, lizards, snakes, and small mammals. It rarely emerges, except after rain. There have been several rapidly fatal snakebites caused by the powerful sarafotoxins (cardiotoxins) in the venom, and there is no antivenom available. This side-stabbing snake cannot be handled safely with bare hands.

The Israeli Burrowing Asp is a stocky snake with a blunt, rounded head, small eyes, and a short tail. It is generally glossy or matte black, sometimes brown, and slightly lighter below.

RELATED SPECIES

In Israel, Jordan, and the Sinai, *Atractaspis engaddensis* may be confused with the Sinai Desert Blacksnake (*Walterinnesia aegyptia*, page 483), while at the southern extent of its range, near Jeddah on the Red Sea coast of Saudi Arabia, it occurs in sympatry with the related Arabian Burrowing Asp (*A. andersonii*), from which it can be separated by its higher dorsal scale count. Both species were formerly subspecies of the Small-scaled Burrowing Asp (*A. microlepidota*, page 358).

Actual size

FAMILY	Lamprophiidae: Atractaspidinae
RISK FACTOR	Horizontal front-fanged, venomous: sarafotoxins, and possibly cytotoxins
DISTRIBUTION	West Africa: Senegal, The Gambia, Mauritania, and Mali
ELEVATION	0–230 ft (0–70 m) asl
HABITAT	Arid Sahel semidesert and scrub
DIET	Lizards, snakes, and toads
REPRODUCTION	Presumed oviparous, clutch size unknown
CONSERVATION STATUS	IUCN Least Concern

ADULT LENGTH
11¾–19¾ in,
rarely 26½ in
(300–500 mm,
rarely 670 mm)

ATRACTASPIS MICROLEPIDOTA
SMALL-SCALED BURROWING ASP
GÜNTHER, 1866

Originally the range of the Small-scaled Burrowing Asp was vast and fragmented, across northern sub-Saharan Africa from coast to coast and into Arabia and the Middle East. However, with the elevation of its subspecies to specific status, the Small-scaled Burrowing Asp is now confined to Senegambia, southern Mauritania, and southwestern Mali. It is closely associated with the sandy soils of the arid Sahel region, south of the Sahara. A fossorial snake, active above ground only at night after rain, it preys on toads, lizards, and other snakes. Its venom contains sarafotoxins, which are cardiotoxins, affecting the heart, and it is believed capable of delivering a fatal snakebite. This snake cannot be handled safely, even gripped behind the head, due to its ability to strike sideways without opening its mouth.

Actual size

The Small-scaled Burrowing Asp is a moderately stout, glossy, jet-black snake, with a head indistinct from the neck, a short tail, and a dark gray venter.

RELATED SPECIES

The taxonomic status and relationships of *Atractaspis microlepidota* are by no means clear. Several former subspecies have been elevated to specific status: Magretti's Burrowing Asp (*A. magretti*), Peter's Burrowing Asp (*A. fallax*), and Phillip's Burrowing Asp (*A. phillipsi*), all from northeast Africa; the Sahelian Burrowing Asp (*A. micropholis*), from West Africa; the Arabian Burrowing Asp (*A. andersonii*), from the southern Arabian Peninsula; and the Israeli Burrowing Asp (*A. engaddensis*, page 357).

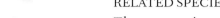

FAMILY	Lamprophiidae: Atractaspidinae
RISK FACTOR	Front-fanged, mildly venomous: venom composition unknown
DISTRIBUTION	Southern Africa: South Africa and Swaziland
ELEVATION	1,720–5,040 ft (525–1,535 m) asl
HABITAT	Wet savanna grasslands
DIET	Threadsnakes
REPRODUCTION	Oviparous, with clutches of 2–4 eggs
CONSERVATION STATUS	IUCN Near Threatened

ADULT LENGTH
8–12½ in
(200–320 mm)

HOMOROSELAPS DORSALIS
STRIPED HARLEQUIN SNAKE
(SMITH, 1949)

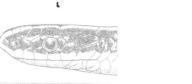

359

The tiny Striped Harlequin Snake inhabits wet savanna grasslands in the highveld of eastern South Africa and Swaziland. It is listed as Near Threatened in South Africa, due to the unregulated burning of its habitat. Rarely encountered, the Striped Harlequin Snake shelters inside termitaria or under logs, and hunts threadsnakes (Leptotyphlopidae), which are killed by its venomous bite. With its small head, narrow mouth gape, low venom yield, and small fangs, it is not considered dangerous to humans, but bites are known from the slightly larger Spotted Harlequin Snake (*Homoroselaps lacteus*) that have resulted in mild hemorrhage, headache, and swollen lymph glands. One of Africa's smallest non-scolecophidian snakes, it produces clutches of two to four extremely small eggs.

Actual size

RELATED SPECIES

At one time this tiny snake, and its close South African relative the Spotted Harlequin Snake (*Homoroselaps lacteus*), were referred to as "dwarf garter snakes" and placed in the genus *Elaps*, the type genus of the Elapidae, because they possessed small front fangs. However, they are more closely related to snakes known as side-stabbing snakes, burrowing asps, or stiletto snakes (*Atractaspis*, pages 355–358) and are now placed in the Atractaspidinae.

The Striped Harlequin Snake is an elongate snake with a small head, which is indistinct from the neck. It is black above and pinkish white below, the two colors separated by a linear demarcation at the midflank. A broad yellow vertebral stripe runs the length of the head, body, and tail.

FAMILY	Lamprophiidae: Lamprophiinae
RISK FACTOR	Nonvenomous, constrictor
DISTRIBUTION	Southern Africa: South Africa, Namibia, Botswana, Zimbabwe, Lesotho, Swaziland, Mozambique; possibly into East Africa
ELEVATION	0–4,430 ft (0–1,350 m) asl
HABITAT	Most habitats, especially around houses
DIET	Small mammals, birds, reptiles, occasionally frogs
REPRODUCTION	Oviparous, with clutches of 8–18 eggs
CONSERVATION STATUS	IUCN not listed

ADULT LENGTH
23¾–35½ in,
rarely 5 ft
(600–900 mm,
rarely 1.5 m)

360

BOAEDON CAPENSIS
COMMON HOUSESNAKE
DUMÉRIL, BIBRON & DUMÉRIL, 1854

The Common Housesnake is a moderately slender snake, with smooth scales, a slightly elongate head, and vertically elliptical pupils. Its dorsal color may vary from yellow to pinkish, or reddish brown to dark brown, with or without broken cross-bands or stripes, but always with a pair of white or yellow stripes on the sides of the head, from the snout, through the eye to the angle of the jaw, and along the supralabials.

The ubiquitous Common Housesnake is one of the most frequently encountered snakes of southern Africa, occurring virtually everywhere, except the highlands of Lesotho. How far it ranges north into East Africa is an open question given the confusion over the identity of these populations. Common Housesnakes inhabit grasslands, forests, semidesert, and even desert, and they are well named because they have adapted well to living in human dwellings, where they hunt and constrict rodents. They will also take birds, bats, lizards, and frogs. Common Housesnakes are docile, and although some specimens will bite, especially if handled roughly, their teeth are small and do little damage. Common Housesnakes make popular first pet snakes for budding herpetologists. Often killed on sight, Common Housesnakes should be encouraged around buildings due to their excellent rodent-hunting abilities.

RELATED SPECIES

Boaedon capensis is a member of the *B. fuliginosus–capensis* species complex, which means it is very close to the Brown Housesnake (*B. fuliginosus*) of sub-Saharan Africa, and the precise division between the two species is blurred, with both possibly containing additional undescribed species. *Boaedon* currently contains 12 species in Africa and one (*B. arabicus*) in Yemen. It is closely related to the seven African housesnake species in genus *Lamprophis* (page 364).

Actual size

FAMILY	Lamprophiidae: Lamprophiinae
RISK FACTOR	Nonvenomous, probably a constrictor
DISTRIBUTION	West and Central Africa: Sierra Leone and Guinea to Uganda, and south to Angola
ELEVATION	0–7,550 ft (0–2,300 m) asl
HABITAT	Dense evergreen forest, rainforest, dry forest, and upland savannas
DIET	Small mammals and lizards
REPRODUCTION	Oviparous, with clutches of 3–5 eggs
CONSERVATION STATUS	IUCN not listed

ADULT LENGTH
23¾–32¾ in,
occasionally 4 ft
(600–830 mm,
occasionally 1.2 m)

BOTHROPHTHALMUS LINEATUS
RED AND BLACK STRIPED SNAKE
PETERS, 1863

The Red and Black Striped Snake is a seldom encountered species across much of its fairly extensive range, through the African Atlantic coastal countries from Guinea to Angola, and eastward to Uganda. It is primarily an inhabitant of dense evergreen forest or rainforest, but it is also found in patches of dry forest, and occurs on savannas at higher elevations. The primary prey of the diurnally active Red and Black Striped Snake are house mice and striped mice, but swamp rats, insectivorous musk shrews, and diurnal skinks are also taken by adults. Juveniles feed exclusively on lizards. It seems likely that this species employs constriction to kill its prey.

The Red and Black Striped Snake is black or dark gray on the back, with or without three to five red or orange longitudinal stripes, and a distinctive black-striped, white head. The venter is pink or red and the head is paler than the body, sometimes even cream, with a dark V-shaped marking on the dorsum.

RELATED SPECIES

The closest relative of *Bothrophthalmus lineatus* is the Brown-bellied Snake (*B. brunneus*) of Cameroon, Gabon, Equatorial Guinea, including Bioko (Fernando Po), Congo, and the Democratic Republic of the Congo. Previous authors have treated *B. brunneus* as a sub-species or a synonym of *B. lineatus*. The patterning of *B. lineatus* also resembles that of Reinhardt's Snake-eater (*Polemon acanthias*).

Actual size

FAMILY	Lamprophiidae: Lamprophiinae
RISK FACTOR	Nonvenomous, constrictor
DISTRIBUTION	Sub-Saharan Africa: Tanzania to Namibia and South Africa
ELEVATION	0–3,280 ft (0–1,000 m) asl
HABITAT	Savanna and coastal forest
DIET	Snakes, lizards, and small mammals
REPRODUCTION	Oviparous, with clutches of 5–13 eggs
CONSERVATION STATUS	IUCN Least Concern

ADULT LENGTH
3 ft 3 in–4 ft,
occasionally 5 ft 9 in
(1.0–1.2 m,
occasionally 1.75 m)

362

GONIONOTOPHIS CAPENSIS
CAPE FILESNAKE
(SMITH, 1847)

The Cape Filesnake is almost triangular in cross section, and heavily built, with a broad, dorsally compressed head. Its gray or purple-brown body scales are strongly keeled and separated from one another by pink or gray interstitial skin. The middorsal scale row is double-keeled and fused to the vertebral column and marked by a broad and distinctive yellow or white longitudinal stripe.

Filesnakes are so-called because they have rough, keeled scales and some species are triangular in cross section, like a metalworker's file. The large, terrestrial, and nocturnal Cape Filesnake occurs from Tanzania, south to KwaZulu-Natal in the east and Namibia in the west. It has a preference for open savanna country, but also inhabits coastal forests. The Cape Filesnake feeds on other snakes, including highly venomous species such as Puff Adders (*Bitis arietans*, page 610), and it is immune to their venom. It also takes lizards and rodents. When handled, the Cape Filesnake is placid and unwilling to bite, but it will void its bowels or the contents of cloacal glands. Despite its gentle nature and its snake-killing abilities, it is much feared as a portent of evil by many Africans.

RELATED SPECIES

The genus *Gonionotophis* contains 15 species of African filesnakes, most of which were previously placed in the genus *Mehelya*. Two former subspecies of *G. capensis* have been elevated to specific status, the Unicolor Filesnake (*G. chanleri*, formerly *unicolor*), from Ethiopia, and the Congo Filesnake (*G. savorgnani*). The African filesnakes should not be confused with the Australasian filesnakes (*Acrochordus*, pages 128–129), which are fully aquatic snakes belonging to the unrelated family Acrochordidae.

Actual size

FAMILY	Lamprophiidae: Lamprophiinae
RISK FACTOR	Nonvenomous, constrictor
DISTRIBUTION	Southern Africa: Swaziland and South Africa
ELEVATION	4,590–6,230 ft (1,400–1,900 m) asl
HABITAT	Rocky outcrops on moist savannas
DIET	Lizards and birds
REPRODUCTION	Oviparous, with clutches of up to 7 eggs
CONSERVATION STATUS	IUCN Near Threatened

ADULT LENGTH
19¾–23¾ in,
rarely 35½ in
(500–600 mm,
rarely 900 mm)

INYOKA SWAZICUS
SWAZI ROCK SNAKE
(SCHAEFER, 1970)

363

Also known as the Swazi Housesnake, the Swazi Rock Snake is only found in Swaziland and South Africa, where it occurs in northern KwaZulu-Natal, Mpumalanga, and Limpopo provinces in the east. It occurs in moist savanna habitats at elevations above 4,590 ft (1,400 m), where it inhabits crevices or shelters under exfoliated flakes of rock, on rocky outcrops, a similar habitat to that adopted by the Broad-headed Snake (*Hoplocephalus bungaroides*, page 505) in Australia. The Swazi Rock Snake's slender body makes it ideally adapted to this saxicolous (rock-dwelling) lifestyle. Its prey comprises small lizards, mostly rock-dwelling geckos and skinks, which are hunted at night. Small birds are also listed as possible prey. This species is considered Near Threatened because of its specialized habitat and limited range.

RELATED SPECIES

The genus *Inyoka* is monotypic, but this species was previously placed in the genus *Lamprophis* (page 364), which contains seven species. The closest relative of the Swazi Rock Snake is believed to be the monotypic Uganda Housesnake (*Hormonotus modestus*), which is also known as the Yellow Forest Snake and is a terrestrial inhabitant of rainforests.

The Swazi Rock Snake is an elongate, slender snake, with a long tail, a narrow head, and bulbous, protruding eyes. It may be dorsally dark red or brown in color, while the venter is white or cream.

Actual size

FAMILY	Lamprophiidae: Lamprophiinae
RISK FACTOR	Nonvenomous, constrictor
DISTRIBUTION	Southern Africa: South Africa and Lesotho
ELEVATION	0–5,580 ft (0–1,700 m) asl
HABITAT	Wet savanna, lowland forest, and fynbos heathland
DIET	Small mammals, lizards, and frogs
REPRODUCTION	Oviparous, with clutches of 8–12 eggs
CONSERVATION STATUS	IUCN Least Concern

ADULT LENGTH
15¾–23¾ in,
rarely 35½ in
(400–600 mm,
rarely 900 mm)

LAMPROPHIS AURORA
AURORA HOUSESNAKE
(LINNAEUS, 1758)

The Aurora Housesnake is endemic to South Africa and Swaziland. It shows a preference for wet savanna habitats but also occurs in low-lying woodlands and forests, and in the South African coastal heathland habitat known as fynbos. It is especially common near watercourses, where it may be discovered sheltering under rocks, and it will occupy old termitaria. This is a secretive species that is not encountered with the same frequency as the Common Housesnake (*Boaedon capensis*, page 360). It emerges at night to hunt rodents, being especially fond of nestlings, but it can also constrict older mice. Lizards and frogs also feature in its diet. It is inoffensive and disinclined to bite.

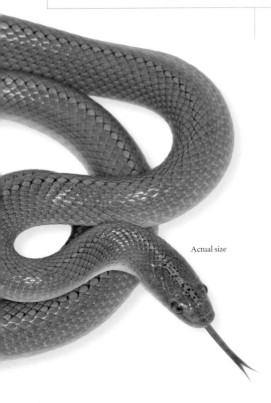

Actual size

RELATED SPECIES
There are currently seven species in genus *Lamprophis*, with the closely related Common Housesnake complex transferred to genus *Boaedon*. Most *Lamprophis* are found in Africa, but one (*L. geometricus*) is endemic to the Seychelles. The Uganda Housesnake (*Hormonotus modestus*) and Swazi Rock Snake (*Inyoka swazicus*, page 363) were also once included in *Lamprophis*.

The Aurora Housesnake is an attractive species. Smooth-scaled, with a rounded head, it is mid-brown to olive-green above, but paler yellow-brown on the lower flanks and yellow or white on the venter. It is quickly recognized by its characteristic yellow or orange vertebral stripe, although small juvenile specimens may be confused with the Striped Harlequin Snake (*Homoroselaps dorsalis*, page 359).

FAMILY	Lamprophiidae: Lamprophiinae
RISK FACTOR	Nonvenomous, constrictor
DISTRIBUTION	Central and East Africa: Lake Tanganyika (Tanzania, Burundi, DRC, and Zambia)
ELEVATION	2,560 ft (780 m) asl
HABITAT	Lake waters and rocky shores
DIET	Fish, and possibly also amphibians
REPRODUCTION	Oviparous, with clutches of 4–8 eggs
CONSERVATION STATUS	IUCN Least Concern

ADULT LENGTH
15¾–19¾ in,
rarely 27½ in
(400–500 mm,
rarely 700 mm)

LYCODONOMORPHUS BICOLOR
LAKE TANGANYIKA WATERSNAKE
(GÜNTHER, 1893)

The genus *Lycodonomorphus* is a genus of African watersnakes that occupies the niche of an aquatic piscivore (fish eater) occupied in Asia, Europe, North America, or Australasia by watersnakes from the Natricidae, a family less well represented in Africa. The Lake Tanganyika Watersnake is endemic to Lake Tanganyika, where it can be very common, floating on the lake surface on calm nights without a moon, or hunting cichlids in the shallows, and it can be attracted toward lantern light. Populations are estimated to range from 3,500 to 14,600 snakes per square mile (9,000 to 38,000 snakes per square kilometer). Small cichlids are captured and constricted, before being swallowed headfirst. By day the snakes shelter in the great rock piles that line the lake shore.

RELATED SPECIES

Lycodonomorphus contains ten species, but the only other aquatic snake in Lake Tanganyika is the endemic Storm's Water Cobra (*Naja annulata stormsi*, page 466), and confusion between those two species is unlikely. The nearest species to *Lycodonomorphus* are the Congo Dark-bellied Watersnake (*L. leleupi*), the Eastern Congo White-bellied Watersnake (*L. subtaeniatus*), and Whyte's Watersnake (*L. whytii*) in southern Tanzania.

The Lake Tanganyika Watersnake is a small snake with a small head and smooth scales. It is usually dark gray or brown above, and yellow-brown on the venter, hence the name *bicolor*. Many specimens have truncated tails, believed to be due to attacks by freshwater crabs.

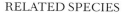

Actual size

FAMILY	Lamprophiidae: Lamprophiinae
RISK FACTOR	Nonvenomous, constrictor
DISTRIBUTION	East and southern Africa: Egypt to South Africa
ELEVATION	0–8,200 ft (0–2,500 m) asl
HABITAT	Grassland, savanna, coastal forest, fynbos, and karoo scrub
DIET	Lizards, occasionally snakes
REPRODUCTION	Oviparous, with clutches of 3–9 eggs
CONSERVATION STATUS	IUCN not listed

ADULT LENGTH
11¾–15¾ in,
occasionally 25¼ in
(300–400 mm,
occasionally 640 mm)

366

LYCOPHIDION CAPENSE
CAPE WOLFSNAKE
(SMITH, 1831)

The Cape Wolfsnake is distributed throughout most of East and southern Africa, except, ironically, the Cape, and the Namib Desert. This is a small snake that is frequently found on roads at night, or sheltering under rocks, boards, or other material during the day. It has a preference for open habitats, grassland, savanna, karoo scrub, or fynbos heathland, but it may also be found in coastal forest or in deserted termitaria. It prefers wet habitats, and it feeds on skinks and geckos, and occasionally small snakes. Wolfsnakes possess specially recurved teeth for gripping smooth-scaled prey such as skinks, and *Lycophidion* is the African equivalent of the Asian colubrid wolfsnakes (*Lycodon*, pages 190–191), or the Australasian groundsnakes (*Stegonotus*, pages 235–236). Generally inoffensive, it flattens its body in defense.

Actual size

The Cape Wolfsnake is a small, smooth-scaled snake with a slightly elongate head and small eyes. It may be uniform black or brown, or have every scale tipped with white, presenting a speckled appearance. The undersides are white or white with black speckling.

RELATED SPECIES

Three subspecies are recognized: the nominate form (*Lycophidion capense capense*) throughout southern Africa, an East African subspecies from the Rift Valley and Great Lakes (*L. c. jacksoni*), and a coastal East African subspecies (*L. c. loveridgei*), which also occurs on Zanzibar. One former subspecies has been elevated to specific status, the Spotted Wolfsnake (*L. multimaculatum*), from the Caprivi Strip and the Namibian–Angolan border. The genus *Lycophidion* contains 20 species, including an endemic species from Pemba Island, Tanzania (*L. pembanum*).

FAMILY	Lamprophiidae: Lamprophiinae
RISK FACTOR	Nonvenomous, constrictor
DISTRIBUTION	Northeast Africa: Ethiopia and Eritrea
ELEVATION	5,250–10,800 ft (1,600–3,300 m) asl
HABITAT	Montane grassland and woodland, lowland forest, and upland moorland
DIET	Amphibians, and possibly small mammals, birds, or lizards
REPRODUCTION	Oviparous, with clutches of up to 21 eggs
CONSERVATION STATUS	IUCN not listed

ADULT LENGTH
38 in
(965 mm)

PSEUDOBOODON LEMNISCATUS

STRIPED ETHIOPIAN MOUNTAIN SNAKE

DUMÉRIL, BIBRON & DUMÉRIL, 1854

367

The genus *Pseudoboodon* is unusual in that its members possess a deep triangular pit on the fifth to sixth supralabials, although the purpose of this feature is unknown. The Striped Ethiopian Mountain Snake inhabits the Ethiopian Highlands, above 5,250 ft (1,600 m) asl, and neighboring Eritrea, on the Red Sea, and is both the commonest and the largest member of the genus. It occupies a variety of habitats ranging from tropical forest at lower elevations, to montane grassland and woodland, and, at the altitudinal extreme of 10,800 ft (3,300 m), alpine moorland. It is primarily nocturnal, and hunts amphibians on the ground. In captivity it has been known to take small mammals, birds, and lizards, but it is not known if these feature in the diet of wild specimens.

The Striped Ethiopian Mountain Snake has a robust, smooth-scaled body, a moderately long tail, and a head that is slightly distinct from the neck. It may be yellow, orange, or brown, with three broad brown or black longitudinal stripes, middorsally and on the lower flanks. The lateral stripes are edged by pale gray stripes, the upper stripe a half to one scale wide, the lower one to two scales wide and merging with the pale venter, which itself often bears two longitudinal rows of dark spots.

RELATED SPECIES

The genus *Pseudoboodon* contains four species, which are all endemic to high elevations in Ethiopia and Eritrea in the Horn of Africa. The other three species are Böhme's Ethiopian Mountain Snake (*P. boehmei*), Gasca's Ethiopian Mountain Snake (*P. gascae*), and Sandford's Ethiopian Mountain Snake (*P. sandfordorum*). *Pseudoboodon boehmei* is striped like *P. lemniscatus*, while *P. gascae* and *P. sandfordorum* are blotched.

Actual size

FAMILY	Lamprophiidae: Psammophiinae
RISK FACTOR	Rear-fanged, mildly venomous: venom composition unknown
DISTRIBUTION	Southern Africa: South Africa, Namibia, and Botswana, to southern Angola
ELEVATION	0–3,610 ft (0–1,100 m) asl
HABITAT	Dry riverbeds, scrubland, rocky slopes, sandveld, and other arid habitats
DIET	Lizards
REPRODUCTION	Oviparous, with clutches of 2–4 eggs
CONSERVATION STATUS	IUCN not listed

ADULT LENGTH
11¾–12½ in,
occasionally 19¾ in
(300–320 mm,
occasionally 500 mm)

DIPSINA MULTIMACULATA
DWARF BEAKED SNAKE
(SMITH, 1847)

368

Actual size

The Dwarf Beaked Snake is a small, smooth-scaled snake with large eyes and a short head, which curves downward to a beak-shaped rostral. Patterning is variable, but usually comprises a patchwork of pastel-colored lozenge shapes arranged as a longitudinal, alternating, vertebral stripe or several series of irregular dark brown spots on a light brown or orange background.

The Dwarf Beaked Snake is endemic to southern Africa, where it occurs through the Northern Cape of South Africa and into Namibia, southwestern Botswana, and possibly southern Angola. It inhabits dry habitats, from dry riverbeds to rocky slopes and sandveld, where it shelters under small bushes or shrubs and waits in ambush for small lizards, such as skinks, lacertids, or geckos. As a small snake it has many enemies itself, from birds to meerkats, other snakes, monitor lizards, and large invertebrates. In defense it will adopt a coiled position in the hopes of mimicking one of the small venomous vipers (*Bitis*, pages 610–617) of southern Africa. Such a small snake is considered incapable of delivering a snakebite, which would mean engaging the fangs at the back of its tiny mouth.

RELATED SPECIES

The tiny *Dipsina multimaculata* resembles the Moila Snake (*Rhagerhis moilensis*, page 375) of North Africa and the Middle East. It may also be mistaken for one of the small southern African vipers such as the Horned Adder (*Bitis caudalis*, page 612), or a juvenile Mole Snake (*Pseudaspis cana*, page 378), which occur in sympatry with this species.

FAMILY	Lamprophiidae: Psammophiinae
RISK FACTOR	Rear-fanged, mildly venomous: venom composition unknown
DISTRIBUTION	Sub-Saharan Africa: South Sudan, Kenya, and Tanzania, east to Togo, and south to Zimbabwe and Limpopo, South Africa
ELEVATION	0–5,250 ft (0–1,600 m) asl
HABITAT	Savanna woodlands, especially mopane woodland, coastal thicket, and semidesert
DIET	Lizards and frogs
REPRODUCTION	Oviparous, with clutches of 2–8 eggs
CONSERVATION STATUS	IUCN not listed

ADULT LENGTH
9¾–13¾ in,
rarely 17 in
(250–350 mm,
rarely 430 mm)

HEMIRHAGERRHIS NOTOTAENIA
MOPANE SNAKE
(GÜNTHER, 1864)

369

Also known as the Eastern Bark Snake, the Mopane Snake is distributed widely through East Africa, from South Sudan, and Kenya, to Zambia, Zimbabwe, Mozambique, and Limpopo Province, South Africa. It also occurs through Central Africa as far as Togo in West Africa. This is a savanna woodland snake particularly associated with mopane woodland. Mopane is a common savanna tree with bilobed leaves. A highly arboreal, diurnal species, it is also found in coastal thicket and semidesert. Being small and slender, it is easily able to secrete itself away in cracks in the tree bark. Prey consists of small skinks, geckos, and gecko eggs. Small frogs are also occasionally taken. The Mopane Snake is inoffensive and is completely unwilling to bite, even when handled.

RELATED SPECIES

A further three species of *Hemirhagerrhis* are recognized in Africa. The Western or Viperine Bark Snake (*H. viperina*), a Namibian endemic, was once treated as a subspecies of *H. nototaenia*, but it is more associated with rocky outcrops. The Striped Bark Snake (*H. kelleri*) and Kenyan Bark Snake (*H. hildebrandtii*) are found in northeast Africa.

Actual size

The Mopane Snake is a small, slender snake with an elongate, flattened head. Most specimens are gray to brown with a distinctive dark vertebral viperine zigzag pattern, but unicolor or striped specimens are not unknown. The zigzag pattern in particular helps this slender, cryptic, and often sedentary snake blend into its tree bark environment.

FAMILY	Lamprophiidae: Psammophiinae
RISK FACTOR	Rear-fanged, mildly venomous: venom composition unknown
DISTRIBUTION	Europe and North Africa: Spain, Portugal, southern France, northwestern Italy, Gibraltar, Morocco, and Western Sahara
ELEVATION	0–9,840 ft (0–3,000 m) asl
HABITAT	Rocky hillsides, boulder-strewn fields, dry stone walls, and forest clearings
DIET	Reptiles, mammals, birds, large insects, and amphibians
REPRODUCTION	Oviparous, with clutches of 4–12 eggs
CONSERVATION STATUS	IUCN Least Concern

ADULT LENGTH
Male
4 ft–4 ft 7 in,
occasionally 6 ft 3 in
(1.2–1.4 m,
occasionally 1.9 m)

Female
6 ft 7 in
(2.0 m)

370

MALPOLON MONSPESSULANUS
WESTERN MONTPELLIER SNAKE
(HERMANN, 1804)

The Western Montpellier Snake is a large snake with a stout, powerful body and a narrow, pointed head with supraocular scales that protrude outward above the large eyes and present a scowling expression. It is a variably patterned snake, being brown, olive, or gray, with or without darker blotches or spots, and sometimes blue on the flanks. The venter is yellow to off-white.

Western Europe's largest snake, the Western Montpellier Snake is found throughout the Iberian Peninsula, southern France, and northern Italy, and across the Gibraltar Strait in Morocco and Western Sahara. A snake of arid habitats, such as boulder-strewn hillsides, fields, or pastures, it will frequent dry stone walls and man-made rock piles. It has an extremely catholic diet, feeding on skinks, anguid lizards, snakes, mammals to the size of rabbits, birds and their eggs, and large insects. Western Montpellier Snakes are highly active, diurnal snakes that are frequently seen rushing across roads, but they are equally frequently seen as roadkill, killed while basking. There are reports of minor envenomings by these rear-fanged snakes, resulting in localized swelling, ptosis (drooping eyelids), shortness of breath, and difficulty in swallowing. Large specimens may be potentially dangerous.

RELATED SPECIES

Two subspecies are recognized, the nominate form (*Malpolon monspessulanus monspessulanus*), from Spain, Portugal, and southern France, and a recently described subspecies (*M. s. saharatlanticus*) from the Atlantic coast of southern Morocco and Western Sahara. Two former subspecies were elevated to specific status, the North African Montpellier Snake (*M. insignitus*), from Morocco to Egypt, on Lampedusa Island, Italy, and through Asia Minor to the Caspian Sea, and the Eastern Montpellier Snake (*M. fuscus*), from Greece, the Balkans, and neighboring Asia Minor. The Moila Snake (*Rhagerhis moilensis*, page 375) was also once included in *Malpolon*.

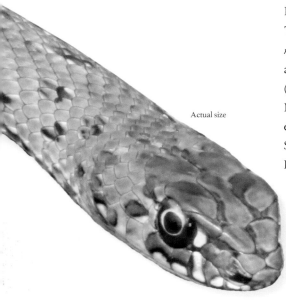

Actual size

FAMILY	Lamprophiidae: Psammophiinae
RISK FACTOR	Rear-fanged, mildly venomous: venom composition unknown
DISTRIBUTION	Madagascar: western, northern, and southern Madagascar
ELEVATION	0–1,970 ft (0–600 m) asl
HABITAT	Dry forest, spiny thornbush savanna, grassland, and humid forest
DIET	Lizards, snakes, frogs, and small mammals
REPRODUCTION	Probably oviparous, clutch size unknown
CONSERVATION STATUS	IUCN Least Concern

ADULT LENGTH
25½–29½ in,
rarely 3 ft 3 in
(650–750 mm,
rarely 1.0 m)

MIMOPHIS MAHFALENSIS
MALAGASY SANDSNAKE
GRANDIDIER, 1867

371

The Malagasy Sandsnake is found throughout the island of
Madagascar, excluding the tropical east. It has a preference for
arid habitats such as dry forest, spiny thornbush savanna, and
open grassland areas, but may be found in humid forests. It is
diurnally active and may even be seen hunting during the heat
of the day. Its prey primarily comprises lizards, but it is also
documented to take small mammals, small snakes, and frogs.
In common with other members of the Psammophiinae it is
probably oviparous, but its reproduction has not been studied.
This species may bite and since it possesses enlarged rear fangs
it should be treated with respect, especially as its venom
composition, yield, and toxicity are unknown. The Mahafaly
are an ethic group native to southwestern Madagascar.

The Malagasy Sandsnake is a slender snake with
a pointed head. It is variably patterned, females
often being uniform brown or gray, while males
are gray to light brown with a dark vertebral
zigzag pattern. Highland specimens possess
broad vertebral and lateral longitudinal brown
stripes. The head is marked by several fine dark
lines that converge into the body pattern. The
undersides are pale brown with two slightly
darker longitudinal stripes.

RELATED SPECIES
Mimophis mahfalensis is the only Malagasy representative of
the subfamily Psammophiinae, and is therefore more closely
related to sandsnakes and their allies from Africa and the
Mediterranean than it is to other Malagasy caenophidian
snakes, which belong to subfamily Pseudoxyrhophiinae.
It is unlikely to be confused with any other species.
Two subspecies are recognized by some authors,
M. m. mahfalensis and *M. m. madagascariensis*,
which was originally described as a separate
species. Other authors do not recognize these.

Actual size

FAMILY	Lamprophiidae: Psammophiinae
RISK FACTOR	Rear-fanged, mildly venomous: venom composition unknown
DISTRIBUTION	Sub-Saharan Africa: Senegal to South Sudan, Kenya, and Tanzania, south to Angola, Namibia, and eastern South Africa and Botswana
ELEVATION	0–4,920 ft (0–1,500 m) asl
HABITAT	Wet savanna, lowland forest, and marshes
DIET	Lizards, snakes, frogs, small mammals, and birds
REPRODUCTION	Oviparous, with clutches of 10–30 eggs
CONSERVATION STATUS	IUCN not listed

ADULT LENGTH
2 ft 7 in–3 ft 3 in,
rarely 6 ft
(0.8–1.0 m,
rarely 1.8 m)

PSAMMOPHIS MOSSAMBICUS
OLIVE GRASS SNAKE
PETERS, 1882

372

The Olive Grass Snake is a stout snake with a long tail and relatively large eyes under shelved supraocular scales, which present a scowling appearance. It is uniform olive-brown above, and white or yellow below, with dark edging to the dorsal scales, forming fine longitudinal stripes in some specimens. The lips, neck, and throat are pale, spotted with darker pigment.

Actual size

Snakes of the genus *Psammophis* are known as sandsnakes, but some species avoid arid, sandy habitats. They are also known as whipsnakes due to their speed. The Olive Grass Snake inhabits savanna, lowland forest, and marshland. Diurnally active and highly alert, it feeds on lizards, frogs, small mammals, birds, and snakes, including venomous species. When on the move *Psammophis* species often elevate the anterior third of the body off the ground. Psammophile (sand-loving) snakes, including *Psammophis*, of all ages and sexes, also anoint and self-rub their bodies with nasal gland secretions. It is thought this helps prevent desiccation in hot weather, but the purpose is not fully understood. Nervous, they bite if handled, and although not considered dangerous, snakebites can cause localized swelling, pain, and nausea.

RELATED SPECIES

There are 23 species in the genus *Psammophis*, including several large species that are related to, and may be confused with, *P. mossambicus*: Phillip's Whipsnake (*P. phillipsii*) and the Western Whipsnake (*P. occidentalis*) from West Africa, the Leopard Whipsnake (*P. leopardinus*) from Namibia and Angola, and the Short-snouted Grass Snake (*P. brevirostris*), from Botswana, northeast, and eastern South Africa and Swaziland. At one time many of these species were treated as subspecies or synonyms of the Hissing Sandsnake (*P. sibilans*), but that species is now limited to northeastern Africa.

FAMILY	Lamprophiidae: Psammophiinae
RISK FACTOR	Rear-fanged, mildly venomous: venom composition unknown
DISTRIBUTION	North Africa and Asia: Morocco south to Mauritania, east to Egypt, Sudan, Eritrea, Somalia, Arabia, Iran, Pakistan, and northwestern India
ELEVATION	0–7,870 ft (0–2,400 m) asl
HABITAT	Rocky or sandy semidesert and desert, or arid savanna, with abundant vegetation
DIET	Lizards, frogs, snakes, small mammals, and birds
REPRODUCTION	Oviparous, with clutches of 5–6 eggs
CONSERVATION STATUS	IUCN not listed

ADULT LENGTH
2 ft 7 in–3 ft 7 in,
rarely 5 ft
(0.8–1.1 m,
rarely 1.5 m)

PSAMMOPHIS SCHOKARI
AFRO-ASIAN SAND RACER
(FORSKÅL, 1775)

373

One of the most widely distributed *Psammophis* species, the slender, arid-adapted Afro-Asian Sand Racer was first described from Yemen but occurs across North Africa and Asia, from Mauritania and Morocco, to Somalia and Eritrea, Israel, and Jordan, UAE and Oman, and Pakistan and northeastern India. It is Arabia's only *Psammophis*. A highly alert, diurnal predator of lizards, it has a preference for arid desert or semidesert habitats with abundant bushes and trees, into which the Afro-Asian Sand Racer may also climb in search of small, sleeping passerine birds. Small mammals and small snakes also feature in its diet. The Afro-Asian Sand Racer is a fast-moving snake that may be active in the heat of midday, rushing between shade patches. A mildly venomous snake, its bites are not dangerous but may cause localized pain.

The Afro-Asian Sand Racer is a slender-bodied, long-tailed, narrow-headed, large-eyed species, and highly variable in color and patterning across its range. Most specimens are olive-brown or yellowish brown above, and white or yellowish below, often with a series of three dorsal body stripes, and a dark stripe through the eye, separating the brown dorsum from the white of the labials and throat.

RELATED SPECIES

A number of slender *Psammophis* species are related to *P. schokari*: the Saharan Sand Racer (*P. aegyptius*) from Algeria to Israel, Spotted Sand Racer (*P. punctulatus*) of northeast Africa, Elegant Sand Racer (*P. elegans*) from West Africa, and Ornate Olympic Snake (*P. praeornatus*) from West and Central Africa. This last species used to belong to genus *Dromophis*, which was recently synonymized with *Psammophis*.

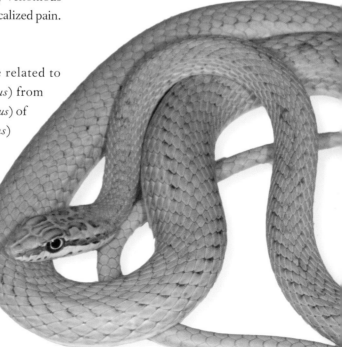

Actual size

FAMILY	Lamprophiidae: Psammophiinae
RISK FACTOR	Rear-fanged, mildly venomous: venom composition unknown
DISTRIBUTION	East and southern Africa: Tanzania, Zambia, Angola, northern Namibia and Botswana, Zimbabwe, and northeastern South Africa
ELEVATION	115–7,220 ft (35–2,200 m) asl
HABITAT	Savanna and grassland, and karoo scrub
DIET	Lizards, frogs, small mammals, and birds
REPRODUCTION	Oviparous, with clutches of 5–18 eggs
CONSERVATION STATUS	IUCN Least Concern

ADULT LENGTH
23¾–25½ in,
rarely 36½ in
(600–650 mm,
rarely 930 mm)

374

PSAMMOPHYLAX TRITAENIATUS
STRIPED SKAAPSTEKER
(GÜNTHER, 1868)

The Striped Skaapsteker has a slender body, long tail, narrow, pointed head, and large eyes. It is boldly marked with three longitudinal dark brown stripes on a gray to pale brown background. The lateral stripes are broad, and continue through the eyes to the snout, while the vertebral stripe is narrow, and fades out on the head. The venter is white with a pale green or yellow midventral stripe.

Skaapstekers are common snakes and most searches within their ranges will turn one up. The name is Afrikaans for "sheep sticker" and is said to have come about because sheep farmers, upon finding a dead snake-bitten sheep, probably killed by a cobra or adder, would search the area and quickly find one of these small, mildly venomous snakes. The skaapsteker is the serpentine equivalent of a "fall guy," getting the blame. The Striped Skaapsteker is found throughout East and southern Africa, from Tanzania to Namibia and northeastern South Africa. It inhabits savanna and grassland habitats and hunts small mammals, lizards, and birds as an adult, and lizards and frogs as a juvenile. Inoffensive and generally unwilling to bite, even when handled, the Striped Skaapsteker is completely harmless to humans, and to sheep.

RELATED SPECIES

The Spotted Skaapsteker (*Psammophylax rhombeatus*) occurs in the Cape, eastern South Africa, Lesotho, and Swaziland, and occurs in sympatry with *P. tritaeniatus* in Limpopo, South Africa. The Beaked Skaapsteker (*P. acutus*) of Zambia, Angola, Tanzania, and Burundi is remarkably similar in appearance to *P. tritaeniatus*. Three other *Psammophylax* species are distributed across Central and East Africa, the Gray-bellied Grass Snake (*P. variabilis*) just entering southern Africa in the Caprivi Strip of Namibia.

Actual size

FAMILY	Lamprophiidae: Psammophiinae
RISK FACTOR	Rear-fanged, venomous: venom composition unknown
DISTRIBUTION	North Africa and Middle East: Western Sahara to Eritrea, Israel, Jordan, Arabia, and western Iran
ELEVATION	0–6,180 ft (0–1,885 m) asl
HABITAT	Arid rocky hills and wadis, gravel pans, coastal dunes, and grassy plains
DIET	Small mammals, birds, lizards, snakes, and large insects
REPRODUCTION	Oviparous, with clutches of 4–18 eggs
CONSERVATION STATUS	IUCN not listed

ADULT LENGTH
27¾–31½ in,
rarely 6 ft 3 in
(700–800 mm,
rarely 1.9 m)

RHAGERHIS MOILENSIS
MOILA SNAKE
(RUESS, 1834)

375

Named for its type locality, Moilah on the Arabian Red Sea coast, the Moila Snake is found from Western Sahara to Eritrea, and in the Arabian Peninsula, excluding the interior. It also occurs in Jordan and Israel, and western Iran. Habitats include rocky wadis and gravel pans, especially those with scrub or bushes, but it is also found in sandy areas. Prey consists of small mammals, such as gerbils, and lizards, snakes, and birds, while juveniles take large insects. Prey may be uncovered by shoveling with the downward-curved snout. The defensive posture involves elevation of the anterior body, flattening of the neck, cobra-style, and curving away from the threat for maximum exposure of the dorsal eyespot pattern. The Moila Snake will bite and its venom is believed to be fairly toxic.

RELATED SPECIES

Once included in the genus *Malpolon*, with the Montpellier Snake (*M. monspessulanus*, page 370), *Rhagerhis moilensis* is more likely to be confused with the Crowned Leafnose Snake (*Lytorhynchus diadema*, page 192, or Common False Smooth Snake (*Macroprotodon cucullatus*, page 193).

The Moila Snake is a slender but muscular snake, with a short head that terminates in a downward-curved snout, and large eyes. It is patterned with pastel desert hues, several irregular series of mid-brown blotches on a sandy to buff background. The only bold markings are a pair of black spots on the neck, contrasting with the white of the labials. These spots may be eyespots to enhance the effect of the neck-hooding display.

Actual size

FAMILY	Lamprophiidae: Psammophiinae
RISK FACTOR	Rear-fanged, mildly venomous: venom composition unknown
DISTRIBUTION	Southeastern and eastern Africa: South Sudan, Ethiopia, Kenya, Tanzania, Zambia, northern Botswana, Mozambique, Zimbabwe, and northeastern South Africa
ELEVATION	1,310–5,250 ft (400–1,600 m) asl
HABITAT	Savanna and grasslands
DIET	Small mammals, lizards, snakes, frogs, birds, and insects
REPRODUCTION	Oviparous, with clutches of 4–18 eggs
CONSERVATION STATUS	IUCN not listed

ADULT LENGTH
3 ft 3 in–4 ft,
rarely 5 ft 2 in
(1.0–1.2 m,
rarely 1.6 m)

376

RHAMPHIOPHIS ROSTRATUS
RUFOUS BEAKED SNAKE
PETERS, 1854

The Rufous Beaked Snake occurs from South Sudan and Ethiopia to northeastern South Africa, in savanna and grassland habitats. It may be active by day or night but is infrequently encountered due to its secretive nature. It spends a lot of time searching burrows for prey, which ranges from naked mole rats and other rodents, to lizards, frogs, snakes, or birds. Juveniles also feed on insects. The beaked snakes have enlarged, downward-pointing rostral scales for excavating burrows. This large snake is also an agile climber. When cornered and unable to escape, it will hiss, and then elevate and flatten its anterior body, and it reacts vigorously to being handled, convulsively jerking its body about, but rarely biting. As it is a large snake with unstudied venom, caution when handling is advisable.

The Rufous Beaked Snake is a large, powerful, stout-bodied snake with a short head that terminates in a large, downward-pointing rostral scale. It is variably colored, being yellow, reddish, brown, or orange above, the scales often bearing dark edges and presenting an overall reticulate appearance. The shape of the head is emphasized by a dark brown stripe from the tip, through the large eye, to the temporal region. The undersides are white to yellow, and juveniles may be speckled.

RELATED SPECIES
Rhamphiophis rostratus was long treated as a subspecies of the Western Beaked Snake (*R. oxyrhynchus*), from West Africa to Uganda, but it is now afforded specific status. A third species, the Red-spotted Beaked Snake (*R. rubropunctatus*), inhabits northeast Africa, in Somalia, Ethiopia, South Sudan, Kenya, and northern Tanzania.

Actual size

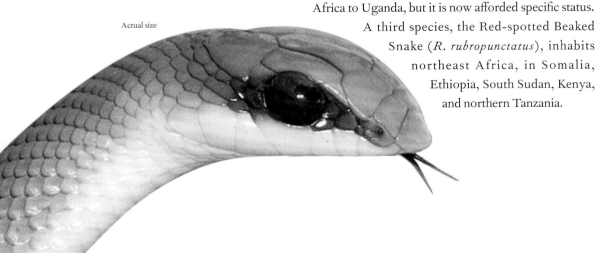

FAMILY	Lamprophiidae: Prosymninae
RISK FACTOR	Nonvenomous
DISTRIBUTION	Southern Africa: South Africa, Lesotho, and Botswana
ELEVATION	0–5,910 ft (0–1,800 m) asl
HABITAT	Savanna, wet or dry grassland, karoo scrub, thicket, and fynbos
DIET	Reptile eggs
REPRODUCTION	Oviparous, with clutches of 3–5 eggs
CONSERVATION STATUS	IUCN not listed

ADULT LENGTH
9¾–14 in
(250–360 mm)

PROSYMNA SUNDEVALLI
SUNDEVALL'S SHOVEL-SNOUT
(SMITH, 1849)

377

Sundevall's Shovel-snout is a southern African species distributed from the Cape to Lesotho, and Limpopo, South Africa, with isolated populations in southern Botswana. It is found in a variety of habitats, ranging from coastal fynbos, to savanna, grassland, and thicket, and may be found sheltering in disused termitaria or under large boulders. It feeds exclusively on reptile eggs, which are slit open with the teeth before being swallowed whole. The fossorial shovel-snouts have a spine on the tip of the tail, which may serve as an aid when forcing themselves along animal burrows or through the soil. When disturbed, Sundevall's Shovel-snout will form a precise body coil and then uncoil again rapidly, then repeat the process, presumably as a form of defense. Carl Jakob Sundevall (1801–75) was a Swedish zoologist who worked in East Africa.

RELATED SPECIES

The genus *Prosymna* is not closely related to any other snake genus and most authors place it in its own subfamily, Prosymninae, while others even elevate it to family status. There are 16 species recognized across sub-Saharan Africa. Three other species occur within the range of *P. sundevalli*: the Two-striped Shovel-snout (*P. bivittata*), East African Shovel-snout (*P. stuhlmanni*), and Lined Shovel-snout (*P. lineata*), which was a former subspecies of *P. sundevalli*.

Actual size

Sundevall's Shovel-snout is a small but stocky-bodied, smooth-scaled snake, with a short tail that ends in a spine, and a narrow head that terminates in an enlarged and slightly upcurved rostral scale, the shovel-snout. The dorsal coloration comprises several rows of dark blotches on a reddish or brown background, with a white to yellow venter. The head is dark brown but the dorsal scales and enlarged rostral are reddish or yellowish in color.

FAMILY	Lamprophiidae: Pseudaspidinae
RISK FACTOR	Nonvenomous, powerful constrictor
DISTRIBUTION	Southern and East Africa: southern Kenya to the Cape, South Africa
ELEVATION	0–8,530 ft (0–2,600 m) asl
HABITAT	Savanna, grassland, thicket woodland, fynbos, karoo, and desert
DIET	Small mammals, lizards, and sometimes birds and their eggs
REPRODUCTION	Viviparous, with litters of 18–50 neonates
CONSERVATION STATUS	IUCN not listed

ADULT LENGTH
3 ft 3 in–4 ft 3 in,
rarely 6–7 ft
(1.0–1.3 m, rarely
1.8–2.1 m)

PSEUDASPIS CANA
MOLE SNAKE
(LINNAEUS, 1758)

The Mole Snake is a stout-bodied, muscular snake with a thick neck, a relatively narrow head that terminates in a pointed snout, and small eyes. Adults may be light or dark brown, orange, reddish brown, or even black. Juveniles exhibit a variety of patterns, from dark transverse cross-bands to scattered spots, or a vertebral zigzag pattern on a pale brown or reddish background. Most specimens lose the juvenile patterning upon reaching maturity, but a few retain some faint markings.

Africa's largest nonvenomous snake after the pythons, the Mole Snake is widely distributed from southern Kenya to the Democratic Republic of the Congo, west to Angola, and throughout most of southern Africa as far south as the Cape. Although large and ubiquitous, the Mole Snake is not frequently encountered due to its secretive fossorial existence, living and hunting in mammal burrow complexes. Juvenile Mole Snakes take lizards but adults are constrictors of rats, moles, and other subterranean mammals, and they also feed on birds, and swallow their eggs whole. The Mole Snake's rodent-hunting abilities make it a valuable snake to have in agricultural areas, but it is still killed on sight, being mistaken for a cobra. When initially handled, Mole Snakes will bite ferociously and cause deep wounds, but they calm down quickly in captivity.

RELATED SPECIES

Juvenile *Pseudaspis cana* may be mistaken for a variety of small patterned snakes, from egg-eaters (*Dasypeltis*, page 157) to the Dwarf Beaked Snake (*Dipsina multimaculata*, page 368), while adults are often mistaken for Snouted Cobras (*Naja annulifera*, page 467). As one of only two species in the Pseudaspidinae, its closest relative would be the Western Keeled Snake (*Pythonodipsas carinata*, page 379).

Actual size

FAMILY	Lamprophiidae: Pseudaspidinae
RISK FACTOR	Nonvenomous, constrictor
DISTRIBUTION	Southern Africa: southwestern Angola and western Namibia
ELEVATION	0–2,850 ft (0–870 m) asl
HABITAT	Rocky desert
DIET	Lizards, small mammals, and birds
REPRODUCTION	Presumed oviparous, clutch size unknown
CONSERVATION STATUS	IUCN not listed

ADULT LENGTH
17¾–24½ in,
occasionally 31½ in
(450–620 mm,
occasionally 800 mm)

PYTHONODIPSAS CARINATA
WESTERN KEELED SNAKE
GÜNTHER, 1868

379

It has not been determined whether the enlarged rear teeth of the Western Keeled Snake are ungrooved palatine fangs or simply enlarged teeth, but this species is not thought to be venomous, even though it appears to have evolved to mimic desert vipers. Nocturnal and terrestrial in habit, it preys on small lizards, such as skinks and geckos, and also small mammals and possibly birds, killing its prey by constriction. Newly captured specimens bite readily but cause little damage and soon become docile. The Western Keeled Snake is confined to the Namib Desert and similar low-elevation desert habitats across the border in southern Angola. Its natural history is poorly documented due to its generally secretive nature, and although it is believed to be oviparous this is not known for certain.

The Western Keeled Snake is a small, slender snake, with weakly keeled or smooth scales, a long tail, a relatively broad, dorsally compressed head, and large, protruding eyes with vertically elliptical pupils. It may be reddish, yellowish, brown, or gray, with a series of darker vertebral blotches that may coalesce to form a zigzag pattern, these colors blending with the pastel shades of its desert habitat.

RELATED SPECIES

Pythonodipsas carinata may be mistaken for a small viper such as the Horned Adder (*Bitis caudalis*, page 612). Its closest relative, by virtue of it being the only species in the Pseudaspidinae, is the Mole Snake (*Pseudaspis cana*, page 378), a widespread and much larger snake.

Actual size

FAMILY	Lamprophiidae: Pseudoxyrhophiinae
RISK FACTOR	Rear-fanged, potentially mildly venomous
DISTRIBUTION	Indian Ocean: northeastern, northern, and western Madagascar
ELEVATION	1,310–2,130 ft (400–650 m) asl
HABITAT	Rainforest and dry forest streams
DIET	Prey preferences unknown
REPRODUCTION	Oviparous, with clutches of up to 5 eggs
CONSERVATION STATUS	IUCN Least Concern

ADULT LENGTH
17½ in
(447 mm)

380

ALLUAUDINA BELLYI
BELLY'S KEELED SNAKE
MOCQUARD, 1894

Belly's Keeled Snake is a nocturnal, terrestrial snake and relatively rare. It is likely to be encountered along streams in wet or dry forests, in the north or northeast of Madagascar. When threatened, specimens will exude a foul-smelling cloacal secretion in defense. Another tactic involves the snake rolling a part of its body upside down to expose its venter. This may be a form of thanatosis (playing dead), a tactic also adopted by other snakes. Belly's Keeled Snake has grooved rear teeth but its capacity regarding snakebite is unknown. The French herpetologist François Mocquard (1834–1917) described this species in honor of the naturalist Charles Alluaud (1861–1949) and his friend M. Belly, who together collected the holotype. Little is known about its natural history; it is oviparous, but its prey preferences are undocumented.

RELATED SPECIES

A second, and much rarer, species, Mocquard's Keeled Snake (*Alluaudina mocquardi*), is known from two specimens from the limestone caves on the Ankarana Massif. It is listed as Endangered by the IUCN. Apart from the leafnosed snakes (*Langaha*, page 386) these are the only colubroid snakes in Madagascar with completely keeled dorsal scales.

Belly's Keeled Snake is a strongly keeled snake with a head slightly broader than the body, and moderate-sized eyes with round pupils. Many of the lateral head scutes are reduced to granular scales. Adults are black or brown above, yellowish on the lower flanks, sometimes with red spots, and white below, while juveniles are mottled and may exhibit a nuchal collar.

Actual size

FAMILY	Lamprophiidae: Pseudoxyrhophiinae
RISK FACTOR	Rear-fanged, mildly venomous: venom composition unknown
DISTRIBUTION	Southern Africa: southern and eastern South Africa, Swaziland, and Zimbabwe
ELEVATION	0–8,500 ft (0–2,590 m) asl
HABITAT	Riparian situations in grassland, montane forest, and fynbos
DIET	Frogs, lizards, and small mammals
REPRODUCTION	Viviparous, with litters of 4–12 neonates
CONSERVATION STATUS	IUCN Least Concern

ADULT LENGTH
15¾–21¾ in,
occasionally 24¾ in
(400–550 mm,
occasionally 630 mm)

AMPLORHINUS MULTIMACULATUS
MANY-SPOTTED SNAKE
SMITH, 1847

381

The Many-spotted Snake occupies a fragmented distribution from the Cape, South Africa, eastward and northward through KwaZulu-Natal, Swaziland, and Mpumalanga, with an isolated population in eastern Zimbabwe, being common near the summit of 8,500 ft (2,590 m) Inyangani Mountain. It is alternatively known as the Cape Reedsnake, as its preferred habitat consists of riparian reedbeds, but it is also found in grassland, coastal fynbos, and montane forest. Diurnal and semi-aquatic, it hunts frogs and lizards, but small mammals also feature in its diet. It is a live-bearer with usual litters of four to eight neonates, but up to 12 have been recorded. A rear-fanged, venomous species, the Many-spotted Snake is not considered dangerous to humans, even though it bites freely with bites causing localized pain and prolonged bleeding.

The Many-spotted Snake is a small snake with a smooth-scaled body, long tail, and rounded head with large eyes. Most specimens are brown with a pair of broad longitudinal dorsolateral stripes and several rows of large dark spots down the body, but specimens from KwaZulu-Natal may be bright green with black spots. The undersides are green or cyan.

RELATED SPECIES

The genus *Amplorhinus* is monotypic, but *A. multimaculatus* could be confused with the Spotted Skaapsteker (*Psammophylax rhombeatus*), or a night adder, the green morphotype especially resembling the Green Night Adder (*Causus resimus*) of East Africa.

Actual size

FAMILY	Lamprophiidae: Pseudoxyrhophiinae
RISK FACTOR	Rear-fanged, mildly venomous: venom composition unknown
DISTRIBUTION	Indian Ocean: eastern Madagascar
ELEVATION	1,970–4,590 ft (600–1,400 m) asl
HABITAT	Rainforest and marshland
DIET	Prey preferences unknown
REPRODUCTION	Reproductive strategy unknown
CONSERVATION STATUS	IUCN Least Concern

ADULT LENGTH
14 in
(353 mm)

382

COMPSOPHIS BOULENGERI
BOULENGER'S MALAGASY FOREST SNAKE
(PERRACA, 1892)

One of many small forest-dwelling snakes found along Madagascar's humid eastern coast, Boulenger's Malagasy Forest Snake is a diurnal and terrestrial inhabitant of rainforest and marshland. Other species are associated with fast-flowing rainforest streams and it is likely that this species also occurs alongside creeks. Little is known about the biology and natural history of Boulenger's Malagasy Forest Snake. The arboreal species *Compsophis infralineatus* preys on frogs and small mammals, while the equally arboreal *C. laphystius* feeds on frogs and their eggs, but the diet of the terrestrial Boulenger's Malagasy Forest Snake is unknown. It was named in honor of George A. Boulenger (1858–1937), a Belgian-born zoologist at the British Museum (Natural History), who made huge contributions to herpetology during the late nineteenth and early twentieth centuries.

RELATED SPECIES

The seven species in *Compsophis* are split between two subgenera: subgenus *Compsophis* containing four small, diurnal, terrestrial species, and subgenus *Geodipsas* comprising three larger, nocturnal, arboreal species. *Compsophis boulengeri* belongs in the subgenus *Geodipsas* and appears closest to the White-bellied Malagasy Forest Snake (*C. albiventris*).

Boulenger's Malagasy Forest Snake is a small snake with smooth body scales, a short tail, and a small, rounded head. Its patterning is brown with dark and light speckling dorsally, and it is bright red underneath. Two pale spots on the neck are suggestive of a broken nuchal collar, and pale spots are also present on the posterior supralabials.

Actual size

FAMILY	Lamprophiidae: Pseudoxyrhophiinae
RISK FACTOR	Rear-fanged, mildly venomous: venom composition unknown
DISTRIBUTION	Indian Ocean: Socotra Island (Yemen)
ELEVATION	33–2,850 ft (10–870 m) asl
HABITAT	Rocky hillsides, scrubby bushland, and stone walls around date palm groves
DIET	Lizards
REPRODUCTION	Reproductive strategy unknown
CONSERVATION STATUS	IUCN Least Concern

ADULT LENGTH
17¾ in
(450 mm)

DITYPOPHIS VIVAX
SOCOTRA NIGHT SNAKE
GÜNTHER, 1881

383

The Socotra Night Snake is endemic to the Yemeni-owned island of Socotra, midway between the Yemen and Somalia, where it has been recorded at 29 localities between 33 and 2,850 ft (10–870 m) asl. It is absent from the neighboring islands of Abd-el-Kuri and the Brothers (Darsa and Samha). This is a nocturnal snake with catlike pupils that adopts a viperid ambush posture among the stones of its rocky hillside habitat. It probably preys on terrestrial geckos, such as nocturnal rock geckos of genus *Pristurus*, of which there are several Socotran species, rather than terrestrial but diurnal skinks. It is also found in scrubby bushlands and in the dry stone walls around date palm groves. Although the Socotra Night Snake is rear-fanged it is small and not considered dangerous to humans.

The Socotra Night Snake is a small snake with smooth body scales, a short tail, an angular head, and relatively large eyes with vertically elliptical pupils. Coloration consists of reddish, brown, or gray with a dorsal pattern of either irregular blurred spots or several rows of larger, more defined spots that may meet and coalesce to form short bands across the back. The lip scales may be cream with a black line through the eye to the angle of the jaw, and the iris is orange.

RELATED SPECIES

Socotra is home to only five other snake species, all endemic to the island, the Socotra Blindsnake (*Xerotyphlops socotranus*), three threadsnakes (*Myriopholis*), and the Socotran Racer (*Hemerophis socotrae*, page 179), the only snake with which *Ditypophis* could be confused, but from which it is easily distinguished by its elliptical pupils. *Ditypophis vivax* is more closely related to the snakes of Madagascar than it is to the Socotra Racer or snakes from neighboring Yemen or Somalia.

Actual size

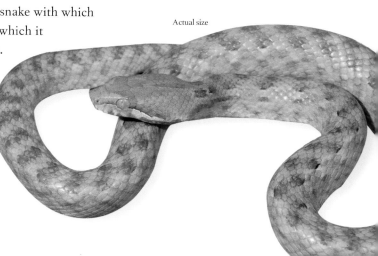

FAMILY	Lamprophiidae: Pseudoxyrhophiinae
RISK FACTOR	Nonvenomous
DISTRIBUTION	Southern and East Africa: southern and eastern South Africa, Swaziland, Lesotho, possibly Kenya, Tanzania, and as far north as Ethiopia
ELEVATION	0–10,700 ft (0–3,250 m) asl
HABITAT	Savanna, coastal fynbos, and lowland forest
DIET	Slugs and snails
REPRODUCTION	Viviparous, with litters of 6–12 neonates
CONSERVATION STATUS	IUCN Least Concern

ADULT LENGTH
11¾–13¾ in,
occasionally 17 in
(300–350,
occasionally 430 mm)

384

DUBERRIA LUTRIX
COMMON SLUG-EATER
(LINNAEUS, 1758)

Actual size

The Common Slug-eater is a small, gracile snake with smooth body scales and a small, narrow, rounded head. Its coloration is usually red to deep brown dorsally, paler brown or even contrasting gray on the flanks, and cream underneath. Some specimens possess a fine black vertebral stripe along the back.

One of the smallest African colubroid snakes, the Common Slug-eater occurs in South Africa. A common and diurnal inhabitant of savanna, grassland, low-lying forest, and coastal fynbos, the Common Slug-eater shelters under any available ground cover and feeds on the slugs and snails that also inhabit these shady, damp microhabitats. It locates its prey by following molluscan slime trails. The Slug-eater should be the gardener's friend, and it should be encouraged. Slug-eaters do not bite; their mouths are very small and their demeanor placid. Their usual defense involves coiling the body into a tight coil with the head hidden inside, and sometimes also emitting a foul-smelling cloacal secretion to deter predation.

RELATED SPECIES

Several subspecies were recognized across the range of *Duberria lutrix*, the South African form being the nominate subspecies (*D. l. lutrix*). The former Zimbabwean subspecies was raised to species status (*D. rhodesiana*). It differs from the South African Slug-eater in the possession of one, rather than two, postocular scales. The same situation exists with regard to the slug-eater found in Tanzania, Kenya, Uganda, and Ethiopia so those populations are also likely different species too. Two other species are recognized, the Shire Slug-eater (*D. shirana*), from Zambia, Malawi, and Tanzania, and the Variegated Slug-eater (*D. variegata*), from South Africa and Mozambique.

FAMILY	Lamprophiidae: Pseudoxyrhophiinae
RISK FACTOR	Rear-fanged, mildly venomous: venom composition unknown
DISTRIBUTION	Indian Ocean: western and northwestern Madagascar, and Comoros
ELEVATION	66–2,300 ft (20–700 m) asl
HABITAT	Dry forest
DIET	Frogs, chameleons, and small mammals
REPRODUCTION	Oviparous, with clutches of up to 5 eggs
CONSERVATION STATUS	IUCN Least Concern

ADULT LENGTH
5 ft 7 in
(1.7 m)

ITHYCYPHUS MINIATUS
CINNABAR VINESNAKE
(SCHLEGEL, 1837)

385

The name Tiny Night Snake has been used for this species, but at 5 ft 7 in (1.7 m) in length and as the longest member of the genus *Ithycyphus*, it is not "tiny," and, being diurnal, the name "night snake" also seems inappropriate. The name *miniatus* means cinnabar or vermillion, not small, and since *Ithycyphus* are slender treesnakes, the name Cinnabar Vinesnake suitably describes this species. The Malagasy call the Cinnabar Vinesnake "Fandrefiala," meaning treesnake. It inhabits dry forests in northern and western Madagascar, and some authors include the Comoros Islands. It preys on frogs, and captive specimens take chameleons and mice, which are quickly subdued by the venom of this rear-fanged species. A human snakebite has been documented; the symptoms were mild, but a large specimen may deliver a more serious bite and caution is advised.

The Cinnabar Vinesnake is a large, slender snake with an elongate tail, and a long head with a pronounced canthus rostralis (a ridge from the eye to above the nostril). The body is gray or brown anteriorly, becoming red-brown posteriorly, and the color of the head is sexually dichromatic, being gray or brown like the body in females but red in males.

RELATED SPECIES
There are four other species in the genus *Ithycyphus*, all endemic to Madagascar: the Perinet Vinesnake (*I. perineti*), Goudot's Vinesnake (*I. goudoti*), and Blanc's Vinesnake (*I. blanci*) in northern and northeastern Madagascar, and Saint-Ours' Vinesnake (*I. oursi*) in southern Madagascar.

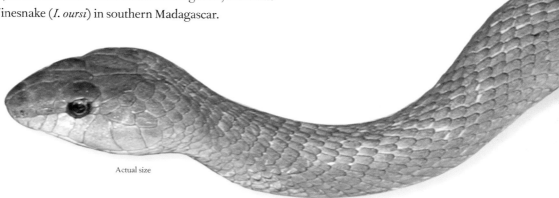

Actual size

FAMILY	Lamprophiidae: Pseudoxyrhophiinae
RISK FACTOR	Rear-fanged, mildly venomous: venom composition unknown
DISTRIBUTION	Indian Ocean: Madagascar
ELEVATION	0–1,770 ft (0–540 m) asl
HABITAT	Lowland wet and dry forest
DIET	Lizards and frogs
REPRODUCTION	Oviparous, with clutches of up to 5 eggs
CONSERVATION STATUS	IUCN Least Concern

ADULT LENGTH
3 ft 3 in
(1.0 m)

LANGAHA MADAGASCARIENSIS
MALAGASY LEAFNOSE SNAKE
BONNATERRE, 1790

The Malagasy Leafnose Snake is one of the strangest-looking snakes in the world, apart for its two relatives, which are somewhat stranger in appearance. This widely distributed species is a slender, diurnal, arboreal inhabitant of lowland wet and dry forests. It possesses a body shape similar to that of the tropical American *Oxybelis* (pages 205–206) or Asian *Ahaetulla* (pages 130–131) vinesnakes, but also possesses a large, fleshy, sexually dimorphic rostral protuberance on the tip of its narrow snout, the purpose of which is unknown. Despite being slender-bodied, the Malagasy Leafnose Snake is capable of catching, killing, and swallowing bulky lizards such as Malagasy iguanids and chameleons, while geckos and frogs are also potential prey. Although rear-fanged and venomous, the leafnose snakes are not considered dangerous to humans.

The Malagasy Leafnose Snake is an extremely slender-bodied vinesnake with a long tail and a long, narrow head. It is both sexually dichromatic and dimorphic. Males are brown above and yellow below, often separated by a white demarcation line along the flank, and have an orange iris to the eye. The male rostral protuberance is a tapering spike that curves downward at the tip (see artwork above). Females are all gray-brown or brown with a brown iris, and a broad, laterally compressed rostral protuberance, which terminates in a fleshy frill like a cock's comb.

RELATED SPECIES

Two other species of *Langaha* are recognized, both of which possess large, raised, fleshy supraocular eyelashes in addition to their nasal protuberances. Neither species is well known. *Langaha alluaudi* occurs in southern and western Madagascar while *L. pseudoalluaudi* is found in southern, western, and northern Madagascar, all localities being near the coast.

Actual size

FAMILY	Lamprophiidae: Pseudoxyrhophiinae
RISK FACTOR	Rear-fanged, mildly venomous: venom composition unknown
DISTRIBUTION	Madagascar: Madagascar, Nosy Be, and Comoros Islands
ELEVATION	33–3,940 ft (10–1,200 m) asl
HABITAT	Most lowland habitats, including anthropogenic localities
DIET	Lizards, frogs, snakes, small mammals, birds, and reptile eggs
REPRODUCTION	Oviparous, with clutches of 10–13 eggs
CONSERVATION STATUS	IUCN Least Concern

ADULT LENGTH
3 ft 3 in–5 ft
(1.0–1.5 m)

LEIOHETERODON MADAGASCARIENSIS
MALAGASY GIANT HOGNOSE SNAKE
(DUMÉRIL & BIBRON, 1854)

387

The Malagasy Giant Hognose Snake is the largest Malagasy snake after the boas (*Acrantophis*, page 124 and *Sanzinia*, page 125). It occurs throughout Madagascar and on the small northern island of Nosy Be. It has been introduced to Grande Comore in the Comoros Islands. The hognose snakes are common, diurnal, and terrestrial snakes, and the Malagasy Giant Hognose Snake is frequently encountered, especially around human dwellings. It has a catholic diet, feeding on frogs, lizards, small mammals, other snakes, and birds. The turned-up, keeled rostral scale on the snout is an excellent implement for excavation, allowing the snakes to dig up and devour entire clutches of iguanine lizard eggs. These snakes can bite, with pain and swelling having been experienced following bites from the related Brown Malagasy Hognose Snake (*Leioheterodon modestus*).

The Malagasy Giant Hognose Snake is a large, stout-bodied snake with a large head, an upturned hognosed-snout, and large eyes with round pupils. Its patterning is black above, becoming browner posteriorly in some specimens, with a pattern of black blotches or cross-bars evident on the posterior body. The undersides are yellow to cream with black spots, this pigment extending onto the flanks, neck, and supralabials, which bear black sutures.

RELATED SPECIES

The genus *Leioheterodon* contains two smaller species, the widespread, arid habitat-dwelling Brown Malagasy Hognose Snake (*L. modestus*), and the Speckled Malagasy Hognose Snake (*L. geayi*) from southern Madagascar.

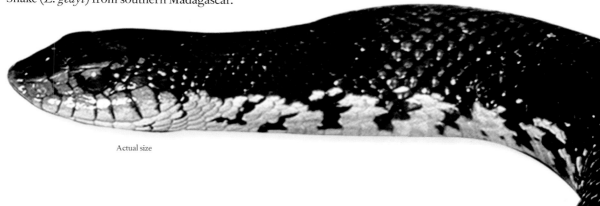

Actual size

FAMILY	Lamprophiidae: Pseudoxyrhophiinae
RISK FACTOR	Rear-fanged, mildly venomous: venom composition unknown
DISTRIBUTION	Indian Ocean: northeastern Madagascar
ELEVATION	0–3,310 ft (0–1,010 m) asl
HABITAT	Primary and secondary rainforest
DIET	Lizards
REPRODUCTION	Reproductive strategy unknown
CONSERVATION STATUS	IUCN Near Threatened

ADULT LENGTH
16½ in
(416 mm)

388

LIOPHIDIUM PATTONI
PATTON'S SPOTTED GROUNDSNAKE
(VIEITES, RATSOAVINA, RANDRIANIAINA, NAGY, GLAW & VENCES, 2010)

Patton's Spotted Groundsnake is only known from a few small reserves in northeastern Madagascar, where it inhabits primary and secondary rainforest. It is diurnal and terrestrial, and the gut of the holotype contained a skink, an abundant diurnal prey type in rainforests. Like many of its congeners this is a poorly known snake species with natural history records largely opportunistic and anecdotal. Its reproductive strategy is unknown. Snakes of the genus *Liophidium* are rear-fanged and mildly venomous, but are placid and rarely bite, and they are not considered dangerous to humans. This species is listed as Near Threatened by the IUCN due to its limited range and threatened rainforest habitat. It was named in honor of the mammalogist and general Malagasy naturalist Jim Patton.

RELATED SPECIES

The genus *Liophidium* contains ten species, nine occurring on Madagascar, with one (*L. torquatum*) extending onto the island of Nosy Be, and another (*L. vaillanti*) also occurring on Réunion Island in the Mascarene Islands. The tenth species is the endemic Mayotte Groundsnake (*L. mayottensis*) from the French department of Mayotte, in the Comoros Archipelago. None could be confused with *L. pattoni*.

Actual size

Patton's Spotted Groundsnake is a small, smooth-scaled snake with a rounded head, indistinct from the neck, and small eyes with round pupils. It exhibits a stunning pattern comprising four longitudinal rows of crimson spots on a black background, the most lateral rows forming a tight zigzag pattern. The spots gradually become blue on the posterior body and tail. The venter of the body is yellow, each scale with a curved black bar, while the subcaudals are pink. The head is black above and yellow on the sides, with a black stripe passing along the supralabials, through the eye to the snout.

FAMILY	Lamprophiidae: Pseudoxyrhophiinae
RISK FACTOR	Nonvenomous
DISTRIBUTION	Indian Ocean: eastern Madagascar
ELEVATION	770–3,280 ft (235–1,000 m) asl
HABITAT	Montane rainforest
DIET	Frogs and their eggs, and fish
REPRODUCTION	Oviparous, clutch size unknown
CONSERVATION STATUS	IUCN Least Concern

ADULT LENGTH
29½ in
(749 mm)

LIOPHOLIDOPHIS RHADINAEA
PINK-BELLIED GROUNDSNAKE
CADLE, 1996

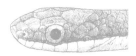

389

The Pink-bellied Groundsnake occurs in mid- to high-elevation montane rainforest on the eastern versant of Madagascar. It is diurnally active and terrestrial, and usually encountered on trails, where it has been observed to freeze and flatten its body to the ground in the hope of avoiding detection. Snakes of the genus *Liopholidophis* prey on lizards and frogs. The Pink-bellied Groundsnake appears to prey on frogs and their eggs, and it is also thought to take fish. Members of the genus are noted for their extreme sexual dimorphism in tail and body length. Males have much longer tails and moderately longer bodies than females, determined by counts of the subcaudal and ventral scales, compared to other Malagasy pseudoxyrhopiine snakes.

RELATED SPECIES

The genus *Liopholidophis* currently contains eight species, all endemic to the rainforests of eastern and central Madagascar. *Liopholidophis rhadinaea* is most closely related to two recently described dwarf species, Bader's Groundsnake (*L. baderi*) and the Short-tailed Groundsnake (*L. oligolepis*).

Actual size

The Pink-bellied Groundsnake is a slender snake with a head only slightly distinct from the neck, and moderately sized eyes with round pupils. It is mid-brown in color, darker dorsally than laterally, with a pair of pale brown dorsolateral stripes. The dorsum of the head is brown, the lips are white, and three yellowish-brown nuchal spots are present behind the head. The venter is bright pink.

FAMILY	Lamprophiidae: Pseudoxyrhophiinae
RISK FACTOR	Rear-fanged, mildly venomous: venom composition unknown
DISTRIBUTION	Indian Ocean: western Madagascar
ELEVATION	410–1,250 ft (125–380 m) asl
HABITAT	Dry forest on limestone karst outcrops
DIET	Possibly lizards
REPRODUCTION	Viviparous, with litters of 2 neonates
CONSERVATION STATUS	IUCN Vulnerable

ADULT LENGTH
18–27¾ in
(460–705 mm)

LYCODRYAS CITRINUS
LEMON TREESNAKE
(DOMERGUE, 1995)

The striking Lemon Treesnake is known from a few scattered localities, including two reserves, in western Madagascar. Its habitat is being degraded by agriculture and cattle grazing, and the species is directly threatened by collection for the pet trade. The Lemon Treesnake inhabits dry forest on limestone karst outcrops. The natural history of *Lycodryas* is largely anecdotal and incompletely known. This species is known to be a live-bearer, one female containing two neonates, but nothing is known of its diet. Other species have been documented feeding on small chameleons or geckos, and it is presumed that this species takes similar prey. It is a mildly venomous rear-fanged snake with unknown venom composition, yield, and toxicity.

The Lemon Treesnake is an extremely slender snake with a laterally compressed body, a long tail, and a rounded head, broader than the neck, with large protruding eyes, a black iris, and vertically elliptical pupils. Its patterning is striking, being bright yellow with a series of large, distinctive black blotches along its back and a black cap over its head. Juveniles are light brown.

RELATED SPECIES

The Lemon Treesnake was originally in genus *Stenophis*, which is now divided into *Lycodryas*, *Phisalixella*, and *Parastenophis*. *Lycodryas* contains ten species, none of which resemble *L. citrinus*. Most *Lycodryas* are Madagascan, but two are from the Comoros Archipelago, the Spotted Treesnake (*L. maculatus*) on Anjouan and Mayotte, and the Coconut Palm Snake (*L. cococola*) on Grande Comore and Mohéli. The name *cococola* is derived from *coco*, meaning "coconut," and *-cola*, meaning "inhabiting," a reference to its presence in coconut groves.

Actual size

FAMILY	Lamprophiidae: Pseudoxyrhophiinae
RISK FACTOR	Rear-fanged, mildly venomous: venom composition unknown
DISTRIBUTION	Indian Ocean: Madagascar
ELEVATION	66–2,300 ft (20–700 m) asl
HABITAT	Most habitats, but especially those near water
DIET	Frogs, lizards, snakes, birds and their eggs, and small mammals
REPRODUCTION	Oviparous, with clutches of 2–6 eggs
CONSERVATION STATUS	IUCN Least Concern

ADULT LENGTH
3 ft 6 in
(1.06 m)

MADAGASCAROPHIS COLUBRINUS
COMMON MALAGASY CATSNAKE
SCHLEGEL, 1837

391

The Common Malagasy Catsnake is a frequently encountered nocturnal, generalist predator of frogs, geckos, chameleons, skinks, other snakes, small mammals, and birds and their eggs. Some specimens exhibit a white tail tip, which could be used to lure frogs or lizards within strike range. This species is distributed island-wide across Madagascar, possibly excluding the arid southwest where two other *Madagascarophis* species occur. It may be found in most lowland habitats, but is especially common around ponds and watercourses. Common Malagasy Catsnakes are terrestrial and arboreal, and they shelter in tree holes, under rocks, or inside buildings during daylight hours. Common Malagasy Catsnakes will bite if handled and there have been instances of mild envenomings following snakebites.

The Common Malagasy Catsnake is a large, somewhat laterally compressed snake with a long tail, a broad head, protruding eyes, and vertically elliptical pupils. There are two main color morphotypes, a brown morphotype and a yellow morphotype, which may be unicolor or speckled, blotched, or spotted with darker pigment. In some specimens the darker pigment coalesces to form a checkerboard pattern. The lips are usually white or yellow.

RELATED SPECIES
Madagascarophis colubrinus is represented by five subspecies across Madagascar. Three other Malagasy catsnake species are also recognized: the Southern Catsnake (*M. meridionalis*), Ocellated Catsnake (*M. ocellatus*), also from the south, and Fuchs' Catsnake (*M. fuchsi*), from northern Madagascar. The Malagasy catsnakes bear a striking resemblance to the catsnakes and cat-eyed snakes of the Colubridae, in the genera *Boiga* (pages 144–146) in Asia and Australasia, *Toxicodryas* (page 247) in Africa, and *Telescopus* (pages 243–244) in Africa and Europe.

Actual size

FAMILY	Lamprophiidae: Pseudoxyrhophiinae
RISK FACTOR	Nonvenomous
DISTRIBUTION	Indian Ocean: northeastern and northern Madagascar, and Nosy Be
ELEVATION	66–2,620 ft (20–800 m) asl
HABITAT	Rainforest
DIET	Snails or lizards
REPRODUCTION	Oviparous, with clutches of up to 10 eggs
CONSERVATION STATUS	IUCN Least Concern

ADULT LENGTH
27½ in
(700 mm)

MICROPISTHODON OCHRACEUS
MALAGASY BLUNT-HEADED SNAKE
MOCQUARD, 1894

The Malagasy Blunt-headed Snake is a slender, smooth-scaled snake with a long tail, a short, chunky head, a short mouth, and large eyes with round pupils. It is dorsally brown, with a series of dark brown transverse bars and chevrons, and one large, distinct chevron on the neck. The interstitial skin is white and may be visible when the snake inflates its body. The venter is brown with black markings.

The Malagasy Blunt-headed Snake is another poorly documented Malagasy species. It is a rainforest inhabitant from the northeast and north of the country, and the small island of Nosy Be. It is both terrestrial and arboreal, one specimen being found 98 ft (30 m) high in a tree, but authors disagree as to whether it feeds on lizards or snails. The narrow body, short, chunky head, large eyes, and short mouth are suggestive of the slug- and snail-eating *Dipsas* (pages 273–274) and *Sibon* (pages 290–291) from South America, or *Pareas* (page 547) from Southeast Asia, and no other Malagasy snake appears to have adapted specifically to feed on terrestrial mollusks, making the niche available. It is believed to be a nonvenomous snake without enlarged rear teeth, despite being related to several opisthoglyphous (rear-fanged) genera.

RELATED SPECIES

The genus *Micropisthodon* is monotypic, and thought to be most closely related to the rear-fanged, mildly venomous Malagasy leafnosed snakes (*Langaha*, page 386) and the Malagasy vinesnakes (*Ithycyphus*, page 385).

Actual size

FAMILY	Lamprophiidae: Pseudoxyrhophiinae
RISK FACTOR	Rear-fanged, mildly venomous: venom composition unknown
DISTRIBUTION	Indian Ocean: eastern Madagascar and Nosy Be
ELEVATION	66–3,280 ft (20–1,000 m) asl
HABITAT	Rainforest, near creeks
DIET	Small mammals, and possibly lizards or frogs
REPRODUCTION	Reproductive strategy unknown
CONSERVATION STATUS	IUCN Least Concern

ADULT LENGTH
39 in
(995 mm)

PSEUDOXYRHOPUS TRITAENIATUS
MALAGASY BLACK-STRIPED GROUNDSNAKE
MOCQUARD, 1894

393

The Malagasy Black-striped Groundsnake is a nocturnal and terrestrial inhabitant of the rainforest on Madagascar's eastern versant. It is also found on Nosy Be. A rare species only known from very few specimens, this snake is usually associated with small streams or creeks, although other species are semi-fossorial. Little is known of its natural history or biology. One specimen was found to have eaten a rodent and other species have been documented to prey on skinks, skink eggs, or chameleon eggs, and frogs are considered another potential prey group. Many specimens possess truncated tails, which are thought to be the result of attacks from carnivorous mammals like tenrecs or civets. Its reproductive strategy is unknown.

RELATED SPECIES

A member of the type genus of the Pseudoxyrhopiinae, *Pseudoxyrhopus tritaeniatus* is one of 11 species of this endemic genus. With its unique and distinctive patterning, it cannot be mistaken for any other *Pseudoxyrhopus* species.

Actual size

The Malagasy Black-striped Groundsnake is a moderately large snake with smooth scales and a head only slightly distinct from the neck, with small eyes, round pupils, and a long tail. Dorsally it is black with usually three longitudinal red stripes. The head is red, with black infusions on most scales.

FAMILY	Lamprophiidae: Pseudoxyrhophiinae
RISK FACTOR	Nonvenomous
DISTRIBUTION	Indian Ocean: Madagascar
ELEVATION	0–5,130 ft (0–1,565 m) asl
HABITAT	Open habitats, including agricultural fields and rice paddies
DIET	Frogs
REPRODUCTION	Oviparous, with clutches of 6–13 eggs
CONSERVATION STATUS	IUCN Least Concern

ADULT LENGTH
Male
30 in
(765 mm)

Female
32½ in
(828 mm)

394

THAMNOSOPHIS LATERALIS
MALAGASY GARTERSNAKE
(DUMÉRIL, BIBRON & DUMÉRIL, 1854)

The Malagasy Gartersnake is a slightly laterally compressed snake with a head distinct from the neck, and large eyes with round pupils. Its patterning comprises an anteriorly white to posteriorly yellow lateral stripe on either side, on a black background. The dorsum is checkerboarded with areas of pure black and black with light blue flecks. The undersides are white.

The Malagasy Gartersnake is one of the commonest snakes of Madagascar. Found across the entire island, it is a diurnal inhabitant of open country rather than closed-canopy rainforest. It is also common in disturbed or anthropogenic habitats such as agricultural fields or rice paddies—anywhere associated with water. Although this species is an excellent swimmer it seems to be found more often in close proximity to water, rather than in the water. This would agree with a preferred diet of frogs, rather than fish. The Malagasy Gartersnake is reported to bask even in the hottest of weather, and to flee to the water or adopt a threatening posture if disturbed. This species is harmless to humans. The local Malagasy name is "Bibilava," which simply means snake.

RELATED SPECIES

The genus *Thamnosophis* contains six endemic Malagasy species that were previously included in the genera *Liopholidophis* and *Bibilava*. *Thamnosophis lateralis* bear a strong resemblance to North American gartersnakes (*Thamnophis*, pages 427–432), and it appears that this species at least occupies their niche in Madagascar.

Actual size

FAMILY	Lamprophiidae: *incertae sedis*
RISK FACTOR	Rear-fanged, mildly venomous: venom composition unknown
DISTRIBUTION	Central Africa: Cameroon, Equatorial Guinea, Gabon, Congo, DRC, Central African Republic, and Uganda
ELEVATION	3,280–7,220 ft (1,000–2,200 m) asl
HABITAT	Montane rainforest
DIET	Frogs
REPRODUCTION	Oviparous, clutch size unknown
CONSERVATION STATUS	IUCN not listed

BUHOMA DEPRESSICEPS

PALE-HEADED FOREST SNAKE

(WERNER, 1897)

ADULT LENGTH
9¾–13¾ in,
occasionally 17¼ in
(250–350 mm,
occasionally 440 mm)

395

The Pale-headed Forest Snake does not always possess a pale head. It is a terrestrial, and probably diurnally active, inhabitant of montane rainforest from Cameroon to Uganda. One specimen was found sheltering inside tussock grass at night, while others have been collected in leaf litter, under ground debris, or in holes in the ground. The only prey on record for this species, and its two congeners, are frogs, but it is possible they may also take small terrestrial lizards or invertebrates. The Pale-headed Forest Snake exudes a noxious-smelling cloacal secretion if handled. Snakes of the genus *Buhoma* are rear-fanged and mildly venomous, but of such small size and gentle disposition that they pose no threat to humans. The genus *Buhoma* is *incertae sedis*, of uncertain placement, within the Lamprophiidae.

The Pale-headed Forest Snake is a small species with smooth scales, a short tail, and a short head with small eyes. It is generally brown, with or without a series of fine longitudinal black lines. The head may be light or dark, but a yellow or white nuchal collar is usually present and serves to distinguish this species from its congeners.

RELATED SPECIES

Formerly contained in the genus *Geophis*, two subspecies of *Buhoma depressiceps* are recognized, a nominate western form (*B. d. depressiceps*) and an eastern form (*B. d. marlieri*). The genus *Buhoma* also contains two other species: the Uluguru Forest Snake (*B. procterae*), and the Usambara Forest Snake (*B. vauerocegae*), both from isolated mountain ranges in Tanzania. The Uluguru Forest Snake is listed as Vulnerable by the IUCN.

Actual size

FAMILY	Lamprophiidae: *incertae sedis*
RISK FACTOR	Rear-fanged, mildly venomous: venom composition unknown
DISTRIBUTION	Southern Africa: South Africa (KwaZulu-Natal), and eastern Lesotho
ELEVATION	6,140–9,400 ft (1,870–2,865 m) asl
HABITAT	High-elevation montane grassland streams
DIET	Frogs
REPRODUCTION	Oviparous, with clutches of up to 6 eggs
CONSERVATION STATUS	IUCN Data Deficient

ADULT LENGTH
11¾–19¾ in
(300–500 mm)

MONTASPIS GILVOMACULATA
CREAM-SPOTTED MOUNTAIN SNAKE
BOURQUIN, 1991

The Cream-spotted Mountain Snake is a small, glossy black snake with smooth body scales and a black head but cream lips, each labial scale edged with black like piano keys. The throat is cream with brown scale edging and the venter of the body is brown with small pale spots.

The Cream-spotted Mountain Snake was discovered in the Cathedral Peak Forest Reserve in 1980 and described in 1991. It is poorly known, with few specimens seen within its small range in the Drakensberg of eastern Lesotho and neighboring KwaZulu-Natal, South Africa. It inhabits high-elevation grasslands, close to streams. Its prey consists of frogs, and it uses its body coils to subdue the prey before swallowing it, possibly because its venom is only mildly toxic. Listed as Data Deficient by the IUCN, the Cream-spotted Mountain Snake should probably be considered potentially threatened, due to its small total range. It will not attempt to bite but will make jerky movements in response to being handled, and exude a noxious-smelling secretion from its cloacal glands. The name *gilvomaculata* means yellow-spotted.

RELATED SPECIES

Montaspis gilvomaculata is *incertae sedis*, of uncertain placement, within the Lamprophiidae, and therefore it is difficult to determine its closest relatives. It does bear a passing resemblance to the Red-lipped Herald Snake (*Crotaphopeltis hotamboeia*, page 156).

Actual size

FAMILY	Lamprophiidae: *incertae sedis*
RISK FACTOR	Nonvenomous, possibly mildly venomous
DISTRIBUTION	South and Southeast Asia: Nepal and northern India to southern China, Taiwan, Thailand, Vietnam, the Philippines, Malaysia, and Indonesia (Sumatra, Java, Borneo, Sulawesi, Lesser Sundas)
ELEVATION	0–6,560 ft (0–2,000 m) asl
HABITAT	Lowland and low to mid-montane evergreen forest and rainforest, also rocky vegetated hillsides and streamsides
DIET	Lizards, frogs, small snakes, reptile eggs
REPRODUCTION	Viviparous, with litters of 3–10 neonates
CONSERVATION STATUS	IUCN Least Concern

ADULT LENGTH
30¼ in
(770 mm)

PSAMMODYNASTES PULVERULENTUS
COMMON MOCK VIPER
(BOIE, 1827)

397

The Common Mock Viper is listed as *incertae sedis* within the Lamprophiidae, a term that means "uncertain placement," indicating that its position in the family has not been determined. It is found across an extremely wide area, from Nepal to Taiwan in the north, and south to the Philippines, Borneo, Sumatra, Java, Sulawesi, and the Lesser Sunda Islands. The Common Mock Viper inhabits forested hills and mountain slopes, rocky hillsides, and riparian (riverine) habitats. It closely resembles a small pitviper and inside its mouth are enlarged, grooved rear teeth that resemble fangs, which this snake uses to chew toxins into its prey, usually lizards such as skinks, frogs, or small snakes. Mock vipers are considered harmless to humans, there being no snakebites on record exhibiting symptoms of envenoming, despite the species' willingness to bite.

The Common Mock Viper is a relatively short snake with an angular, viperine head, and eyes with vertically elliptical pupils. Unlike many vipers, it has smooth scales. It is a highly variable species, being brown, reddish, yellow, or pale gray, with or without darker flecks or streaks. Males are believed to be lighter colored than females, but this may not differentiate them across the entire range. The head usually bears four longitudinal stripes that run onto the neck, and the undersides are gray, pink, or brown, spotted with black or striped.

RELATED SPECIES

Two subspecies of *Psammodynastes* are recognized, the nominate form (*P. pulverulentus pulverulentus*) throughout most of the range, and a Taiwanese population that is treated as a separate subspecies (*P. p. papenfussi*). A second species of *Psammodynastes* also exists, the Painted Mock Viper (*P. pictus*), which occurs in sympatry with *Psammodynastes pulverulentus* in Peninsular Malaysia, Sumatra, and Borneo, but has a much smaller total range.

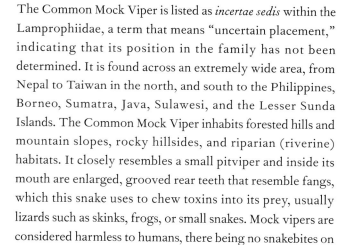

Actual size

FAMILY	Natricidae
RISK FACTOR	Nonvenomous
DISTRIBUTION	South Asia: Pakistan, India, Nepal, Sri Lanka, Bangladesh, Bhutan, Taiwan, and southern China
ELEVATION	0–6,560 ft (0–2,000 m) asl
HABITAT	Paddy fields, gardens, grasslands, agricultural fields, streams, and ponds
DIET	Frogs, toads, tadpoles, insects, scorpions, fish, lizards, and small mammals
REPRODUCTION	Oviparous, with clutches of 5–15 eggs
CONSERVATION STATUS	IUCN not listed, CITES Appendix III (India)

ADULT LENGTH
15¾–31½ in
(400–800 mm)

AMPHIESMA STOLATUM
BUFF-STRIPED KEELBACK
(LINNAEUS, 1758)

Actual size

One of the most commonly encountered diurnal snakes of South Asia, the Buff-striped Keelback inhabits wetlands, gardens, grasslands, agricultural fields, and rice paddies, living alongside humans at elevations from sea-level to 6,560 ft (2,000 m). An adept swimmer, it is often seen hunting in the reeds or grass of the shallows. Its prey comprises mostly amphibians, frogs and toads for adults, and tadpoles for juveniles, but animals as diverse as large insects, scorpions, fish, lizards, and rodents may also be taken. Buff-striped Keelbacks are most active in the morning or evening, sheltering through the hot day or during the night inside termite mounds and animal burrows. This is an inoffensive snake that rarely bites, preferring body-flattening, to expose the bright colors of its neck skin, or adopting thanatosis to deter further interference.

RELATED SPECIES

The genus *Amphiesma* once contained 44 species, but following a recent molecular study and taxonomic revision, these were split between the genera *Hebius* (41 species, pages 404–405) and *Herpetoreas* (two species, page 407), with only *A. stolatum* remaining. *Amphiesma stolatum* is now thought to be more closely related to keelbacks in the genera *Xenochrophis* (page 439), *Atretium* (page 400), and *Rhabdophis* (pages 423–424).

The Buff-striped Keelback has keeled body scales and large eyes. Its patterning is complex, being olive-green to brown with a series of dark transverse bars, each bearing a lateral white or yellow spot, which graduates to form a bold longitudinal buff or yellow stripe by the midbody, extending onto the tail. The undersides are white, with a small black spot on either side of each ventral scale, while the lips and throat are yellow.

FAMILY	Natricidae
RISK FACTOR	Nonvenomous
DISTRIBUTION	South Asia: Sri Lanka
ELEVATION	2,460–6,890 ft (750–2,100 m) asl
HABITAT	Wet lowland and hill forest and rainforest, and tea plantations
DIET	Earthworms
REPRODUCTION	Oviparous, with clutches of 4–12 eggs
CONSERVATION STATUS	IUCN not listed

ADULT LENGTH
15 in
(380 mm)

ASPIDURA TRACHYPROCTA

COMMON ROUGHSIDE

COPE, 1860

399

The roughsides are endemic to Sri Lanka. All are small, diurnal, fossorial snakes that feed on earthworms. The common name originates from the spinous, keeled scales found on the flanks of males, in the cloacal region and on the tail. Roughsides inhabit humid lowland and hillside forest and rainforest, and also plantations and gardens, in the southwest and south-center of Sri Lanka, and are absent from the arid east and north. The Common Roughside occurs from 2,460 to 6,890 ft (750–2,100 m) asl, while some of its congeners are more restricted in their vertical distribution. The Common Roughside can be very common where it occurs, often being discovered in small colonies in piles of decaying leaf litter. It breeds all year round, laying eggs under rotten logs. Roughsides are inoffensive, never attempting to bite.

Actual size

The Common Roughside is a small snake with a narrow, pointed head, a cylindrical body, and a short tail, males bearing spinous scales on the posterior body and tail. It is orange to dark brown above with two darker longitudinal stripes and several rows of dark spots. The underside may be yellow or heavily infused with black.

RELATED SPECIES

There are six other species in *Aspidura*: Boie's Roughside (*A. brachyorrhos*); the Sri Lankan Roughside (*A. ceylonensis*); Cope's Roughside (*A. copei*); Deraniyagala's Roughside (*A. deraniyagalae*); Drummond-Hay's Roughside (*A. drummondhayi*); and Günther's Roughside (*A. guentheri*). *Aspidura trachyprocta* is the second largest species after Cope's Roughside, which measures 25 in (635 mm), while Günther's Roughside is the smallest at 6⅓ in (160 mm).

FAMILY	Natricidae
RISK FACTOR	Nonvenomous
DISTRIBUTION	South Asia: India, Sri Lanka, and Nepal
ELEVATION	245–5,510 ft (75–1,680 m) asl
HABITAT	Freshwater ponds, streams, and rivers, rice paddies, and brackish creeks
DIET	Frogs, tadpoles, fish, crustaceans, and insect larvae
REPRODUCTION	Oviparous, with clutches of 10–30 eggs
CONSERVATION STATUS	IUCN Least Concern

ADULT LENGTH
17¾–19¾ in,
rarely 3 ft 3 in
(450–500 mm,
rarely 1.0 m)

400

ATRETIUM SCHISTOSUM
OLIVE KEELBACK
(DAUDIN, 1803)

The Olive Keelback is a common aquatic snake of ponds, rivers, and streams from Nepal, peninsular India, and Sri Lanka, although its range may not be continuous throughout. It is also known to occur in brackish tidal creeks and is common in rice paddies. Its prey is also aquatic, comprising frogs, tadpoles, fish, crabs, and reportedly even mosquito larvae. The Olive Keelback is diurnal, swims and climbs well, and often burrows into soft mud or hides in crab holes around watercourses. In common with many other Asian keelbacks, it is oviparous, mating in the wet season, and producing clutches of up to 30 eggs. Although some keelbacks bite easily, this species is relatively calm and inoffensive, its defense being generally limited to flattening its neck.

RELATED SPECIES

The closest relative of *Atretium schistosum* is the Yunnan Olive Keelback (*A. yunnanensis*) from Yunnan Province, western China, and possibly also Myanmar.

Actual size

The Olive Keelback is a stocky snake with a rounded head and keeled scales. It is usually uniform olive-green to olive-brown above, although some specimens exhibit a double row of black spots, and dark yellow, orange, or white below and on the supralabials, with a stark delineation, often a reddish line, between the two on the second dorsal scale row.

FAMILY	Natricidae
RISK FACTOR	Rear-fanged, mildly venomous
DISTRIBUTION	South Asia: Sri Lanka
ELEVATION	3,000–4,000 ft (915–1,220 m) asl
HABITAT	Rainforest, and wet low montane forest
DIET	Frogs and orthopterans
REPRODUCTION	Oviparous, with clutches of up to 7 eggs
CONSERVATION STATUS	IUCN Near Threatened

ADULT LENGTH
17¾–19¾ in
(450–500 mm)

BALANOPHIS CEYLONENSIS
BLOSSOM KRAIT
(GÜNTHER, 1858)

401

A Sri Lankan endemic, the Blossom Krait gets its name from the Sinhalese name *Mal Karawala*, translating as "flower krait." Sri Lanka is home to two true kraits, including the lethal Common Krait (*Bungarus caeruleus*, page 444), but the Blossom Krait is not a krait (Elapidae) but a keelback (Natricidae). However, it is related to the genus *Rhabdophis* (pages 423–424) and there is a snakebite documented from this rare snake that caused headache, photophobia, fainting, blurred vision, vomiting, and bleeding. There is no antivenom. The Blossom Krait inhabits forest leaf litter in Sri Lanka's humid south, and feeds on frogs, and occasionally grasshoppers. When disturbed it raises its body to expose a red patch under its neck, but it cannot bite unless it is handled.

The Blossom Krait has strongly keeled scales, a large head, and bulbous eyes. The dorsum of the head is orange-brown, extending as a stripe onto the anterior of the gray-brown body, which has patterned black transverse cross-bars at intervals, the meeting of these stripes being marked by bright yellow eyespots, presumably the blossom marking of its name.

RELATED SPECIES

The genus *Balanophis* is monotypic, but it is closely related to the genus *Rhabdophis* (pages 423–424), which contains two species that have caused serious snakebites, the Red-necked Keelback (*R. subminiatus*) and the Tiger Keelback, or Yamakagashi (*R. tigrinus*, page 424).

Actual size

FAMILY	Natricidae
RISK FACTOR	Rear-fanged, mildly venomous
DISTRIBUTION	North America: southern Great Lakes, USA (Illinois, Indiana, Michigan, Ohio, and Pennsylvania; also Missouri and Kentucky)
ELEVATION	295–2,200 ft (90–670 m) asl
HABITAT	Marshes, canals, creeks, pastures, and woodland
DIET	Earthworms, crayfish, and fish; possibly leeches
REPRODUCTION	Viviparous, with litters of 7–15 neonates
CONSERVATION STATUS	IUCN Least Concern

ADULT LENGTH
Male
13 in
(330 mm)

Female
24½ in
(625 mm)

402

CLONOPHIS KIRTLANDII
KIRTLAND'S SNAKE
(KENNICOTT, 1856)

The diminutive Kirtland's Snake, named for the naturalist Jared Potter Kirtland (1793–1877), occurs across the southern Great Lakes in Illinois and Indiana, southern Michigan, Ohio, and Pennsylvania, with small incursions into Missouri and Kentucky. Although aquatic and found in marshes, creeks, and canals, this small snake also inhabits damp woodland and open pastures. Kirtland's Snake preys on earthworms, adopting a Z-shaped hunting posture before attacking its prey. Other documented prey includes crayfish and minnows, and it is reported captive specimens fed on leeches. Females are considerably larger than males, and whereas most Old World keelbacks and watersnakes are oviparous, those from North America are viviparous, large female Kirtland's Snakes producing litters of up to 15 neonates. These snakes are completely harmless.

Kirtland's Snake is relatively stout-bodied, with a smallish black head, relatively long tail, and keeled body scales. Patterning comprises a double series of large black spots down either side, on a gray to brown background, and a reddish-brown vertebral stripe, at least anteriorly. The undersides are orange to red with a series of small black spots along the ventral margins.

RELATED SPECIES

Clonophis is a monotypic genus that may be most closely related to other small North American natricids such as the brownsnakes (*Storeria*, page 426), crayfish snakes (*Liodytes*, page 411 and *Regina*, page 422), Smooth Earthsnake (*Virginia valeriae*, page 438), and Rough Earthsnake (*Haldea striatula*, page 403).

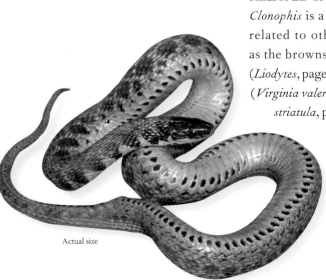

Actual size

FAMILY	Natricidae
RISK FACTOR	Rear-fanged, mildly venomous
DISTRIBUTION	North America: southeastern USA from Virginia to northern Florida, west to Texas, and north to Missouri
ELEVATION	0–1,150 ft (0–350 m) asl
HABITAT	Swamps, marshes, woodlands, edge situations, and urban areas
DIET	Earthworms, also possibly mollusks, isopods, insects, and small amphibians
REPRODUCTION	Viviparous, with litters of 2–13 neonates
CONSERVATION STATUS	IUCN Least Concern

ADULT LENGTH
11¾–13¾ in
(300–350 mm)

HALDEA STRIATULA
ROUGH EARTHSNAKE
(LINNAEUS, 1766)

The Rough Earthsnake is found across much of the eastern and southeastern USA, but this distribution is broken by the broad swath of the Mississippi Valley, from where it is absent. It is believed this is a prehistoric artifact from when the Mississippi Valley was a lengthy saltwater inlet. The Rough Earthsnake inhabits marshes, swamps, and woodlands, hunting under leaf litter or logs, although it is also encountered in urban environments under household debris. It feeds, primarily or exclusively, on earthworms. Other prey suggested includes mollusks, slaters, insects, and small frogs, but earthworms were the only prey found in the guts of museum specimens. This snake is an expert at tracking earthworms back to their burrows.

RELATED SPECIES

Previously included in the genus *Virginia*, with the Smooth Earthsnake (*V. valeriae*, page 438), the monotypic genus *Haldea* is probably also closely related to other small North American watersnakes, including the brownsnakes (*Storeria*, page 426), Kirtland's Snake (*Clonophis kirtlandii*, page 402), and the crayfish snakes (*Liodytes*, page 411 and *Regina*, page 422).

The Rough Earthsnake is a small snake with keeled dorsal body scales, and a slightly pointed head. Coloration is reddish brown to dark brown, although juveniles, and some adults, exhibit a pale band across the back of the head.

Actual size

FAMILY	Natricidae
RISK FACTOR	Nonvenomous
DISTRIBUTION	South Asia: southern India
ELEVATION	1,640–6,890 ft (500–2,100 m) asl
HABITAT	Rainforest creeks
DIET	Fish, also toads
REPRODUCTION	Oviparous, clutch size unknown
CONSERVATION STATUS	IUCN Least Concern

ADULT LENGTH
22¾ in
(580 mm)

HEBIUS MONTICOLA
HILL KEELBACK
(JERDON, 1853)

404

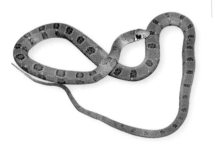

The Hill Keelback is a small snake with very large eyes and keeled scales. It is red-brown or olive above and white below, with a pattern of large dark brown blotches arranged in three longitudinal rows, although these markings may be faint in some specimens. A conspicuous and characteristic white or yellow line crosses the head behind the eyes. Juveniles are often banded.

Also known as the Wayanad Keelback after its Kerala type locality, the Hill Keelback is endemic to the Western Ghats of southwestern India, as far north as Goa. This is a high rainfall area, clothed in dense montane rainforest, the perfect habitat for this semi-aquatic species. Hill Keelbacks may be found along rainforest streams, and also occasionally on paths, although they are not commonly encountered and might even be considered rare. Hill Keelbacks are active diurnally, when they hunt the frogs that constitute their primary prey, although they are also reported to feed on toads. In common with other Asian keelbacks, this species is oviparous, but its egg-clutch size in unrecorded.

RELATED SPECIES

Most of the species formerly contained in genus *Amphiesma* are now placed in genus *Hebius*. The genus contains 41 species, from southern India's *H. monticola* to the Japanese Keelback (*H. vibakari*) in Amur (Russia), Korea, and Japan, and south to the Sulawesi Keelback (*H. celebicum*) in eastern Indonesia. *Hebius monticola* most resembles Beddome's Keelback (*H. beddomei*), which also occurs in the Western Ghats of India.

Actual size

FAMILY	Natricidae
RISK FACTOR	Nonvenomous
DISTRIBUTION	East Asia: southern China, Hainan Island, Hong Kong, northern Vietnam, and Taiwan
ELEVATION	1,640–4,920 ft (500–1,500 m) asl
HABITAT	Montane forest, grassy valleys, and meadows near streams
DIET	Earthworms, slugs, and tadpoles
REPRODUCTION	Oviparous, with clutches of up to 5 eggs
CONSERVATION STATUS	IUCN Least Concern

ADULT LENGTH
14–27½ in
(360–700 mm)

HEBIUS SAUTERI
KOSEMPO KEELBACK
(BOULENGER, 1909)

405

Also known as Sauter's Keelback after the German naturalist Hans Sauter (1871–1943), who lived much of his life on Taiwan, the Kosempo Keelback is named for its Taiwanese type locality. It is now known to be more widely distributed than just Taiwan, being recorded from southern China, Hainan Island, Hong Kong, and also northern Vietnam, but it is nowhere a common snake. It is a small keelback that is associated with low montane forest, grassy valleys, and meadows, usually near to small watercourses, and it is believed to prey on earthworms, slugs, and tadpoles. Being a small snake it only produces small clutches of eggs, with five eggs being recorded. The Kosempo Keelback is totally inoffensive, not even biting when handled, and is harmless to humans.

The Kosempo Keelback is a small snake with a slender body, keeled scales, a head broader than the neck, and a moderately long tail. It is olive to gray in color with markings consisting of a weakly defined reddish longitudinal stripe, which is interrupted by a series of black spots with orange centers, and extensive black flecking. The lips are black and white barred, the throat is white, and the remainder of the undersides are dark yellowish to tan.

RELATED SPECIES

The large genus *Hebius* contains 41 species formerly included in genus *Amphiesma*. Some of these species exist in sympatry with *H. sauteri*, including Pope's Keelback (*H. popei*) in southern China and on Hainan Island, and Maki's Keelback (*H. miyajimae*) on Taiwan.

Actual size

FAMILY	Natricidae
RISK FACTOR	Nonvenomous
DISTRIBUTION	Central Africa: western Democratic Republic of the Congo
ELEVATION	605–1,480 ft (185–450 m) asl
HABITAT	Riverine forest and swamp forest
DIET	Not known, but presumed fish or frogs
REPRODUCTION	Reproductive strategy unknown
CONSERVATION STATUS	IUCN not listed

ADULT LENGTH
Male
22¾ in
(580 mm)

Female
25½ in
(650 mm)

HELOPHIS SCHOUTEDENI
SCHOUTEDEN'S SUN SNAKE
(WITTE, 1922)

One of the least known snakes of Africa, Schouteden's Sun Snake is known from approximately 30 specimens collected over the past century from the Congo River in the western Democratic Republic of the Congo. It is an aquatic snake with dorsally positioned nostrils for breathing at the surface, but otherwise its natural history is unknown. A specimen kept in captivity for a short time preferred to remain in the water, suggesting a very aquatic habit. Its prey preferences are unknown but will likely include fish and/or frogs. Its reproductive strategy is also undocumented. This snake alone is proof that there is still a great deal to be learned about the natural history of snakes. It is named in honor of Henri Schouteden (1881–1972), a zoologist who had a specialist interest in the Congo.

RELATED SPECIES

The genus *Helophis* is monotypic. *Helophis schoutedeni* is very poorly known, but is believed most closely related to the Smooth-scaled Watersnake (*Hydraethiops laevis*) and Black-bellied Watersnake (*H. melanogaster*), both from Central Africa. This relationship is based on morphology rather than molecular data as no DNA is yet available from this species.

Schouteden's Sun Snake is a stout-bodied, cylindrical, smooth-scaled snake with a short tail, a chunky head that is just distinct from the thick neck, and small eyes with round pupils. Dorsally it is black with at least 80 irregular, deep red or orange transverse bars, ventrally it is black, and the head, dorsally and ventrally, is black with red infusions on every scale.

Actual size

FAMILY	Natricidae
RISK FACTOR	Nonvenomous
DISTRIBUTION	South and Central Asia: India, Pakistan, Nepal, Tibet, Bhutan, and Bangladesh
ELEVATION	4,100–12,000 ft (1,250–3,657 m) asl
HABITAT	Montane forests, cultivated fields, and streams
DIET	Frogs, toads, tadpoles, fish, reptiles, and small mammals
REPRODUCTION	Oviparous, with clutches of 5–15 eggs
CONSERVATION STATUS	IUCN not listed

ADULT LENGTH
27½–37 in,
rarely 4 ft
(700–940 mm,
rarely 1.2 m)

HERPETOREAS PLATYCEPS
HIMALAYAN KEELBACK
(BLYTH, 1854)

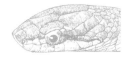

407

Arguably the keelback occurring at the greatest altitude, with a record of 12,000 ft (3,657 m), the Himalayan Keelback inhabits all the continental countries of South Asia, and also occurs in Tibet, but is not known to occur below 4,100 ft (1,250 m), which is above the maximum elevation for many lowland keelbacks. Active in the early evening, it hunts a wide range of prey, which may be in low population densities at altitude. Apart from frogs, toads, and tadpoles, the Himalayan Keelback will tackle fish, skinks, other snakes, and small mammals. Even snake eggs are eaten. It prefers small streams in montane forest, but may also be found in cultivated fields and gardens. Its threats include thrashing about and mouth gaping, but it rarely bites and is harmless to humans.

The Himalayan Keelback is slender-bodied, with a relatively broad head and keeled scales. Coloration is usually brown to gray, with patterning limited to black and white flecking on the flanks, black stripes from the snout, through the eye to the angle of the jaw, and sometimes a fine white, dark-edged collar across the neck or white lips. Some specimens are more boldly striped. The undersides are white or pale yellow.

RELATED SPECIES

Herpetoreas platyceps used to be in the genus *Amphiesma* but is now placed in the genus *Herpetoreas*, which is also home to two other high-elevation species, the Sikkim Keelback (*H. sieboldii*), from elevations of 4,000–11,800 ft (1,220–3,600 m) in the eastern Himalayas, and the recently described Burbrink's Keelback (*H. burbrinki*), from an elevation of 6,200 ft (1,890 m) in Tibet.

Actual size

FAMILY	Natricidae
RISK FACTOR	Rear-fanged mildly venomous
DISTRIBUTION	Philippines: Panay
ELEVATION	1,480–4,950 ft (450–1,510 m) asl
HABITAT	Lowland dipterocarp forest, submontane forest, and upper mossy forest
DIET	Not known
REPRODUCTION	Not known, presumed oviparous
CONSERVATION STATUS	IUCN Endangered

ADULT LENGTH
11½–14 in
(290–359 mm)

408

HOLOGERRHUM DERMALI
DERMAL'S CYLINDRICAL SNAKE
BROWN, LEVITON, FERNER & SISON 2001

Dermal's Cylindrical Snake is a small, smooth-scaled snake with a brown dorsum and yellow-white venter. The head is hardly distinct from the neck but the eyes are prominent. Patterning consists of a stark white stripe along the lips and onto the neck, a dark brown stripe running off the back of the head, and a series of short black crossbars that become broken into paravertebral spots in the anterior body and eventually peter out at midbody. The throat is brown with a series of white spots, black edged, and a dark midventral line runs the length of the undersides, with black spots either side.

This is a very rare snake, known only from five specimens collected in the West Panay Mountain Range, on the island of Panay in the central Philippines. During one three-year survey in part of its range only a single specimen was discovered. Secretive in nature, it inhabits lowland dipterocarp forest, submontane forest, and upper mossy forest, where it is usually found along streams. It appears to be diurnal but nothing is known of its prey preferences. Its reproductive strategy is also unknown, although it may be presumed to be oviparous. This species is listed as Endangered by the IUCN, due to its small range. The specific epithet *dermali* honors the American herpetologist Ronald "Dermal" Crombie, who has worked intensively in the Philippines.

RELATED SPECIES

The genus *Hologerrhum* contains a second species, the Philippine Stripe-lipped snake (*H. philippinum*), which is known from Luzon and Polillo islands. It seems highly likely that more species from this secretive genus remain to be discovered in the Philippines. The closest relative of *Hologerrhum* has not been determined, but it does share some characteristics with the Philippine triangle-spotted snakes (*Cyclocorus*), which are currently *insertae sedis* within the Colubridae.

Actual size

FAMILY	Natricidae
RISK FACTOR	Nonvenomous
DISTRIBUTION	Southeast Asia: Borneo (Sabah, Sarawak, Brunei, and Kalimantan)
ELEVATION	490–1,970 ft (150–600 m) asl
HABITAT	Lowland and low montane primary rainforest near streams and rivers
DIET	Dietary preferences unknown
REPRODUCTION	Probably oviparous, but with clutch size unknown
CONSERVATION STATUS	IUCN Least Concern

ADULT LENGTH
20 in
(530 mm)

HYDRABLABES PERIOPS

BORNEO DWARF WATERSNAKE

(GÜNTHER, 1872)

409

Also sometimes known as the Yellow-spotted Watersnake, or confusingly the Olive Small-eyed Snake, the Borneo Dwarf Watersnake is endemic to the island of Borneo, being found in the Malaysian states of Sabah and Sarawak, the Indonesian provinces that comprise Kalimantan, and the independent sultanate of Brunei. Its generic name, *Hydrablabes*, comes from *Hydra*, meaning water, and *-ablabes*, meaning not causing harm, a reference to the harmless nature of the snake. This keelback is highly aquatic, being associated with rivers, creeks, and streams in low-lying or low montane pristine rainforest. Despite being relatively common, it has not been studied in the wild and is underrepresented in museum collections, so its dietary and reproductive strategies are unknown.

RELATED SPECIES

The genus *Hydrablabes* contains one other species, the similar-sized Mount Kinabalu Watersnake (*H. praefrontalis*), which is endemic to the Malaysian state of Sabah, northeastern Borneo, but is poorly documented in the wild and even less well known than *H. periops*. Some authors believe *H. praefrontalis* is a synonym of *H. periops*. The former has only a single prefrontal scale compared to two in *H. periops*.

Actual size

The Borneo Dwarf Watersnake has a slender body that tapers smoothly into the small head and short tail, and which is covered in strongly keeled scales. There is a ring of small scales around the eye, from which this species gets the name *peri*, meaning around, and *-ops*, meaning eye. It is usually uniform gray or brown, although some specimens exhibit brown lateral or vertebral stripes, or spots on the flanks. The undersides are off-white to yellow.

FAMILY	Natricidae
RISK FACTOR	Rear-fanged, mildly venomous
DISTRIBUTION	Southern Africa: DRC, Zambia, Angola, Zimbabwe, Botswana, and Namibia
ELEVATION	3,120–5,540 ft (950–1,690 m) asl
HABITAT	Slow-moving rivers and lakes, wetlands, and swamps
DIET	Fish, including spiny eels
REPRODUCTION	Oviparous, with clutches of 5 eggs
CONSERVATION STATUS	IUCN not listed

ADULT LENGTH
Male
18¼ in
(465 mm)

Female
22¾ in,
rarely 24½ in
(580 mm,
rarely 625 mm)

LIMNOPHIS BANGWEOLICUS
BANGWEULU WATERSNAKE
(MERTENS, 1936)

The distinctive, narrow, and slightly pointed head of the Bangweulu Watersnake, a fish specialist, is probably an adaptation for searching underwater crevices for prey, particularly spiny eels. Named after Lake Bangweulu in Zambia, it occurs in both the Okavango and Zambezi river systems, in swamps, marshes, lakes, or rivers. It has been recorded from Botswana, Zimbabwe, and Zambia, southern Democratic Republic of the Congo, and westward to Angola and Namibia. The only report of its reproduction involved a large female that contained five eggs.

The Bangweulu Watersnake is a robustly built snake with smooth body scales, and a narrow, slightly pointed head with large eyes. It may be dorsally uniform black or mid-brown with a broad, light brown, pale-edged longitudinal stripe down either flank. The scales below this stripe are black-edged, forming a series of fine black longitudinal stripes. Anteriorly there may also be a whitish stripe, which begins on the supralabials. The undersides are yellowish or red, the scales under the chin and tail being black-edged.

RELATED SPECIES

The closest relative to *Limnophis bangweolicus* is the Angolan Watersnake (*L. bicolor*), also from the Okavango and Zambezi drainages, and of which it was once a subspecies (*L. bicolor bangweolicus*), but the two species are sympatric without integration. *Limnophis* can be distinguished from *Natriciteres* by its single, triangular internasal scale.

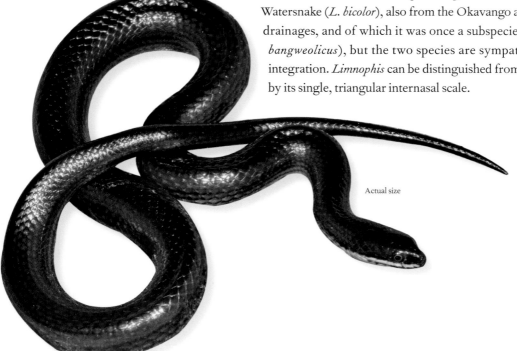

Actual size

FAMILY	Natricidae
RISK FACTOR	Nonvenomous
DISTRIBUTION	North America: USA (Florida Peninsula and southern Georgia)
ELEVATION	0–245 ft (0–75 m) asl
HABITAT	Freshwater marshes, swamps, ponds, streams, and occasionally in brackish water
DIET	Crayfish, shrimps, dragonfly nymphs, and amphibians
REPRODUCTION	Viviparous, with litters of 4–12 neonates
CONSERVATION STATUS	IUCN Least Concern

ADULT LENGTH
19¾–23¾ in,
rarely 27½ in
(500–600 mm,
rarely 700 mm)

LIODYTES ALLENI
STRIPED CRAYFISH SNAKE
(GARMAN, 1874)

411

The Striped Crayfish Snake is an inhabitant of still or slow-moving watercourses, preferably those with muddy bottoms, including lakes, marshes, swamps, sloughs, and slow-moving streams. It shelters in the muddy bank or under littoral or floating vegetation such as grasses or water hyacinth. In coastal situations it has been known to enter brackish watercourses. This small snake occurs throughout the Florida Peninsula, but not the Panhandle, and it is also present in southern Georgia. The Striped Crayfish Snake possesses a specially adapted skull to enable it to feed on hard-shelled crayfish, its primary prey. Adults also take shrimps, juveniles feed on dragonfly nymphs, and frogs, tadpoles, and sirens (aquatic salamanders) are also occasional prey. Joel Asaph Allen (1838–1921) was an American mammalogist, ornithologist, and museum curator.

The Striped Crayfish Snake is a glossy, smooth, or weakly keel-scaled snake with a slender head. It is olive to brown above with three broad, black longitudinal stripes, vertebral and lateral, and a broad, dark yellow stripe running along the lower flanks. This stripe joins with the ventral pigment, which may be yellow or pink, spotted occasionally with black.

RELATED SPECIES

Liodytes alleni was, until recently, included in the genus *Regina* with the Queen Snake (*R. septemvittata*, page 422), Graham's Crayfish Snake (*R. grahamii*), and the Glossy Crayfish Snake (*R. rigida*), which has three subspecies. A revision of these small aquatic snakes resulted in the resurrection of genus *Liodytes*, for *L. alleni*, its close relative the Glossy Crayfish Snake, and also the Black Swampsnake (*L. pygaea*), formerly known as *Seminatrix pygaea*.

Actual size

FAMILY	Natricidae
RISK FACTOR	Nonvenomous
DISTRIBUTION	Indian Ocean: Seychelles
ELEVATION	0–3,000 ft (0–915 m) asl
HABITAT	Wet and dry tropical forest, and secondary forest
DIET	Lizards
REPRODUCTION	Oviparous, clutch size unknown
CONSERVATION STATUS	IUCN Endangered

ADULT LENGTH
2 ft 5 in–3 ft 3 in
(0.75–1.0 m)

LYCOGNATHOPHIS SEYCHELLENSIS
SEYCHELLES WOLFSNAKE
(SCHLEGEL, 1837)

The Seychelles Wolfsnake is a small, smooth-scaled snake with a slightly pointed head, barely distinct from the neck, and moderate-sized eyes with vertically elliptical pupils. Dorsally it is brown or yellow with or without a pattern of dorsolateral pale spots, which are linked transversely by faint dark crossbars. The undersides are yellow, flecked with dark pigment, with a row of black spots along the outer edges of the scales. The most noticeable characteristic is a wavy white stripe that runs along the lips, flanked above by a dark brown or black stripe.

Lycognathophis means "wolf-jawed snake," but the Seychelles Wolfsnake is not related to Asian wolfsnakes (*Lycodon*, pages 190–191) or African wolfsnakes (*Lycophidion*, page 366), but is rather a member of the watersnake family Natricidae. One previous author suggested it should be called the Seychelles Cliffsnake instead, but this suggestion does not appear to have been popular. It is endemic to the Seychelles, where it occurs on Mahé, Silhouette, Praslin and Aride, La Digue and Frègate Islands, and inhabits wet and dry tropical forest. Although it may also occur in secondary habitats it has not adapted to plantations. The Seychelles Wolfsnake preys on geckos and skinks, and lays eggs but its clutch size is unknown. This species is listed as Endangered by the IUCN due to its small range.

RELATED SPECIES

Lycognathophis is a monotypic genus with no relatives in the Seychelles. Recent molecular analysis suggests its closest known relative may be the equally monotypic African Brown Watersnake (*Afronatrix anascopus*) from West Africa. The only other snakes to occur in the Seychelles are the introduced Brahminy Blindsnake (*Indotyphlops braminus*, page 57) and the endemic Seychelles Housesnake (*Lamprophis geometricus*). There are no dangerous land snakes in the Seychelles.

Actual size

FAMILY	Natricidae
RISK FACTOR	Possibly rear-fanged, mildly venomous
DISTRIBUTION	Southeast Asia: southern Thailand, Peninsular Malaysia, Singapore, Sumatra, Java, and Borneo
ELEVATION	490–4,360 ft (150–1,330 m) asl
HABITAT	Primary and secondary rainforest, and wetland habitats, in association with streams and creeks
DIET	Frogs, toads, and tadpoles
REPRODUCTION	Oviparous, with clutches of up to 25 eggs
CONSERVATION STATUS	IUCN Least Concern

ADULT LENGTH
23¾–29½ in
(600–750 mm)

MACROPISTHODON RHODOMELAS
BLUE-NECKED KEELBACK
(BOIE, 1827)

413

The Blue-necked Keelback is a striking snake, especially when it adopts its defensive posture, which involves raising its anterior body and spreading a hood like a cobra, a practice of several harmless Asian snakes, such as the mock cobras (*Plagiopholis*, page 440 and *Pseudoxenodon*, page 441). Although this keelback is inoffensive, it does possess enlarged rear teeth, and other keelbacks with similar dentition (*Rhabdophis*, pages 423–424) have proved to be dangerous, so it should be treated with respect. The Blue-necked Keelback inhabits rainforests and wetland areas close to streams and rivers, from southern Thailand to Sumatra, Java, and Borneo. Adults feed on amphibians, frogs and toads, and juveniles on tadpoles and froglets. Where it occurs, the Blue-necked Keelback may be fairly abundant.

RELATED SPECIES

The genus *Macropisthodon* also contains two other species, the Olive Keelback (*M. plumbicolor*), from India, Sri Lanka, and Myanmar, and the Orange-lipped Keelback (*M. flaviceps*), which mirrors the distribution of *M. rhodomelas*, with the exception of Java.

The Blue-necked Keelback is a slender, keel-scaled snake with a moderately broad head and large eyes. It has a very characteristic dorsal pattern comprising a red-brown ground color overlain by a broad black vertebral stripe, which forks to form a distinctive black chevron on the back of the neck. The undersides are pink, each ventral scale spotted with black, but the throat is pale blue, this color extending to the black chevron. These markings are especially obvious when the snake displays defensively.

Actual size

FAMILY	Natricidae
RISK FACTOR	Nonvenomous
DISTRIBUTION	West and Central Africa: Guinea-Bissau to Central African Republic, south to Congo and DRC
ELEVATION	0–2,620 ft (0–800 m) asl
HABITAT	Rainforest, and wet low montane forests
DIET	Froglets, tadpoles, and invertebrates
REPRODUCTION	Oviparous, clutch size not known
CONSERVATION STATUS	IUCN Near Threatened

ADULT LENGTH
11¾–19¾ in
(300–500 mm)

414

NATRICITERES FULIGINOIDES
COLLARED FOREST MARSHSNAKE
(GÜNTHER, 1858)

The Collared Forest Marshsnake is small and slender with a drab olive-brown body, often with faint bands, especially on the neck, and speckled with black and yellow, while its head is darker olive with black and white lip scales. The undersides are pale.

A widespread diurnal African watersnake, occurring from Guinea-Bissau to the Central African Republic and as far south as Gabon, Congo, and the Democratic Republic of the Congo, the Collared Forest Marshsnake is associated with rainforest and low montane forests, usually in close vicinity to watercourses. It is a small snake that feeds on small frogs, tadpoles, and forest-floor invertebrates in the leaf litter. This species has an unusual defensive tactic usually associated with small lizards: when accosted by a predator it will shed part of its tail and escape, leaving a wriggling tail section to occupy its would-be predator. The blood vessels in the stump cauterize and prevent excessive blood loss, but unlike the skinks and geckos that practice caudal autotomy, the Collared Forest Marshsnake is not known to regenerate a new tail, and because of this the tactic is called pseudoautotomy.

RELATED SPECIES

There are five other species in the genus *Natriciteres*, including the Southwestern Forest Marshsnake (*N. bipostocularis*), from Angola and Zambia, the Southeastern Forest Marshsnake (*N. sylvatica*), from Tanzania to Mozambique, and the Variable Marshsnake (*N. variegata*), from Guinea to Mozambique. The Olive Marshsnake (*N. olivacea*) occurs throughout most of sub-Saharan Africa, except the south, while the species with the smallest range is the endemic Pemba Island Marshsnake (*N. pembana*) from Tanzania.

Actual size

FAMILY	Natricidae
RISK FACTOR	Nonvenomous
DISTRIBUTION	Europe: southern Scotland, western Germany, Italy and southern France, Sicily, Sardinia, and Corsica
ELEVATION	0–8,200 ft (0–2,500 m) asl
HABITAT	Water meadows, streams, rivers, lakes, marshes, damp woodland, and canals
DIET	Frogs, and fish, small mammals, and earthworms on occasion
REPRODUCTION	Oviparous, with clutches of 5–25 eggs; record 64 eggs
CONSERVATION STATUS	IUCN Low Risk/Least Concern

ADULT LENGTH
Male
19⅔ in–27½ in
(500–700 mm)

Female
3 ft 3 in–4 ft,
British record 6 ft
(1.0–1.2 m,
British record 1.8 m)

NATRIX HELVETICA
WESTERN GRASS SNAKE
(LACÉPÈDE, 1789)

415

The Western Grass Snake is the species occurring in the British Isles, where it is widely distributed through England and Wales and ventures into southern Scotland. On mainland Europe it occurs through France, the Low Countries, western Germany, Italy, and on the islands of Sicily, Corsica, and Sardinia. It inhabits wetlands areas, including rivers, lakes, marshes, and man-made canals, reservoirs, and garden ponds, and also wet woodland, railway embankments, and vineyards. Its main prey comprises frogs, but fish are also taken and some populations prey on small mammals, while juveniles may eat earthworms. Females can reach almost 6 ft 7 in (2 m), but males are much smaller. They are completely harmless, despite their loud hissing. As a defense they adopt thanatosis, combined with the exudation of foul-smelling cloacal gland secretions. Females often lay their eggs in communal sites such as compost heaps.

The Western Grass Snake may range from green to dark olive or brown, with black flecks, or totally black in melanistic specimens. The undersides are off-white with black spotting. They are also known as Ringed Snakes, a name that translates into many European languages, and which is a reference to the broad, lobed, yellow and black collar many exhibit, although this marking may become obscured in older specimens.

RELATED SPECIES

Following the recent elevation of this taxon to specific status there are now four grass snake species. The other three are the Eastern Grass Snake (*N. natrix*) from Germany and Scandinavia to Russia, the Iberian Grass Snake (*N. astreptophora*) on the Iberian Peninsula and northwest Africa, and the Big-headed Grass Snake (*N. melanocephalus*) in the Caucasus. The Western Grass Snake contains five subspecies, the nominate form throughout most of its range; an Italian subspecies (*N. h. lanzai*), south Italian and Sicilian subspecies (*N. h. sicula*), and Corsican (*N. h. corsa*) and Sardinian (*N. h. cetti*) subspecies.

Actual size

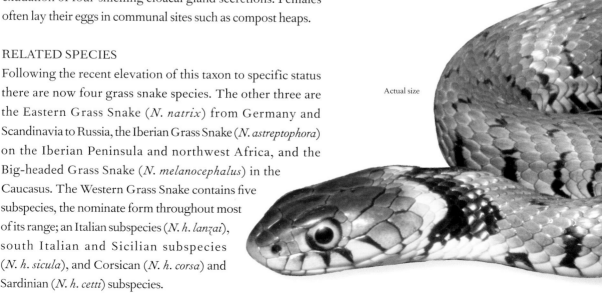

FAMILY	Natricidae
RISK FACTOR	Nonvenomous
DISTRIBUTION	Southwestern Europe: Spain, Portugal, France, northern Italy, the Balearics, Corsica, and Sardinia
ELEVATION	0–8,530 ft (0–2,600 m) asl
HABITAT	Slow-moving creeks, mountain streams, lakes, ponds, marshes, canals, fish farms, and livestock watering holes
DIET	Fish, tadpoles, and occasionally frogs, toads, newts, or invertebrates
REPRODUCTION	Oviparous, with clutches of 3–16 eggs
CONSERVATION STATUS	IUCN Least Concern

ADULT LENGTH
23¾–31½ in,
rarely 3 ft 3 in
(600–800 mm,
rarely 1.0 m)

NATRIX MAURA
VIPERINE WATERSNAKE
(LINNAEUS, 1758)

The Viperine Watersnake is well named, as its appearance and display do resemble those of a viper such as Lataste's Viper (*Vipera latastei*, page 642) or the Asp Viper (*V. aspis*, page 639). It is olive to brown in color, often with broken black zigzags and even a "V" on the top of its head. Some specimens have a series of white-centered black spots on the flanks, while others are strongly striped.

Actual size

Often mistaken for a venomous viper, the harmless Viperine Watersnake is the master of huff and bluster, hissing loudly, making mock strikes, and even mimicking the dangerous snakes' patterning and posture, but it is all bluff as these snakes are harmless. The Viperine Watersnake occurs through the Iberian Peninsula, into France and northern Italy, and also on the Balearic Islands, Corsica, and Sardinia. In the south of its range (Iberia) the Viperine Watersnake is commonly encountered hunting along mountain streams or in livestock watering holes, but in France it is more associated with lowland habitats, from marshes and lakes to man-made canals. It is much smaller than its relatives, the grass snakes, and prefers small fish or tadpoles to larger, stronger frogs and toads. Where they occur, Viperine Watersnakes are often very common.

RELATED SPECIES

Natrix maura occurs in sympatry with several other *Natrix*: the Iberian Grass Snake (*N. n. astreptophora*) in Spain and Portugal, the Western Grass Snake (*N. h. helvetica*) in France, and the Corsican Grass Snake (*N. h. corsica*) and Sardinian Grass Snake (*N. h. cetti*). The related Dice or Tessellated Snake (*N. tessellata*) occurs from Italy to Ukraine. All three species may occur together in northwestern Italy.

FAMILY	Natricidae
RISK FACTOR	Nonvenomous
DISTRIBUTION	North America and West Indies: USA, coast of Florida Peninsula and Gulf of Mexico, and north coast of Cuba
ELEVATION	0–82 ft (0–25 m) asl
HABITAT	Coastal saltmarshes, mangrove swamps, and estuaries; also coastal freshwater habitats
DIET	Inshore marine fish, crayfish, and shrimps
REPRODUCTION	Viviparous, with litters of 1–24 neonates
CONSERVATION STATUS	IUCN Least Concern, locally threatened

ADULT LENGTH
15–29½ in,
rarely 3 ft 3 in
(380–750 mm,
rarely 1.0 m)

NERODIA CLARKII
SALTMARSH SNAKE
(BAIRD & GIRARD, 1853)

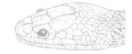

417

The most marine-adapted North American snake, the Saltmarsh Snake lives in the coastal strip from Florida's Atlantic coast to the Gulf of Mexico and the northern coast of Cuba, in saltmarshes and also mangrove swamps and estuaries, where it feeds on mullet, killifish, and other fish, and also crayfish and shrimps. It has therefore had to adapt to living in saline conditions. To avoid high levels of salinity it drinks pooling rainwater, because it lacks the salt-excretory glands of the seasnakes. Saltmarsh Snakes also inhabit coastal freshwater habitats. Although this species is not considered endangered by the IUCN, the Atlantic coastal population is considered locally threatened and is protected by both federal and Florida state laws. This species is named for the surveyor and naturalist Lieutenant John Henry Clark (1830–85).

RELATED SPECIES

Nerodia clarkii was once a subspecies of the Southern or Banded Watersnake (*N. fasciata*). Three subspecies are recognized, the Atlantic Saltmarsh Snake (*N. c. taeniata*), from the northern Florida coast, the Mangrove Saltmarsh Snake (*N. c. compressicauda*), of the Florida Peninsula and Cuba, and the Gulf Saltmarsh Snake (*N. c. clarkii*), from the Florida Panhandle to Texas.

The Saltmarsh Snake is a variable snake. Gulf Saltmarsh Snakes (*Nerodia clarkii clarkii*) are gray with four dark longitudinal stripes, a reddish or black venter, and a central row of yellow spots. In the Atlantic Saltmarsh Snake (*N. c. taeniata*) the longitudinal stripes often form blotches or transverse bands. Mangrove Saltmarsh Snakes (*N. c. compressicauda*) are the most variable, some specimens being gray with dark cross-bands, while others are uniform reddish or orange. Litters often contain both color morphotypes.

Actual size

FAMILY	Natricidae
RISK FACTOR	Nonvenomous
DISTRIBUTION	North America: USA (Mississippi River from Illinois to Louisiana, west to Texas, east to Florida)
ELEVATION	0–490 ft (0–150 m) asl
HABITAT	Slow-moving rivers, shallow lakes, wooded pools, bayous, sloughs, inundated woodland, canals, and brackish waters
DIET	Fish, also amphibians and crayfish
REPRODUCTION	Viviparous, with litters of 7–37 neonates
CONSERVATION STATUS	IUCN Least Concern

ADULT LENGTH
2 ft 4 in–3 ft 3 in,
rarely 4 ft
(0.7–1.0 m,
rarely 1.27 m)

418

NERODIA CYCLOPION
MISSISSIPPI GREEN WATERSNAKE
(DUMÉRIL, BIBRON & DUMÉRIL, 1854)

The Mississippi Green Watersnake is a stout-bodied snake with a relatively narrow, pointed head, and keeled scales. It is dark olive-green in color, with a heavy infusion of black pigment that largely obscures the background color. The underside is yellow anteriorly, and gray with yellow half-moon markings posteriorly.

One of the larger North American watersnakes, the Mississippi Green Watersnake occurs the length of the Mississippi Valley, from Illinois to Louisiana, and both east and west along the coast to the Florida Panhandle and Texan coast respectively. It shows a preference for quiet watercourses, slow-moving rivers, still lakes and pools, inundated woodland, sloughs, and bayous. It also occurs in man-made habitats such as canals, and enters brackish water along the coast. The primary prey of the Mississippi Green Watersnake comprises fish, a wide variety of species being taken. A very small proportion of its diet also comprises crayfish, frogs, and sirens and amphiumas (aquatic salamanders). Although nonvenomous, this snake is powerful and will bite, drawing blood, if mishandled.

RELATED SPECIES

The most closely related watersnake to *Nerodia cyclopion* is the Florida Green Watersnake (*N. floridana*), which was formerly a subspecies. It may also be confused with the highly venomous Cottonmouth (*Agkistrodon piscivorus*, page 554).

Actual size

FAMILY	Natricidae
RISK FACTOR	Nonvenomous
DISTRIBUTION	North America: USA (Brazos River system, central Texas)
ELEVATION	820–1,800 ft (250–550 m) asl
HABITAT	Rocky shorelines of swift-flowing rivers and streams
DIET	Fish, frogs, salamanders, and crayfish
REPRODUCTION	Viviparous, with litters of 4–24 neonates
CONSERVATION STATUS	IUCN Near Threatened

ADULT LENGTH
23¾–35½ in
(600–900 mm)

NERODIA HARTERI
BRAZOS RIVER WATERSNAKE
(TRAPEDO, 1941)

419

The Brazos River Watersnake is confined to the Brazos River drainage of central Texas, and as such is vulnerable to habitat changes. It prefers rocky shores along swift-flowing rivers to slow-moving rivers or lakes, and it hunts fish, as well as frogs, salamanders, and crayfish. It is listed as Near Threatened by the IUCN because of its small range and because it was believed that the construction of dams along the Brazos River would lead to large areas of flooded habitat to which the Brazos River Watersnake might not adapt, and that this would impact adversely on its survival. This small snake has persevered through the development but it still requires protection given the vulnerability of existing in such a small range. Philip Harter (d. 1971) was the herpetologist who collected the holotype.

The Brazos River Watersnake is a small, slender snake with an olive-green or brown dorsum marked by a darker zigzag pattern, formed from the coalescence of upper lateral rows of blotches. A lower longitudinal row of dark blotches is also present on the midflank area. The throat and chin are off-white, while the venter of the body varies from pinkish to orange-brown.

RELATED SPECIES

Nerodia harteri is not the only small, threatened watersnake in Texas. The Concho River Watersnake (*N. paucimaculata*), which was formerly treated as a subspecies of *N. harteri* and appears to prefer similar habitats and feed on similar prey, inhabits the Concho–Colorado drainages. Another species with a small total range, it was also feared threatened by the extensive development of reservoirs. This small watersnake also seems to have survived the development of its habitat, but both are still extremely vulnerable.

Actual size

FAMILY	Natricidae
RISK FACTOR	Nonvenomous
DISTRIBUTION	North America: central USA (Iowa and Illinois to Texas) and eastern Mexico (Tamaulipas to Tabasco)
ELEVATION	0–7,330 ft (0–2,235 m) asl
HABITAT	Lakes, ponds, slow-moving rivers, wooded swamps, bayous, and marshes
DIET	Fish, also frogs
REPRODUCTION	Viviparous, with litters of 8–35, occasionally 2, neonates
CONSERVATION STATUS	IUCN Least Concern

ADULT LENGTH
2 ft 7 in–3 ft 7 in,
rarely 5 ft 9 in
(0.8–1.1 m,
rarely 1.75 m)

420

NERODIA RHOMBIFER
DIAMOND-BACKED
WATERSNAKE
(HALLOWELL, 1852)

The Diamond-backed Watersnake is a large aquatic snake that is often killed in the belief it is a rattlesnake (*Crotalus*, pages 572–581) or Cottonmouth (*Agkistrodon piscivorus*, page 554), yet although large specimens can inflict a bloody bite they are nonvenomous and harmless to humans. An inhabitant of slow-moving or still watercourses, from river to lakes, wooded swamps, and bayous, the Diamond-backed Watersnake is primarily a predator of fish, taking a wide variety of species, with frogs the next most popular prey group. This snake is distributed from Iowa and Illinois in central USA, to Mississippi, Louisiana, Texas, and over the border into Mexico and down the Gulf of Mexico coast to Tabasco. Unlike Eurasian watersnakes, North American watersnakes are live-bearers, with large female Diamond-backed Watersnakes producing large litters.

The Diamond-backed Watersnake is a stout-bodied snake with strongly keeled scales, a long tail, and a broad head. It is usually olive-brown in color with three series of black cross-bands, on the dorsum and each flank, which may coalesce at the corners to create a continuous dorsal pattern.

RELATED SPECIES

In appearance this watersnake resembles the Brown Watersnake (*Nerodia taxispilota*) of the southeastern USA, and small specimens may also be confused with Southern Watersnakes (*N. fasciata*) or Northern Watersnakes (*N. sipedon*), with which this species occurs in sympatry in parts of its range. Three subspecies are recognized: the nominate Northern Diamond-backed Watersnake (*N. rhombifer rhombifer*) in the USA, the Tampico Diamond-backed Watersnake (*N. r. blanchardi*) in Tamaulipas and northern Veracruz, Mexico, and the Tabasco Diamond-backed Watersnake (*N. r. werleri*) in southern Veracruz and Tabasco.

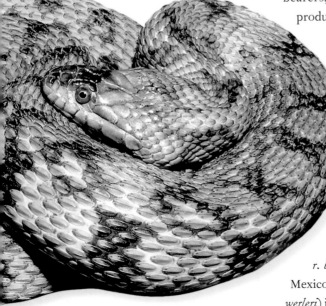

Actual size

FAMILY	Natricidae
RISK FACTOR	Nonvenomous
DISTRIBUTION	Southeast and East Asia: southern China, Hong Kong, and northern Vietnam
ELEVATION	560–3,940 ft (170–1,200 m) asl
HABITAT	Rocky, forested, montane streams and pools
DIET	Freshwater shrimps, also tadpoles and fish
REPRODUCTION	Oviparous, with clutches of 2–5 eggs
CONSERVATION STATUS	IUCN Least Concern

ADULT LENGTH
15¾–19¾ in
(400–500 mm)

OPISTHOTROPIS LATERALIS
MAU SON MOUNTAIN-STREAM SNAKE
BOULENGER, 1903

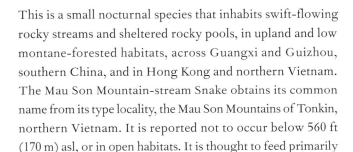

421

This is a small nocturnal species that inhabits swift-flowing rocky streams and sheltered rocky pools, in upland and low montane-forested habitats, across Guangxi and Guizhou, southern China, and in Hong Kong and northern Vietnam. The Mau Son Mountain-stream Snake obtains its common name from its type locality, the Mau Son Mountains of Tonkin, northern Vietnam. It is reported not to occur below 560 ft (170 m) asl, or in open habitats. It is thought to feed primarily on freshwater shrimps, but tadpoles and small fish are also believed to feature in its diet. Oviparous females lay from two to five eggs, often among stones near to a watercourse, but not in the water.

The Mau Son Mountain-stream Snake is uniform dark brown or gray-brown above and pale yellow on the venter, with a dark brown lateral stripe separating the two colors on the flanks. The narrow head is the same color as the body, although the supralabials and infralabials are yellow and there is a dark brown flash behind the eye, which may coalesce with the lateral stripe.

RELATED SPECIES

The genus *Opisthotropis* contains 22 species of mountain-stream snakes, from southern China, Vietnam, Laos, Thailand, Borneo, Sumatra, the Philippines, and the Ryukyu Islands. Several species have been described only recently, including the Tam Dao Mountain-stream Snake (*O. tamdaoensis*), from the Tam Dao Mountains of northern Vietnam, which may occur in sympatry with *O. lateralis*.

Actual size

FAMILY	Natricidae
RISK FACTOR	Nonvenomous
DISTRIBUTION	North America: eastern Canada and USA, from southern Great Lakes to Gulf of Mexico
ELEVATION	0–2,490 ft (0–760 m) asl
HABITAT	Brooks, streams, and marshes, usually with rocky bottoms
DIET	Crayfish, fish, amphibians, insect nymphs
REPRODUCTION	Viviparous, with 4–39 neonates
CONSERVATION STATUS	IUCN Least Concern

ADULT LENGTH
14–23¼ in,
rarely 35½ in
(355–590 mm,
rarely 900 mm)

REGINA SEPTEMVITTATA
QUEEN SNAKE
(SAY, 1825)

The Queen Snake demonstrates a preference for clean, unpolluted, quiet aquatic habitats, being found in streams, brooks, and marshes, under tree cover or in the open, but usually with rocky bottoms. It is found from the southern Great Lakes in southwestern Ontario, Canada, and Wisconsin, USA, to New York state, south to Mississippi, Alabama, and the Florida Panhandle, with a western population in Missouri and Arkansas. The Queen Snake is believed to feed almost exclusively on crayfish, with only occasional records of small fish, frogs, or insect nymphs being taken. The name Queen Snake refers to the generic name *Regina*, meaning queen.

The Queen Snake is slender-bodied with keeled scales. It is olive-green to brown above with a series of three black longitudinal stripes and a broad, distinctive yellow stripe along the lower lateral surfaces. The underside is yellow with four fine brown longitudinal stripes, while the head is green like the body, but with a yellow throat and labials. The name *septemvittata* is a reference to the dark stripes, *septem* meaning seven, and *-vittata* meaning stripes.

RELATED SPECIES

The genus *Regina* contained four species of crayfish-eating snakes, but two of these were recently moved to the genus *Liodytes* (page 411), leaving only *R. septemvittata* and its closest relative, Graham's Crayfish Snake (*R. grahamii*). *Regina* is closely related to *Liodytes*, and also the brownsnakes (*Storeria*, page 426), Kirtland's Snake (*Clonophis kirtlandii*, page 402), and the earthsnakes (*Haldea*, page 403 and *Virginia*, page 438).

Actual size

FAMILY	Natricidae
RISK FACTOR	Rear-fanged, venomous: possibly procoagulants, anticoagulants, or hemorrhagins
DISTRIBUTION	Southeast Asia: southern China and Myanmar, Thailand, Laos, Vietnam, and Cambodia
ELEVATION	1,310–7,910 ft (400–2,410 m) asl
HABITAT	Rainforest and monsoon forest, especially close to rivers
DIET	Frogs and fish
REPRODUCTION	Oviparous, clutch size unknown
CONSERVATION STATUS	IUCN not listed

ADULT LENGTH
37½ in
(950 mm)

RHABDOPHIS NIGROCINCTUS

BLACK-RINGED KEELBACK

(BLYTH, 1856)

The genus *Rhabdophis* contains 22 species of Asian keelbacks that possess enlarged rear teeth and toxin-secreting Duvernoy's glands. They are arguably the most dangerous members of the Natricidae, with serious snakebites and even fatalities attributed to the genus. There are no bites on record from the Black-ringed Keelback, though it should be treated with caution given its close relationship to proven dangerous species. This attractive rainforest and monsoon forest-dwelling snake occurs throughout mainland Southeast Asia, from China to Thailand and Cambodia, and is usually found in forest bordering rivers in undisturbed, upland habitats. It is both nocturnal and diurnal, and it feeds on fish and frogs, which succumb quickly to its bite. Although the Black-ringed Keelback is oviparous its clutch size remains undocumented.

The Black-ringed Keelback is a striking snake, being bright green anteriorly and olive-green posteriorly, with a series of fragmented black bars or rings across the dorsum, and a head that is olive to copper above and pinkish on the sides, with stark black stripes from the eye to the lip and the rear of the eye to the angle of the jaw, while another black stripe runs back onto the side of the neck. The eyes are large, and the scales strongly keeled.

RELATED SPECIES

Of the 22 species of *Rhabdophis*, several occur in sympatry with *R. nigrocinctus*, including the Red-necked Keelback (*R. subminiatus*), which has been implicated in serious snakebites, but no deaths thus far.

Actual size

FAMILY	Natricidae
RISK FACTOR	Rear-fanged, venomous: procoagulants, and possibly also anticoagulants and hemorrhagins
DISTRIBUTION	Southeast and East Asia: eastern Russia, China, Korea, Taiwan, and Japan
ELEVATION	0–7,220 ft (0–2,200 m) asl
HABITAT	Paddy fields, streams, and wooded or shrubby hillsides
DIET	Toads, frogs, fish, tadpoles, other snakes, and possibly beetles
REPRODUCTION	Oviparous, with clutches of 9–27 eggs
CONSERVATION STATUS	IUCN not listed

ADULT LENGTH
2 ft–3 ft 7 in,
rarely 5 ft 7 in
(0.6–1.1 m,
rarely 1.7 m)

424

RHABDOPHIS TIGRINUS
TIGER KEELBACK
(BOIE, 1826)

The Tiger Keelback is olive-brown or green above, but red on the neck, this pigment continuing along the flanks as a lateral stripe. Several rows of black spots form irregular bands across the body. The nape of the neck bears a pair of black half-moons, backed by yellow or red pigment, presumably as a warning to indicate to predators that it is toxic to eat. The undersides are white.

The Tiger Keelback, also called the Japanese Keelback or Yamakagashi, is a dangerous snake with at least one proven human fatality recorded. Its venom contains procoagulants, and possibly also anticoagulants and hemorrhagins, which contribute to intravascular coagulation, leading to incoagulable blood and potential cerebral hemorrhage. This venomous snake is also poisonous because it preys on toads and sequesters their bufotoxins (toxins found in the parotoid glands of toxic toads) into its own skin, presumably as an antipredator tactic. Despite this, the Tiger Keelback is a fairly inoffensive species and was long believed to be nonvenomous due to its disinclination to bite. When threatened it raises its anterior body and arches its neck in warning, a display enhanced by its red coloration. It is found in a wide variety of habitats, from rice paddies to forested hillsides, and can be common.

RELATED SPECIES

There are a further 20 species in the genus *Rhabdophis*, distributed through Southeast and East Asia. Up to three subspecies of *R. tigrinus* may be recognized, the nominate form (*R. t. tigrinus*) from Japan, a mainland Asian form (*R. t. lateralis*), and an endemic Taiwanese subspecies (*R. t. formosanus*). The Red-necked Keelback (*R. subminiatus*), of Southeast Asia, is a close relative.

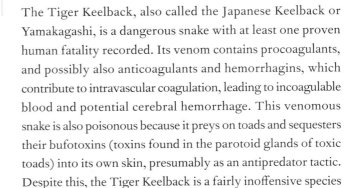

Actual size

FAMILY	Natricidae
RISK FACTOR	Nonvenomous
DISTRIBUTION	East Asia: eastern and southeastern China, Taiwan
ELEVATION	985–6,560 ft (300–2,000 m) asl
HABITAT	Rice paddies and hill streams
DIET	Small fish, frogs, and tadpoles
REPRODUCTION	Viviparous, with litters of 9–13 neonates
CONSERVATION STATUS	IUCN not listed

ADULT LENGTH
19¾–31½ in
(500–800 mm)

SINONATRIX ANNULARIS
RINGED KEELBACK
(HALLOWELL, 1899)

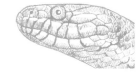

425

The Ringed Keelback is a snake of the hill streams and rice paddies of southeastern China, including Hainan Island, and also Taiwan. It occurs north to the Ningpo Mountains, near Shanghi, and westward along the Yangtse River valley. The Ringed Keelback is the only Old World natricid keelback snake known to give birth to live young, although early reports also suggest that some specimens may also lay eggs. It is a relatively small aquatic species that preys on small fish, such as the loaches that are common in fast-flowing streams, but frogs have also been recorded in the diet of the Ringed Keelback. Unlike some Asian keelbacks, this species is not dangerous to humans.

The Ringed Keelback is a robustly built snake with a relatively large head and small eyes. In coloration it is olive above, red on the flanks, and red or white on the undersides, with a series of broad transverse black bands, sometimes edged with yellow scales, that cross the belly and run up the flanks, and may meet over the center of the back. The lips are yellow or white with black sutures.

RELATED SPECIES

There are three other species of *Sinonatrix* recognized: the Chinese Keelback (*S. percarinata*), Chinese Spotted Keelback (*S. aequifasciata*), and Yunnan Keelback (*S. yunnanensis*). These other three *Sinonatrix* species are all oviparous species, as are all other Asian keelbacks for which reproductive strategies are known.

Actual size

FAMILY	Natricidae
RISK FACTOR	Nonvenomous
DISTRIBUTION	North and Central America: southeastern Canada, eastern USA, eastern Mexico, Guatemala, and Honduras
ELEVATION	0–6,680 ft (0–2,035 m) asl
HABITAT	Almost all terrestrial and semi-wetland habitats from marshes to urban areas, especially woodland
DIET	Earthworms, slugs, insects, small frogs, and tadpoles
REPRODUCTION	Viviparous, with litters of 3–31, occasionally 40 neonates
CONSERVATION STATUS	IUCN Least Concern

ADULT LENGTH
8–15¾ in,
rarely 20½ in
(200–400 mm,
rarely 520 mm)

426

STORERIA DEKAYI
DEKAY'S BROWNSNAKE
(HALLOWELL, 1839)

Actual size

DeKay's Brownsnake is a small gray or brown snake with markings comprising four rows of dark brown spots or two fine longitudinal stripes. Some specimens possess a white collar on the nape.

The petite DeKay's Brownsnake is a very widely distributed species, occurring from Quebec and Ontario, Canada, through eastern and central USA, to eastern Mexico, with an isolated population from southern Mexico to Honduras. DeKay's Brownsnake is found in almost every terrestrial or semi-wet habitat within its range, from marshland to urban areas, but especially woodland, and while it is often found around small lowland ponds it may also be found in montane cloud forest. It may occur anywhere where there is ground cover, whether fallen logs or urban trash. DeKay's Brownsnakes feed on the earthworms and slugs that abound under such cover, but also take soft-bodied insects, small frogs, and tadpoles. The live-bearing females may produce more than one litter of neonates a year.

RELATED SPECIES

Seven subspecies of *Storeria dekayi* are recognized: the Northern Brownsnake (*S. d. dekayi*), Midland Brownsnake (*S. d. wrightorum*), Western Brownsnake (*S. d. texana*), Marsh Brownsnake (*S. d. limnetes*) on the Gulf of Mexico, Tamaulipas Brownsnake (*S. d. temporalineata*) and Veracuz Brownsnake (*S. d. anomala*) in Mexico, and the Tropical Brownsnake (*S. d. tropica*) from Chiapas to Honduras. The Florida Brownsnake (*S. victa*) was a former subspecies. There are also three other brownsnake species: the Red-bellied Snake (*S. occipitomaculata*), Mexican Brownsnake (*S. storerioides*), and Mexican Yellow-bellied Brownsnake (*S. hidalgoensis*).

FAMILY	Natricidae
RISK FACTOR	Nonvenomous
DISTRIBUTION	North America: Great Lakes of USA and Canada
ELEVATION	490–1,510 ft (150–460 m) asl
HABITAT	Marshland, Great Lakes coastal plains, open grasslands, and vacant urban lots or abandoned industrial sites
DIET	Earthworms, and also leeches, small frogs, toads, and salamanders
REPRODUCTION	Viviparous, with litters of 8–11 neonates
CONSERVATION STATUS	IUCN Least Concern, locally endangered

ADULT LENGTH
15–20 in,
rarely 27 in
(380–510 mm,
rarely 690 mm)

THAMNOPHIS BUTLERI
BUTLER'S GARTERSNAKE
(COPE, 1889)

427

Named for the Indiana naturalist Amos William Butler (1860–1937), the diminutive Butler's Gartersnake is found in the vicinity of the Great Lakes, in southeastern Ontario, Michigan, Ohio, and Indiana, with an isolated population in Wisconsin. It is an inhabitant of marshland, open grasslands, pastures, and shoreline areas, but is also found in urban areas where it may shelter under boards on abandoned industrial sites or vacant lots. It feeds primarily on earthworms but also eats leeches. Small frogs, toads, and salamanders are also reported as occasional prey. It is often found in sympatry with the Eastern Gartersnake (*Thamnophis sirtalis sirtalis*, page 432) and the Northern Ribbon Snake (*T. sauritus septentrionalis*). Although not thought threatened by the IUCN, Butler's Gartersnake is considered Endangered in Ontario and Indiana.

Butler's Gartersnake is a small, slender, gracile snake with a small head. It may be brown to black in color, with three yellow longitudinal stripes, and a double row of black spots between the dorsal and lateral stripes.

RELATED SPECIES

The slightly smaller Short-headed Gartersnake (*Thamnophis brachystoma*), which also eats earthworms, occurs to the east of *T. butleri*, in the Allegheny Highlands of New York State and Pennsylvania.

Actual size

FAMILY	Natricidae
RISK FACTOR	Nonvenomous
DISTRIBUTION	North America: western USA, southwestern Canada, northwestern Mexico
ELEVATION	0–12,000 ft (0–3,660 m) asl
HABITAT	Marshes, meadows, mountain lakes, streams, springs, semidesert, and coastal scrub
DIET	Fish, slugs, leeches, earthworms, frogs, salamanders, lizards, small mammals, and birds
REPRODUCTION	Viviparous, with litters of 4–27 neonates
CONSERVATION STATUS	IUCN Least Concern

ADULT LENGTH
15¾–35½ in,
rarely 3 ft 7 in
(400–900 mm,
rarely 1.1 m)

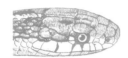

428

THAMNOPHIS ELEGANS
WESTERN TERRESTRIAL GARTERSNAKE
(BAIRD & GIRARD, 1853)

The Western Terrestrial Gartersnake is fairly variably patterned, ranging from black to dark gray, olive, pale gray, or even red on the flanks, with two rows of black spots visible on paler specimens, and three longitudinal yellow stripes. The Coast Gartersnake (*Thamnophis elegans terrestris*) is especially variable, occurring as black, brown, or red morphotypes, which could be mistaken for different species.

The Western Terrestrial Gartersnake is widely distributed across western USA and southwestern Canada, with small populations in northwestern Mexico. It is one of the more terrestrial gartersnakes, hence its common name, having adapted to live in xeric semidesert habitats and cool montane habitats up to 12,000 ft (3,660 m) asl. Habitats occupied by Western Terrestrial Gartersnakes range from lowland marshes to mountain lakes and coastal scrub, and the diets of the different populations also vary. Some populations of Wandering Gartersnake (*Thamnophis elegans vagrans*) are primarily piscivorous, while others are mainly amphibiophagous (amphibian eaters), and some populations of Coast Gartersnake (*T. e. terrestris*) are believed to prefer slugs, and to completely avoid amphibians. Other occasional prey items include leeches, small mammals, and small birds. Females may produce large litters of up to 27 neonates.

RELATED SPECIES

Across western USA and southwestern Canada, five subspecies are recognized: the Mountain Gartersnake (*Thamnophis elegans elegans*), Arizona Gartersnake (*T. e. ariςonae*), Upper Basin Gartersnake (*T. e. vascotanneri*), Coast Gartersnake (*T. e. terrestris*), and Wandering Gartersnake (*T. e. vagrans*). The San Pedro Martir Gartersnake (*T. e. hueyi*) is a localized subspecies from northern Baja California. The Mexican Wandering Gartersnake (*T. errans*) of northwest Mexico is a former subspecies.

Actual size

FAMILY	Natricidae
RISK FACTOR	Nonvenomous
DISTRIBUTION	North America: western USA (northern and central California)
ELEVATION	0–400 ft (0–122 m) asl
HABITAT	Marshes, lakes, ponds, and rice fields
DIET	Fish and frogs
REPRODUCTION	Viviparous, with litters of up to 24 neonates
CONSERVATION STATUS	IUCN Vulnerable

ADULT LENGTH
2 ft 7 in–4 ft,
rarely 5 ft 4 in
(0.8–1.2 m,
rarely 1.62 m)

THAMNOPHIS GIGAS
GIANT GARTERSNAKE
FITCH, 1940

The largest gartersnake, the Giant Gartersnake was previously treated as a subspecies of the Aquatic Gartersnake (*T. couchii*). The Giant Gartersnake is confined to the Sacramento and San Joaquin valleys of northern and central California, where it inhabits marshes, ponds, sloughs, and lakes, but is rarely encountered in large rivers. It is now considered Vulnerable by the IUCN because it has been extirpated from a large part of its historical range, by habitat alteration and destruction. In places Giant Gartersnakes have been forced to inhabit rice fields that replaced their former habitats. It primarily preys on fish and frogs, but has adapted to feeding on introduced fish such as carp, with the loss of the native fish and frogs that made up its original diet.

The Giant Gartersnake occurs as two different color morphotype. The striped morphotype, which is the pattern of Sacramento Valley specimens, is black with three longitudinal yellow stripes. A spotted morphotype, found alongside the striped pattern in San Joaquin Valley specimens, is olive-green with a faint yellow-green dorsal stripe and two rows of black spots on the flanks.

RELATED SPECIES
Thamnophis gigas is related to the Aquatic Gartersnake (*T. couchii*), Santa Cruz Gartersnake (*T. atratus atratus*), and Oregon Gartersnake (*T. a. hydrophilus*), all of which occur in northern California.

Actual size

FAMILY	Natricidae
RISK FACTOR	Nonvenomous
DISTRIBUTION	North and Central America: central USA and western Mexico to Costa Rica
ELEVATION	0–8,000 ft (0–2,438 m) asl
HABITAT	Ponds, lakes, swamps, marshes, creeks, and desert springs
DIET	Frogs, toads, occasionally fish or lizards
REPRODUCTION	Viviparous, with litters of 8–12 neonates
CONSERVATION STATUS	IUCN Least Concern

ADULT LENGTH
19¾–29½ in,
rarely 4 ft
(500–750 mm,
rarely 1.23 m)

430

THAMNOPHIS PROXIMUS
WESTERN RIBBONSNAKE
(SAY, 1823)

The Western Ribbonsnake is a slender snake, which may be olive-green to black, with overlain rows of black spots, and three bright yellow stripes, except in the Red-striped Ribbonsnake of the Midwest, which has a dark red dorsal stripe.

Ribbonsnakes are more gracile than gartersnakes. They also feed on more amphibians, primarily frogs and toads, than their gartersnake relatives, which exhibit primarily fish-orientated diets, although the Western Ribbonsnake is also documented to occasionally take fish, or lizards such as skinks. This species has an extensive distribution, from the southern Great Lakes, through the Midwest and western Mexico to the Yucatán Peninsula, and, as a series of small, isolated populations, from southern Mexico to Costa Rica. The Western Ribbonsnake is an aquatic snake associated with bushy habitats that offer escape routes from predators. It even inhabits desert watercourses provided there is cover available. Marshes and swamps with dense growths of reeds or grasses are also popular. Being slender snakes, females produce relatively small litters of 8–12 neonates.

RELATED SPECIES

Despite its extensive and fragmented distribution only six subspecies are recognized, including the Western Ribbonsnake (*Thamnophis proximus proximus*) in the north, Red-striped Ribbonsnake (*T. p. rubrilineatus*) in the Midwest, and Gulf Coast Ribbonsnake (*T. p. orarius*) and Arid Land Ribbonsnake (*T. p. diabolicus*) in Texas and northern Mexico. The Lowland Tropical Ribbonsnake (*T. p. rutiloris*) accounts for all the scattered southern Mexico and Central American populations, except a montane population of Alpine Ribbonsnake (*T. p. alpinus*), in Chiapas. The Eastern Ribbonsnake (*T. saurita*) occurs in eastern USA and Canada, with four subspecies.

Actual size

FAMILY	Natricidae
RISK FACTOR	Nonvenomous
DISTRIBUTION	North America: southwestern USA (Arizona) and New Mexico
ELEVATION	2,300–7,970 ft (700–2,430 m) asl
HABITAT	Lakes and streams in rocky, wooded habitats
DIET	Fish, salamander larvae, frogs, and tadpoles
REPRODUCTION	Viviparous, with litters of 8–17 neonates
CONSERVATION STATUS	IUCN Least Concern

ADULT LENGTH
18–34 in,
occasionally 37½ in
(460–860 mm,
occasionally 950 mm)

THAMNOPHIS RUFIPUNCTATUS
NARROW-HEADED GARTERSNAKE
(COPE, 1875)

431

This Arizona gartersnake resembles a watersnake (*Nerodia*, pages 417–420) more than a gartersnake, given its complete lack of the yellow stripes that give gartersnakes their common name. Since no watersnakes occur as far west as Arizona, confusion is unlikely. The Narrow-headed Gartersnake is one of the most aquatic of the gartersnakes and may occupy the niche left vacant by the absent watersnakes. It prefers wooded, rocky lakes and shallow, fast-flowing streams with rocky bottoms. It is encountered more often in the water than out of it, but may be found sheltering under boulders on the bank or basking in overhanging riparian vegetation. In keeping with its aquatic lifestyle, this diurnal snake feeds primarily on small fish, but the larvae of tiger salamanders, small frogs, and tadpoles are also reported in its diet.

RELATED SPECIES

The closest relative to *Thamnophis rufipunctatus* is probably its former subspecies, the Mexican Narrow-headed Gartersnake (*T. unilabialis*) from Chihuahua and Coahuila. Other stripeless Mexican gartersnakes include the Mexican West Coast Gartersnake (*T. valida*), Durango Spotted Gartersnake (*T. nigronuchalis*), Black-bellied Gartersnake (*T. melanogaster*), Tamaulipan Montane Gartersnake (*T. mendax*), and Sumichrast's Gartersnake (*T. sumichrasti*).

Actual size

The Narrow-headed Gartersnake is a slender species with an elongate, slightly pointed head. Its patterning is much more like a watersnake than a gartersnake, being dorsally brown or gray, with five to six rows of black or dark brown spots but no distinctive yellow stripes.

FAMILY	Natricidae
RISK FACTOR	Nonvenomous
DISTRIBUTION	North America: Canada, USA, and northern Mexico
ELEVATION	0–8,330 ft (0–2,540 m) asl
HABITAT	Lakes, ponds, rivers, wooded swamps, bayous, marshes, woodland, prairies, and grasslands
DIET	Amphibians, fish, invertebrates, earthworms, small mammals, and birds
REPRODUCTION	Viviparous, with litters of 7–36 neonates
CONSERVATION STATUS	IUCN Least Concern

ADULT LENGTH
17¾–26 in,
females occasionally 4 ft 3 in
(450–660 mm,
females occasionally 1.3 m)

432

THAMNOPHIS SIRTALIS
COMMON GARTERSNAKE
(LINNAEUS, 1758)

The Common Gartersnake is extremely variable in both coloration and patterning, ranging from olive-green with faint yellow-green stripes and black lateral spotting, to black with three bright yellow stripes and small red markings on the flanks. The spectacular western forms are dominated by red, from scattered spots between the yellow stripes to red stripes between white dorsal and pale blue lateral stripes.

The most widely distributed gartersnake, and the northernmost snake in the Americas, the Common Gartersnake occurs from Nova Scotia to British Columbia in Canada. It is also found throughout much of the northwestern, eastern, and central United States, with scattered Midwest and northern Mexico populations. It inhabits both aquatic and terrestrial habitats, western populations being more aquatic than those in the east. Prey is equally varied, from frogs, tadpoles, earthworms, and fish, to occasional rodents or birds. Common Gartersnakes can be extremely common, and the Manitoba "snake pits," with carpets of writhing Red-sided Gartersnakes (*Thamnophis sirtalis parietalis*, page 26), are a phenomenon of the natural world.

RELATED SPECIES
Eleven subspecies are recognized, the largest ranges being the Eastern Gartersnake (*Thamnophis sirtalis sirtalis*) in the east, Red-sided Gartersnake (*T. s. parietalis*) in the Midwest, Maritime Gartersnake (*T. s. pallidulus*) in the extreme northeast, and Valley Gartersnake (*T. s. fitchi*) in the northwest. Localized subspecies include the Blue-striped Gartersnake (*T. s. similis*) from the Florida Panhandle, Texas Gartersnake (*T. s. annectens*), Puget Sound Gartersnake (*T. s. pickeringii*), New Mexico Gartersnake (*T. s. dorsalis*), and Chicago Gartersnake (*T. s. semifasciatus*). Some western races are stunning, i.e. the Red-spotted Gartersnake (*T. s. concinnus*) and Californian Red-sided Gartersnake (*T. s. infernalis*).

Actual size

FAMILY	Natricidae
RISK FACTOR	Nonvenomous
DISTRIBUTION	South Asia: northern Pakistan, India, and Nepal
ELEVATION	3,020–8,500 ft (920–2,590 m) asl
HABITAT	Deciduous forested rocky hillsides and mountains, rice paddies, and cow dung heaps
DIET	Earthworms, and possibly soft-bodied insects or larvae
REPRODUCTION	Oviparous, with clutches of 3–6 eggs
CONSERVATION STATUS	IUCN not listed

ADULT LENGTH
17¾–26 in,
females occasionally 4 ft 3 in
(450–660 mm,
females occasionally 1.3 m)

TRACHISCHIUM FUSCUM
DARJEELING
WORM-EATING SNAKE
(BLYTH, 1854)

A Himalayan species, the Darjeeling Worm-eating Snake occurs from the Kashmir regions of Pakistan and India, through Nepal, to Darjeeling, Assam, and northeastern India. It is primarily an inhabitant of rocky hillsides and mountains with deciduous woodland cover, but is also found alongside humans, inhabiting rice paddies and even cow-dung heaps, drawn by the large numbers of earthworms in these habitats. Earthworms are the only prey known for this species, and its relatives, but the possibility remains that it may also take soft-bodied insects or their larvae. It is active during the evening and at night, sheltering under stones during the day. All the Himalayan worm-eating snakes are inoffensive and do not bite when handled.

RELATED SPECIES

There are four other species in the genus *Trachischium*: the Rosebelly Worm-eating Snake (*T. guentheri*), Olive Worm-eating Snake (*T. laeve*), Orange-bellied Worm-eating Snake (*T. monticola*), and Yellow-bellied Worm-eating Snake (*T. tenuiceps*). The alternative name of *T. fuscum* is Black-bellied Worm-eating Snake.

The Darjeeling Worm-eating Snake is a small snake with a cylindrical body and a small head, which is indistinct from the neck. Its scales are mostly smooth, though males may have keels on the scales of the posterior body. The tail terminates as a short spine. This snake is dark brown to black, sometimes with fine black longitudinal stripes. The undersides are often black, resulting in its alternative name.

Actual size

FAMILY	Natricidae
RISK FACTOR	Nonvenomous
DISTRIBUTION	North America: USA (South Dakota and Minnesota to Texas and New Mexico)
ELEVATION	0–6,610 ft (0–2,015 m) asl
HABITAT	Prairie, grassland, woodland, creeks, ponds, parks and gardens, cemeteries, and abandoned lots
DIET	Earthworms and insect larvae
REPRODUCTION	Viviparous, with litters of 2–17 neonates
CONSERVATION STATUS	IUCN Least Concern

ADULT LENGTH
8¾–15 in,
rarely 21½ in
(224–380 mm,
rarely 544 mm)

434

TROPIDOCLONION LINEATUM
LINED SNAKE
(HALLOWELL, 1856)

The diminutive Lined Snake is found in the USA from South Dakota and Minnesota, south to the Gulf of Mexico coast of Texas, and as far east as Illinois and west to New Mexico. It was originally a member of the prairie fauna and is still found in grassland, woodland, and along creeks, but it has also adapted to live alongside humans in parks, gardens, cemeteries, and abandoned lots—anywhere providing ground cover, from rotten logs to urban trash. A nocturnal species, it feeds almost exclusively on earthworms, especially after rain when they are on the surface, but it also takes soft-bodied insect larvae. Lined Snakes are inoffensive and do not bite when handled.

Actual size

The Lined Snake is small and slender, and olive-brown above with three longitudinal stripes, those on the flanks being orange or yellow, while the dorsal stripe may be white. This species is often distinguished by a double row of black half-moon markings on the white underbelly.

RELATED SPECIES

The only snakes with which *Tropidoclonion lineatum* may be confused would be juvenile gartersnakes, such as the Plains Gartersnake (*Thamnophis radix*). Some authors recognize four subspecies: Northern Lined Snake (*T. l. lineatum*), Central Lined Snake (*T. l. annectens*), Texas Lined Snake (*T. l. texanum*), and New Mexico Lined Snake (*T. l. mertensi*).

FAMILY	Natricidae
RISK FACTOR	Nonvenomous
DISTRIBUTION	Melanesia: throughout New Guinea, except the west, and Aru Islands (Indonesia)
ELEVATION	0–4,270 ft (0–1,300 m) asl
HABITAT	Rainforest rivers
DIET	Frogs and their eggs, and fish
REPRODUCTION	Oviparous, with clutches of 2–8 eggs
CONSERVATION STATUS	IUCN not listed

ADULT LENGTH
3 ft–3 ft 7 in
(0.9–1.1 m)

TROPIDONOPHIS DORIAE
BARRED KEELBACK
(BOULENGER, 1897)

435

The Barred Keelback is probably the largest of the 11 keelback species known to occur in New Guinea, and it is the only mainland species possessing 17 scale rows at midbody, with other mainland species having 15 scale rows, which means it is more heavily built than other mainland keelbacks. It inhabits rainforest rivers, and while it is not frequently encountered in the southern Trans-Fly, it is common along forested rivers in Central Province. It hunts fish, frogs, and frogs' eggs. Along the Brown River, Central Province, this species can be found living in sympatry with the Common Keelback (*Tropidonophis mairii*, page 436), Long-tailed Keelback (*T. multiscutellatus*, page 437), and Painted Keelback (*T. picturatus*), which raises an interesting question as to how four species of diurnal keelbacks partition the resources and avoid competition.

The Barred Keelback is the most ruggedly built of the New Guinea keelbacks, and also the most variable in coloration. It may be brown, pink, orange, or yellow, and either unicolor, faintly banded, or strongly banded with darker pigment, which may also vary from brown to orange. Several distinct color morphotypes may be found within a short distance on the same river.

RELATED SPECIES
The only other keelbacks in the region with 17 scale rows at midbody are two species from New Britain, to the east of New Guinea (*Tropidonophis dahlii* and *T. hypomelas*). *Tropidonophis doriae* is unlikely to be confused with any other New Guinea species.

Actual size

FAMILY	Natricidae
RISK FACTOR	Nonvenomous
DISTRIBUTION	Australasia: northern Australia (Arnhem Land to New South Wales) and southern New Guinea (Trans-Fly and Central Province)
ELEVATION	0–4,920 ft (0–1,500 m) asl
HABITAT	Shallow forest rivers, savanna creeks, swamps, and marshes
DIET	Fish, frogs, and tadpoles
REPRODUCTION	Oviparous, with clutches of 3–18 eggs
CONSERVATION STATUS	IUCN Least Concern

ADULT LENGTH
31½–36½ in
(800–930 mm)

436

TROPIDONOPHIS MAIRII
COMMON KEELBACK
(GRAY, 1841)

The Common Keelback is a relatively stout snake, with a rounded head and large eyes. It may be olive-brown to gray throughout, or gray on the head and neck, with a light brown body, flecked with black and white. The undersides are immaculate white.

The Common Keelback is the only member of the genus *Tropidonophis* in Australia, occurring from Arnhem Land, Northern Territory, to northern New South Wales. In New Guinea it occurs in the southern Trans-Fly region and the coastal strip either side of Port Moresby. It inhabits watercourses from forested rivers to savanna creeks, flooded grassland, and swamps, and feeds on fish, frogs, and tadpoles. When disturbed it will dive into the water and swim to the bottom or hide in yabbie (freshwater crayfish) holes in the bank. If handled, Australian specimens shed their tails, but it is not known if Papuan specimens also do this. This is one of the few Australian reptiles able to eat introduced cane toads, being immune to the bufotoxins (toxins found in the parotoid glands of toxic toads), because its Asiatic ancestor's diet included toads.

RELATED SPECIES

Two subspecies are recognized, the nominate form (*Tropidonophis mairii mairii*) occurring in northern Australia and the southern coast of the Papuan Peninsula, with a Trans-Fly form (*T. m. plumbea*) occurring in Western Province, Papua New Guinea, and over the border into Indonesian New Guinea. In Australia the Common Keelback may be confused with the highly venomous Rough-scaled Snake (*Tropidechis carinatus*, page 534).

Actual size

FAMILY	Natricidae
RISK FACTOR	Nonvenomous
DISTRIBUTION	New Guinea: throughout New Guinea
ELEVATION	50–4,720 ft (15–1,440 m) asl
HABITAT	Shallow forest rivers, savanna creeks, forest, and plantations
DIET	Frogs and fish
REPRODUCTION	Oviparous, with clutches of 2–7 eggs
CONSERVATION STATUS	IUCN Least Concern

ADULT LENGTH
31½–37½ in
(800–950 mm)

TROPIDONOPHIS MULTISCUTELLATUS
LONG-TAILED KEELBACK
(BRONGERSMA, 1948)

437

The Long-tailed Keelback occurs throughout the island of New Guinea, from the southern tip of the Papuan Peninsula in the southeast, to the Vogelkop Peninsula in the northwest. It is also found both north and south of the Central Cordillera, and on islands to the west of New Guinea. The Long-tailed Keelback may eventually turn out to be more than one species, a species complex, but only molecular analysis will confirm this. An alert and fast-moving snake, this species feeds on frogs, which are captured during the day by hunting along rivers and creeks, both in the water and in the riparian vegetation. It also takes fish, but may not be as aquatic as some other Australo-Papuan keelbacks, since it is also found in piles of oil-palm debris in plantations or along forest or plantation tracks.

The Long-tailed Keelback is a slender, gracile snake with an elongate head, large eyes, and a long tail. It is usually uniform light brown, dark brown, or reddish brown, but some specimens are slightly greenish on the neck and anterior body. Faint transverse bands may be just visible. The head is the same color as the body, but it may have fine black suturing or flecking on the scales. The underside is immaculate white.

RELATED SPECIES

A slender species, *Tropidonophis multiscutellatus* may be mistaken for other slender snakes such as the East Papuan Keelback (*T. aenigmaticus*) or the Montane Keelback (*T. statisticus*), or possibly the slender, venomous Müller's Crowned Snake (*Aspidomorphus muelleri*, page 490). The genus *Tropidonophis* currently contains 19 species in New Guinea, eastern Indonesia, the Philippines, and northern Australia.

Actual size

FAMILY	Natricidae
RISK FACTOR	Nonvenomous
DISTRIBUTION	North America: eastern and southern USA (New Jersey and Delaware to Iowa, Florida, and Texas)
ELEVATION	0–2,900 ft (0–885 m) asl
HABITAT	Woodland and open areas shaded by trees, and also occasionally suburban habitats
DIET	Earthworms, and occasionally slugs or insect larvae
REPRODUCTION	Viviparous, with litters of 2–18 neonates
CONSERVATION STATUS	IUCN Least Concern

ADULT LENGTH
7–13 in,
occasionally 15 in
(180–330 mm,
occasionally 380 mm)

438

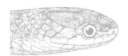

VIRGINIA VALERIAE
SMOOTH EARTHSNAKE
BAIRD & GIRARD, 1853

The Smooth Earthsnake is a smooth-scaled snake with a narrow, slightly pointed head that is only slightly wider than the neck. Coloration ranges from dirty orange to reddish brown or olive, the pigment being darkest dorsally and lighter on the lower flanks, and patterning, if present, confined to a faint vertebral stripe.

Primarily a woodland species, the Smooth Earthsnake may also be found in open meadows close to woodland. It has adapted to living in suburban habitats too, sheltering under domestic trash just as it would under natural woodland cover. It is an inoffensive species that feeds almost exclusively on earthworms, although slugs and insect larvae have been reported in its diet. The Smooth Earthsnake has a wide range across eastern USA, from New Jersey and Delaware in the northeast, to Florida in the south, Iowa in the west, and Texas in the southwest, but it is curiously absent from much of the Mississippi Valley. Earthsnakes are live-bearing, like other American natricid snakes, and in contrast to Eurasian watersnakes.

RELATED SPECIES

The genus *Virginia* used to contain *Virginia valeriae* and also the Rough Earthsnake, but the latter is now placed in its own genus as *Haldea striatula*. There are three subspecies of *V. valeriae*: the nominate Eastern Earthsnake (*V. v. valeriae*), the Mountain Earthsnake (*V. v. pulchra*) from the Appalachians, and the Western Earthsnake (*V. v. elegans*).

Actual size

FAMILY	Natricidae
RISK FACTOR	Nonvenomous
DISTRIBUTION	South and Southeast Asia: Afghanistan to Sri Lanka, China, Myanmar, Thailand, and Laos
ELEVATION	1,640–6,890 ft (500–2,100 m) asl
HABITAT	Ponds, lakes, rice paddies, marshes, and rivers in lowland and low montane habitats
DIET	Fish and frogs
REPRODUCTION	Oviparous, with clutches of 4–100 eggs
CONSERVATION STATUS	IUCN not listed

ADULT LENGTH
2 ft–3 ft 3 in,
occasionally 5 ft 9 in
(0.6–1.0 m,
occasionally 1.75 m)

XENOCHROPHIS PISCATOR
CHECKERED KEELBACK
(SCHNEIDER, 1799)

439

Also known as the Olive Keelback, the Checkered Keelback is a large snake, and relatively common in parts of its range, which covers a considerable area from Afghanistan to Sri Lanka, southern China, Laos, and Thailand. Its habitat preferences comprise watercourses, from rivers to rice paddies. Its scientific name, *piscator*, means "fisherman," an apt summation of this species' prey preferences of fish and frogs. It hunts by both day and night. Checkered Keelbacks are large snakes and they react quickly to being handled, delivering painful bites that bleed profusely, though they are not dangerous to humans. Large females may lay as many as 100 eggs.

RELATED SPECIES

The genus *Xenochrophis* contains 13 species. The closest relatives of *Xenochrophis piscator* are its former subspecies, the Yellow-spotted Keelback (*X. flavipunctatus*), from India and Southeast Asia, and St. John's Keelback (*X. sanctijohannis*), from the southern Himalayas from Pakistan to Myanmar. Other species include several island endemics, the Sri Lankan Keelback (*X. asperrimus*), the Andaman Keelback (*X. tytleri*), and the Javan Keelback (*X. melanzostus*).

The Checkered Keelback is a stockily built snake, one of the largest Asian keelbacks, with a stout body and large head, although the head is often less broad than the body in large adults. Patterning usually comprises a multi-rowed checkerboard of black squares, hence the common name, over an olive to brown background. There is often a diagonal black stripe from the eye to the angle of the jaw, and the undersides are white or pale yellow.

Actual size

FAMILY	Pseudoxenodontidae
RISK FACTOR	Rear-fanged, mildly venomous
DISTRIBUTION	Southeast Asia: southern China, eastern Myanmar, northwestern Thailand, and Vietnam; possibly Laos
ELEVATION	1,970–5,310 ft (600–1,620 m) asl
HABITAT	Submontane forest
DIET	Earthworms
REPRODUCTION	Oviparous, with clutches of 5–11 eggs
CONSERVATION STATUS	IUCN Least Concern

ADULT LENGTH
Male
22 in
(560 mm)

Female
20¾ in
(525 mm)

440

PLAGIOPHOLIS NUCHALIS
COMMON MOCK COBRA
(BOULENGER, 1893)

The Common Mock Cobra is dorsally dark brown to red-brown, with a series of dark spots and light cross-bars that break up the background color, but vary between specimens. A dark brown inverted chevron is usually present on the nape of the neck, a marking further emphasized when the hood is spread, and the white venter is boldly marked with transverse black spots, which are also more in evidence when the snake rears up and mimics a venomous cobra.

Often referred to as the Assam Mountain Snake, this species does not appear to occur in Assam, its range being southern China, and neighboring Myanmar, Thailand, Vietnam, and possibly Laos. The Common Mock Cobra is a small snake that defends itself by mimicking a hooding cobra. Although rear-fanged and mildly venomous, it is not considered dangerous to humans. It inhabits submontane forest, where it occupies a terrestrial, possibly semi-fossorial niche, feeding on earthworms. The genus *Plagiopholis* is one of only two genera in the Pseudoxenodontidae. It can be distinguished from the related *Pseudoxenodon* by its lower midbody scale count (15), entire anal plate, and smaller size, the Common Mock Cobra being the largest member of the genus.

RELATED SPECIES
There are three other species in the genus *Plagiopholis*, all of which occur in the same geographical area as, and occupy similar montane forest habitats to, *Plagiopholis nuchalis*: Blakeway's Mock Cobra (*P. blakewayi*), Delacour's Mock Cobra (*P. delacouri*), and the Fujian Mock Cobra (*P. styani*).

Actual size

FAMILY	Pseudoxenodontidae
RISK FACTOR	Rear-fanged, mildly venomous
DISTRIBUTION	Southeast Asia: northeast India, Nepal, China, Myanmar, Thailand, Vietnam, Laos, and West Malaysia
ELEVATION	1,640–10,800 ft (500–3,300 m) asl
HABITAT	Evergreen forest, and montane and submontane forest
DIET	Frogs and lizards
REPRODUCTION	Oviparous, with clutches of 6–10 eggs
CONSERVATION STATUS	IUCN Least Concern

ADULT LENGTH
3 ft 3 in–4 ft 7 in
(1.0–1.4 m)

PSEUDOXENODON MACROPS
LARGE-EYED MOCK COBRA
(BLYTH, 1855)

441

Also known as the Large-eyed Mountain or Bamboo Snake, this mock cobra occurs from northeast India to China, inhabiting evergreen forest and montane forest at elevations up to 10,800 ft (3,300 m), the highest elevation recorded within the Pseudoxenodontidae. Whereas the smaller *Plagiopholis* species feed on earthworms, the generally larger *Pseudoxenodon*, which often exceed 3 ft 3 in (1 m), prey on frogs and lizards. *Pseudoxenodon* can also be distinguished from *Plagiopholis* by its higher midbody scale count (17–19) and divided anal plate. Being large and rear-fanged these snakes are capable of delivering a snakebite, and although their venom is not known to be harmful to humans, large specimens should be treated with respect.

The Large-eyed Mock Cobra may be brown, red, or yellowish, with a patterning of darker cross-bars and spots, and a dark inverted chevron on the nape of the neck. The venter is white with broken black cross-bars and spots. The markings further enhance the mock cobra display when this species is defending itself. As its name suggests, the eyes are large and distinctive.

RELATED SPECIES

Most of the range is occupied by the nominate subspecies (*Pseudoxenodon macrops macrops*), but populations from southern and northern China are treated as separate subspecies (*P. m. fukienensis* and *P. m. sinensis*, respectively). Six congeners are recognized: the Bamboo Mock Cobra (*P. bambusicola*) and Karl Schmidt's Mock Cobra (*P. karlschmidti*) from mainland Southeast Asia; the Sumatran Mock Cobra (*P. jacobsonii*); the Javan Mock Cobra (*P. inornatus*); the Baram Mock Cobra (*P. baramensis*) from Sarawak, Borneo; and the Taiwan Mock Cobra (*P. stejnegeri*).

Actual size

FAMILY	Elapidae: Elapinae
RISK FACTOR	Highly venomous: presynaptic neurotoxins, but otherwise unknown
DISTRIBUTION	Southern Africa: southern Angola, Namibia, and southern and eastern South Africa (Namaqualand and Cape)
ELEVATION	0–4,640 ft (0–1,415 m) asl
HABITAT	Semiarid scrubland and desert edges
DIET	Small mammals, lizards, reptile eggs
REPRODUCTION	Oviparous, with clutches of 3–11 eggs
CONSERVATION STATUS	IUCN not listed

ADULT LENGTH
23¾–31½ in
(600–800 mm)

ASPIDELAPS LUBRICUS
AFRICAN CORALSNAKE
(LAURENTI, 1768)

442

The African Coralsnake may be banded black and salmon-pink, and pale yellow to white below with black bands across the throat, but its patterning varies considerably across its range (see above). When it feels threatened it may adopt a defensive display, elevating its anterior body like a true cobra, although the hood is much more narrowed.

The African Coralsnake, sometimes called the Coral Cobra, can be a very attractive serpent, the nominate subspecies, from South Africa and southern Namibia, being strikingly banded. However, this is a highly variable species and the western form is weakly banded with a dull yellow background, while the Angolan form is uniform gray-brown with a black head. This small elapid will raise its anterior body and hood like a cobra; its venom is less toxic than that of the true cobras (*Naja*), although bites to children can be serious and fatalities are known. Inhabiting semiarid scrub habitats, it emerges from cover on cool nights to feed on rodents and reptiles.

RELATED SPECIES
Three subspecies are recognized: the Cape Coralsnake (*Aspidelaps lubricus lubricus*), the Western Coralsnake (*A. l. infuscatus*), and the Angolan Coralsnake (*A. l. cowlesi*). The only close relative of *A. lubricus* is the Shieldnose Snake (*A. scutatus*, page 443), also of southern Africa.

Actual size

FAMILY	Elapidae: Elapinae
RISK FACTOR	Highly venomous: presynaptic neurotoxins, but otherwise unknown
DISTRIBUTION	Southern Africa: Namibia, Botswana, Zimbabwe, Zambia, Mozambique, and northeastern South Africa
ELEVATION	0–5,480 ft (0–1,670 m) asl
HABITAT	Savanna, sandveld, and shallow pans
DIET	Frogs, lizards, small mammals, reptile eggs, and snakes
REPRODUCTION	Oviparous, with clutches of 4–11 eggs
CONSERVATION STATUS	IUCN not listed

ADULT LENGTH
23¾–29½ in
(600–750 mm)

ASPIDELAPS SCUTATUS
SHIELDNOSE SNAKE
SMITH, 1849

443

Shieldnose Snakes are stout, but relatively short, snakes of sandveld and savanna habitats across the north of southern Africa. They hide in burrows or under fallen logs during the day and venture out at night in search of amphibians around waterholes, but they will also take small mammals, lizards, and occasionally other snakes. The Shieldnose Snake is an accomplished burrower, using its enlarged rostral scale to excavate the loose soil or sand under logs or rocks in search of prey. When disturbed, a Shieldnose Snake will mimic a cobra, raising its body and hissing loudly. Little is known of its venom; some bites exhibit only mild symptoms, but there are also records of children having died from Shieldnose Snake bites.

The Shieldnose Snake is stout-bodied and brown, orange, or pinkish, with black mottling to every scale and black bands around the throat and neck. The short, rounded head may be black or brown, and terminates in a large shield-shaped rostral scale that earns the snake its common name. The scales of the tail may be strongly keeled.

RELATED SPECIES

Aspidelaps scutatus is related to the African Coralsnake (*A. lubricus*, page 442). Three subspecies are recognized. The nominate form (*A. s. scutatus*) occurs across most of the range, while the Eastern Shieldnose (*A. s. fulafula*) occurs in southern Mozambique. The Intermediate Shieldnose (*A. s. intermedius*) is found between these two ranges, in northeastern South Africa.

Actual size

FAMILY	Elapidae: Elapinae
RISK FACTOR	Highly venomous: pre- and postsynaptic neurotoxins
DISTRIBUTION	South Asia: Pakistan, India, Nepal, Bangladesh, and Sri Lanka
ELEVATION	0–5,580 ft (0–1,700 m) asl
HABITAT	Low-lying and low montane open habitats and woodland, especially rice paddies and villages
DIET	Snakes, including other kraits
REPRODUCTION	Oviparous, with clutches of 8–12 eggs
CONSERVATION STATUS	IUCN Least Concern

ADULT LENGTH
3 ft 3 in–5 ft 9 in
(1.0–1.75 m)

BUNGARUS CAERULEUS
COMMON KRAIT
(SCHNEIDER, 1801)

444

The Common Krait is either black with narrow or broad white broken bands around the body, or uniform gray to black, with a pale underside. The absence of a loreal scale and the presence of an enlarged vertebral scale row distinguish it from most nonvenomous mimics.

Kraits are highly venomous snakes. Of the 14 known species, half are implicated in the very high annual death rate across Asia. Although shy and retiring during the day, often hiding their heads under their coils, they become lethal and dangerous at night, a true serpentine "Jekyll and Hyde." The Common Krait is found in many habitats but is especially at home in habitats altered by humans, such as rice paddies, where it hunts other snakes. It is one of the "Big Four" lethal snakes of South Asia, entering houses in search of prey, and biting sleeping villagers at night, many of its victims never waking up again. Because kraits rarely climb, many krait bites could be avoided by simply sleeping off the ground or using a mosquito net.

RELATED SPECIES

Bungarus caeruleus appears most similar to the Andaman Krait (*B. andamanensis*), the Sri Lankan Krait (*B. ceylonicus*), the Black Krait (*B. niger*), and the Sind Krait (*B. sindanus*), but many harmless snakes mimic its patterning for protection, such as the Travanacore Wolfsnake (*Lycodon travancoricus*) and Zaw's Wolfsnake (*L. zawi*).

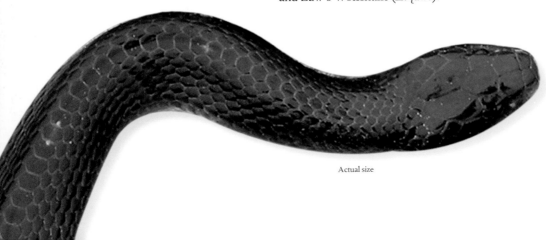

Actual size

FAMILY	Elapidae: Elapinae
RISK FACTOR	Highly venomous: pre- and postsynaptic neurotoxins
DISTRIBUTION	South and Southeast Asia: India, Nepal, and China to Malaysia, Sumatra, Java, and Borneo
ELEVATION	0–8,200 ft (0–2,500 m) asl
HABITAT	Coastal lowland and low montane forest, swamps, and cultivated habitats
DIET	Snakes, lizards, small mammals, frogs, and fish, including eels
REPRODUCTION	Oviparous, with clutches of 3–12 eggs
CONSERVATION STATUS	IUCN Least Concern

ADULT LENGTH
5–7 ft 5 in
(1.5–2.25 m)

BUNGARUS FASCIATUS

BANDED KRAIT

(SCHNEIDER, 1801)

445

Many kraits have enlarged scales along the vertebral row, presenting a ridge down the back, but this appearance reaches its most extreme in the Banded Krait, which is almost triangular in cross section. Although a large and dangerous snake, the Banded Krait features less frequently in snakebites than its relatives the Common Krait (*Bungarus caeruleus*, page 444) or the Malayan Krait (*B. candidus*). In common with other kraits, it primarily preys on elongate vertebrates such as snakes or eels, but frogs, lizards, other fish, and small mammals are also taken. It spends the day in mammal burrows or termite mounds, emerging to hunt after dark. Despite its less truculent attitude, compared to its congenerics, bites are serious and there are fatalities on record.

The Banded Krait is broadly banded with yellow, or cream, and black and is triangular in cross section with an enlarged, raised vertebral scale row. The head is broad and flattened. Its tail terminates in a blunt, rounded tip and the subcaudal scales are undivided.

RELATED SPECIES

The patterning of *Bungarus fasciatus* is so characteristic it is unlikely to be confused with any other species, and even though the mildly venomous Mangrove Snake (*Boiga dendrophila*, page 145) is also banded black and yellow, it is obviously cylindrical, rather than triangular, in cross section, and has a long prehensile tail with divided subcaudal scales.

Actual size

FAMILY	Elapidae: Elapinae
RISK FACTOR	Highly venomous: pre- and postsynaptic neurotoxins
DISTRIBUTION	Southeast Asia: Myanmar, Thailand, Cambodia, Vietnam, Malaysia, Indonesia, Sumatra, Java, and Borneo
ELEVATION	1,800–5,090 ft (550–1,550 m) asl
HABITAT	Lowland and low montane rainforest
DIET	Snakes and lizards
REPRODUCTION	Oviparous, clutch size unknown
CONSERVATION STATUS	IUCN Least Concern

ADULT LENGTH
3 ft 3 in–6 ft 7 in
(1.0–2.0 m)

446

BUNGARUS FLAVICEPS
RED-HEADED KRAIT
REINHARDT, 1843

The Red-headed Krait is triangular in cross section and brightly colored. The head is red or yellow, the anterior half of the body is blue-black with three longitudinal white stripes either present or absent, and the posterior body and short tail are bright coral red and ringed with black and white in Mt. Kinabalu specimens.

The Red-headed Krait is also known as the Yellow-headed Krait, *flaviceps* meaning "yellow-headed." Populations in mainland Southeast Asia are usually red-headed, while those from Borneo are often yellow-headed, so either name is applicable. Although highly dangerous it features in snakebite statistics much less frequently than other species of krait, and it is relatively understudied in nature. It is a forest species that inhabits lowland and low montane rainforest, especially close to water, where it shelters in leaf litter or under logs during the day. At night it emerges to hunt and it is believed to prey on other snakes, and elongate lizards such as skinks. Although it is believed to be oviparous, like other kraits, its clutch size is unknown.

RELATED SPECIES

Bungarus flaviceps is unlike any other krait species, but its patterning is remarkably similar to that of the Blue Long-glanded Coralsnake (*Calliophis bivirgata*, page 448), with which it occurs in sympatry. Two subspecies are recognized, the nominate form, which has an unmarked red tail, and the Mt. Kinabalu Krait (*B. f. baluensis*) from Sabah, Borneo, which exhibits a series of black and white rings around the red tail.

Actual size

FAMILY	Elapidae: Elapinae
RISK FACTOR	Highly venomous: pre- and postsynaptic neurotoxins
DISTRIBUTION	Southeast Asia: southern China, Taiwan, Myanmar, Vietnam, and Laos
ELEVATION	0–4,920 ft (0–1,500 m) asl
HABITAT	Low-lying wetland areas, including rice paddies
DIET	Snakes, lizards, small mammals, frogs, and eels
REPRODUCTION	Oviparous, with clutches of 3–12 eggs
CONSERVATION STATUS	IUCN Least Concern

ADULT LENGTH
3 ft 3 in–4 ft 5 in
(1.0–1.35 m)

BUNGARUS MULTICINCTUS
MANY-BANDED KRAIT
BLYTH, 1861

447

A highly dangerous snake, it was a Many-banded Krait that claimed the life of the respected American herpetologist Joe Slowinski in 2001, when he mistook a juvenile krait for a harmless wolfsnake (*Lycodon*, pages 190–191), one of several mimics of this lethal species. The Many-banded Krait is found in lowland habitats over a wide range, from southern and southeastern China to Vietnam, Laos, and Myanmar, and it is also present on the island of Taiwan. Within its range this species, which feeds on snakes and other vertebrates, is a serious public health risk because it is very commonly encountered in agricultural habitats, such as rice paddies, or around buildings.

The Many-banded Krait is blue-black to brown with a series of regularly spaced, broad pale bands around the body and tail, a common pattern mimicked by many harmless wolfsnakes (*Lycodon*).

RELATED SPECIES

Many snakes resemble *Bungarus multicinctus*, including several harmless mimics and a new species of krait, the Red River Krait from Vietnam, which was named *B. slowinskii*, in honor of the fallen herpetologist. Wanghaoting's Krait (*B. wanghaotingi*) was formerly a subspecies of *B. multicinctus*.

Actual size

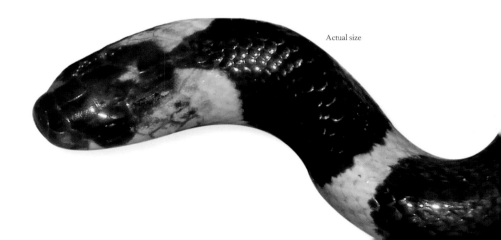

FAMILY	Elapidae: Elapinae
RISK FACTOR	Highly venomous: composition unknown
DISTRIBUTION	Southeast Asia: southern Thailand, Malaysia, Sumatra, Borneo, and Java
ELEVATION	0–4,510 ft (0–1,375 m) asl
HABITAT	Lowland and low montane rainforest, also edges of cultivated land
DIET	Small snakes
REPRODUCTION	Oviparous, with clutches of 1–3 eggs
CONSERVATION STATUS	IUCN Least Concern

ADULT LENGTH
3 ft 3 in–6 ft
(1.0–1.85 m)

448

CALLIOPHIS BIVIRGATA
BLUE LONG-GLANDED CORALSNAKE
(BOIE, 1827)

The Blue Long-glanded Coralsnake is an elongate snake that is blue or blue-black above with a series of longitudinal stripes, which may be white or pale blue. The head, entire tail, and underbelly are bright coral red.

Blue Long-glanded Coralsnakes possess extremely long venom glands that extend for one-third of the body length, and given that a bite may deliver a large quantity of venom of unknown composition, they must be considered highly dangerous. A Southeast Asian lowland and low montane rainforest species, it may also be encountered on cultivated land. Nocturnal and semi-fossorial, it hunts other snakes. If disturbed it will bury its head under its coils, and raise its bright red tail in the air as a decoy, inviting attack away from the head. If handled it will bite. At least two deaths are thought to have been caused by this species, with death occurring in under two hours.

RELATED SPECIES

Calliophis bivirgata comprises three subspecies depending on the presence or absence of stripes, from mainland Southeast Asia and Sumatra (*C. b. flaviceps*), Borneo (*C. b. tetrataenia*), and Java (*C. b. bivirgata*). This coralsnake may easily be confused with the sympatric Red-headed Krait (*Bungarus flaviceps*, page 446) but its closest relative is the Striped Long-glanded Coralsnake (*C. intestinalis*, page 449). These two species were previously placed in genus *Maticora*. *Calliophis bivirgatus* is also closely related to the recently described Dinagat Island Banded Coralsnake (*C. salitan*) from the Philippines.

Actual size

FAMILY	Elapidae: Elapinae
RISK FACTOR	Venomous: composition unknown
DISTRIBUTION	Southeast Asia: southern Thailand, Malaysia, Sumatra, Borneo, and the Philippines
ELEVATION	0–5,000 ft (0–1,525 m) asl
HABITAT	Lowland and low montane rainforest, also parks and gardens
DIET	Small snakes
REPRODUCTION	Oviparous, with clutches of 1–3 eggs
CONSERVATION STATUS	IUCN Least Concern

ADULT LENGTH
23¾–28 in
(600–710 mm)

CALLIOPHIS INTESTINALIS
STRIPED LONG-GLANDED CORALSNAKE
(LAURENTI, 1768)

449

The Striped Long-glanded Coralsnake is a small, secretive snake that hides in leaf litter or under logs, and burrows in the subsoil. Although a lowland and low montane rainforest species, it also evades detection in city parks and gardens. Its prey consists of the many small snake species that occur in similar habitats across its wide Southeast Asian range. The Striped Long-glanded Coralsnake is an inoffensive species. If uncovered it will raise its tail to expose the red underside, to distract attention from the head, or roll its entire body upside down to expose the checkerboard patterning underneath. Despite its small mouth, this species has extremely long venom glands that extend one-third of the length of the body, and it demands respect. Serious snakebites are known.

The Striped Long-glanded Coralsnake is an extremely elongate snake with a rounded head, which is indistinct from the neck, and a short tail. The dorsum is brown, with or without a longitudinal orange or scarlet vertebral stripe, while the underside is black and white, except for the tail, which is red underneath.

RELATED SPECIES
Six subspecies are recognized, from Sumatra and Java (*Calliophis intestinalis intestinalis*), mainland Southeast Asia and Sumatra (*C. i. lineata*), Borneo (*C. i. thepassi*), Palawan (*C. i. bilineata*), the Philippines (*C. i. philippina*), and the Sulu Archipelago (*C. i. suluensis*).

Actual size

FAMILY	Elapidae: Elapinae
RISK FACTOR	Highly venomous: presynaptic neurotoxins
DISTRIBUTION	East and southeast Africa: Kenya, Tanzania, Zimbabwe, and Mozambique
ELEVATION	0–5,580 ft (0–1,700 m) asl
HABITAT	Coastal bush, woodlands, thickets, and hill forests
DIET	Birds, lizards, bats, and other small mammals
REPRODUCTION	Oviparous, with clutches of up to 10 eggs
CONSERVATION STATUS	IUCN not listed

ADULT LENGTH
4 ft 3 in–6 ft 7 in
(1.3–2.0 m)

450

DENDROASPIS INTERMEDIUS
EASTERN GREEN MAMBA
(GÜNTHER, 1865)

The Eastern Green Mamba is bright green above, sometimes with every scale of the head edged with yellow, and yellow-green on the belly. The eye may have an orange or green iris.

The Eastern Green Mamba inhabits coastal scrub and forests, but has a patchy distribution from Kenya to Mozambique. Until recently it was included in the species *Dendroaspis angusticeps*, which is now confined to KwaZulu-Natal, South Africa. A highly alert, diurnal snake, it relies on camouflage to avoid detection, and prefers flight to fight, but if cornered will defend itself. Its venom contains highly toxic presynaptic neurotoxins (see pages 34–35), and fatalities have occurred, albeit rarely. Unusually for an elapid, Eastern Green Mamba bites cause rapid swelling of the bitten limb. Eastern Green Mambas prey on birds and their nestlings, and also small mammals, ranging from bats to rats. Chameleons feature in the diet of juveniles. In pristine habitat this species may be very common, with up to five sometimes being found in the same tree.

RELATED SPECIES

There are three other green mambas: the Southern Green Mamba (*Dendroaspis angusticeps*) from KwaZulu-Natal, South Africa; West African Green Mamba (*D. viridis*) from Senegal to Benin; and Jameson's Mamba (*D. jamesoni*) from Guinea to Sudan and Kenya, and south to Angola and Zambia. Green mambas are often confused with the common and totally harmless bushsnakes (*Philothamnus*, pages 211–212), which are often killed as a result.

Actual size

FAMILY	Elapidae: Elapinae
RISK FACTOR	Highly venomous: powerful presynaptic neurotoxins
DISTRIBUTION	East and southern Africa: Eritrea and Ethiopia, to Angola, Namibia, and South Africa, with isolated West African records from Senegal, The Gambia, and Burkina Faso
ELEVATION	6,000 ft (1,830 m) asl
HABITAT	Coastal bush, savanna woodland, and riverine forest
DIET	Small mammals, and occasionally birds or other snakes
REPRODUCTION	Oviparous, with clutches of 6–17 eggs
CONSERVATION STATUS	IUCN Least Concern

ADULT LENGTH
9 ft–11 ft 6 in
(2.7–3.5 m)

DENDROASPIS POLYLEPIS

BLACK MAMBA

GÜNTHER, 1864

451

This is probably the most feared snake in Africa, yet from its vantage point in the trees the alert Black Mamba usually avoids confrontations with humans. However, if it feels threatened it will advance and deliver several bites in quick succession, the highly toxic presynaptic neurotoxic venom causing rapid respiratory arrest and death unless antivenom is administered. The venom is designed to kill rats, squirrels, hyraxes, and elephant shrews. Black Mambas inhabit open wooded habitats across sub-Saharan Africa, but they do not enter rainforest. Popular habitats include overgrown rocky outcrops with an abundance of prey. Mambas are diurnal, but crepuscular in hot weather, and although primarily arboreal are equally at home on the ground. They move quickly, often in a straight line with the anterior body raised, but stories of them overtaking horses are exaggerations.

The Black Mamba is not generally black, it is more often gunmetal gray or brown. The inside of its mouth is black and this can be seen during the threat display, which involves neck-flattening and gaping widely before the strike. This is a slender-bodied snake with a long tail and a long, somewhat aptly coffin-shaped head.

RELATED SPECIES

There are four other mambas, but none resemble *Dendroaspis polylepis*, which is more likely to be mistaken for the Yellow-throated Black Treesnake (*Thrasops flavigularis*) or a black Boomslang (*Dispholidus typus*, page 160). The northeast African *D. polylepis* is sometimes treated as a separate subspecies, *D. p. antinorii*.

Actual size

FAMILY	Elapidae: Elapinae
RISK FACTOR	Venomous: composition unknown
DISTRIBUTION	East and southeast Africa: Tanzania, Burundi, eastern DRC, Botswana, and South Africa
ELEVATION	0–4,920 ft (0–1,500 m) asl
HABITAT	Wet savannas
DIET	Snakes, lizards, frogs, and small mammals
REPRODUCTION	Oviparous, with clutches of 4–8 eggs
CONSERVATION STATUS	IUCN not listed

ADULT LENGTH
19¾–27½ in
(500–700 mm)

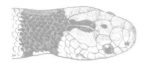

452

ELAPSOIDEA BOULENGERI
ZAMBESI GARTERSNAKE
BOETTGER, 1895

The Zambesi Gartersnake is a stocky, glossy-scaled snake with a short tail. Patterning is variable, from uniform black above and white below, to a ringed pattern comprising pairs of fine pale bands. Banding is more distinctive in juveniles, with broad off-white bands alternating with the black bands.

The Zambesi Gartersnake is a slow-moving, wet savanna inhabitant that preys on small snakes, including its own species, and smooth-scaled lizards such as skinks, but it will also take frogs, geckos, and small mammals. Inactive in animal burrows or under logs during the day, it generally emerges at night, especially after rain. This species occurs from southern Tanzania, to Limpopo and KwaZulu-Natal, South Africa, and it has been found at high elevations (4,920 ft / 1,500 m). The usual defense adopted by the Zambesi Gartersnake is to inflate and compress its body, hiss loudly, and jerk suddenly, especially if touched. Generally unwilling to bite, it will if molested, and although its venom composition is a mystery, snakebites are on record as only causing localized pain and swelling, and, curiously, nasal congestion.

RELATED SPECIES

This species closely resembles the Angolan Gartersnake (*Elapsoidea semiannulata*), which occurs farther west, but it may also be confused with Sundevall's Gartersnake (*E. sundevalli*) of southern Africa.

Actual size

FAMILY	Elapidae: Elapinae
RISK FACTOR	Venomous, composition unknown
DISTRIBUTION	East Africa: northeast Tanzania (Usambara and Uluguru mountains)
ELEVATION	985–6,230 ft (300–1,900 m) asl
HABITAT	Montane evergreen forest
DIET	Caecilians
REPRODUCTION	Oviparous, with clutches of 2–5 eggs
CONSERVATION STATUS	IUCN Endangered

ADULT LENGTH
15¾–23¾ in
(400–600 mm)

ELAPSOIDEA NIGRA

USAMBARA GARTERSNAKE

GÜNTHER, 1888

453

Also known as the Black Gartersnake, the Usambara Gartersnake is endemic to the Usambara and Uluguru mountains of northeastern Tanzania. It may also inhabit the Udzungwa Mountains of Tanzania, or the Taita Hills, Kenya. A secretive inhabitant of leaf litter and rotten logs in montane evergreen forests, it burrows in soft soil or under logs, assisted by a spike on its tail that serves as an anchorage for pushing the body forward. It hunts primarily caecilians (legless elongate amphibians), but whether it takes other cylindrical prey like amphisbaenians (worm-lizards), limbless skinks, or small snakes is unknown. Usambara Gartersnakes are active at night, especially after rain, but may be diurnal on overcast days. Generally inoffensive, this species will bite if handled, but nothing is known of its venom composition.

The Usambara Gartersnake is a small snake with a short head and tail and smooth, glossy scales. Adults may be uniform gray or exhibit a series of alternating white-edged black bands and light gray interspaces, but juveniles are more vividly patterned with a pale orange head and the anterior gray interspaces replaced by pale orange.

RELATED SPECIES

Elapsoidea nigra is most likely to be confused with other banded gartersnakes, such as the Zambesi Gartersnake (*E. boulengeri*, page 452), Central African Gartersnake (*E. laticincta*), or East African Gartersnake (*E. loveridgei*), although only the last of these occurs close to the range of the Usambara Gartersnake.

Actual size

FAMILY	Elapidae: Elapinae
RISK FACTOR	Highly venomous: possibly neurotoxins, hemorrhagins, or cytotoxins
DISTRIBUTION	Southern Africa: South Africa and Zimbabwe
ELEVATION	0–8,200 ft (0–2,500 m) asl
HABITAT	Lowveld and highveld grasslands
DIET	Toads and small mammals
REPRODUCTION	Viviparous, with litters of 20–30 neonates
CONSERVATION STATUS	IUCN Least Concern

ADULT LENGTH
2 ft 7 in–3 ft 9 in
(0.8–1.15 m)

HEMACHATUS HAEMACHATUS
RINKHALS
(GÜNTHER, 1865)

454

The Rinkhals is usually banded black and pale yellow or gray, although specimens from higher elevations are often uniform gray, brown, or black. Rinkhals have several broad black cross-bands on the throat, visible when they rise up to hood, and it is these markings that earn them the Afrikaans name of "rinkhals," meaning "ring neck."

The Rinkhals is a spitting cobra that differs from the true cobras (*Naja*, pages 466–479) in possessing keeled scales and giving birth to live young. It inhabits the grasslands of South Africa, to elevations of 8,200 ft (2,500 m), with an isolated population on the Zimbabwean border with Mozambique. Rinkhals are nocturnal and terrestrial, although they may be active on overcast or wet days, venturing out to hunt toads, which comprise the majority of their diet. If confronted, a Rinkhals will spread its hood and spit venom into the eyes of the enemy. The venom causes pain and temporary blindness, during which time the cobra will attempt to flee. Its final defense is "thanatosis," rolling onto its back and playing dead, like the hognose snakes (*Heterodon*, page 264) or Western Grass Snake (*Natrix helvetica*, page 415).

RELATED SPECIES

Hemachatus haemachatus could be confused with its closest relatives, the true cobras (*Naja*, pages 466–479), but for its strongly keeled scales.

Actual size

FAMILY	Elapidae: Elapinae
RISK FACTOR	Venomous: composition unknown
DISTRIBUTION	Southeast Asia: Philippines
ELEVATION	0–2,620 ft (0–800 m) asl
HABITAT	Lowland and low to mid-montane rainforest
DIET	Small snakes
REPRODUCTION	Presumed oviparous, clutch size unknown
CONSERVATION STATUS	IUCN Least Concern

ADULT LENGTH
15¾–21¾ in
(400–550 mm)

HEMIBUNGARUS CALLIGASTER
PHILIPPINE BARRED CORALSNAKE
(WIEGMANN, 1834)

455

The Philippine Barred Coralsnake is an inhabitant of low- to medium-elevation rainforest in the northern and central Philippines, but is most common at lower elevations. It shelters in leaf litter or under logs, where it also hunts other snakes, although whether it takes other prey is unknown. Neither is it known how many eggs it lays, although it is assumed to be oviparous, in common with other terrestrial Asian elapids. If uncovered, the Philippine Barred Coralsnake will expose the red on the underside of its tail, drawing attention away from its head. A reported snakebite caused intense pain, swelling, blistering, fever, and vomiting, but the victim eventually recovered without treatment.

RELATED SPECIES
Three subspecies are recognized, from Luzon (*Hemibungarus calligaster calligaster*), Polillo (*H. c. mcclungi*), and Negros, Masbate, Cebu, and Panay (*H. c. gemianulis*). *Hemibungarus* may be more closely related to African gartersnakes (*Elapsoidea*, pages 452–453) than other Asian coralsnakes.

The Philippine Barred Coralsnake is brown above, with broad black bands, edged with white, while the underside is bright red with black bands that are contiguous with the dorsal bands, each with a white center. The head is black above with a pale snout and a pale or red band on the nape.

Actual size

FAMILY	Elapidae: Elapinae
RISK FACTOR	Highly venomous: postsynaptic neurotoxins, and probably myotoxins
DISTRIBUTION	North America: southwest USA (Arizona, New Mexico) and northwest Mexico (Sonora, Sinaloa)
ELEVATION	0–5,910 ft (0–1,800 m) asl
HABITAT	Desert, thorn scrub, tropical dry forest, and rocky areas with sandy soil, such as arroyos
DIET	Small snakes
REPRODUCTION	Oviparous, with clutches of 2–6 eggs
CONSERVATION STATUS	IUCN Least Concern

ADULT LENGTH
11¾–15¼ in,
rarely 22 in
(300–400 mm,
rarely 560 mm)

MICRUROIDES EURYXANTHUS
SONORAN CORALSNAKE
(KENNICOTT, 1860)

One of the smallest American coralsnakes, the Sonoran Coralsnake is the only coralsnake in southwestern USA and northwestern Mexico, although there are a number of nonvenomous snakes in the genera *Chilomeniscus* (page 148), *Chionactis* (page 149), *Lampropeltis* (pages 182–187), and *Rhinocheilus* (page 225) that may mimic its patterning for protection. Unlike most other coralsnakes, this is a desert species that is most common in sandy areas around rocky outcrops, such as arroyos. It feeds on a variety of small snakes. It is generally inoffensive, being a small species with a small mouth. Snakebites are said to have produced only local pain and mild effects. When molested, the Sonoran Coralsnake will invert its tail and produce a curious sound, known as "cloacal popping." How this is accomplished is not known, but it may involve forced expulsion of air from the gut.

The Sonoran Coralsnake is a small, banded snake with a black snout and a white band across the back of the head. The bands follow the pattern red, white, black, white, red, which means the famous rhyme "red to yellow [or white], kill a fellow, red to black, venom lack" does work for this species.

RELATED SPECIES

Micruroides euryxanthus is placed in its own genus because it exhibits differences in scalation that separate it from all other American coralsnakes (*Micrurus*). Three subspecies are recognized, from Arizona to Sonora and Isla Tiburon (*Micruroides e. euryxanthus*), southern Sonora (*M. e. australis*), and Sinaloa (*M. e. neglectus*).

Actual size

FAMILY	Elapidae: Elapinae
RISK FACTOR	Highly venomous: postsynaptic neurotoxins, and probably myotoxins
DISTRIBUTION	South America: southern and eastern Brazil, Uruguay, Paraguay, and northern Argentina
ELEVATION	0–1,970 ft (0–600 m) asl
HABITAT	Tropical and subtropical deciduous forest, and evergreen forest
DIET	Amphisbaenians, caecilians, lizards, and other snakes
REPRODUCTION	Oviparous, with clutches of 4–10 eggs
CONSERVATION STATUS	IUCN not listed

ADULT LENGTH
23¾–33½ in,
rarely 38½ in
(600–850 mm,
rarely 980 mm)

MICRURUS CORALLINUS
PAINTED CORALSNAKE
(MERREM, 1820)

457

The Painted Coralsnake occurs down South America's Atlantic coast from Rio Grande do Norte, Brazil, to Uruguay, and northern Argentina, and on a number of Brazilian islands, including São Sebastião, Alcatrazes, and Vitória, in a variety of lowland and low montane forest habitats. Like other coralsnakes it feeds on elongate vertebrates, and seems to prefer amphisbaenians (worm-lizards) and caecilians (legless amphibians), but the Painted Coralsnake also takes leaf-litter-dwelling lizards and other snakes, and cannibalism has been documented. They appear very good at locating and excavating their fossorial prey. Painted Coralsnakes are said to react nervously to human contact, and given that they are large enough to deliver a snakebite they must be considered extremely dangerous.

The Painted Coralsnake exhibits a monadal (single black rings) pattern of black rings, each of which is sandwiched between a pair of white rings, between broad red interspaces. The head is black-snouted, with a white band over the nape that extends forward onto the lips. The tail is banded black and white without any red.

RELATED SPECIES

Three coralsnakes occur in sympatry with *Micrurus corallinus*, the South American Coralsnake (*M. lemniscatus*, page 460), Decorated Coralsnake (*M. decoratus*), and Cerrado Coralsnake (*M. frontalis*), but they are unlikely to be confused with *M. corallinus* as it is a monadal (single black ring) species, while the other three species exhibit triad patterns of three black rings, separated from each other by white rings, between each pair of red interspaces.

Actual size

FAMILY	Elapidae: Elapinae
RISK FACTOR	Highly venomous: postsynaptic neurotoxins, and probably myotoxins
DISTRIBUTION	North America: southeastern USA (Florida to Carolinas, Louisiana, and eastern Texas)
ELEVATION	0–1,310 ft (0–400 m) asl
HABITAT	Scrub and live oak hammocks, flatwoods, and pinelands
DIET	Small snakes, lizards, and amphisbaenians
REPRODUCTION	Oviparous, with clutches of 1–13 eggs
CONSERVATION STATUS	IUCN Least Concern

ADULT LENGTH
23¾–31½ in,
rarely 4 ft
(600–800 mm,
rarely 1.2 m)

MICRURUS FULVIUS
EASTERN CORALSNAKE
(LINNAEUS, 1766)

458

The Eastern Coralsnake is a relatively small, slender snake with a black snout and a broad yellow band over the back of the head. The body bands follow the order black, yellow, red, yellow, black, yellow, red, with the "red to yellow, kill a fellow" combination indicating a dangerous species.

The only coralsnake endemic to the USA, the Eastern Coralsnake occurs throughout the Florida Peninsula, north to the Carolinas, and west as far as Louisiana, and possibly extreme eastern Texas. It inhabits dry hammocks in the Everglades and dry woodland habitats throughout the southeast. A secretive, nocturnal species, it may be encountered abroad after rain. Prey comprises elongate vertebrates, such as small snakes, including its own species, the Floridian amphisbaenian, and small lizards. Although most Eastern Coralsnakes are small, they are capable of delivering a serious and fast-acting snakebite and there are deaths on record. Before the advent of antivenom there was a 20 percent death rate following Eastern Coralsnake bites. This species is reported to have caused the first Confederate casualty of the American Civil War.

RELATED SPECIES

A close relative is the Texas Coralsnake (*Micrurus tener*), which was formerly a subspecies of *M. fulvius*. The only snakes within its range with which *M. fulvius* may be confused are the harmless Scarlet Kingsnake (*Lampropeltis elapsoides*) and Scarletsnake (*Cemophora coccinea*, page 147).

Actual size

FAMILY	Elapidae: Elapinae
RISK FACTOR	Highly venomous: postsynaptic neurotoxins, and probably myotoxins
DISTRIBUTION	South America: Amazonia, in the Guianas, Brazil, Venezuela, Colombia, Ecuador, Peru, and Bolivia
ELEVATION	0–3,940 ft (0–1,200 m) asl
HABITAT	Primary lowland and low montane rainforest and gallery forest
DIET	Onychophorans, also amphisbaenians and small snakes
REPRODUCTION	Oviparous, with clutches of 1–8 eggs
CONSERVATION STATUS	IUCN not listed

ADULT LENGTH
19¾–23¾ in,
rarely 35½ in
(500–600 mm,
rarely 900 mm)

MICRURUS HEMPRICHII
HEMPRICH'S CORALSNAKE
(JAN, 1858)

459

An alternative name for Hemprich's Coralsnake is Worm-eating Coralsnake, but this is inaccurate. The primary prey sought by Hemprich's Coralsnake are not earthworms, but "velvet worms" (Onychophora), a curious group of tube-footed invertebrates that inhabit rotten logs and are closely related to the Arthropoda. Although Hemprich's Coralsnake specializes in velvet worms it will also prey on elongate vertebrates such as amphisbaenians (worm-lizards) and small snakes. It occurs throughout the Amazon Basin countries from French Guiana to Bolivia, in lowland and low montane tropical forests, and riverine forest. It lives and hunts in the damp leaf litter or under fallen logs, and reacts quickly to being uncovered by writhing and burrowing down into the substrate. Wilhelm Friedrich Hemprich (1796–1825) was a German naturalist.

RELATED SPECIES

Three subspecies of *Micrurus hemprichii* are recognized, from the lower and middle Amazon (*Micrurus h. hemprichii*), upper Amazon (*M. h. ortonii*), and Rôndonia (*M. h. rondonianus*). No other coralsnake resembles this species, with its broad triads, orange rings, and undivided cloacal plate.

Hemprich's Coralsnake has a black snout and dorsum to its head, followed by an orange ring across the nape and onto the lips. The body is marked with a repeating pattern of three extremely broad black bands, separated from each other by fine white bands that form a "triad," each triad being separated from the next by an orange band. Not all coralsnakes are black, red, and yellow.

Actual size

FAMILY	Elapidae: Elapinae
RISK FACTOR	Highly venomous: postsynaptic neurotoxins, and probably myotoxins
DISTRIBUTION	South America: across the Amazon, in Trinidad, the Guianas, Brazil, Venezuela, Colombia, Ecuador, Peru, Bolivia, Paraguay, and Argentina
ELEVATION	0–1,460 ft (0–445 m) asl
HABITAT	Lowland and low montane rainforest, savannas, floodplains, rocky habitats, and cultivated areas
DIET	Swamp eels and knifefish, small snakes, amphisbaenians, elongate lizards, and caecilians
REPRODUCTION	Oviparous, with clutches of 3–8 eggs
CONSERVATION STATUS	IUCN not listed

ADULT LENGTH
19¾–35½ in,
rarely 4 ft 7 in
(500–900 mm,
rarely 1.4 m)

460

MICRURUS LEMNISCATUS
SOUTH AMERICAN CORALSNAKE
(LINNAEUS, 1758)

The South American Coralsnake has a black head with a white band over the snout. The body is marked with wide black triads, with white rings, between narrow red interspaces. The contact between red and black might suggest "venom lack," but this species is dangerous and the rhyme (see page 456), useful in the USA, should not be used in South America.

The South American Coralsnake is found from Trinidad and Ilha de Marajó in the mouth of the Amazon to the headwaters in Peru, Ecuador, and Bolivia, but it is absent from upper Amazonian Brazil. Populations of South American Coralsnakes also occur on the Brazilian Atlantic coast and in Paraguay. This may be the most widely distributed coralsnake in the Americas. Its prey preferences are equally wide, ranging from swamp eels and knifefish to small snakes, leaf-litter lizards, amphisbaenians (worm-lizards), and caecilians. The South American Coralsnake has also been documented as cannibalistic. Its habitat preferences are also broad, from lowland rainforest to rocky habitats, open savanna, and disturbed areas with cultivation. Documented snakebites are few and symptoms have been mild, but as a coralsnake it should not be underestimated.

RELATED SPECIES

With its wide but punctuated distribution, several subspecies are recognized: in eastern Brazil and the Guianas (*Micrurus lemniscatus lemniscatus*); Venezuela, Trinidad, and the Guianas (*M. l. diutius*); upper Amazon (*M. l. helleri*); Bolivia (*M. l. frontifasciatus*); and the Brazilian Atlantic Forest, Argentina, and Paraguay (*M. l. carvalhoi*). Many other triad-marked coralsnakes resemble this species, but so also do several mildly venomous mimics such as the Aesculapian False Coralsnake (*Erythrolamprus aesculapii*, page 311).

Actual size

FAMILY	Elapidae: Elapinae
RISK FACTOR	Highly venomous: postsynaptic neurotoxins, and probably myotoxins
DISTRIBUTION	North and Central America: southern Mexico, Guatemala, El Salvador, Honduras, Nicaragua, Costa Rica, and Panama, to northwest Colombia
ELEVATION	0–5,250 ft (0–1,600 m) asl
HABITAT	Lowland and low to mid-montane rainforest, and lowland dry forest
DIET	Snakes, lizards, reptile eggs, and caecilians
REPRODUCTION	Oviparous, with clutches of 5–15, occasionally 23, eggs
CONSERVATION STATUS	IUCN Least Concern, CITES Appendix III (Honduras)

ADULT LENGTH
23¾–29½ in,
rarely 3 ft 7 in
(600–750 mm,
rarely 1.1 m)

MICRURUS NIGROCINCTUS
CENTRAL AMERICAN CORALSNAKE
(GIRARD, 1854)

461

The Central American Coralsnake is the most widely distributed coralsnake in Central America, with six subspecies recognized from southern Mexico to northwest Colombia, and on Big Corn Island, Nicaragua, in the Caribbean and Coiba Island, Panama, in the Pacific. This moderately large coralsnake inhabits lowland and montane rainforest and dry forests, where it feeds on a wide range of prey species, including snakes, lizards, caecilians, and even reptile eggs. Snakebites from this species are thought to have caused human fatalities. Although it is not considered threatened by the IUCN, Honduras have placed the Central American Coralsnake on CITES Appendix III to control trade.

The Central American Coralsnake is a monadal species with single black bands between narrow white or yellow bands, between broad red interspaces. Its head is black with a white band across the center. Beyond this description this is a very variable species, sometimes even lacking any white or yellow bands.

RELATED SPECIES

Micrurus nigrocinctus possesses four mainland subspecies, and two island endemics, on Big Corn Island (*M. n. babaspul*) and Coiba Island (*M. n. coibensis*). Several harmless mimics resemble this species, including the Variegated False Coralsnake (*Pliocercus elapoides*, page 285) and the Central America Milksnake (*Lampropeltis polyzona*). Several other coralsnakes also occur in sympatry with this species, but the species believed closely related is the Endangered Roatán Coralsnake (*M. ruatanus*, page 463), formerly treated as a subspecies, from the Islas de la Bahía off the Caribbean coast of Honduras.

Actual size

FAMILY	Elapidae: Elapinae
RISK FACTOR	Highly venomous: postsynaptic neurotoxins, and probably myotoxins
DISTRIBUTION	South America: eastern Venezuela, Guyana, Suriname, French Guiana, and northeastern Brazil
ELEVATION	0–1,640 ft (0–500 m) asl
HABITAT	Low-lying tropical rainforest, low montane wet forest, gallery forest, and savanna–forest edge situations
DIET	Small gymnophthalmid lizards
REPRODUCTION	Oviparous, clutch size unknown
CONSERVATION STATUS	IUCN not listed

ADULT LENGTH
21¾–23¾ in,
rarely 35¾ in
(550–600 mm,
rarely 910 mm)

462

MICRURUS PSYCHES
CARIB CORALSNAKE
(DAUDIN, 1803)

The Carib Coralsnake is a monadal species with single black rings alternating with slightly wider red rings and separated by fine white rings only one scale wide. On the dorsum the red rings are so heavily infused with black they are almost black themselves, but they are usually visible on the venter. Not all coralsnakes have visible red rings.

At first glance the Carib Coralsnake might not be thought a coralsnake, so dark are its red rings, but this species demonstrates the diversity of patterning found through the approximately 80 species of *Micrurus*. It occurs from eastern Venezuela, through the Guianas to Amapá, Brazil, in lowland and low montane rainforest, gallery forest, and along the edges of savannas. The only prey recorded for the Carib Coralsnake are the small, slender, short-legged, leaf-litter-dwelling gymnophthalmid lizards of the genus *Bachia*. The Carib Coralsnake is a poorly known species; even the number of eggs it produces is unknown. At one time a number of isolated populations of coralsnakes across northern South America were treated as subspecies of *M. psyches*, but they have since been elevated to valid species in their own rights.

RELATED SPECIES
South American coralsnakes that exhibit similar patterning to *Micrurus psyches* include Medem's Coralsnake (*Micrurus medemi*) from Colombia, and the Speckled Coralsnake (*M. margaritiferus*) from Peru, while the Red-tailed Coralsnake (*M. mipartitus*) of Colombia is banded black and white, but with a distinct red head and tail.

Actual size

FAMILY	Elapidae: Elapinae
RISK FACTOR	Highly venomous: probably postsynaptic neurotoxins, and myotoxins
DISTRIBUTION	Central America: Roatán (Islas de la Bahía, Honduras)
ELEVATION	0–66 ft (0–20 m) asl
HABITAT	Low-lying tropical moist forest and disturbed areas
DIET	Whiptail lizards and small snakes
REPRODUCTION	Presumed oviparous, otherwise undocumented
CONSERVATION STATUS	IUCN Endangered

ADULT LENGTH
23¾–29½ in,
rarely 3 ft 7 in
(600–750 mm,
rarely 1.1 m)

MICRURUS RUATANUS
ROATÁN CORALSNAKE
(GÜNTHER, 1895)

463

The Roatán Coralsnake is endemic to the island of Roatán, in the Honduran Islas de la Bahía, 40 miles (64 km) off the Caribbean coast. Due to habitat destruction and alteration it is now considered Endangered by the IUCN. Roatán is a relatively small island (60 sq miles/156 sq km) with a high point of 1,100 ft (335 m) asl, the coralsnake being found between sea-level and 66 ft (20 m), where there has been much disturbance. Semi-fossorial and both diurnal and nocturnal, this snake feeds on diurnal whiptail lizards (*Cnemidophorus*) and small snakes. Local people believe venomous snakes sequester the poison of toads taken as prey, and since toads are absent from Roatán they believe the coralsnake is nonvenomous. There is an anecdotal account of a fatality following a snakebite from this species.

The Roatán Coralsnake is usually bicolored red and black, lacking white or yellow bands. Its head is bright red with a black snout tip.

RELATED SPECIES
The closest relative of *Micrurus ruatanus* is the widely distributed mainland Central American Coralsnake (*M. nigrocinctus*, page 461), of which the Roatán Coralsnake was formerly a subspecies. Other insular endemics include coralsnakes on Nicaragua's Isla del Maíz (*M. n. babaspul*) and Panama's Coiba Island (*M. n. coibensis*). There are no snakes on Roatán with which *M. ruatanus* could be confused.

Actual size

FAMILY	Elapidae: Elapinae
RISK FACTOR	Highly venomous: postsynaptic neurotoxins, and probably myotoxins
DISTRIBUTION	South America: Amazonian Brazil, Venezuela, Colombia, Ecuador, Peru, Bolivia, and Paraguay
ELEVATION	165–3,940 ft (50–1,200 m) asl
HABITAT	Primary rainforest, secondary growth, gallery forest, and savanna
DIET	Lizards and snakes, including pitvipers, and caecilians
REPRODUCTION	Oviparous, with clutches of 6–12 eggs
CONSERVATION STATUS	IUCN not listed

ADULT LENGTH
2 ft 7 in–3 ft 7 in,
rarely 4 ft 7 in
(0.8–1.1 m,
rarely 1.4 m)

464

MICRURUS SPIXII
AMAZONIAN CORALSNAKE
WAGLER, 1824

The Amazonian Coralsnake has a gray and red banded head, the pattern heavily infused with black. Its body exhibits typical coralsnake triads, but the usual narrow white bands are broad, with every scale tipped with black. The scales of the red bands may also be black tipped.

The Amazonian Coralsnake is also known as Spix's Coralsnake, named in honor of the German naturalist Johann Baptist von Spix (1781–1826). It is distributed through the Amazonian countries, but is absent from the Rio Negro watershed of Venezuela and northeastern Brazil, and the Guianas. It is equally at home in primary rainforest and secondary growth and may be found in cultivated gardens or around houses. It is reportedly often associated with leaf-cutter ant nests but the reasons why are unclear. It preys on a wide range of lizards and snakes and even preys on dangerous species like lancehead pitvipers (*Bothrops*, pages 560–569). Caecilians are also taken in the upper Amazon. Although reportedly inoffensive, such a large species must be considered extremely dangerous, even though there are apparently no snakebites on record.

RELATED SPECIES

Four subspecies are recognized, from the lower Amazon (*Micrurus spixii martiusi*), middle Amazon south to Paraguay (*M. s. spixii*), upper Amazon (*M. s. obscurus*), and central Bolivia (*M. s. princeps*). *Micrurus spixii* closely resembles the Venezuelan Coralsnake (*M. isozonus*), although that species occurs in the Orinoco watershed of Colombia and Venezuela to the north.

Actual size

FAMILY	Elapidae: Elapinae
RISK FACTOR	Highly venomous: postsynaptic neurotoxins, and probably myotoxins
DISTRIBUTION	South America: upper Orinoco in Venezuela, and Amazonian Venezuela, Colombia, the Guianas, Brazil, Ecuador, Peru, and Bolivia
ELEVATION	98–1,890 ft (30–575 m) asl
HABITAT	Lowland and low montane rainforest, especially near water
DIET	Swamp eels, knifefish, catfish, and other bony fish
REPRODUCTION	Oviparous, with clutches of 5–13 eggs
CONSERVATION STATUS	IUCN not listed

ADULT LENGTH
2 ft 7 in–3 ft 3 in,
rarely 4 ft
(0.8–1.0 m,
rarely 1.2 m)

MICRURUS SURINAMENSIS
AQUATIC CORALSNAKE
(CUVIER, 1817)

465

Found throughout the Amazonian countries, the Aquatic Coralsnake is one of the largest coralsnake species. Although it is a snake of lowland and lower montane rainforest, the distribution of the most aquatic of all coralsnakes is dominated by creeks and rivers. It swims well and can remain submerged for long periods. Clues to its aquatic habits can be seen in its relatively dorsally positioned nostrils and eyes. Prey comprises mostly fish, from swamp eels and knifefish to catfish and other bony fish. The Aquatic Coralsnake is unlikely to be confused with any other coralsnake, and it is the only species where only the fourth supralabial scale contacts the eye. Several harmless aquatic snakes in the genera *Hydrops* (page 317), *Helicops* (pages 314–315), and *Hydrodynastes* (page 316) may mimic the Aquatic Coralsnake's patterning for protection.

The Aquatic Coralsnake is characterized by its red head, each scale edged with black. The body is patterned with black-white-black-white-black triads between red interspaces, each scale of the interspaces being black-edged.

RELATED SPECIES
Two subspecies of *Micrurus surinamensis* are recognized: the widespread Amazonian–Guianan form (*M. surinamensis surinamensis*), and an upper Orinoco and Rio Negro form (*M. s. nattereri*).

Actual size

FAMILY	Elapidae: Elapinae
RISK FACTOR	Highly venomous: postsynaptic neurotoxins
DISTRIBUTION	Central and East Africa: Cameroon and Gabon to Rwanda, Tanzania, and Zambia
ELEVATION	0–3,490 ft (0–1,065 m) asl
HABITAT	Lakes and rivers, fringing woodland, or rocky shores
DIET	Fish, and possibly amphibians
REPRODUCTION	Oviparous, with clutches of 22–24 eggs
CONSERVATION STATUS	IUCN Least Concern

ADULT LENGTH
4 ft 7 in–7 ft 3 in,
rarely 9 ft
(1.4–2.2 m,
rarely 2.7 m)

NAJA ANNULATA
BANDED WATER COBRA
PETERS, 1876

466

The Banded Water Cobra is pale brown, orange, or reddish brown, with a series of bold black rings around the body, these rings becoming more expanded on the flanks to encompass areas of paler background color. The rings are especially in evidence on the throat when the cobra rears up and spreads its narrow hood. In the Lake Tanganyika subspecies (*Naja annulata stormsi*) the rings are present only on the neck and anterior body.

A truly aquatic snake, the Banded Water Cobra is more at home in the water than on land, swimming gracefully, diving rapidly to 82 ft (25 m), and remaining submerged for 20 minutes at a time. It is believed to subsist entirely on a diet of fish, although amphibians may also be taken. Active by day or night, the Banded Water Cobra sleeps in caves under boulders on rocky shores, emerging with first light to hunt. At night it explores underwater crevices for sleeping fish. Although Banded Water Cobras are not considered endangered in general, Storm's Water Cobra (*Naja annulata stormsi*), the subspecies endemic to Lake Tanganyika, may be threatened because of local fishermen who set their gill nets along rocky shores and many cobras become ensnared and drown as a result. Human snakebites are rare.

RELATED SPECIES

For a long time placed in the genus *Boulengerina*, *Naja annulata* is closely related to the Congo Water Cobra (*N. christyi*), a little-known species from the mouth of the Congo River, and the Forest Cobra (*N. melanoleuca*, page 470).

Actual size

FAMILY	Elapidae: Elapinae
RISK FACTOR	Highly venomous: postsynaptic neurotoxins, cytotoxins, and possibly cardiotoxins
DISTRIBUTION	Southern Africa: southern Zambia, Zimbabwe, eastern Botswana, southern Mozambique, northwestern South Africa, and Swaziland
ELEVATION	0–4,510 ft (0–1,375 m) asl
HABITAT	Savanna, bushveld, and lowveld
DIET	Small mammals, snakes, toads, and birds' eggs
REPRODUCTION	Oviparous, with clutches of 8–33 eggs
CONSERVATION STATUS	IUCN not listed

NAJA ANNULIFERA
SNOUTED COBRA
PETERS, 1876

ADULT LENGTH
Male
7 ft
(2.1 m)

Female
6 ft 3 in
(1.9 m)

467

The Snouted Cobra is an inhabitant of arid grassland habitats across southeastern Africa, from southern Zambia to KwaZulu-Natal. This species often spends the day sleeping inside termitaria, before emerging at night to hunt toads, other snakes, including venomous species like Puff Adders (*Bitis arietans*, page 610), and small mammals, or to raid chicken coops for eggs. If confronted, the Snouted Cobra will spread a broad hood and hiss loudly, and as a non-spitting species its other options include escape, thanatosis (playing dead), or biting. Snakebites from this species, like those from any cobra, are serious medical emergencies; death within hours due to respiratory paralysis is a likely outcome without treatment with antivenom.

RELATED SPECIES

The Angolan cobra (*Naja anchietae*) was a former subspecies of *N. annulifera*, and both were originally subspecies of the northeast African Egyptian Cobra (*N. haje*, page 469), but physically *N. annulifera* is very different from the Egyptian Cobra, with different head and hood shapes.

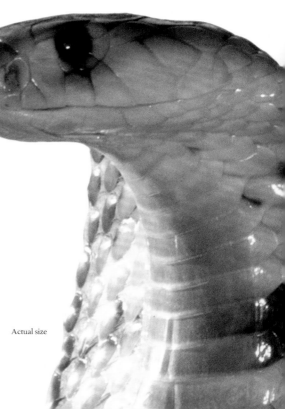

The Snouted Cobra may be uniform brown, light brown with darker brown speckling, or it may occur as a banded morphotype, with distinctive broad pale brown and dark purple-brown bands along the body. It has a large rostral scale on the snout and may possess a pointed snout in profile, though not as pointed as in the related Angolan Cobra.

Actual size

FAMILY	Elapidae: Elapinae
RISK FACTOR	Highly venomous: postsynaptic neurotoxins
DISTRIBUTION	Arabia: southeastern Oman, Yemen, and southeastern Saudi Arabia
ELEVATION	3,280–7,870 ft (1,000–2,400 m) asl
HABITAT	Highland wadis and streams, especially in monsoon areas
DIET	Toads, snakes, lizards, fish, birds, and small mammals
REPRODUCTION	Oviparous, with clutches of 8–20 eggs
CONSERVATION STATUS	IUCN Least Concern

ADULT LENGTH
4–5 ft,
rarely 6 ft 3 in
(1.2–1.5 m,
rarely 1.9 m)

468

NAJA ARABICA
ARABIAN COBRA
SCORTECCI, 1932

The Arabian Cobra may be yellow, brown, reddish, or jet black. When it raises a hood it lacks any black collar or bands across the neck.

The Arabian Cobra is the only true cobra on the Arabian Peninsula, a representative of an African clade of cobras in western Asia. It is primarily an inhabitant of highland areas, away from the lowland sandy desert, where it adopts a diurnal lifestyle, but avoids the heat of the day, and is most often encountered in habitats near water or within monsoon areas. Although it shows a preference for feeding on toads, it will take other snakes, lizards, birds, small mammals, and even fish as prey. Snakebites to humans are rare but are serious medical emergencies due to the high risk of neuromuscular paralysis and death. In common with its close relative, the Egyptian Cobra (*Naja haje*, page 469), this cobra does not spit venom.

RELATED SPECIES
A former subspecies of the Egyptian Cobra (*Naja haje*, page 469) of northeast Africa, *N. arabica* is most likely to be confused with racers of the genus *Platyceps* (pages 216–217), which share its body shape, coloration, patterning, and activity cycle, though they do not grow as large as the cobra.

Actual size

FAMILY	Elapidae: Elapinae
RISK FACTOR	Highly venomous: postsynaptic neurotoxins, cytotoxins, and possibly cardiotoxins
DISTRIBUTION	North and East Africa: Western Sahara to Egypt north of the Sahara, and Senegal to Somalia, Kenya, and Tanzania south of the Sahara
ELEVATION	0–6,560 ft (0–2,000 m) asl
HABITAT	Arid savanna, woodland, and semidesert
DIET	Small mammals, birds and their eggs, toads, and snakes
REPRODUCTION	Oviparous, with clutches of 8–20 eggs
CONSERVATION STATUS	IUCN not listed

NAJA HAJE
EGYPTIAN COBRA
(LINNAEUS, 1758)

ADULT LENGTH
4 ft 3 in–6 ft,
rarely 8 ft 2 in
(1.3–1.8 m,
rarely 2.5 m)

469

The probable means of Cleopatra's suicide, and the cobra that adorned the headdress of the Pharaohs, the Egyptian Cobra is not confined to Egypt. It has an extremely fragmented distribution around but not into the Sahara, with populations from Western Sahara, Morocco, and Algeria; along the coast from Tunisia to Egypt; in the Sahel from Senegal to Cameroon; South Sudan; and Uganda, Eritrea, Ethiopia, Somalia, Kenya, and Tanzania. It inhabits arid grassland, woodland, and semidesert, but never enters rainforest. It feeds on a variety of vertebrates, from other snakes and toads, to small mammals, birds, and their eggs. A terrestrial species, it is active by day or night. A non-spitter, when confronted it will hood and hiss, before biting. Snakebites from this species cause death through respiratory paralysis.

The Egyptian Cobra is usually light brown in color, often with dorsal speckling of dark brown or black. It has a narrow hood with a broad black band across the throat. Moroccan specimens are often jet black.

RELATED SPECIES

Naja haje was once a pan-African species, but most former subspecies are now treated as distinct species, including the Arabian Cobra (*N. arabica*, page 468), Snouted Cobra (*N. annulifera*, page 467), and Angolan Cobra (*N. anchietae*) from southwestern Africa, and a recently described West African species (*N. senegalensis*) that could be called the Sahel Cobra because it occurs between the West African rainforests and the Sahara. The Moroccan population of *N. haje*, sometimes referred to as *N. h. legionis*, is no longer recognized as distinct enough to warrant taxonomic recognition.

Actual size

FAMILY	Elapidae: Elapinae
RISK FACTOR	Highly venomous: postsynaptic neurotoxins, cytotoxins, and possibly cardiotoxins
DISTRIBUTION	Sub-Saharan Africa: Senegal to Ethiopia, south to Angola, Zambia, Mozambique, Zimbabwe, and South Africa
ELEVATION	0–8,860 ft (0–2,700 m) asl
HABITAT	Tropical and subtropical rainforest, gallery forest, coastal thicket, arid woodland, and oil-palm plantations
DIET	Small mammals, birds, snakes, amphibians, and fish
REPRODUCTION	Oviparous, with clutches of 15–26 eggs
CONSERVATION STATUS	IUCN not listed

ADULT LENGTH
Male
7 ft 3 in
(2.2 m)

Female
5 ft 2 in
(1.6 m)

470

NAJA MELANOLEUCA
FOREST COBRA
HALLOWELL, 1857

The Forest Cobra of West and Central Africa is a glossy black snake with a pale yellow underbelly and black bands across the throat, although some specimens are broadly banded on the back, with the lips banded black and white, the source of the old name Black and White-lipped Cobra. Southern and eastern populations are matte brown on the head and anterior body and matte black on the posterior body and tail.

The Forest Cobra is an arboreal, non-spitting cobra that inhabits rainforest, woodland, and oil-palm plantations. It climbs with ease, but is also active on the ground. It has a wide distribution, from Senegal to Uganda and Zambia, with isolated populations in Ethiopia, Kenya, Tanzania, Malawi, and Zimbabwe to South Africa. With this distribution, and differences in coloration between the West African rainforest specimens and the matte brown, black-tailed specimens from southeastern Africa, it is probable that *N. melanoleuca* is a species complex, containing several species that are yet to be defined. Its prey includes rodents, elephant shrews, birds, snakes, toads, and fish. Alert Forest Cobras avoid confrontation with humans and snakebites are rare, but fatalities are known.

RELATED SPECIES

The closest relative of the Forest Cobra is probably the Banded Water Cobra (*Naja annulata*, page 466) of Central Africa. The recently described São Tomé cobra (*N. peroescobari*) from the Gulf of Guinea is also a close relative.

Actual size

FAMILY	Elapidae: Elapinae
RISK FACTOR	Highly venomous: postsynaptic neurotoxins, cytotoxins, and possibly cardiotoxins
DISTRIBUTION	Southeastern Africa: Tanzania to South Africa, west to northeastern Namibia
ELEVATION	0–5,910 ft (0–1,800 m) asl
HABITAT	Savanna, woodland, coastal forest, semidesert, and highland areas (Zimbabwe)
DIET	Toads, snakes, lizards, and small mammals
REPRODUCTION	Oviparous, with clutches of 10–22 eggs
CONSERVATION STATUS	IUCN not listed

ADULT LENGTH
2 ft 7 in–4 ft 3 in,
rarely 5 ft
(0.8–1.3 m,
rarely 1.5 m)

NAJA MOSSAMBICA
MOZAMBIQUE SPITTING COBRA
PETERS, 1854

471

The Mozambique Spitting Cobra is a common cause of snakebites, and although deaths are rare, the cytotoxic effects of the venom are severe and can lead to necrosis (tissue death), loss of tissue, limb deformity, and long recovery periods. An inhabitant of lowland and coastal savanna, woodland, and semidesert, it hunts toads and other vertebrates by day or night. This cobra is often associated with termitaria, in which it shelters. Spitting cobras have evolved a form of defense from a distance, propelling jets of venom up to 6 ft 7 in (2 m) into the face of a perceived enemy, causing intense pain and temporary blindness if it contacts the eyes, at which point the cobra makes its escape. Spitting cobras can spit many times in quick succession. The Zulu name for this cobra is "M'fezi."

The Mozambique Spitting Cobra is a relatively small species that is greenish brown dorsally, with black interstitial skin presenting a lightly reticulate appearance, and white, pale yellow, or pinkish below, with heavy black infusions on all the ventral scales, and a broad black band across the throat when it hoods.

RELATED SPECIES

This cobra is a member of a clade that also contains the Zebra Spitting Cobra (*Naja nigricincta*, page 474) of southwest Africa, Black-necked Spitting Cobra (*N. nigricollis*) of West and Central Africa, the recently described Ashe's Spitting Cobra (*N. ashei*) of northeast Africa, and possibly the West African Spitting Cobra (*N. katiensis*).

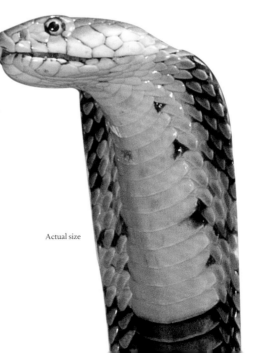

Actual size

FAMILY	Elapidae: Elapinae
RISK FACTOR	Highly venomous: venom composition unknown
DISTRIBUTION	Central Africa: southern Cameroon, Equatorial Guinea, Gabon, Congo, and DRC
ELEVATION	985–2,620 ft (300–800 m) asl
HABITAT	Tropical forest and savanna woodland
DIET	Unknown, possibly snakes or amphibians
REPRODUCTION	Unknown, probably oviparous; clutch size unknown
CONSERVATION STATUS	IUCN Least Concern

ADULT LENGTH
19¾–31½ in
(500–800 mm)

472

NAJA MULTIFASCIATA
BURROWING COBRA
(WERNER, 1902)

One of the least known African snake species, the Burrowing Cobra is a terrestrial or semi-fossorial inhabitant of tropical forest and savanna woodland from Cameroon and Gabon to the Democratic Republic of the Congo. Like most other elapids it is probably oviparous, though there are no records of its reproductive strategy. Even its diet is unknown, and although it is expected to feed on either amphibians or small snakes, there is no data to support either possibility. There is also no data regarding the venom of the Burrowing Cobra, and no snakebites on record. This is a rarely encountered species in dire need of study in nature.

RELATED SPECIES

In the past this species was placed in the monotypic genus *Paranaja* and three dubious subspecies were recognized. Although now in *Naja*, it does not resemble other cobras but in appearance being more like the African gartersnakes (*Elapsoidea*, pages 452–453).

The Burrowing Cobra is a small snake that is pale green with every scale tipped with black, presenting a reticulate appearance. The head is yellow with a black cap, the darker pigment extending downward through the eye and to the angle of the jaw, with a black band across the nape of the neck.

Actual size

FAMILY	Elapidae: Elapinae
RISK FACTOR	Highly venomous: postsynaptic neurotoxins, cytotoxins, and possibly cardiotoxins
DISTRIBUTION	South Asia: India (except the northeast), Pakistan, Nepal, Bangladesh, Bhutan, and Sri Lanka
ELEVATION	0–6,560 ft (0–2,000 m) asl
HABITAT	Forests, woodland, open country, and especially cultivated rice paddies
DIET	Small mammals, amphibians, snakes, and birds
REPRODUCTION	Oviparous, with clutches of 12–30 eggs
CONSERVATION STATUS	IUCN not listed, CITES Appendix II

NAJA NAJA

INDIAN COBRA
(LINNAEUS, 1758)

ADULT LENGTH
3 ft–3 ft 3 in,
rarely 7 ft 3 in
(0.9–1.0 m,
rarely 2.2 m)

473

The Indian Cobra, also called the Spectacled Cobra, bears a "spectacle" marking on its hood. Sri Lankan Buddhists believe this is the mark of Buddha's two fingers, made after the cobra sheltered him from the rain. Thai Buddhists have the Monocled Cobra (*Naja kaouthia*), which has a single round "monocle" marking, made by Buddhas's thumb. Even though Hindus and Buddhists revere the Indian Cobra, it is still a major contributory factor in thousands of snakebite fatalities each year, being one of the "Big Four" snakebite killers. But it also plays an important role, ridding rice paddies of rats that would spoil the crop and spread disease, so while cobras take lives, they also save them. Found in most habitats, with a catholic vertebrate diet, Indian Cobras are one of South Asia's commonest snakes.

RELATED SPECIES

The closest relatives of *Naja naja* are likely to be the non-spitting Asian cobras, the Transcaspian Cobra (*N. oxiana*, page 476), Monocled or Thai Cobra (*N. kaouthia*), Andaman Cobra (*N. sagittifera*), and Chinese Cobra (*N. atra*). At one time all Asian *Naja* were treated as subspecies of *Naja naja*.

The Indian Cobra is a fairly variable snake. Indian specimens are usually brown with the characteristic spectacle marking on the hood, while Sri Lankan specimens are often banded. Specimens from Nepal and Pakistan are often melanistic, being all black with only the faintest of hood markings visible.

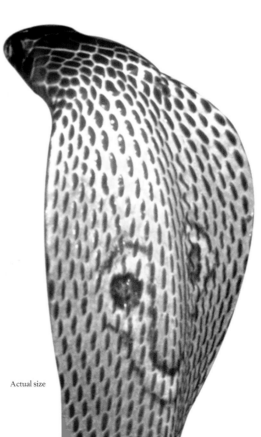

Actual size

FAMILY	Elapidae: Elapinae
RISK FACTOR	Highly venomous: postsynaptic neurotoxins, cytotoxins, and possibly cardiotoxins
DISTRIBUTION	Southwest Africa: Angola, Namibia, and northwest South Africa
ELEVATION	0–5,350 ft (0–1,630 m) asl
HABITAT	Arid rocky habitats, and semidesert
DIET	Toads, lizards, and small mammals
REPRODUCTION	Oviparous, clutch size unknown
CONSERVATION STATUS	IUCN not listed

ADULT LENGTH
2 ft 7 in–4 ft 3 in,
rarely 6 ft
(0.8–1.3 m,
rarely 1.8 m)

NAJA NIGRICINCTA
ZEBRA SPITTING COBRA
BOGERT, 1940

474

The Zebra Spitting Cobra is usually banded with alternating black and white bands, with a totally black head, hood, neck, and throat. Its southern subspecies, the Black Spitting Cobra (*Naja nigricincta woodi*), is jet black throughout. These cobras spread a broad hood.

Also known as the Western Barred Spitting Cobra, this is a relatively small cobra that occurs as two subspecies, the southern subspecies achieving a slightly larger size (6 ft/ 1.8 m) than the northern subspecies (5 ft/1.5 m). Inhabiting semidesert, with a preference for rocky outcrops, the two subspecies both hunt toads, lizards, and small mammals by day. Although most specimens are small, they can produce large venom yields and are potentially capable of causing a fatality, but tissue loss due to necrosis is the more likely result of a snakebite and even bites treated with antivenom require a long recovery period and skin grafts. The preferred defense of these snakes is to hood as a warning, and if that is not heeded, to spit venom into the face of their enemy. The cobra then drops its hood and escapes.

RELATED SPECIES

The closest relatives of *Naja nigricincta* are the Black-necked Spitting Cobra (*N. nigricollis*) of West and Central Africa, of which it was once a subspecies, the Mozambique Spitting Cobra (*N. mossambica*, page 471) of southeast Africa, and Ashe's Spitting Cobra (*N. ashei*) of northeast Africa. Two subspecies are recognized, the zebra-striped nominate form (*N. n. nigricincta*) from Namibia and southern Angola, and the Black Spitting Cobra (*N. n. woodi*) from southern Namibia and Namaqualand, South Africa.

Actual size

FAMILY	Elapidae: Elapinae
RISK FACTOR	Highly venomous: postsynaptic neurotoxins, cytotoxins, and possibly cardiotoxins
DISTRIBUTION	Southern Africa: eastern, southern, and central South Africa, Lesotho, southern Namibia, and Botswana
ELEVATION	0–8,200 ft (0–2,500 m) asl
HABITAT	Arid grassland, desert, semidesert, rocky hills, riverbeds, and coastal fynbos
DIET	Small mammals, snakes, lizards, amphibians, and weaverbirds
REPRODUCTION	Oviparous, with clutches of 8–20 eggs
CONSERVATION STATUS	IUCN not listed

NAJA NIVEA
CAPE COBRA
(LINNAEUS, 1758)

ADULT LENGTH
Male
5 ft 7 in
(1.7 m)

Female
4 ft 7 in
(1.4 m)

475

The Cape Cobra is also known as the Yellow Cobra, although not all specimens are yellow. It is an inhabitant of relatively arid open habitats throughout southwestern Africa, from the Cape, South Africa, to Namaqualand and Namibia, Botswana, and Lesotho. It is also found in southern coastal fynbos habitats. Although a terrestrial species, the Cape Cobra is an extremely agile climber, and it is known to climb into weaver bird nests to steal eggs and chicks. Small mammals, snakes, lizards, frogs, and toads are also eaten. Cape Cobras possess one of the most toxic of all African cobra venoms and in some parts of the Cape they are responsible for killing sheep and other domesticated animals. Human snakebites, though rare, are extremely serious and may end tragically without antivenom therapy.

The Cape Cobra is a variable species that may be bright yellow, orange, brown, or reddish in color, often speckled with black. Some specimens are heavily mottled with black, almost obscuring the paler pigment, and totally black specimens are not rare, especially at higher elevations.

RELATED SPECIES
Naja nivea is not closely related to other African cobras and is unlikely to be confused with any other species. It occurs in sympatry with the Black Spitting Cobra (*N. nigricincta woodi*, page 474) and the Rinkhals (*Hemachatus haemachatus*, page 454), both of which spit venom, while the Cape Cobra is a non-spitter.

Actual size

FAMILY	Elapidae: Elapinae
RISK FACTOR	Highly venomous: postsynaptic neurotoxins, cytotoxins, and possibly cardiotoxins
DISTRIBUTION	Western Asia: Turkmenistan, Uzbekistan, Tajikistan, Kyrgyzstan, northern Iran, Afghanistan, northern Pakistan, and northeast India
ELEVATION	805–6,890 ft (245–2,100 m) asl
HABITAT	Arid hills, mountains, and rocky valleys
DIET	Amphibians, lizards, snakes, birds, and small mammals
REPRODUCTION	Oviparous, clutch size unknown
CONSERVATION STATUS	IUCN Data Deficient, CITES Appendix II

ADULT LENGTH
3 ft 3 in–5 ft,
rarely 5 ft 7 in
(1–1.5 m, rarely
1.7 m)

NAJA OXIANA
TRANSCASPIAN COBRA
(EICHWALD, 1831)

The Transcaspian Cobra is a fairly uniform snake that is sometimes called the Black Cobra, not to be confused with melanistic specimens of the Indian Cobra (*Naja naja*, page 473), which may be much darker in Pakistan. It is found in the states to the east of the Caspian Sea, from Turkmenistan and northern Iran, through Afghanistan and northern Pakistan, to northwest India (Jammu and Kashmir). Although there are few records of snakebites it must be assumed to possess venom equally as toxic as the Indian Cobra, and therefore to be a snakebite risk. Transcaspian Cobras inhabit arid, rugged hillsides, mountains, and valleys and exhibit a very catholic vertebrate diet.

The Transcaspian Cobra is usually gray or brown, often banded with subdued bands. Unlike its Asian relatives, it does not usually bear any hood markings, but it does expose broad dark gray bands across its throat when it hoods.

RELATED SPECIES

The closest relative to *Naja oxiana* is probably the Indian or Spectacled Cobra (*N. naja*) of South Asia, with which it may occur in sympatry in northwest India and Pakistan.

Actual size

FAMILY	Elapidae: Elapinae
RISK FACTOR	Highly venomous: postsynaptic neurotoxins
DISTRIBUTION	Northeast Africa: Somalia, Djibouti, Ethiopia, Kenya, and northern Tanzania
ELEVATION	0–4,920 ft (0–1,500 m) asl
HABITAT	Savanna, semidesert, and desert
DIET	Amphibians, small mammals, birds, and reptiles
REPRODUCTION	Oviparous, with clutches of 6–15 eggs
CONSERVATION STATUS	IUCN not listed

ADULT LENGTH
3 ft–3 ft 3 in,
rarely 5 ft
(0.9–1.0 m,
rarely 1.5 m)

NAJA PALLIDA
RED SPITTING COBRA
BOULENGER, 1896

477

This small cobra is found in the Horn of Africa, from Somalia to Djibouti, and south through eastern Ethiopia and Kenya to northern Tanzania. It inhabits savanna habitats but also occurs in desert and semidesert. Red Spitting Cobras are mostly active at night, when they emerge to hunt toads, frogs, small mammals, birds, and reptiles. Although terrestrial, this cobra is an agile climber, an individual being observed aloft in a thorn tree eating foam-nest tree frogs. It shelters in brush piles or under logs during the day and if disturbed will defend itself by spitting venom, in common with other spitting cobras. The venom is not thought to be as toxic as that of other African species and snakebites are rare.

The Red Spitting Cobra is salmon red or pink, or sometimes reddish brown, with a broad black band across its throat.

RELATED SPECIES
The closest relative of *Naja pallida* is the Nubian Spitting Cobra (*N. nubiae*), which occurs in isolated pockets of distribution in Eritrea, Egypt, Sudan, South Sudan, Chad, and Niger, and is distinguished from *N. pallida* by the presence of more than one black band across the throat.

Actual size

FAMILY	Elapidae: Elapinae
RISK FACTOR	Highly venomous: postsynaptic neurotoxins, cytotoxins, and possibly cardiotoxins
DISTRIBUTION	Southeast Asia: southeastern Philippines (Mindanao, Bohol, Leyte, and Samar)
ELEVATION	0–3,280 ft (0–1,000 m) asl
HABITAT	Lowland grassland and cultivated areas to low montane rainforest
DIET	Snakes, lizards, amphibians, and small mammals
REPRODUCTION	Oviparous, with clutches of up to 8 eggs
CONSERVATION STATUS	IUCN Least Concern, CITES Appendix II

ADULT LENGTH
31½–36½ in,
rarely 4 ft 7 in
(800–930 mm,
rarely 1.4 m)

478

NAJA SAMARENSIS
SOUTHEASTERN PHILIPPINE COBRA
PETERS, 1861

Also known as the Samar or Visayan Cobra, the Southeastern Philippine Cobra is a common spitting cobra from Mindanao, Bohol, Leyte, and Samar, where it occurs at low or medium elevations. It inhabits rainforest and open grasslands, including cultivated rice paddies, where it comes into contact with humans. It feeds on snakes, lizards, frogs, toads, and small mammals, and is drawn toward human habitations by high populations of rodents. Although a nervous snake, observers suggest it is more reluctant to spit venom than its relative, the Northern Philippine Cobra (*Naja philippinensis*). It is a highly venomous snake that has been documented killing large pigs within a few hours.

RELATED SPECIES

The closest relative of *Naja samarensis* is the Northern Philippine Cobra (*N. philippinensis*) from Luzon, Mindoro, Masbate, and Catanduanes. The yellow and black *N. samarensis* may be more slender than this brown northern relative, having only 17–19 scale rows at midbody, compared to 21–23 in the northern species.

The Southeastern Philippine Cobra is usually an iridescent, black-scaled snake with yellow interstitial skin, combining to present a reticulate pattern. The undersides of the head and throat are also yellow, while the venter is slate-gray or black.

Actual size

FAMILY	Elapidae: Elapinae
RISK FACTOR	Highly venomous: postsynaptic neurotoxins, cytotoxins, and possibly cardiotoxins
DISTRIBUTION	Southeast Asia: Java, Bali, and Lesser Sunda Islands
ELEVATION	0–2,300 ft (0–700 m) asl
HABITAT	Monsoon forest, tropical dry forest, grassland, and cultivated rice paddies
DIET	Small mammals, snakes, lizards, frogs, and toads
REPRODUCTION	Oviparous, with clutches of 6–25 eggs
CONSERVATION STATUS	IUCN Least Concern, CITES Appendix II

ADULT LENGTH
3 ft–3 ft 3 in,
rarely 5 ft
(0.9–1.0 m,
rarely 1.55 m)

NAJA SPUTATRIX
INDONESIAN SPITTING COBRA
BOIE, 1827

479

The Indonesian Spitting Cobra is found in Java and Bali, and through the Inner Banda Arc of the Lesser Sunda Islands, from Lombok to Alor. The name *sputatrix* means "the spitter." It inhabits low-lying monsoon forest, tropical dry forest, grassland, and cultivated rice paddies, feeding on rodents, frogs, toads, snakes, and lizards. This species is quick to raise a hood and spit venom at its enemy, making a quick getaway while the recipient is struggling with the pain and temporary blindness. Few human fatalities are on record, but this snake must be considered a highly dangerous species. Although not considered threatened by the IUCN, hundreds of thousands of Asian cobras are collected for the skin trade, and are placed on Appendix II of CITES to monitor and control the trade.

RELATED SPECIES

Naja sputatrix is the southernmost in a series of Southeast Asian spitting cobras, including the Equatorial Spitting Cobra (*N. sumatrana*) in Sumatra, Borneo, and Malaysia, the Thai Spitting Cobra (*N. siamensis*) of Thailand, Cambodia, and Vietnam, and the Burmese Spitting Cobra (*N. mandalayensis*), from Myanmar.

The Indonesian Spitting Cobra is light brown, reddish brown, or dark brown, without any distinctive hood markings.

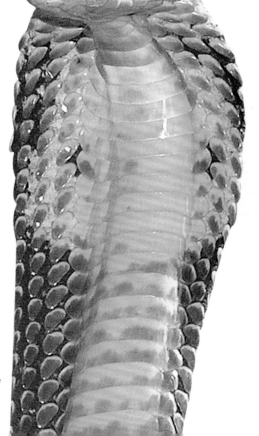

Actual size

FAMILY	Elapidae: Elapinae
RISK FACTOR	Highly venomous: postsynaptic neurotoxins, and cardiotoxins
DISTRIBUTION	Asia: northeast and southwest India to south China, Southeast Asia, the Philippines, Malaysia, and Indonesia, to Sulawesi
ELEVATION	0–7,150 ft (0–2,180 m) asl
HABITAT	Pristine rainforest, plantations, mangrove swamps
DIET	Snakes, monitor lizards, mammals
REPRODUCTION	Oviparous, with clutches of 20–50 eggs
CONSERVATION STATUS	IUCN Vulnerable, CITES Appendix II

ADULT LENGTH
Male
16 ft 5 in
(5.0 m)

Female
10 ft
(3.0 m)

OPHIOPHAGUS HANNAH
KING COBRA
(CANTOR, 1836)

480

The King Cobra is an imposing sight when it spreads its hood, raising one third of its length off the ground. King Cobras may be black or brown, and faintly banded, and they often have a series of inverted chevrons on the rear of their relatively narrow hoods. The narrower hood and the presence of two scales called "occipitals" on the back of the head help distinguish this species from common cobras (*Naja*, pages 466–479).

The King Cobra is the longest venomous snake, with large males achieving almost 16 ft 5 in (5 m) in length. A snake of undisturbed rainforest, it also enters plantations and mangrove swamps. King Cobras are primarily snake-eaters, *Ophiophagus* meaning "snake-swallower," and favored species include keelbacks (*Xenochrophis*, page 439), Dharman Ratsnakes (*Ptyas mucosa*, page 221), and Reticulated Pythons (*Malayopython reticulatus*, page 90), although some eat monitor lizards or mammals. Female King Cobras are the only snakes to build a nest for their eggs, gathering leaves with their coils, and they will guard them against all threats, even confronting and killing elephants. This species' venom is not as toxic as that of other cobras (*Naja*, pages 466–479); it is quantity, not quality, that kills, and they can inject a very large amount, possibly more than ⅓ fl oz (10 ml).

RELATED SPECIES

There are numerous cobra species in Asia but *Ophiophagus hannah* is not closely related to any of them, being placed in its own genus, and recent research suggests it may be closer, in relationship, to African mambas. *Ophiophagus hannah* is probably a species complex containing several species.

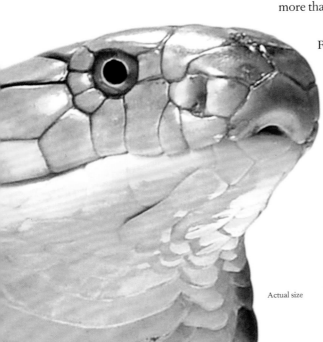

Actual size

FAMILY	Elapidae: Elapinae
RISK FACTOR	Highly venomous, presumed to contain neurotoxins
DISTRIBUTION	West Africa: Sierra Leone, Liberia, Ivory Coast, Ghana, and Togo, possibly Nigeria
ELEVATION	Elevation unknown
HABITAT	Pristine rainforest, riverine forest, and savanna woodland
DIET	Amphibians, possibly small arboreal mammals
REPRODUCTION	Oviparous, clutch size unknown
CONSERVATION STATUS	IUCN not listed

ADULT LENGTH
5 ft 2 in – 7 ft 3 in
(1.6 – 2.2 m)

PSEUDOHAJE NIGRA
BLACK TREE COBRA
GÜNTHER, 1858

481

Due to its arboreal canopy-dwelling habits, the Black Tree Cobra is very poorly known and rarely seen, and it has only been recorded from Sierra Leone to Togo, with the Nigeria record considered dubious. It inhabits a variety of forested habitats, but is mostly associated with pristine rainforest and riverine forests, although it is also found in isolated hill forests and savanna woodland. The Black Tree Cobra is believed to feed on amphibians, but it is also possible that it takes small arboreal mammals. Almost nothing is known about its venom, but it is thought to contain a very powerful neurotoxin similar to that found in its closest relative, Goldie's Tree Cobra (*Pseudohaje goldii*). Tree cobras may be the most venomous cobras in Africa, and there is no antivenom.

The Black Tree Cobra has a short head and very large eyes. It scales are smooth, the body being glossy black above and yellow below, and on the side of the head, where the supralabials (upper lip scales) are edged with black.

RELATED SPECIES

The closest relative is Goldie's Tree Cobra (*Pseudohaje goldii*), which also occurs in West Africa, although the two species only occur in sympatry in Ghana. Black Tree Cobras may be confused with the Black Treesnakes (*Thrasops*), male Blanding's Treesnakes (*Toxicodryas blandingii*, page 247), or the glossy black morphotype of the Forest Cobra (*Naja melanoleuca*, page 470).

Actual size

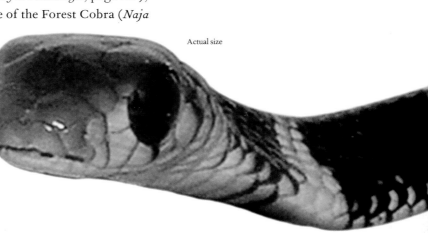

FAMILY	Elapidae: Elapinae
RISK FACTOR	Venomous: venom composition unknown, presumed neurotoxic
DISTRIBUTION	Asia: India, Nepal, Bangladesh, Myanmar, Thailand, Vietnam, Laos, China, Taiwan, and the Ryukyu Islands
ELEVATION	150–8,200 ft (45–2,500 m) asl
HABITAT	Lowland and low montane rainforest, and evergreen forest
DIET	Snakes and lizards
REPRODUCTION	Oviparous, with clutches of 6–14 eggs
CONSERVATION STATUS	IUCN Least Concern

ADULT LENGTH
27½–33 in
(700–840 mm)

SINOMICRURUS MACCLELLANDI
MACCLELLAND'S CORALSNAKE
(REINHARDT, 1844)

MacClelland's Coralsnake is a very slender, red or red-brown species with a pattern comprising thin black bands, edged with even narrower yellow or white bands. The head is white with a black snout tip and nape band.

MacClelland's Coralsnake, the most widely distributed member of its genus, is found in lowland and low montane rainforest and evergreen forest, across a large part of Asia from India to the Ryukyu Islands. It inhabits leaf litter and subsoil, where it adopts a semi-fossorial existence, hunting small snakes and lizards, especially skinks, at night. Although it is generally inoffensive, with a small head, the toxicity of its venom and its ability to bite should not be underestimated. Fatalities have been recorded, including that of the Swiss herpetologist Hans Schnurrenberger (1925–64), who ignored a bite from a 11¾ in (300 mm) specimen and died eight hours later.

RELATED SPECIES

Four subspecies are recognized, the nominate form (*Sinomicrurus macclellandi macclellandi*) occurring across China to Bangladesh and south to Vietnam and Laos, and others in northern India and Nepal (*S. m. univirgatus*), on Taiwan (*S. m. swinhoei*), and on the southern Ryukyu Islands (*S. m. iwasakii*). In Vietnam and Laos it occurs in sympatry with an extremely similar-looking species (*S. kelloggi*), on Taiwan it occurs with two other species (*S. sauteri* and *S. hatori*), and in the Ryukyus with a fourth species (*S. japonicus*).

Actual size

FAMILY	Elapidae: Elapinae
RISK FACTOR	Highly venomous: postsynaptic neurotoxins
DISTRIBUTION	Middle East: northeast Egypt, Israel, Jordan, and northwestern Saudi Arabia
ELEVATION	1,640–3,940 ft (500–1,200 m) asl
HABITAT	Desert and semidesert, arid hills, gravel pans, scrubland, and also in cultivated areas
DIET	Lizards, small mammals, and toads
REPRODUCTION	Oviparous, clutch size unknown
CONSERVATION STATUS	IUCN Least Concern

ADULT LENGTH
3 ft–3 ft 3 in,
rarely 4 ft 7 in
(0.9–1.0 m,
rarely 1.4 m)

WALTERINNESIA AEGYPTIA
SINAI DESERT BLACKSNAKE
LATASTE, 1887

483

The Sinai Desert Blacksnake is glossy black to black-brown, without patterning. It has a relatively long head with a large rostral scale for burrowing.

Also known as the Sinai Desert Cobra, the Sinai Desert Blacksnake occurs from Israel to the Sinai of Egypt, and northwestern Saudi Arabia. It inhabits desert or semidesert habitats, in wadis, on gravel pans, and on rocky hillsides, but it may also be found around cultivated areas and habitations. Although it has been known for around 130 years it is still largely unstudied in the wild, possibly because it spends much of its time underground in animal burrows, and is nocturnal. Prey includes lizards, including dabb lizards, small mammals, and toads. Although sometimes called a "cobra" the Sinai Desert Blacksnake does not spread a hood, its defensive posture involving turning its snout to the ground, from where it will strike sideways if its loud hissing goes unheeded. Fatalities have been recorded.

RELATED SPECIES

The more widely distributed eastern population of desert blacksnake from Iraq, Iran, and central Saudi Arabia is treated as a separate species (*Walterinnesia morgani*). Sympatric snake species with which *W. aegyptia* could be confused include the highly venomous Israeli Burrowing Asp (*Atractaspis engaddensis*, page 357).

Actual size

FAMILY	Elapidae: Hydrophiinae
RISK FACTOR	Highly venomous: presynaptic and postsynaptic neurotoxins, also myotoxins and anticoagulants
DISTRIBUTION	Indo-Australia: New Guinea, Australia (northern Torres Strait Islands), and eastern Indonesia (Seram, Maluku)
ELEVATION	0–6,560 ft (0–2,000 m) asl
HABITAT	Tropical lowland and montane rainforest, kunai grasslands, cultivated gardens, monsoon forest, sago swamps, and coffee plantations
DIET	Lizards, small mammals, and frogs
REPRODUCTION	Viviparous, with litters of 8–15 neonates
CONSERVATION STATUS	IUCN Least Concern

ADULT LENGTH
11¾–19¾ in,
rarely 23¼ in
(300–500 mm,
rarely 590 mm)

ACANTHOPHIS LAEVIS
SMOOTH-SCALED DEATH ADDER
MACLEAY, 1878

Vipers are absent from Australo-Papua, so the death adders, which closely resemble them as short, stout, sit-and-wait ambushers, with vertical pupils, nocturnal habits, live-bearing, and the caudal luring of prey, have evolved to occupy their vacant niche, a classic case of convergent evolution. The Smooth-scaled Death Adder is the most widely distributed venomous snake in New Guinea and eastern Indonesia, being present almost everywhere, in all habitats, including cultivated gardens and highland coffee plantations. Prey comprises lizards, small mammals, and probably frogs, which are ambushed at night. During the day death adders sleep under trash or along trails and bite anyone who steps on them. Fortunately, the effects of bites can be reversed with antivenom. Stories of 3 ft 3 in (1 m) death adders are probably misidentified New Guinea Ground Boas (*Candoia aspera*, page 114).

The Smooth-scaled Death Adder may be gray, olive, brown, or reddish, with darker banding, a fine yellow tail tip ending in a spine, for luring prey, and bold black and white lip scales. The body is short and squat like that of a viper, and the head angular with raised, hornlike supraocular scales, and vertically elliptical pupils.

RELATED SPECIES

Acanthophis laevis is most similar in appearance to the equally smooth-scaled Common Death Adder (*A. antarcticus*) of Australia, but the only other death adder occurring in New Guinea is the easily distinguished Rough-scaled Death Adder (*A. rugosus*), from the southern Trans-Fly region.

Actual size

FAMILY	Elapidae: Hydrophiinae
RISK FACTOR	Highly venomous: presynaptic and postsynaptic neurotoxins, and also anticoagulants
DISTRIBUTION	Australia: Western and Central Australia
ELEVATION	0–2,940 ft (0–895 m) asl
HABITAT	Desert, spinifex sandplains, acacia scrub, sand ridges, stony flats, and rocky outcrops
DIET	Lizards and small mammals
REPRODUCTION	Viviparous, with litters of 10–13 neonates
CONSERVATION STATUS	IUCN not listed, considered Vulnerable in South Australia

ADULT LENGTH
15¾–23¾ in,
rarely 29½ in
(400–600 mm,
rarely 750 mm)

ACANTHOPHIS PYRRHUS
DESERT DEATH ADDER
BOULENGER, 1898

485

The Desert Death Adder inhabits arid habitats across Western and Central Australia, in contrast to moister habitat preferences exhibited by the Smooth-scaled Death Adder (*Acanthophis laevis*, page 484) of New Guinea and eastern Indonesia, or the Common Death Adder (*A. antarcticus*) in eastern Australia. Being keel-scaled is useful for a desert snake because the ridges on the scales collect early morning dew and provide a drink. Habitats range from sandy desert to rocky outcrops and spinifex grasslands, in which this desert-camouflaged snake appears to vanish. Prey comprises lizards and small mammals, ambushed at night when the snake lies in wait on deserted sandy roads and tracks. Its defense involves flattening the body, hiding the head under the coils, hissing, and, if still approached, launching a rapid jabbing strike.

The Desert Death Adder has strongly keeled scales, a squat body shape, a thin, tapering tail, and an angular head. Its banded red or orange dorsal coloration enables it to blend in perfectly with its desert habitat.

RELATED SPECIES

The closest relatives of *Acanthophis pyrrhus* are the Pilbara Death Adder (*A. wellsi*) and the recently described Kimberley Death Adder (*A. cryptamydros*). These three species can be distinguished by their differing midbody scale row count and the condition of their prefrontal scales (divided or entire).

Actual size

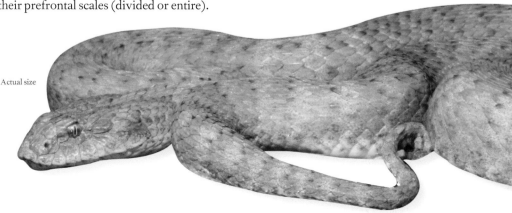

FAMILY	Elapidae: Hydrophiinae
RISK FACTOR	Highly venomous: postsynaptic neurotoxins, and possibly also myotoxins, with secondary nephrotoxicity and cardiotoxicity
DISTRIBUTION	Timor Sea: Western Australia (Ashmore and Hibernia reefs, possibly Western Australia coastline)
ELEVATION	Sea-level to 33 ft (10 m) bsl
HABITAT	Shallow coral reef flats and coral sand
DIET	Fish (eels)
REPRODUCTION	Viviparous, litter size unknown
CONSERVATION STATUS	IUCN Critically Endangered

ADULT LENGTH
19¾–23¾ in
(500–600 mm)

486

AIPYSURUS APRAEFRONTALIS
SHORT-NOSED SEASNAKE
SMITH, 1926

The Short-nosed Seasnake is believed to be endemic to Ashmore and Hibernia reefs, in the Timor Sea, north of Western Australia, although it may also occur in suitable shallow coral reef flats on the Western Australian coast. Since 2000 there has been a steep decline in the number of seasnakes occurring on these reefs, with the frequency of encountered snakes dropping from at least 40 snakes of nine species a day, to under seven snakes of two species a day. This decline has raised fears for those species believed endemic to these reefs, resulting in this species and others being listed as Critically Endangered by the IUCN. Its biology is almost unknown, other than that it is a live-bearer, and there is one record of a specimen containing an eel.

RELATED SPECIES

Two of the closest relatives of *Aipysurus apraefrontalis*, the Leaf-scaled Seasnake (*A. foliosquama*) and Dusky Seasnake (*A. fuscus*), are also endemic to Ashmore and Hibernia Reefs, and are respectively listed as Critically Endangered and Endangered due to the sudden and inexplicable decline in seasnake numbers and diversity.

The Short-nosed Seasnake is dark brown or purple-brown with pale bands of cream or olive-brown, which may be obscured dorsally. It has a small, short, pointed head and small eyes. All seasnakes have paddle-shaped tails.

Actual size

FAMILY	Elapidae: Hydrophiinae
RISK FACTOR	Highly venomous: postsynaptic neurotoxins, and possibly also myotoxins, with secondary nephrotoxicity and cardiotoxicity
DISTRIBUTION	Timor, Arafura, and Coral seas: Australia, New Guinea, and New Caledonia
ELEVATION	Sea-level to 260 ft (80 m) bsl
HABITAT	Shallow reef flats, coral bommies, and seaweed beds
DIET	Fish (moray eels, gobies, blennies, parrotfish, and surgeonfish)
REPRODUCTION	Viviparous, with litters of 2 neonates
CONSERVATION STATUS	IUCN Least Concern

ADULT LENGTH
2 ft 7 in–3 ft 3 in,
rarely 5 ft
(0.8–1.0 m,
rarely 1.5 m)

AIPYSURUS DUBOISII
REEF SHALLOWS SEASNAKE
BAVAY, 1869

487

The Reef Shallows Seasnake, also known as Dubois' Seasnake, is a small species with short fangs and a low venom yield, yet its venom is extremely toxic, comparable to that of the Eastern Brownsnake (*Pseudonaja textilis*, page 529) or Inland Taipan (*Oxyuranus microlepidotus*, page 521), making it one of the most venomous snakes in the world. It occurs from the Timor Sea, west of Australia, to New Caledonia in the Coral Sea, east of Australia, where it inhabits shallow coral flat platforms and the upper levels of coral bommies, diving to depths of 165 or 260 ft (50 or 80 m). It also inhabits seaweed beds and is often camouflaged by a growth of seaweed on its scales. It is a generalist feeder, feeding on a variety of fish that are ambushed or actively hunted on the reef.

The Reef Shallows Seasnake may be either uniform dark brown, black-brown with pale markings, or pale yellow-brown with darker bands, although this patterning may be obscured by a growth of seaweed.

RELATED SPECIES
Several related seasnakes occur in the same locality as *Aipysurus duboisii* on Ashmore and Hibernia Reefs in the Timor Sea: the Short-nosed Seasnake (*Aipysurus apraefrontalis*, page 486), Leaf-scaled Seasnake (*A. foliosquama*), and Dusky Seasnake (*A. fuscus*).

Actual size

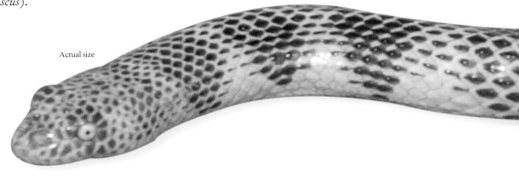

FAMILY	Elapidae: Hydrophiinae
RISK FACTOR	Highly venomous: postsynaptic neurotoxins, and possibly also myotoxins, with secondary nephrotoxicity and cardiotoxicity
DISTRIBUTION	Timor, Arafura, and Coral seas: eastern Indonesia, Timor-Leste, northern and eastern Australia, New Guinea, and New Caledonia
ELEVATION	Sea-level to 165 ft (50 m) bsl
HABITAT	Coral flats and bommies
DIET	Fish (snappers, catfish, surgeonfish, cardinalfish, damselfish, groupers, sweepers, parrotfish, and stonefish), and shrimps
REPRODUCTION	Viviparous, with litters of 2–5 neonates
CONSERVATION STATUS	IUCN Least Concern

ADULT LENGTH
2 ft 7 in–4 ft,
occasionally 6 ft 7 in
(0.8–1.2 m,
occasionally 2.0 m)

AIPYSURUS LAEVIS
OLIVE SEASNAKE
LACÉPÈDE, 1804

488

The Olive Seasnake is a large, stocky snake, with variable patterning, from dark brown to purple-brown or olive-brown markings that fade on the flanks, and with cream or yellow undersides and lower flanks.

The nine species of seasnakes in genus *Aipysurus* are known as the pipe seasnakes. The Olive Seasnake is the largest species in the genus. It is widely distributed from the Timor Sea to the Coral Sea, where it inhabits coral flats and bommies (coral stacks), to depths of 165 ft (50 m) bsl. Olive Seasnakes are generalist feeders, taking a variety of fish, including highly venomous stonefish, and crustaceans such as shrimps. Olive Seasnakes demonstrate "site fidelity," being associated with one coral bommie for long periods. They are also curious snakes that investigate divers closely. Although inoffensive and disinclined to bite, unless roughly handled, this is one of the few seasnakes capable of biting through a wetsuit. As it is highly venomous, this presents a serious scenario.

RELATED SPECIES
Probably the closest relative of *Aipysurus laevis* is its former subspecies, the Shark Bay Seasnake (*A. pooleorum*), from southwestern Australia.

Actual size

FAMILY	Elapidae: Hydrophiinae
RISK FACTOR	Venomous: postsynaptic neurotoxins, and possibly also myotoxins, but venom almost depleted
DISTRIBUTION	Arafura and Coral seas: northern Australia and southern Papua
ELEVATION	Sea-level to 165 ft (50 m) bsl
HABITAT	Deep water, river mouths, and turbid estuaries over mangrove mud
DIET	Benthic goby eggs
REPRODUCTION	Viviparous, with litters of fewer than 4 neonates
CONSERVATION STATUS	IUCN Least Concern

ADULT LENGTH
19¼–36 in
(490–915 mm)

AIPYSURUS MOSAICUS

MOSAIC SEASNAKE

SANDERS, RASMUSSEN, ELMBERG, MUMPUNI, GUINEA, BLIAS, LEE & FRY, 2012

489

While other members of the genus *Aipysurus* inhabit the clear waters found over coral reefs and substrates of coral sand, the Mosaic Seasnake, and the related Eydoux's Seasnake (*A. eydouxii*), occur in the turbid water found offshore and at estuary mouths, over mangrove mud substrates that cloud the water considerably. Also in common with Eydoux's Seasnake, the Mosaic Seasnake feeds on the eggs of benthic gobies, extracted from their seabed burrows. Neither species has any requirement for venom, and both are evolving toward a potentially nonvenomous state, through the atrophy of the venom glands and reduction in fang size. Mosaic Seasnakes are encountered at night, resting on the surface in the waters offshore from large estuaries, or are trawled up from deeper waters during the day.

RELATED SPECIES

Aipysurus mosaicus was only recently described for the southern population of the widely distributed Eydoux's Seasnake (*A. eydouxii*), which is its closest relative. *Aipysurus eydouxii* is the only known *Aipysurus* species to occur in Southeast Asian waters.

The Mosaic Seasnake is cream or salmon-colored dorsally, with irregular dark brown cross-bands, while the head is brown. The hexagonal scales are black-edged, presenting a mosaic pattern throughout.

Actual size

FAMILY	Elapidae: Hydrophiinae
RISK FACTOR	Venomous: venom composition unknown
DISTRIBUTION	New Guinea: New Ireland, New Britain, New Guinea and its satellite islands, and Seram
ELEVATION	0–3,280 ft (0–1,000 m) asl
HABITAT	Lowland and montane rainforest, plantations, and gardens
DIET	Lizards
REPRODUCTION	Oviparous, with clutches of 3–5 eggs
CONSERVATION STATUS	IUCN Least Concern

ADULT LENGTH
23¾–28½ in
(600–725 mm)

ASPIDOMORPHUS MUELLERI
MÜLLER'S CROWNED SNAKE
(SCHLEGEL, 1837)

Müller's Crowned Snake can be quite variable in patterning, most specimens being light or dark brown, with white lips and collar, a dark throat, and a "humbug" marbled pattern on the crown, which earns them their common name. Some specimens lack any markings.

The crowned snake genus *Aspidomorphus* is endemic to the New Guinea region. Müller's Crowned Snake is the most widely distributed species, from New Ireland and New Britain in the Bismarck Archipelago, throughout New Guinea, onto the islands of Milne Bay Province, and in Indonesian New Guinea, on the Schouten and Raja Ampat islands, and as far west as Seram. These relatively small elapids can be fairly common, hiding under logs in rainforest, in plantations, or in garden trash, where they hunt primarily skinks. Although small, these snakes are tenacious and will bite easily. Fortunately those few bites on record have resulted in only localized pain and nausea, but like any small elapid this species should be treated with respect. Salomon Müller (1804–64) was a German naturalist who collected in the Dutch East Indies.

RELATED SPECIES

The Striped Crowned Snake (*Aspidomorphus lineaticollis*) occurs in northern Papua New Guinea, east of the Sepik River and onto the islands of Milne Bay Province, while Schlegel's Crowned Snake (*A. schlegelii*) inhabits northwestern New Guinea, and Papua New Guinea west of the Sepik. Both have much shorter tails than *A. muelleri*.

Actual size

FAMILY	Elapidae: Hydrophiinae
RISK FACTOR	Highly venomous: postsynaptic neurotoxins, also cytotoxins, and possibly myotoxins and anticoagulants
DISTRIBUTION	Australia: Victoria, Tasmania, and Bass Strait islands
ELEVATION	0–6,970 ft (0–2,125 m) asl
HABITAT	Moist habitats, marshes, woodland, coastal heath, and tussock grassland
DIET	Lizards and frogs
REPRODUCTION	Viviparous, with litters of up to 15 neonates
CONSERVATION STATUS	IUCN not listed

ADULT LENGTH
4 ft–5 ft 7 in,
rarely 6 ft
(1.2–1.7 m,
rarely 1.8 m)

AUSTRELAPS SUPERBUS
LOWLANDS COPPERHEAD
(GÜNTHER, 1858)

491

Australian copperheads are not related to the American Copperhead (*Agkistrodon contortrix*, page 553), which is a pitviper. The Lowlands Copperhead occurs in southern Victoria and on both Tasmania and the islands of the Bass Strait, the largest specimens occurring on King Island. Tasmania only has three species of snakes, and they have two things in common: they are all elapids, and they are all live-bearers, an adaptation for living in cooler conditions. Lowlands Copperheads are powerful snakes that inhabit swamps, marshes, damp areas in woodlands, coastal heathland, and tussock grassland. In Tasmania they are often found in sympatry with Tigersnakes (*Notechis scutatus*, page 519). They are active both by day and night and hunt lizards and frogs. Snakebites are rare occurrences, but are serious given the high toxicity of the venom.

The Lowlands Copperhead is a powerful, muscular snake with strong neck muscles and a pointed snout. It varies from uniform light brown to red-brown or chocolate-brown in color, and white or pale brown below.

RELATED SPECIES

Southeastern Australia is home to both of the other Australian copperheads, the Highlands Copperhead (*Austrelaps ramsayi*), which also occurs in New South Wales, and the Pygmy Copperhead (*A. labialis*), which is confined to Kangaroo Island and extreme southern South Australia and is listed as Vulnerable by the IUCN.

Actual size

FAMILY	Elapidae: Hydrophiinae
RISK FACTOR	Venomous: venom composition unknown
DISTRIBUTION	Australia: Western Australia and South Australia
ELEVATION	0–2,640 ft (0–805 m) asl
HABITAT	Arid and mesic habitats, hummock grassland, scrub, coastal heath, and dunes
DIET	Reptile eggs
REPRODUCTION	Oviparous, with clutches of 3 eggs
CONSERVATION STATUS	IUCN not listed

ADULT LENGTH
11¾–13¾ in,
rarely 15¾ in
(300–350 mm,
rarely 400 mm)

492

BRACHYUROPHIS SEMIFASCIATUS
SOUTHERN SHOVEL-NOSED SNAKE
GÜNTHER, 1863

The Southern Shovel-nosed Snake is a smooth-scaled snake with a pointed, shovel-shaped snout for burrowing. The head bears a black cap while the body is orange to red with regular dark brown bands that extend halfway down the flanks.

Shovel-nosed snakes are nocturnal and semi-fossorial, with a wedge-shaped snout specially adapted for burrowing. The Southern Shovel-nosed Snake, also known as the Half-girdled Snake because the markings only extend part way down its flanks, inhabits both arid and mesic habitats, ranging from coastal dunes and heaths to inland grassland and scrubland, where it shelters during the day under logs or stumps. Some species feed on skinks but the Southern Shovel-nosed Snake feeds entirely on reptile eggs. It has specialized bladelike teeth at the rear of the jaw with which it slits the soft-shelled eggs open, before devouring both contents and shell. Although an elapid, this is an inoffensive snake. Shovel-nosed snakes are an example of an elapid evolving to occupy a niche that elsewhere would be occupied by a nonvenomous colubrid snake, like a kukri snake (*Oligodon*, pages 198–200).

RELATED SPECIES

There are eight species of shovel-nosed snakes distributed across most of Australia, except the southeast. Some are banded like *Brachyurophis semifasciatus*, including the related Einasleigh Shovel-nosed Snake (*B. campbelli*) and the Northern Shovel-nosed Snake (*B. roperi*), but the Unbanded Shovel-nosed Snake (*B. incinctus*) and Arnhem Shovel-nosed Snake (*B. morrisi*) have unicolor bodies and only neck and nape markings.

Actual size

FAMILY	Elapidae: Hydrophiinae
RISK FACTOR	Venomous: venom composition unknown
DISTRIBUTION	Australia: eastern tropical Queensland
ELEVATION	0–2,810 ft (0–855 m) asl
HABITAT	Rainforest and eucalypt forest
DIET	Lizards
REPRODUCTION	Oviparous, with clutches of up to 5 eggs
CONSERVATION STATUS	IUCN not listed

ADULT LENGTH
11¾–17¾ in
(300–450 mm)

CACOPHIS CHURCHILLI
NORTHERN DWARF CROWNED SNAKE
WELLS & WELLINGTON, 1985

493

The Northern Dwarf Crowned Snake is one of the smallest members of its genus, and also the one with the smallest range, being confined to the coastal strip of tropical and subtropical Queensland, from Mossman to Townsville, inland to the Atherton Tableland and Bluewater Range, and offshore on Lindeman and Magnetic Islands. This diminutive snake feeds on skinks, which it hunts in the leaf litter or under logs in its rainforest or eucalypt forest habitat. Secretive and nocturnal, the Northern Dwarf Crowned Snake defends itself vigorously when uncovered, elevating its body and making wild strikes or thrashing its body about, but strikes are often with the mouth closed, and it is reluctant to bite. However, as its venom composition remains unstudied, caution is advised and bites should be avoided.

The Northern Dwarf Crowned Snake is a small, smooth-scaled snake, blue-gray to brown above, pale gray underneath, and with a narrow cream, light gray, or yellowish collar.

RELATED SPECIES

Three other species of small crowned snakes are found along the coast of southern Queensland and New South Wales, the White-crowned snake (*Cacophis harriettae*), the Southern Dwarf Crowned Snake (*C. krefftii*), and, at up to 29½ in (750 mm), the largest species, the Golden-crowned Snake (*C. squamulosus*). The fact that they share a common name does not mean that they are closely related to the New Guinea crowned snakes (*Aspidomorphus*, page 490).

Actual size

FAMILY	Elapidae: Hydrophiinae
RISK FACTOR	Venomous: venom composition unknown
DISTRIBUTION	Australia: Queensland (Prince of Wales Island, Torres Strait)
ELEVATION	Sea-level
HABITAT	Open eucalypt and paperbark woodland
DIET	Unknown, presumably lizards
REPRODUCTION	Viviparous, litter size unknown
CONSERVATION STATUS	IUCN not listed

ADULT LENGTH
15¾ in
(400 mm)

494

CRYPTOPHIS INCREDIBILIS
PINK SNAKE
(WELLS & WELLINGTON, 1985)

The Pink Snake is a small, uniform coral pink snake with no additional markings. Its small, protruding black eyes stand out against this background color.

One of the least known Australian elapids, and also one with an extremely limited distribution, the Pink Snake is only recorded from Prince of Wales Island in the southern Torres Strait, Queensland. It is possible that a search of similar habits on other islands in the southern Torres Strait may extend its known range. The Pink Snake is thought to inhabit open eucalypt and paperbark woodlands, on sandy soils, but the few specimens known were found under flotsam in the upper strand zone between the woodland and the beach. It is probably nocturnal, and a live-bearer that feeds on skinks, like other members of the Australian small-eyed snake genus *Cryptophis*. This is the smallest species known within the genus.

RELATED SPECIES
The ranges of two Australo-Papuan relatives, the Carpentaria Snake (*Cryptophis boschmai*) and the Black-striped Snake (*C. nigrostriatus*, page 495), overlap the range of *C. incredibilis*, but it is not known if they occur in sympatry with those species.

Actual size

FAMILY	Elapidae: Hydrophiinae
RISK FACTOR	Venomous: myotoxins
DISTRIBUTION	Australo-Papua: Australia (Queensland) and Papua New Guinea (Western Province)
ELEVATION	0–2,810 ft (0–855 m) asl
HABITAT	Sclerophyll forest, woodland, savanna woodland, and gardens
DIET	Lizards
REPRODUCTION	Viviparous, with litters of 4–9 neonates
CONSERVATION STATUS	IUCN not listed

ADULT LENGTH
11¾–22 in,
rarely 24¼ in
(300–560 mm,
rarely 615 mm)

CRYPTOPHIS NIGROSTRIATUS
BLACK-STRIPED SNAKE
(KREFFT, 1864)

495

The Black-striped Snake occurs in coastal Queensland, and
also in southern Papua New Guinea, in the southern Trans-
Fly region of Western Province, although it is relatively rare
here and much more commonly encountered in Queensland.
In Queensland it inhabits sclerophyll forest and woodland,
but in Papua New Guinea it has been collected in savanna
woodland in association with termite mounds, and also in
cleared gardens. This is a semi-fossorial species, secretive and
usually active nocturnally, although specimens are also found
in the daytime. Prey seems to consist entirely of skinks, which
abound in all these habitats. Although the Black-striped Snake
is inoffensive and disinclined to bite, it should be noted that
other *Cryptophis* species have caused bites, with the similar-
sized Eastern Small-eyed Snake (*C. nigrescens*) causing a
fatality, due to kidney failure.

The Black-striped Snake is a slender snake with
smooth scales and small, protruding eyes. It is
red in color with a black cap to its head and a
broad black stripe down the back to the tail.
The undersides are pale cream.

RELATED SPECIES
The genus *Cryptophis*, sometimes called *Rhinoplocephalus*,
contains five species. Most species occur in eastern Australia,
although two also occur in southern New Guinea, including
the Carpentaria Snake (*C. boschmai*), and one species, the
Northern Small-eyed Snake (*C. pallidiceps*), occurs from
the Kimberley of Western Australia to Arnhem Land,
Northern Territory.

Actual size

FAMILY	Elapidae: Hydrophiinae
RISK FACTOR	Venomous: venom composition unknown
DISTRIBUTION	Australia: southern Australia, north to southern Cape York, and also Pilbara, Western Australia
ELEVATION	0–3,590 ft (0–1,095 m) asl
HABITAT	Coastal heath and forest, eucalypt forest, hummock grassland, dry woodland, and arid scrubland
DIET	Lizards
REPRODUCTION	Oviparous, with clutches of 5–20 eggs
CONSERVATION STATUS	IUCN not listed, Near Threatened (Victoria)

ADULT LENGTH
2 ft 7 in–3 ft 3 in
(0.8–1.0 m)

496

DEMANSIA PSAMMOPHIS
YELLOW-FACED WHIPSNAKE
(SCHLEGEL, 1837)

The Yellow-faced Whipsnake varies across its range. The brightest patterned western specimens have brown heads, yellow cheeks with a black spot under the large eye, greenish to olive bodies, every scale edged black, and a reddish-brown tail. Eastern specimens are usually olive to gray, often with two brown stripes down the back.

The Yellow-faced Whipsnake has a fragmented distribution from coast to coast across much of southern and central Australia, ranging north to the Pilbara, Western Australia, and southeastern Cape York Peninsula, Queensland. It inhabits a wide variety of habitats from eucalypt forest to coastal heath, and grasslands to woodland and scrub. Fast-moving diurnal predators of lizards such as agamids and skinks, whipsnakes are frequently seen dashing about in the middle of the day. The eastern population of the Yellow-faced Whipsnake is known to nest communally, with nest sites containing several hundred eggs. The small and localized Victorian population is considered Near Threatened, although this species is not currently listed by the IUCN.

RELATED SPECIES

The eastern population is the nominate subspecies (*Demansia psammophis psammophis*), while the western population is treated as a separate subspecies (*D. p. cupreiceps*). The Reticulated Whipsnake (*D. reticulata*), from southwestern Western Australia, was once also a subspecies of *D. psammophis*.

Actual size

FAMILY	Elapidae: Hydrophiinae
RISK FACTOR	Venomous: weakly coagulant, otherwise venom composition unknown
DISTRIBUTION	Australo-Papua: Australia (Western Australia and Northern Territory) and southern New Guinea
ELEVATION	0–1,670 ft (0–510 m) asl
HABITAT	Savanna and savanna woodland
DIET	Lizards and frogs
REPRODUCTION	Oviparous, with clutches of 4–13, rarely 20, eggs
CONSERVATION STATUS	IUCN not listed

ADULT LENGTH
2 ft 7 in–3 ft 3 in,
rarely 4 ft
(0.8–1.0 m,
rarely 1.2 m)

DEMANSIA VESTIGIATA
LESSER BLACK WHIPSNAKE
(DE VIS, 1884)

497

The Lesser Black Whipsnake is a highly alert diurnal predator of lizards, from skinks to agamids, and also frogs. It inhabits savanna and savanna woodland in southern New Guinea, and northern Australia from the Kimberley to Arnhem Land. It is one of the few snakes seen abroad during the heat of the day, when it carries itself with head and anterior body raised, large eyes alert for prey and predators alike. It moves extremely quickly, rushing across roads in the blink of an eye. Whipsnakes avoid humans and escape rather than bite, but if handled they will bite freely. Although the venom is believed to be weakly coagulant, rather than neurotoxic, it causes localized pain and should be treated with caution as there is at least one fatality on record.

RELATED SPECIES

Demansia vestigiata is just one of 14 species of Australian whipsnake, but the only species known to occur in New Guinea. It is related to the larger Greater Black Whipsnake (*D. papuensis*), which, despite its scientific name, is an Australian endemic.

The Lesser Black Whipsnake is a slender-bodied, black or olive-brown snake, each scale edged with dark pigment to present a reticulate pattern, and with a very long, reddish-brown, whiplike tail. The undersides are white anteriorly and gray posteriorly, and the head is elongate and the eyes are large.

Actual size

FAMILY	Elapidae: Hydrophiinae
RISK FACTOR	Venomous: venom composition unknown
DISTRIBUTION	Australia: southeastern South Australia, New South Wales, southern Victoria, and Tasmania
ELEVATION	0–6,970 ft (0–2,125 m) asl
HABITAT	Tussock grassland and mossy wetlands
DIET	Skinks, and also skink eggs, frogs, and small mammals
REPRODUCTION	Viviparous, with litters of 2–10 neonates
CONSERVATION STATUS	IUCN not listed

ADULT LENGTH
15¾–17¾ in
(400–450 mm)

498

DRYSDALIA CORONOIDES
WHITE-LIPPED SNAKE
(GÜNTHER, 1858)

The White-lipped Snake is a small snake, brown to black above, and cream to yellow or pink below. It has a black stripe along the side of the head, and underneath the bold white stripe that earns it its common name.

One of only three snakes occurring on Tasmania, the White-lipped Snake, which also inhabits the southeastern corner of mainland Australia, is a cold-adapted species. Being a live-bearer is an adaptation that permits the species to inhabit latitudes where eggs would perish. Being active during the day is also important as night-time temperatures may be too cold for reptiles to function. Being small is also useful, as the snake can warm quickly in the sun and easily hide away. Feeding primarily on diurnal skinks, but also taking skink eggs, frogs, and occasionally small mammals, the White-lipped Snake inhabits tussock grasslands and mossy habitats, often in relatively wet areas, where it shelters under fallen logs or within the tussocks. Its venom composition is unknown, but bites are known to cause pain.

RELATED SPECIES

There are two other species of *Drysdalia*, Master's Snake (*D. mastersii*), found along the south coast of Australia, and the Rose-bellied Snake (*D. rhodogaster*) from southeastern New South Wales.

Actual size

FAMILY	Elapidae: Hydrophiinae
RISK FACTOR	Venomous: probably neurotoxins, otherwise unknown
DISTRIBUTION	Australia: southern Western Australia, South Australia, New South Wales, and Victoria
ELEVATION	0–2,100 ft (0–640 m) asl
HABITAT	Heathland, mallee woodland, and spinifex grassland
DIET	Lizards, frogs, birds, small mammals, and insects
REPRODUCTION	Viviparous, with litters of 3–14 neonates
CONSERVATION STATUS	IUCN Near Threatened, Vulnerable (Victoria), Endangered (New South Wales)

ADULT LENGTH
15¼–23¾ in,
rarely 28 in
(400–600 mm,
rarely 710 mm)

ECHIOPSIS CURTA
BARDICK
(SCHLEGEL, 1837)

499

The Bardick, an Aboriginal name, is a cool-adapted, live-bearing snake. It occurs in three separate southern populations from Western Australia to New South Wales. It inhabits heathland, mallee woodland, and spinifex grassland, a combination of semiarid eucalypt woodland with grassland that experiences seasonal rain. A primarily nocturnal, occasionally diurnal, species, it shelters under logs, emerging to hunt a wide variety of vertebrate and invertebrate prey, which is largely captured from ambush rather than actively foraged. In many ways it is similar in its biology to the death adders (*Acanthophis*, pages 484–485). The venom composition of the Bardick is poorly known, but is probably neurotoxic and may be potentially dangerous. Threats to the Bardick include fires, overgrazing, and heathland clearance.

The Bardick is short and stout, hence the name *curta*. It is a uniform reddish-brown or gray-brown snake with a broad head, sometimes with white lateral flecks, and vertically elliptical pupils.

RELATED SPECIES

The closest relative of *Echiopsis curta* is probably the rare Lake Cronin Snake (*Paroplocephalus atriceps*, page 525) from south-central Western Australia.

Actual size

FAMILY	Elapidae: Hydrophiinae
RISK FACTOR	Venomous: postsynaptic neurotoxins, and possibly myotoxins, but venom almost depleted
DISTRIBUTION	Timor and Coral seas: Ashmore and Hibernia Reefs (Western Australia), eastern Queensland, and New Caledonia
ELEVATION	Sea-level to 82 ft (25 m) bsl
HABITAT	Shallow, clear coral reefs
DIET	Eggs of coral-dwelling and benthic fish
REPRODUCTION	Viviparous, with litters of 2–5 neonates
CONSERVATION STATUS	IUCN Least Concern

ADULT LENGTH
28–35¾ in
(715–910 mm)

EMYDOCEPHALUS ANNULATUS
SOUTHERN TURTLE-HEADED SEASNAKE
KREFFT, 1869

The Southern Turtle-headed Seasnake is sexually dichromatic and dimorphic. Females are yellow or banded with irregular black or brown rings, while males are melanistic jet black. The male also has a spine on its snout that it uses to stroke the female's back during courtship (see page 27), similar to the way male pythons and boas use their spurs. The second supralabial scale is greatly enlarged for scraping fish eggs from coral.

The Southern Turtle-headed Seasnake gets its name from the enlarged second supralabial scale on the lip, which resembles the sharp lip of a sea turtle. This scale is used to scrape the eggs of coral-nesting gobies, blennies, clownfish, and damselfish off the coral. This species is a diurnal inhabitant of shallow coral reefs on Ashmore Reef and neighboring Timor Sea reefs in Western Australia, and eastern Queensland to New Caledonia in the Coral Sea. It can be very common, with males often seen following and courting females. Although the Southern Turtle-headed Seasnake is technically venomous, its diet has led to atrophy of its venom glands, low venom yield and toxicity, and the shrinking of the fangs as the species slowly evolves toward becoming nonvenomous. These snakes are extremely inoffensive.

RELATED SPECIES

Turtle-headed seasnakes are related to the pipe seasnakes (*Aipysurus*, pages 486–489). Two other species are known, the Japanese Turtle-headed Seasnake (*Emydocephalus ijimae*) and the Vietnamese Turtle-headed Seasnake (*E. szczerbaki*).

Actual size

FAMILY	Elapidae: Hydrophiinae
RISK FACTOR	Venomous: postsynaptic neurotoxins, and possibly myotoxins
DISTRIBUTION	Timor Sea: Western Australia
ELEVATION	Sea-level
HABITAT	Mangrove swamps and mudflats
DIET	Mudskippers and other gobies
REPRODUCTION	Viviparous, litter size unknown
CONSERVATION STATUS	IUCN Least Concern

ADULT LENGTH
19¾–26 in
(500–660 mm)

EPHALOPHIS GREYAE

NORTHWESTERN MANGROVE SEASNAKE

SMITH, 1921

501

One of the most terrestrial of seasnakes, the Northwestern Mangrove Seasnake inhabits mudflats and mangrove swamps, where it hunts mudskippers and other gobies on the mud or in their burrows. It easily moves on the mud because it retains the broad ventral scales of terrestrial snakes, unlike more marine-adapted seasnakes. A small species, confined to coastal Western Australia, between Shark Bay and King Sound, it is very poorly known, and although a live-bearer, like other true seasnakes, its litter size is unknown. Its scientific name was originally written *greyi* by the describer, Malcolm Smith, but some authorities believe it should be corrected to *greyae* as it was named in honor of its collector, Beatrice Grey. The suffix *-i* or *-ii* is for a male while *-ae* is for a female.

RELATED SPECIES

Other mangrove and mudflat-dwelling seasnakes include the Port Darwin Seasnake (*Hydrelaps darwiniensis*, page 506) and Arafura Smooth Seasnake (*Parahydrophis mertoni*, page 523) from the northern Australian coastline.

The Northwestern Mangrove Seasnake is a small snake that is cream to pale gray dorsally, with irregular dark gray blotches, which may form a zigzag, and bands on its tail.

Actual size

FAMILY	Elapidae: Hydrophiinae
RISK FACTOR	Venomous: possibly neurotoxins and procoagulants, otherwise unknown
DISTRIBUTION	Australia: northern and central Western Australia, Northern Territory, and northern South Australia, to Cape York Peninsula, Queensland
ELEVATION	0–3,330 ft (0–1,015 m) asl
HABITAT	Tropical woodland, savannas and grasslands, floodplains, semidesert, and sandplains
DIET	Lizards
REPRODUCTION	Oviparous, with clutches of up to 1–6 eggs
CONSERVATION STATUS	IUCN not listed

ADULT LENGTH
23½–27½ in
(650–700 mm)

FURINA ORNATA
MOON SNAKE
(GRAY, 1842)

The Moon Snake is red-brown dorsally, often with dark edges to the scales, presenting a reticulate appearance. The head and anteriormost body is dark brown with a broad orange or yellow nape band across the neck. This species is smooth-scaled and glossy in appearance. Its eyes are small, dark, and bead-like. Males are slightly smaller and more brightly colored than females.

Also known as the Orange-naped Snake, the Moon Snake is the most widely distributed member of its genus, being found across most of northern, central, and western Australia, in habitats ranging from tropical woodland to grassland, sandplains, and semidesert. It may be seen crossing roads, and if confronted it will raise the anterior portion of its body off the ground in threat. It does not flatten its neck like cobras (*Naja*, pages 466–479) or blacksnakes (*Pseudechis*, pages 526–527), or form a fast-moving S-shape, like brownsnakes (*Pseudonaja*, pages 528–529), but it will sway and make mock closed-mouth strikes. Despite its display this is a relatively inoffensive species, preferring to thrash about rather than bite. Although nocturnal, the Moon Snake feeds on diurnal skinks, which are presumably sought while they are asleep.

RELATED SPECIES

Furina ornata is most closely related to the Yellow-naped snake (*F. barnardi*), Red-naped snake (*F. diadema*), and Dunmall's Snake (*F. dunmalli*).

Actual size

FAMILY	Elapidae: Hydrophiinae
RISK FACTOR	Venomous: possibly neurotoxins and procoagulants, otherwise unknown
DISTRIBUTION	Australo-Papua: southern New Guinea, Aru Islands (Indonesia), and Cape York Peninsula (Queensland, Australia)
ELEVATION	0–260 ft (0–80 m), possibly 1,640 ft (500 m) asl
HABITAT	Monsoon forest, rainforest, savanna woodland, gardens, plantations, and even into cities
DIET	Lizards
REPRODUCTION	Oviparous, with clutches of up to 6 eggs
CONSERVATION STATUS	IUCN not listed

ADULT LENGTH
27½–31½ in,
rarely 3 ft 3 in
(700–800 mm,
rarely 1.0 m)

FURINA TRISTIS
BROWN-HEADED SNAKE
(GÜNTHER, 1858)

503

An alternative name for this species is Gray-naped Snake, and although rare in Australia, being confined to northern Cape York Peninsula, it is common in most lowland habitats in southern New Guinea, from rainforest to savanna woodland, and into village gardens and coconut plantations, where it shelters under trash and logs. The Brown-headed Snake is small enough to remain hidden even in relatively large towns, including the Papua New Guinea capital, Port Moresby. Prey consists of the abundant skinks and geckos that occur in these habitats. Although it is not generally considered dangerous to humans, there is a snakebite on record that caused some unpleasant symptoms, including abdominal pain, headache, respiratory distress, nausea, and diarrhea. As with any relatively unknown small elapid, respect is required and handling should be avoided.

The Brown-headed Snake is a smooth-scaled, slender snake with an elongate, rounded head and small, bead-like eyes. It may be glossy black, gray, or brown, either unmarked or with light edges to the scales, presenting a reticulate appearance. The head is often dark brown with a broad yellowish or grayish nape marking across the neck, though not all specimens possess this.

RELATED SPECIES

At one time this species was placed in the genus *Glyphodon*. It is now in genus *Furina* with four other Australian species: the Yellow-naped snake (*F. barnardi*), Red-naped snake (*F. diadema*), Dunmall's Snake (*F. dunmalli*), and Moon Snake or Orange-naped Snake (*F. ornata*, page 502).

Actual size

FAMILY	Elapidae: Hydrophiinae
RISK FACTOR	Venomous: procoagulants, otherwise unknown
DISTRIBUTION	Australia: coastal New South Wales and Queensland
ELEVATION	0–4,330 ft (0–1,320 m) asl
HABITAT	Swamps and creeks, rocky outcrops, coastal heathland, tropical rainforest, and dry sclerophyll forest
DIET	Lizards and frogs
REPRODUCTION	Viviparous, with litters of 4–20 neonates
CONSERVATION STATUS	IUCN not listed

ADULT LENGTH
23¾–27½ in
(600–700 mm)

504

HEMIASPIS SIGNATA

BLACK-BELLIED
SWAMP SNAKE

(JAN, 1859)

An eastern coastal Australian species, the Black-bellied Swamp Snake is found in a wide variety of habitats, from woodland bordering beaches to freshwater swamps and creeks, heathland, and both wet and dry forest. It is found along the New South Wales–Queensland coastline, with two isolated populations farther north on the Queensland coast, and a population on Fraser Island. The prey of this generally diurnal snake consists of skinks and frogs. In hot weather it may become more crepuscular or nocturnal. Bites reportedly cause pain, but this is not considered a dangerous species to humans. Even so, caution with all small elapids is always recommended.

The Black-bellied Swamp Snake is uniform brown, olive, or even black above, paler on the lower flanks, with two bold cream or yellow stripes on each side of the head, and usually a black belly, although the northern Queensland populations are reddish above and cream or pink below, which rather belies the common name.

RELATED SPECIES

The only other species in genus *Hemiaspis* is Damel's Snake, also known as the Gray Snake (*H. damelii*), a gray snake with a black nape band, occurring in woodland habitats in southern Queensland and central New South Wales.

Actual size

FAMILY	Elapidae: Hydrophiinae
RISK FACTOR	Highly venomous: procoagulants, and also possibly neurotoxins and myotoxins
DISTRIBUTION	Australia: Blue Mountains of coastal New South Wales
ELEVATION	0–2,430 ft (0–740 m) asl
HABITAT	Hawkesbury Sandstone outcrops
DIET	Geckos
REPRODUCTION	Viviparous, with litters of 8–20 neonates
CONSERVATION STATUS	IUCN Vulnerable, CITES Appendix II, Endangered in New South Wales

ADULT LENGTH
19¾–21¾ in,
rarely 35½ in
(500–550 mm,
rarely 900 mm)

HOPLOCEPHALUS BUNGAROIDES
BROAD-HEADED SNAKE
SCHLEGEL, 1837

505

One of the most distinctive New South Wales snakes, the Broad-headed Snake is only found on the Hawkesbury Sandstone ridges of the Blue Mountains, from Sydney to Nowra. It is only found under slabs of eroded sandstone, in rock-on-rock situations, never in rock-on-soil, which means it is generally found at the very edges of the rocky outcrops. With such a precise and limited distribution and habitat, this small snake is severely threatened by habitat loss due to the illegal collection of "bushrock" to enhance Sydney gardens. The Broad-headed Snake feeds primarily on Lesueur's Velvet Gecko (*Amalosia lesueurii*) but will take other geckos and sometimes skinks. This snake is considered highly venomous, the venom inducing coagulopathy (inability of the blood to clot) and uncontrolled bleeding, similar to that caused by the brownsnakes (*Pseudonaja*, pages 528–529).

RELATED SPECIES

Two other species of *Hoplocephalus* are found farther north in coastal New South Wales and Queensland, the Pale-headed Snake (*H. bitorquatus*) and Stephen's Banded Snake (*H. stephensii*), Australia's only truly arboreal elapids.

The Broad-headed Snake is a relatively slender, boldly patterned snake, being yellow with distinct black blotches and saddles dorsally, and off-white with black scale edging below. Its defensive posture involves raising its anterior body in an S-shape and flattening its already broad head in order to emphasize the threat. From this position it will make rapid open-mouthed strikes with a considerable reach.

Actual size

FAMILY	Elapidae: Hydrophiinae
RISK FACTOR	Venomous: postsynaptic neurotoxins, and possibly myotoxins
DISTRIBUTION	Timor and Arafura Seas: Northern Territory and Queensland (Australia), and southern New Guinea
ELEVATION	Sea-level
HABITAT	Mangrove swamps and mudflats
DIET	Mudskippers and other gobies
REPRODUCTION	Believed viviparous, litter size unknown
CONSERVATION STATUS	IUCN not listed

ADULT LENGTH
17–17½ in
(435–445 mm)

506

HYDRELAPS DARWINIENSIS
PORT DARWIN SEASNAKE
BOULENGER, 1896

The Port Darwin Seasnake is small, and patterned with dark rings on a pale white to yellow background, the rings either completely encircling the body or breaking to form staggered Y-shaped markings along the back.

This is a small, intertidal species from northern Australia, and also southern New Guinea, where it is rare. It lives in sympatry with several rear-fanged mudsnakes, including the Crab-eating Mangrove Snake (*Fordonia leucobalia*, page 541) and Richardson's Mangrove Snake (*Myron richardsonii*, page 544). It forages for mudskippers and gobies on the mudflats, searching for them in their burrows. Unlike other intertidal seasnakes, such as the Northwestern Mangrove Seasnake (*Ephalophis greyae*, page 501) and the Arafura Smooth Seasnake (*Parahydrophis mertoni*, page 523), the Port Darwin Seasnake hunts on the drying mudflats, while its relatives prefer to hunt along the creeks and remain close to, or in, the water. As the mudflats dry the Port Darwin Seasnake will retreat to a burrow to await the incoming tide. It has neurotoxic venom but is not considered dangerous to humans.

RELATED SPECIES

Hydrelaps darwiniensis is related to the Arafura Smooth Seasnake (*Parahydrophis mertoni*), and together they form a sister group to the large seasnake genus *Hydrophis* (pages 507–513).

Actual size

FAMILY	Elapidae: Hydrophiinae
RISK FACTOR	Highly venomous: postsynaptic neurotoxins, and possibly myotoxins
DISTRIBUTION	Timor and Arafura Seas: northern Australia, including Torres Strait, and southern New Guinea
ELEVATION	Sea-level to 260 ft (80 m) bsl
HABITAT	Estuaries, tidal creeks, and river mouths over muddy or sandy bottoms, and also in clear coral reef waters
DIET	Slender fish, including eels
REPRODUCTION	Viviparous, with litters of up to 23 neonates
CONSERVATION STATUS	IUCN Least Concern

ADULT LENGTH
5 ft 2 in–6 ft 3 in,
rarely 7 ft
(1.6–1.9 m,
rarely 2.1 m)

HYDROPHIS ELEGANS
ELEGANT SEASNAKE
(GRAY, 1842)

The Elegant Seasnake is one of the longest seasnake species within its range. It occurs in the Timor and Arafura Seas, off northern Australia and southern New Guinea. It is a relatively common turbid estuarine and tidal river-dwelling species that is found over both muddy and sandy bottoms, and although encountered on the surface at night it is usually caught in deepwater trawl nets. Occasionally this species is also found on coral reefs. It is active by day or night, and feeds on slender benthic (bottom-dwelling) fish, especially eels, as necessitated by its small head and correspondingly small mouth gape. A highly venomous species, it is easily capable of causing a fatal snakebite.

RELATED SPECIES

According to a recent phylogenetic study the closest relative of *Hydrophis elegans* is Stokes' Seasnake (*H. stokesii*, page 512), then the Olive-headed Seasnake (*H. major*). As currently recognized, the genus *Hydrophis* contains 44 of the 63 species of true seasnakes.

The Elegant Seasnake is a large seasnake with a small head and slender anterior body, and a much broader posterior body and flattened, paddle-shaped tail. This species' patterning comprises a black or gray head and a yellow or gray body, with black rings on the anterior body, becoming oval saddles with vertical rows of spots in the interspaces on the posterior body and tail. Juveniles are more boldly patterned than adults. The large nostrils are positioned dorsally for breathing at the surface.

Actual size

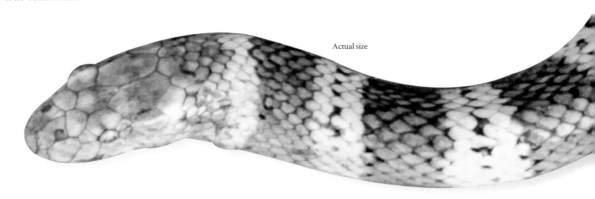

FAMILY	Elapidae: Hydrophiinae
RISK FACTOR	Highly venomous: postsynaptic neurotoxins, and possibly myotoxins
DISTRIBUTION	Indian and Pacific Oceans: Myanmar to Japan, the Philippines, Indonesia, Australia, New Guinea, and New Caledonia
ELEVATION	Sea-level to 50 ft (15 m) bsl
HABITAT	Estuaries and river mouths over muddy bottoms, and also offshore in clear water over sand
DIET	Fish
REPRODUCTION	Viviparous, with litters of 1–15 neonates
CONSERVATION STATUS	IUCN Least Concern

ADULT LENGTH
2 ft 7 in–3 ft,
rarely 4 ft 3 in
(0.8–0.9 m,
rarely 1.3 m)

508

HYDROPHIS HARDWICKII
HARDWICKE'S
SPINE-BELLIED SEASNAKE
(GRAY, 1834)

Hardwicke's Spine-bellied Seasnake is so named because males possess spinous ventral scales. It is a stocky snake with a large head and the usual seasnake paddle-shaped tail. Coloration is pale gray to off-white with dark olive-gray saddles that narrow and fade on the flanks.

This species is very common in the turbid waters of muddy river mouths and estuaries, often floating on the surface at night, but it is also found in clear water, though not over reefs. It occurs from Myanmar to Japan, the Philippines, and Indonesia, and south to Australia and New Caledonia. It is a generalist feeder, taking a wide variety of fish at all levels in the water column. Hardwicke's Spine-bellied Seasnake is an aggressive species that will bite with little provocation. Its high venom yield and attitude make it a highly dangerous species, suspected to have caused three human fatalities. It has few enemies, large sharks being its only threat, although even tiger sharks seem to prefer other seasnakes to this pugnacious species.

RELATED SPECIES

Until recently, this snake occupied is own genus, *Lapemis*, but a phylogenetic study revealed a rapid radiation within the true seasnakes, leading to many dissimilar species that are more closely related than was formerly believed. Twelve genera have now been synonymized within *Hydrophis*, including *Lapemis*. Hardwicke's Spine-bellied Seasnake has also been treated as a subspecies of the Short or Spine-bellied Seasnake (*H. curtus*) from the Persian Gulf but current opinion treats them as separate species.

Actual size

FAMILY	Elapidae: Hydrophiinae
RISK FACTOR	Highly venomous: postsynaptic neurotoxins, and possibly myotoxins
DISTRIBUTION	Pacific Ocean: Australia and New Caledonia, Thailand to Taiwan
ELEVATION	Sea-level to 195 ft (60 m) bsl
HABITAT	Coral reefs, coral sand or seagrass bottoms, and turbid estuaries
DIET	Fish and shrimps
REPRODUCTION	Viviparous, with litters of 4–10 neonates
CONSERVATION STATUS	IUCN Least Concern

ADULT LENGTH
3 ft 3 in–4 ft
(1.0–1.2 m)

HYDROPHIS PERONII
SPINY-HEADED SEASNAKE
(DUMÉRIL, 1853)

Primarily found from Australia to New Caledonia, this seasnake is also found from Thailand to Taiwan and it is also called Peron's Seasnake in honor of the French naturalist-explorer François Peron (1775–1810), who collected the holotype. The Spiny-headed Seasnake is instantly recognizable due to its uniquely spiny scalation. It inhabits coral reefs and seabeds of coral sand or seagrass, and may gather a covering of seaweed or barnacles. It is also found in turbid estuarine waters. Adults prey on small fish, especially gobies, while juveniles take shrimps. The golden trevally fish has been observed swimming unmolested with this seasnake, suggesting some sort of symbiotic relationship. Predators include tiger and bull sharks. Though shy and reluctant to bite, with no snakebites on record, it is a highly venomous species certainly capable of causing a fatal snakebite.

The Spiny-headed Seasnake can be recognized instantly from the strongly keeled dorsal body scales and the spinous scales on the head, especially over the eyes. Its coloration is pale yellow to gray with regular, broad, olive-gray bands around the body.

RELATED SPECIES

This species was placed in the monotypic genus *Acalyptophis*, until a recent study moved it to *Hydrophis* (see *H. hardwickii*, page 508). Its closest relatives include the Eyed Seasnake (*H. ocellatus*), which inhabits reefs and estuaries and also enters tidal river systems in Australia, New Guinea, and eastern Indonesia.

Actual size

FAMILY	Elapidae: Hydrophiinae
RISK FACTOR	Highly venomous: postsynaptic neurotoxins, and possibly myotoxins
DISTRIBUTION	Indian and Pacific Oceans: South Africa to the Americas
ELEVATION	Sea-level to 66 ft (20 m) bsl
HABITAT	Open ocean, especially where conflicting currents meet
DIET	Small pelagic fish
REPRODUCTION	Viviparous, with litters of 2–6 neonates
CONSERVATION STATUS	IUCN Least Concern

ADULT LENGTH
2 ft 7 in–3 ft
(0.8–0.9 m)

510

HYDROPHIS PLATURA
PELAGIC SEASNAKE
(LINNAEUS, 1766)

The Pelagic Seasnake is black and yellow, or occasionally completely yellow, with an extremely elongate head, and tiny ventral scales that enable it to flatten its entire body like a ribbon for effortless swimming. Its scales are non-overlapping, to deny anchorage for marine parasites, but this also means a Pelagic Seasnake on land is completely helpless. However, the warning colors prevent even vultures from feeding on individuals cast ashore by storms.

Also known as the Black and Yellow Seasnake, this is the most marine-adapted, and the most widely distributed, naturally occurring snake in the world. It is found from California to Ecuador, across the Pacific to Japan, Korea, and New Zealand, then the Indian Ocean to the Persian Gulf and South Africa, and around the Cape to Namibia, although the geographical extremes comprise records of waif specimens. Thousands of Pelagic Seasnakes gather in floating aggregations where outflowing river currents meet oceanic currents, and rafts of floating debris collect. The rafts and the snakes offer the only shade and are attractive to pelagic (surface-dwelling) fish, which provide the snakes with an easy meal. The Pelagic Seasnake has oral salt-excretory glands that expel salt with the flickering tongue. Deaths from bites by this species are known.

RELATED SPECIES

Until recently, this ultimate seasnake occupied its own monotypic genus, *Pelamis*, but a recent study placed it in *Hydrophis*. Its closest relative is reportedly the Spine-bellied Seasnake (*H. curtus*) and Hardwick's Spine-bellied Seasnake (*H. hardwickii*, page 508).

Actual size

FAMILY	Elapidae: Hydrophiinae
RISK FACTOR	Highly venomous, venom composition unknown but probably postsynaptic neurotoxins, possibly myotoxins
DISTRIBUTION	Southeast Asia: Borneo, Kalimantan, Sibau Kecil River
ELEVATION	Sea-level
HABITAT	Freshwater river system
DIET	Prey preferences unknown, presumed to include fish
REPRODUCTION	Viviparous, with a litter size of 7 neonates
CONSERVATION STATUS	IUCN Data Deficient

ADULT LENGTH
19¾–29 in
(500–735 mm)

HYDROPHIS SIBAUENSIS
SIBAU RIVER SEASNAKE
(RASMUSSEN, AULIYA & BÖHME, 2001)

511

The Sibau River Seasnake is only known from the Sibau Kecil River, in the Kapuas Basin, West Kalimantan, Indonesian Borneo. It is a land-locked seasnake that does not occur in the ocean, but rather occurs 620 miles (1,000 km) inland. It was described in 2001, but originally placed in the genus *Chitulia*, now a subgenus of *Hydrophis*. This species has been little studied in nature, even its prey preferences are unknown, though it might be expected to feed on fish like its congenerics. Like all true seasnakes the Sibau River Seasnake is live-bearing with a litter size of seven neonates on record.

The Sibau River Seasnake is a small snake with a slender body, keeled scales, flattened paddle-shaped tail, a moderately elongate head, slightly broader than the neck, and small eyes. It is gray-brown in color with a series of short, transverse orange bands dorsally.

RELATED SPECIES

The genus *Hydrophis* contains a further 46 species, all but one being oceanic in their distribution, but the Lake Taal Seasnake (*H. semperi*) is endemic to a single lake on the island of Luzon, in the Philippines. Crocker's Sea Krait (*Laticauda crockeri*) is also an endemic land-locked lake inhabitant from Lake Te-Nggano, on Rennell Island in the Solomons. The species deemed closest to the Sibau River Seasnake is the Garlanded Seasnake (*H. torquatus*), which is widely distributed in Southeast Asia.

Actual size

FAMILY	Elapidae: Hydrophiinae
RISK FACTOR	Highly venomous: postsynaptic neurotoxins, and possibly myotoxins
DISTRIBUTION	Indian and Pacific Oceans: Arabian Peninsula to northern Australia
ELEVATION	Sea-level to 72 ft (22 m) bsl
HABITAT	Turbid estuaries, coral reefs, and shelf edges
DIET	Fish
REPRODUCTION	Viviparous, with litters of 1–14 neonates
CONSERVATION STATUS	IUCN Least Concern

ADULT LENGTH
4 ft–5 ft 2 in
(1.2–1.6 m)

HYDROPHIS STOKESII
STOKES' SEASNAKE
(GRAY, 1846)

Stokes' Seasnake is a large, stocky-bodied, large-headed seasnake with a gray to yellow background overlain by widely spaced dark, circular blotches, with transverse bars or spots in the interspaces, although unpatterned specimens are not uncommon.

Stokes' Seasnake is a massively built species, and one of the more dangerous seasnake species because its large head and mouth conceal fangs up to ¼ in (6.7 mm) in length. This is one of the few species that can bite through a ⅕ in (5 mm) neoprene wetsuit. Combined with its relatively high venom yield and truculent nature, this makes it a species to avoid when diving. Serious snakebites have occurred, although no fatalities are on record. Stokes' Seasnake may be found in turbid estuaries, coral reefs, and oceanic shelf edges, over muddy or sandy bottoms, on the coasts of the Arabian Peninsula and South and Southeast Asia, to northern Australia. Stokes' Seasnake is a generalist piscivore, and given its large head it can feed on armored fish, like frogfish, that smaller-mouthed seasnakes would find difficult to swallow. John Lort Stokes (1811–85) was an Admiral in the Royal Navy who explored Australian waters.

RELATED SPECIES

Until recently placed in the monotypic genus *Astrotia*, this species was moved to *Hydrophis* by a recent study (see *H. hardwickii*, page 508). The Elegant Seasnake (*H. elegans*, page 507) is a close relative.

Actual size

FAMILY	Elapidae: Hydrophiinae
RISK FACTOR	Highly venomous: postsynaptic neurotoxins, and possibly myotoxins
DISTRIBUTION	Arafura and Coral seas: northern Australia and New Guinea
ELEVATION	Sea-level to 16 ft (5 m), sometimes 98 ft (30 m), bsl
HABITAT	Estuaries and river mouths over muddy bottoms, and also offshore in clear water over sand
DIET	Fish
REPRODUCTION	Viviparous, with litters of 3–34 neonates
CONSERVATION STATUS	IUCN Least Concern

ADULT LENGTH
27–31 in
(690–790 mm)

HYDROPHIS ZWEIFELI
SOUTHERN BEAKED SEASNAKE
(KHARIN, 1985)

513

The Southern Beaked Seasnake is a nocturnally active inhabitant of turbid, muddy-bottomed estuaries and river mouths. It lives in a "cotton candy" world of soft mangrove mud and never meets a straight line or hard surface. This means that despite being highly venomous and a very dangerous snake to humans, and a successful generalist piscivorous predator, it is extremely susceptible to damage, captured snakes quickly developing edema of the head, even if handled gently. The beaked seasnakes are the species used in seasnake antivenom production, but this vulnerability may make them difficult to maintain in captivity and handle for venom extraction. The prey preferences of the Southern Beaked Seasnake are unknown, but the Northern Beaked Seasnake (*Hydrophis schistosus*) feeds on a variety of estuarine fish, from moray eels to catfish and pufferfish.

The Southern Beaked Seasnake is a relatively slender species, although the posterior body may be fairly broad. It is patterned with 45–55 dark to mid-gray or olive cross-bars spaced regularly on a pale gray to cream dorsal background, the bars not extending fully down the flanks, leaving the venter uniform whitish. The name "beaked seasnake" come from a downward-pointing fleshy protuberance from the front of the rostral scale. Another distinguishing characteristic is an elongate "daggerlike" mental scale under the throat.

RELATED SPECIES

This species represents the southern population of what was once a widely distributed Indo-Australian species, the Beaked or Common Seasnake (*Enhydrina schistosa*). *Enhydrina zweifeli* was first described from the Sepik River mouth, Papua New Guinea, but a recent molecular study expanded the use of this name to encompass all Australo-Papuan specimens previously treated as *E. schistosa*, reserving that name for Asian populations. Both *Enhydrina* are now included in *Hydrophis* (see *H. hardwickii*, page 508).

Actual size

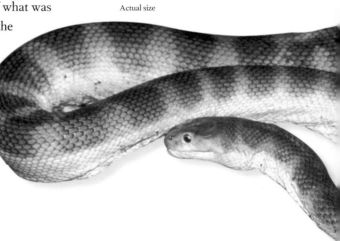

FAMILY	Elapidae: Hydrophiinae
RISK FACTOR	Venomous: possibly postsynaptic neurotoxins or myotoxins
DISTRIBUTION	Indian and Pacific Oceans: Bangladesh to Japan, south to northern Australia, east to New Caledonia, Solomon Islands, New Zealand, and Samoa
ELEVATION	Sea-level to 150 ft (45 m) bsl
HABITAT	Coral reefs, small rocky islands, and mangrove entanglements
DIET	Fish, including eels
REPRODUCTION	Oviparous, with clutches of 4–10 eggs
CONSERVATION STATUS	IUCN Least Concern

ADULT LENGTH
3 ft 3 in–4 ft 7 in,
rarely 5 ft 5 in
(1.0–1.4 m,
rarely 1.65 m)

514

LATICAUDA COLUBRINA
COLUBRINE SEA KRAIT
(SCHNEIDER, 1799)

The Colubrine Sea Krait is strongly and regularly banded blue-gray and black or gunmetal on the body and tail. The head is black above with two black postocular stripes on the sides, and yellow lips that earn this species the alternative name of Yellow-lipped Sea Krait.

Sea kraits differ from true seasnakes in several important ways. They have broad, overlapping ventral scales and are as able to move on land as well as any terrestrial snake, and have even been found in trees, or climbing cliffs. They are also oviparous, which means they must come onto land to lay their eggs. But they do possess flattened, paddle-shaped tails and, being strongly banded, are frequently mistaken for seasnakes. The Colubrine Sea Krait occurs across a wide swath of the Indian and Pacific Oceans, from Bangladesh to Samoa, but usually occurs close to shore. Occasional waifs have been found in Japan and New Zealand. The primary prey comprises moray eels, particularly the Zebra Eel (*Gymnomuraena ʒebra*), which resembles the sea krait, but occasionally other fish are eaten. Inoffensive, sea kraits are extremely reluctant to bite humans, even when handled.

RELATED SPECIES

There are eight species of sea kraits, but they are not closely related to kraits of genus *Bungarus*. *Laticauda colubrina* is related to the Papuan Sea Krait (*L. guineai*) from southern New Guinea, a dwarf species from Vanuatu (*L. frontalis*), and the New Caledonian Sea Krait (*L. saintgironsi*, page 515). Through much of its range, *L. colubrina* occurs in sympatry with the Blue-ringed or Brown-lipped Sea Krait (*L. laticaudata*), which is related to an endemic species (*L. crockeri*) from the freshwater Lake Te-Nggano, on Rennell Island in the Solomons.

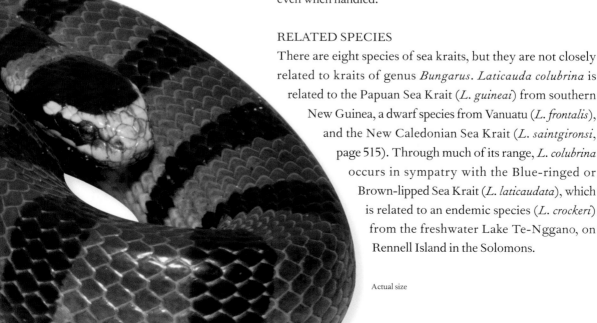

Actual size

FAMILY	Elapidae: Hydrophiinae
RISK FACTOR	Venomous: possibly postsynaptic neurotoxins or myotoxins
DISTRIBUTION	Coral Sea: New Caledonia and the Loyalty Islands
ELEVATION	Sea-level to 16 ft (5 m), sometimes 98 ft (30 m), bsl
HABITAT	Coral reefs, lagoons, and small rocky or sandy islands
DIET	Eels
REPRODUCTION	Oviparous, with clutches of 4–19 eggs
CONSERVATION STATUS	IUCN Least Concern

ADULT LENGTH
3 ft 3 in–4 ft 3 in
(1.0–1.3 m)

LATICAUDA SAINTGIRONSI
NEW CALEDONIAN SEA KRAIT
COGGER & HEATWOLE, 2005

515

The New Caledonian Sea Krait is endemic to the French territory of New Caledonia, 1,000 miles (1,616 km) east of Australia, around Grande Terre, the Île des Pins, and many smaller islands, and the Loyalty Islands to the east. Aggregations of sea kraits are often found on small islands at particular times of the year, it not being rare to count 30–60 on a single small island. They can climb cliffs and travel overland. When this species occurs in sympatry with the Brown-lipped Sea Krait (*Laticauda laticaudata*) it outnumbers the other by ten to one. Like other sea kraits, this is an inoffensive species. It feeds on eels, the larger females feeding farther offshore than the males. This species was named in honor of the French herpetologist Hubert Saint Girons (1926–2000).

The New Caledonian Sea Krait is an attractive snake, being regularly and broadly banded black and gold, with a black cap on the rear of the head and a black stripe through the eye, joining the band at the rear of the cap.

RELATED SPECIES
Laticauda saintgironsi is related to the widespread Colubrine Sea Krait (*Laticauda colubrina*, page 514), the dwarf Vanuatu Sea Krait (*L. frontalis*), and the Papuan Sea Krait (*L. guineai*) from southern New Guinea.

Actual size

FAMILY	Elapidae: Hydrophiinae
RISK FACTOR	Venomous: venom composition unknown
DISTRIBUTION	Melanesia: Solomon Islands
ELEVATION	0–2,300 ft (0–700 m) asl
HABITAT	Rainforests and overgrown creeks
DIET	Lizards and snakes
REPRODUCTION	Oviparous, with clutches of up to 9 eggs
CONSERVATION STATUS	IUCN not listed

ADULT LENGTH
Male
2–3 ft
(0.6–0.9 m)

Female
2 ft 7 in–4 ft 3 in
(0.8–1.3 m)

516

LOVERIDGELAPS ELAPOIDES
SOLOMONS
SMALL-EYED SNAKE
(BOULENGER, 1890)

The Solomons Small-eyed Snake occurs on many islands and exhibits several different island color morphotypes, but usually the body is banded black and orange or yellow, the lighter pigment giving way to white on the underbelly. The head may be all white or black and white.

This stunning snake is poorly known in the wild. Found on most of the islands of the Solomon Island Archipelago, but not on Bougainville, it is generally feared by locals, even though it is generally placid and inoffensive. Bites are rare and no fatalities are known. On Malaita Island it is known as *baekwa i tolo*, which translates as "Shark of the Jungle," but the reasoning behind this name is not known. A nocturnal, secretive, semifossorial species, the Solomons Small-eyed Snake is not commonly encountered in the field. Its prey consists of forest floor lizards, such as skinks of the genus *Sphenomorphus*, and blindsnakes (Typhlopidae and Gerrhopilidae), but larger snakes are also taken as food, including the Solomons Coralsnake (*Salomonelaps par*, page 530).

RELATED SPECIES

In the past this snake was included in the genus *Micropechis* with the New Guinea Small-eyed Snake (*M. ikaheka*, page 518), but it was moved to its own genus *Loveridgelaps*, named for the famous British herpetologist Arthur Loveridge (1891–1980). Within its range there is no other snake that resembles *L. elapoides*. It is probably most closely related to the Solomons Coralsnake (*Salomonelaps par*, page 530).

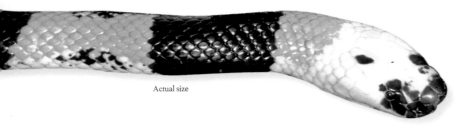

Actual size

FAMILY	Elapidae: Hydrophiinae
RISK FACTOR	Venomous: postsynaptic neurotoxins, and possibly myotoxins
DISTRIBUTION	Indian and Pacific Oceans: Persian Gulf to South China Sea, New Guinea, and northern Australia
ELEVATION	3 ft 3 in–98 ft (1–30 m) bsl
HABITAT	Turbid deepwater bays and gulfs with muddy bottoms
DIET	Eels
REPRODUCTION	Viviparous, with litters of 1–6 neonates
CONSERVATION STATUS	IUCN Least Concern

ADULT LENGTH
3 ft–3 ft 3 in
(0.9–1.0 m)

MICROCEPHALOPHIS GRACILIS
GRACEFUL SMALL-HEADED SEASNAKE
(SHAW, 1802)

517

The Graceful Small-headed Seasnake is a rarely encountered species in Australo-Papuan waters, being more common in Asian waters, from the Persian Gulf to the South China Sea. It inhabits muddy, deepwater bays and gulfs where it swims along the bottom hunting benthic snake-eels, especially the Oriental Worm-eel (*Lamnostoma orientalis*). The extremely small head and neck permits the seasnake to hunt eels by probing their muddy burrows on the seabed. Little is known of the venom composition or toxicity for this seasnake, but given that *Microcephalophis* is closely related to *Hydrophis* (pages 507–513), which contains many highly venomous species, it must be considered potentially dangerous, despite the small size of its head. As with all true seasnakes, the Graceful Small-headed Seasnake is a live-bearer.

The Graceful Small-headed Seasnake is an unusual shape. The posterior body is vast, broader even than the paddle-shaped tail and four to five times broader than the anterior body. The neck and head are slender, the head being minuscule in comparison with the body. Patterning comprises a series of faint gray bands, darkest on the anterior body, on a pale gray background. The bands fade on the flanks and with maturity. The undersides are white while the head is black.

RELATED SPECIES

The closest relative of *Microcephalophis gracilis* is the similarly proportioned Cantor's Small-headed Seasnake (*M. cantoris*) from South Asia. Some authors synonymize *Microcephalophis* with *Hydrophis* (pages 507–513).

Actual size

FAMILY	Elapidae: Hydrophiinae
RISK FACTOR	Highly venomous: postsynaptic neurotoxins, myotoxins, anticoagulants, and possibly also hemorrhagins
DISTRIBUTION	New Guinea: Papua New Guinea, Indonesian west New Guinea, and associated satellite islands (Aru, Karkar, Manam, Waigeo, and Batanta)
ELEVATION	0–5,580 ft (0–1,700 m) asl
HABITAT	Swamps, rainforest, plantations, and overgrown creeks
DIET	Snakes, lizards, small mammals, and eels
REPRODUCTION	Oviparous, with clutches of 4–6 eggs
CONSERVATION STATUS	IUCN not listed

ADULT LENGTH
5 ft–7 ft 7 in
(1.5–2.3 m)

518

MICROPECHIS IKAHEKA
NEW GUINEA
SMALL-EYED SNAKE
LESSON, 1830

The New Guinea Small-eyed Snake is a New Guinea endemic. It occurs in damp areas such as freshwater swamps and creeks and in tropical rainforest, where it adopts a semi-fossorial, nocturnal existence. It is commonly encountered in coconut or oil-palm plantations, living inside accumulated piles of coconut husks or palm fronds. It poses a serious risk to plantation workers, biting and chewing without the slightest provocation. There is no specific antivenom and human fatalities are on record. The New Guinea Small-eyed Snake has quite a catholic diet, preying on lizards such as skinks, small mammals, ranging from rats to bandicoots, and other snakes—including the New Guinea Ground Boa (*Candoia aspera*, page 114). It is even known to resort to cannibalism. The most unusual prey taken by a New Guinea Small-eyed Snake was a freshwater eel.

RELATED SPECIES

Unlike many of the other large, venomous New Guinea snakes, *Micropechis ikaheka* does not have any close Australian relatives. At one time the Solomons Small-eyed Snake (*Loveridgelaps elapoides*, page 516) was treated as its sister species, before being transferred to its own genus.

The New Guinea Small-eyed Snake is generally a banded snake with regular wide brown or red bands on a yellow or white background, and a grayish head. Specimens from the Vogelkop of west New Guinea are often all yellow without bands, and specimens from Batanta and Waigeo islands are totally black.

Actual size

FAMILY	Elapidae: Hydrophiinae
RISK FACTOR	Highly venomous: pre- and postsynaptic neurotoxins, myotoxins, and procoagulants
DISTRIBUTION	Australia: southwestern and southeastern Australia, Tasmania, and offshore islands
ELEVATION	0–4,820 ft (0–1,470 m) asl
HABITAT	Rainforest, sclerophyll woodland, coastal dunes, marshes, heaths, floodplains, and islands
DIET	Small mammals, frogs, lizards, snakes, and birds
REPRODUCTION	Viviparous, with litters of 17–23, occasionally > 100, neonates
CONSERVATION STATUS	IUCN Least Concern

ADULT LENGTH
3–5 ft,
occasionally 6 ft 7 in
(0.9–1.5 m,
occasionally 2.0 m)

NOTECHIS SCUTATUS
TIGERSNAKE
(PETERS, 1861)

519

In the early years of European colonization of Australia, Tigersnakes were responsible for many snakebite fatalities, but today the effectiveness of antivenom has greatly reduced the death rate. Cool-adapted and live-bearing, Tigersnakes prefer damp habitats, including wet forest, coastal dunes, marshes, and floodplains. They occur in southwestern and southeastern mainland Australia, throughout Tasmania, and on the islands of the Bass Strait and Spencer Gulf, south of Adelaide. The prey of mainland Tigersnakes consists of small mammals, frogs, lizards, and snakes, with cannibalism of smaller conspecifics not being uncommon. On the islands prey is less abundant. On Mount Chappell Island, Bass Strait, adult Tigersnakes, which grow very large, survive entirely on a diet of muttonbird chicks, gorging themselves over the six-week breeding period and then fasting until the following year, while juveniles eat skinks.

The Tigersnake is a smooth-bodied snake, with a large, bulbous head. In coloration it is very variable, specimens being brown, yellow, gray, unicolor, or broadly banded, with certain patterns being more common in certain populations. Island populations are often melanistic jet black, an adaptation that enables the snakes to warm up more quickly when basking.

RELATED SPECIES

At one time two species were recognized, the Eastern Tigersnake (*Notechis scutatus*) from mainland southeastern Australia, and the Black Tigersnake (*N. ater*) with five subspecies: *N. a. ater* in South Australia, *N. a. occidentalis* in Western Australia, *N. a niger* on Kangaroo Island, *N. a. humphreysi* on Tasmania and King Island, and *N. a. serventyi* on Mount Chappell Island. Today only one species is recognized, with eastern (*N. s. scutatus*) and western (*N. s. occidentalis*) subspecies. Some authors do recognize the other subspecies, but as a subspecies of *N. scutatus*.

Actual size

FAMILY	Elapidae: Hydrophiinae
RISK FACTOR	Venomous: venom composition unknown
DISTRIBUTION	Fiji: Viti Levu
ELEVATION	33–490 ft (10–150 m) asl
HABITAT	Lowland rainforest and yam gardens, associated with termite mounds
DIET	Earthworms, and possibly also soft-bodied insects
REPRODUCTION	Oviparous, with clutches of 2–3 eggs
CONSERVATION STATUS	IUCN Endangered

ADULT LENGTH
4–9¾ in,
rarely 12½ in
(100–250 mm,
rarely 320 mm)

520

OGMODON VITIANUS
FIJI SNAKE
PETERS, 1864

The Fiji Snake, alternatively known as the Fiji Burrowing Snake, Bola, or *Gata ni Balabala*, which means "snake of the mountain ferns," is a small, secretive, fossorial species that feeds on earthworms, possibly exclusively. It inhabits lowland rainforest and yam gardens and is endemic to the main island of Fiji, Viti Levu, where it appears to occur in very small, localized areas in the south. Habitat destruction and foraging groups of semi-feral pigs, which will eat small snakes, are major threats to its survival. Although venomous, this tiny snake poses no threat to humans. It is even featured on Fijian postage stamps. The only other terrestrial snakes on Fiji are the introduced Brahminy Blindsnake (*Indotyphlops braminus*, page 57) and the Pacific Boa (*Candoia bibroni*).

RELATED SPECIES

Ogmodon vitianus occurs over 1,240 miles (2,000 km) from the Solomon Islands, home of the nearest terrestrial elapids, and over 1,680 miles (2,700 km) from a snake with a similar biology, the Bougainville Coralsnake (*Parapistocalamus hedigeri*). Its natural history is also similar to the worm-eating snakes (*Toxicocalamus*, page 533) of New Guinea.

The Fiji Snake is a small, short-tailed, smooth-scaled, drab gray, brown, or black snake, with its markings confined to a white collar across the nape of the neck, although this may be indistinct in adults.

Actual size

FAMILY	Elapidae: Hydrophiinae
RISK FACTOR	Highly venomous: pre- and postsynaptic neurotoxins, myotoxins, and procoagulants
DISTRIBUTION	Australia: eastern-central Australia
ELEVATION	0–425 ft (0–130 m) asl
HABITAT	Black soil river floodplains and lateritic gibber plains
DIET	Mammals
REPRODUCTION	Oviparous, with clutches of 12–20 eggs
CONSERVATION STATUS	IUCN not listed

ADULT LENGTH
5 ft–6 ft 7 in,
occasionally 8 ft 2 in
(1.5–2.0 m,
occasionally 2.5 m)

OXYURANUS MICROLEPIDOTUS
INLAND TAIPAN
(MCCOY, 1879)

521

Its alternative name, Fierce Snake, does not do this highly venomous snake justice; it is usually a fairly placid snake, and is certainly not fierce. The Inland Taipan inhabits eastern-central Australia, where the states of Queensland, the Northern Territory, New South Wales, and South Australia meet. It occurs in black soil habitats, on riverine floodplains, and open desert stony pans known as gibber plains, where it hunts rats in the large fissures that develop in the substrate when it has been dry for a long period. Inland Taipans are not encountered with any frequency, rarely coming into contact with humans, and snakebites are therefore extremely rare, but not unknown. Although still rated as the most venomous snake in the world, the Inland Taipan is not yet known to have claimed a human life.

The Inland Taipan is a large, brown snake, with every scale edged with darker pigment and dark flecking on its head scales. Some specimens have entirely black heads and necks. The undersides are yellowish, and are clearly visible if the snake raises itself up in a threat posture. The head is more rounded than that of its relative, the Coastal Taipan (*Oxyuranus scutellatus*, page 522).

RELATED SPECIES

There are two other taipans, the Coastal Taipan (*Oxyuranus scutellatus*, page 522) of northern Australia and southern New Guinea, and the recently discovered and described Western Desert Taipan (*O. temporalis*) from the border where Western Australia meets South Australia and the Northern Territory.

Actual size

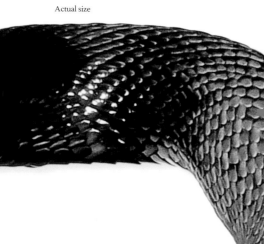

FAMILY	Elapidae: Hydrophiinae
RISK FACTOR	Highly venomous: pre- and postsynaptic neurotoxins, myotoxins, and procoagulants
DISTRIBUTION	Australo-Papua: northern and eastern coastal Australia, and southern New Guinea
ELEVATION	0–2,620 ft (0–800 m) asl
HABITAT	Savanna woodland, kunai grassland, oil-palm plantations, sugarcane fields, and urban areas
DIET	Mammals
REPRODUCTION	Oviparous, with clutches of 5–22 eggs
CONSERVATION STATUS	IUCN not listed

ADULT LENGTH
6 ft 7 in–8 ft 2 in,
possibly 10 ft
(2.0–2.5 m,
possibly 3.0 m)

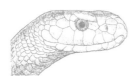

OXYURANUS SCUTELLATUS
COASTAL TAIPAN
PETERS, 1867

522

The Coastal Taipan is a large, slender snake with a long, coffin-shaped head, large eyes, and protruding supraocular scales that give the impression the snake is scowling. Australian specimens are usually brown, paler on the lower flanks, with a pale brown head, which may be yellowish in juveniles. Specimens from Western Province, Papua New Guinea are also brown, while those from Central Province are black or gunmetal, with a broad orange stripe down the back.

Australo-Papua's largest venomous snake, the Coastal Taipan inhabits northern Australia and the southern coast of New Guinea, in savanna and savanna woodland habitats. It has adapted well to living in oil-palm plantations in Papua New Guinea and sugar-cane fields in Queensland. In both it is a serious threat to plantation workers. The fast-striking Coastal Taipan is a predator of bandicoots and rats, which are rapidly killed by its highly toxic venom. Although at one time it was a much-feared snake, today in Australia there are probably few human snakebites caused by this species, but in southern New Guinea it is the most medically important species, causing many deaths. Snakebites even occur in urban areas; Port Moresby is in prime Coastal Taipan habitat. Death is due to respiratory paralysis, but patients can bleed to death if traditional razor-cutting treatment is adopted.

RELATED SPECIES

Oxyuranus scutellatus occurs as two subspecies, the nominate Australian Coastal Taipan (*O. s. scutellatus*), and the Papuan Taipan (*O. s. canni*) in southern New Guinea. There is some doubt whether the subspecies are truly different taxa.

Actual size

FAMILY	Elapidae: Hydrophiinae
RISK FACTOR	Venomous: postsynaptic neurotoxins, and possibly myotoxins
DISTRIBUTION	Arafura Sea: Northern Territory (Australia), the Aru Islands (Indonesia), and possibly New Guinea
ELEVATION	Sea-level
HABITAT	Estuaries, mudflats, and mangrove swamps
DIET	Not known, but presumed gobies such as mudskippers
REPRODUCTION	Viviparous, with litters of 3 neonates
CONSERVATION STATUS	IUCN Data Deficient

ADULT LENGTH
11¾–19¾ in
(300–500 mm)

PARAHYDROPHIS MERTONI
ARAFURA SMOOTH SEASNAKE
(ROUX, 1910)

523

Little is known about this relatively rare inshore, mangrove- and mudflat-dwelling seasnake. It occurs around the Arnhem Land coast of the Northern Territory, Australia, from Darwin to the Gulf of Carpentaria, but the type locality is the Indonesian Aru Islands, in the Arafura Sea, so it may also be expected to occur on the southern coast of at least western New Guinea. There are no data on prey preferences, but the options of intertidal snakes are gobies such as mudskippers, shrimps, or crabs. It is likely that this species preys on mudskippers, like other intertidal seasnakes. The only reproductive data involves a litter of three neonates from one female. Hugo Merton (1879–1939) was a zoologist who visited the Kai and Aru Islands.

RELATED SPECIES

The relatives of *Parahydrophis mertoni* are the other semi-terrestrial inshore species, the Northwestern Mangrove Seasnake (*Ephalophis greyae*, page 501) and the Port Darwin Seasnake (*Hydrelaps darwiniensis*, page 506).

The Arafura Smooth Seasnake is a cylindrical, smooth-scaled seasnake with a shallow, flattened tail and overlapping scales. It is blue-gray or brown with a patterning of black bands, often with centers the same color as the dorsal color. It has relatively broad ventral scales that permit locomotion over mudflats.

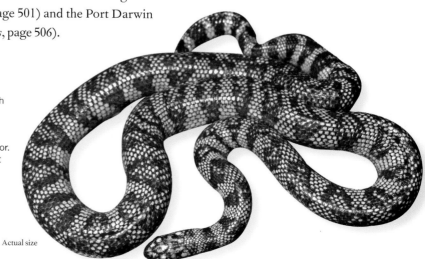

Actual size

FAMILY	Elapidae: Hydrophiinae
RISK FACTOR	Venomous: venom composition unknown
DISTRIBUTION	Australia: southern Australia, around the Great Australian Bight to New South Wales and Victoria
ELEVATION	0–3,590 ft (0–1,095 m) asl
HABITAT	Mallee, dry woodland, acacia scrub, and shrubland
DIET	Lizards
REPRODUCTION	Viviparous, with litters of 3 neonates
CONSERVATION STATUS	IUCN not listed

ADULT LENGTH
15¾ in
(400 mm)

PARASUTA SPECTABILIS
SPECTACLED HOODED SNAKE
(KREFFT, 1869)

The Spectacled Hooded Snake is brown above, paler on the back, with every scale having a black base and often a black edge, presenting a reticulate appearance. The head is black, giving the "hood" of the common name, with a pale stripe across the snout, and pale lips. The undersides are white.

Also known as the Mallee Black-headed Snake or Port Lincoln Snake, this species is an inhabitant of dry woodland, mallee, and other well-drained habitats across southern Australia, from the Great Australian Bight, to Victoria and New South Wales. A terrestrial species, it is found under logs or rocks or in small burrows during the day. It emerges at night to hunt lizards, primarily geckos, but skinks are also taken. Its defense involves body-flattening and thrashing about if handled, but nothing is known of its venom composition. There are no bites on record, but bites from related species have caused unpleasant symptoms and, as with all small elapids, caution is advised.

RELATED SPECIES

Three subspecies are recognized: the nominate subspecies (*Parasuta spectabilis spectabilis*) in the east, a Nullarbor Plain subspecies (*P. s. nullarbor*), and a western subspecies (*P. s. bushi*), named after the Western Australian herpetologist Brian Bush. Five other species are known: Dwyer's Snake (*P. dwyeri*), a former eastern subspecies; Mitchell's Short-tailed Snake (*P. nigriceps*), which occurs in sympatry with *P. spectabilis*; the Monk Snake (*P. monachus*) from central Australia; Gould's Hooded Snake (*P. gouldii*) from southwestern Australia; and the Little Whipsnake (*P. flagellum*) from Victoria.

Actual size

FAMILY	Elapidae: Hydrophiinae
RISK FACTOR	Venomous: possibly neurotoxins, otherwise venom composition unknown
DISTRIBUTION	Australia: Lake Cronin and Peak Eleanora, southern Western Australia
ELEVATION	885–1,440 ft (270–440 m) asl
HABITAT	Ephemeral lake in semiarid eucalypt woodland, mallee, and shrubland, and rocky outcrops
DIET	Lizards
REPRODUCTION	Presumed to be viviparous, litter size unknown
CONSERVATION STATUS	IUCN Vulnerable

ADULT LENGTH
19¾–23¾ in
(500–600 mm)

PAROPLOCEPHALUS ATRICEPS
LAKE CRONIN SNAKE
(STORR, 1980)

525

The Lake Cronin Snake may be the Australian elapid with the most localized distribution. It was originally thought to be confined to this small, semi-permanent freshwater lake, but specimens have also been obtained from the granite Peak Eleanora, 90 miles (145 km) east, in the Eastern Goldfields area of southern Western Australia. The habitat comprises semiarid eucalypt woodland, mallee, shrubland, and granite outcrops. The Lake Cronin Snake is both terrestrial and arboreal, and diurnal and nocturnal, and it feeds on small lizards. It is presumed to be a live-bearer, in common with other cold-adapted, southern elapids. It is considered potentially dangerous, with one bite recorded as causing serious symptoms. However, this is a rare snake, with fewer than ten specimens known.

The Lake Cronin Snake is a small snake with smooth scales, a brown body and tail, and a distinctive black head and neck, against which the yellow irises of the eyes and the vertical pupils stand out. The black of the head is also broken by a fine white stripe along the lips.

RELATED SPECIES

Paroplocephalus atriceps was once included in the genus *Echiopsis* with the Bardick (*E. curta*, page 499), which it resembles, but it is now known to be more closely related to the genus *Hoplocephalus*, which contains the Broad-headed Snake (*H. bungaroides*, page 505). The smooth scales of the Lake Cronin Snake distinguish it from the similar-looking, dangerously venomous Rough-scaled Snake (*Tropidechis carinatus*, page 534), and the non-venomous Common Keelback (*Tropidonophis mairii*, page 436), both of which have strongly keeled scales.

Actual size

FAMILY	Elapidae: Hydrophiinae
RISK FACTOR	Highly venomous: myotoxins and procoagulants, and possibly neurotoxins
DISTRIBUTION	Australia: continent-wide, excluding extreme south and southeast
ELEVATION	0–3,260 ft (0–995 m) asl
HABITAT	Arid and humid woodland, monsoon forest, rainforest, grassland, semidesert, and desert
DIET	Small mammals, lizards, and snakes
REPRODUCTION	Oviparous, with clutches of 4–19 eggs
CONSERVATION STATUS	IUCN not listed

ADULT LENGTH
3–5 ft,
occasionally 6 ft 7 in
(0.9–1.5 m,
occasionally 2.0 m)

526

PSEUDECHIS AUSTRALIS
KING BROWNSNAKE
(GRAY, 1842)

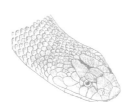

The King Brownsnake is a large, stocky brown snake with shades varying from yellow-brown to red-brown. It is paler on the flanks and underbelly, and has a broad, rounded head.

The King Brownsnake is actually a "blacksnake" (*Pseudechis*) rather than a "brownsnake" (*Pseudonaja*, pages 528–529). It is also known as the Mulga Snake after the *Acacia* savanna woodland in which it occurs, but this species also inhabits almost every continental Australian habitat, from rainforest to desert. It is distributed throughout most of Australia, except the south and the state of Victoria. Its prey comprises mostly small mammals but it also feeds on reptiles, including snakes, and cannibalism has been recorded. When disturbed, the King Brownsnake raises its anterior body and flattens its neck in warning, although this defense is far more impressive in the cobras (*Naja*, pages 466–479). King Brownsnakes are highly venomous, easily capable of causing a snakebite fatality.

RELATED SPECIES

The closest relatives of *Pseudechis australis* are snakes that were formerly included within this widespread species: the Papuan Pygmy Mulga Snake (*Pseudechis rossignolii*) of New Guinea, Western Pygmy Mulga Snake (*P. weigeli*) from the Kimberley, Western Australia, and Eastern Pygmy Mulga Snake (*P. pailsei*) from near Mt. Isa, Queensland.

Actual size

FAMILY	Elapidae: Hydrophiinae
RISK FACTOR	Highly venomous, postsynaptic neurotoxins, procoagulants, myotoxins, and hemolysins
DISTRIBUTION	Australia: eastern Queensland, New South Wales, Victoria, and southeast South Australia
ELEVATION	0–2,620 ft (0–800 m) asl
HABITAT	Lowland wet habitats, lagoons, swamps, and flooded grasslands
DIET	Frogs, small mammals, lizards, and birds
REPRODUCTION	Viviparous, with litters of 5–18, possibly 40 neonates
CONSERVATION STATUS	IUCN not listed

ADULT LENGTH
5 ft–6 ft 7 in
(1.5 – 2.0 m)

PSEUDECHIS PORPHYRIACUS
RED-BELLIED BLACKSNAKE
(SHAW, 1794)

527

The Red-bellied Blacksnake inhabits low-lying damp habitats such as swamps and lagoons, from eastern coastal and southeastern Queensland, through eastern New South Wales and Victoria, to southeastern South Australia. This is the only viviparous member of the Australo-Papuan genus *Pseudechis*, with litter sizes of 5–18, possibly up to 40 neonates, a reflection of its primarily southerly distribution. Diurnal in habit, it preys principally on frogs, but small mammals, lizards, and birds are also taken. Some populations suffered declines due to them feeding on the poisonous introduced cane toad. Generally an inoffensive snake that relies on its neck-spreading defensive posture to deter interference, the venom of the Red-bellied Blacksnake is considered less toxic than that of other blacksnakes, and whether it has caused any fatalities is questionable since bites usually result in minor symptoms, but the possibility exists for a fatal envenoming.

The Red-bellied Blacksnake is a relatively stocky-bodied, broad-headed, glossy, smooth-scaled snake with small eyes. It is dorsally jet black in color with a vivid red belly.

RELATED SPECIES

The genus *Pseudechis* contains nine species, mostly endemic to Australia; only two species, the Papuan Blacksnake (*P. papuanus*) and Pygmy Mulga Snake (*P. rossignolii*), occur in New Guinea. Although called "blacksnakes," several other species are brown in color, such as the widely distributed King Brownsnake (*P. australis*, page 526), with which the Red-bellied Blacksnake occurs in sympatry. It also occurs in sympatry with the Spotted Blacksnake (*P. guttatus*).

Actual size

FAMILY	Elapidae: Hydrophiinae
RISK FACTOR	Highly venomous: pre- and postsynaptic neurotoxins, and procoagulants
DISTRIBUTION	Australia: western and central Australia, excluding the extreme north and south
ELEVATION	0–1,280 ft (0–390 m) asl
HABITAT	Sclerophyll woodland, arid scrubland, grasslands, sand dunes, sandy and stony desert, and semidesert
DIET	Small mammals, lizards, snakes, and birds
REPRODUCTION	Oviparous, with clutches of up to 12 eggs
CONSERVATION STATUS	IUCN not listed

ADULT LENGTH
3 ft 3 in–4 ft,
occasionally 6 ft 7 in
(1.0–1.2 m,
occasionally 2.0 m)

528

PSEUDONAJA MENGDENI
GWARDAR
WELLS & WELLINGTON, 1985

The Gwardar is extremely variably patterned. It may be orange, reddish, brown, or even black, dorsally, with fine black edging to the scales of lighter specimens and a range of dorsal bands on the body. The head is often black, this pigment extending onto the neck and anterior body. While some specimens are uniform in color, others may be boldly banded with broad, alternating yellow and black bands. The undersides may be yellow, cream, gray, or orange. Identification of brownsnakes by color pattern can be misleading.

The Gwardar is also known as the Western Brownsnake. This species used to be included with the population in the Top End of the Northern Territory, and was known as *Pseudonaja nuchalis*. However, the common names Gwardar and Western Brownsnake are now confined to the western and central Australian species *P. mengdeni*, with *P. nuchalis* becoming the Northern Brownsnake. The Gwardar inhabits arid habitats from woodland to grassland and sandy or stony deserts. It preys on small mammals, lizards, other snakes, and ground-dwelling birds. The defensive posture of many brownsnakes involves raising the anterior body off the ground and forming a moving S-shape, whereupon they will sometimes advance on the threat. This is a highly venomous snake, responsible for serious snakebites and fatalities across its range.

RELATED SPECIES

The wide range of *Pseudonaja mengdeni* overlaps with that of the much smaller Ringed Brownsnake (*Pseudonaja modesta*). To the east it occurs in sympatry with the Eastern Brownsnake (*P. textilis*, page 529), Speckled Brownsnake (*P. guttata*), Ingram's Brownsnake (*P. ingrami*), and Strap-snouted Brownsnake (*P. aspidorhyncha*). To the north is found the Northern Brownsnake (*P. nuchalis*), and to the south, the Dugite (*P. affinis*) and Peninsula Brownsnake (*P. inframacula*).

Actual size

FAMILY	Elapidae: Hydrophiinae
RISK FACTOR	Highly venomous: pre- and postsynaptic neurotoxins, myotoxins, and procoagulants
DISTRIBUTION	Australo-Papua: eastern and central Australia, and southern New Guinea
ELEVATION	0–4,050 ft (0–1,235 m) asl
HABITAT	Savanna woodland, sclerophyll forests, coastal heathland, grassland, parks, and oil-palm plantations
DIET	Small mammals, lizards, and snakes
REPRODUCTION	Oviparous, with clutches of 10–35 eggs
CONSERVATION STATUS	IUCN not listed

ADULT LENGTH
3 ft 3 in–5 ft,
occasionally 6 ft 7 in
(1.0–1.5 m,
occasionally 2.0 m)

PSEUDONAJA TEXTILIS
EASTERN BROWNSNAKE
(DUMÉRIL, BIBRON & DUMÉRIL, 1854)

529

The snake most often associated with serious and fatal snakebites in eastern Australia, the Eastern Brownsnake is a fast-moving, rapid-striking, and commonly encountered species that occurs in a wide variety of dry or semi-dry habitats from grassland to woodland, and even in parks and gardens. It is found throughout eastern Australia, excluding Cape York Peninsula, but is also known from several locations in New Guinea, where it also inhabits oil-palm plantations. Originally it was believed that the brownsnake had been accidentally introduced during, or after, World War II, but it has now been established that the Papuan populations are naturally occurring. Prey consists of small mammals, lizards, and snakes, including its own species. Snakebites to humans are extremely dangerous, with death often caused by cerebral hemorrhage.

The Eastern Brownsnake is usually dark or light brown but specimens from the Papuan Peninsula are often black in color. Juveniles are light brown to yellow, with a black cap marking on the head, a broad black nape band, and a series of black bands around the body and tail.

RELATED SPECIES

Brownsnakes as a group are found throughout mainland Australia. The range of *Pseudonaja textilis* overlaps with those of the Gwardar (*P. mengdeni*, page 528), Ringed Brownsnake (*P. modesta*), Ingram's Brownsnake (*P. ingrami*), Speckled Brownsnake (*P. guttata*), and Strap-snouted Brownsnake (*P. aspidorhyncha*). The identity of sympatric brownsnakes often relies upon scale counts as coloration and patterning are highly variable.

Actual size

FAMILY	Elapidae: Hydrophiinae
RISK FACTOR	Venomous: venom composition unknown
DISTRIBUTION	Melanesia: Solomon Islands, and Buka Island, Papua New Guinea
ELEVATION	0–2,300 ft (0–700 m) asl
HABITAT	Rainforests and overgrown creeks
DIET	Lizards, frogs, and snakes
REPRODUCTION	Oviparous, with clutches of 3–7 eggs
CONSERVATION STATUS	IUCN Least Concern

ADULT LENGTH
Male
2 ft–2 ft 5 in
(0.6–0.75 m)

Female
2 ft 7 in–4 ft
(0.8–1.18 m)

530

SALOMONELAPS PAR
SOLOMONS CORALSNAKE
(BOULENGER, 1884)

The Solomons Coralsnake occurs in at least five color patterns ranging from uniform gray, brown, red, or orange, to banded or reticulate morphotypes, to two-tone specimens, lighter anteriorly and darker posteriorly, and specimens with contrasting lighter heads. Unsurprisingly, in the past, various island populations with distinctive patterns were treated as different species.

The Solomons Coralsnake is found throughout the Solomon Island Archipelago. However, it is curiously absent from the island of Bougainville, part of Papua New Guinea, despite occurring again on Buka Island to the north of Bougainville, the only elapid on Bougainville being the tiny endemic Bougainville Coralsnake (*Parapistocalamus hedigeri*). The Solomons Coralsnake inhabits rainforest and is especially common along overgrown creeks where it feeds on skinks, frogs, and occasionally agamid lizards, geckos, and blindsnakes (Typhlopidae and Gerrhopilidae). The Solomons Coralsnake will flatten its body and hiss when disturbed. Little is known of its venom, although villagers report it has claimed lives. The only known documented snakebite was symptomless.

RELATED SPECIES

The only other terrestrial elapid within the range of *Salomonelaps par* is the similar-sized Solomons Small-eyed Snake (*Loveridgelaps elapoides*, page 516), a species that sometimes preys on *S. par*.

Actual size

FAMILY	Elapidae: Hydrophiinae
RISK FACTOR	Venomous: venom composition unknown
DISTRIBUTION	Australia: southern Western Australia and South Australia
ELEVATION	0–2,660 ft (0–810 m) asl
HABITAT	Coastal plains, dunes, spinifex plains, semidesert, and desert
DIET	Lizards
REPRODUCTION	Oviparous, with clutches of 1–8 eggs
CONSERVATION STATUS	IUCN not listed

ADULT LENGTH
11¾ in
(300 mm)

SIMOSELAPS BERTHOLDI
SOUTHERN DESERT BANDED SNAKE
(JAN, 1859)

531

The Southern Desert Banded Snake occurs in southern Western Australia and South Australia, in habitats ranging from coastal dunes to tussock grasslands and desert. It is an adept burrower and sand-swimmer, due to the presence of a broad, shovel-like snout, and it can disappear quickly if uncovered. The prey of the Southern Desert Banded Snake comprises mostly skinks, especially the genus *Lerista*, known as sliders, which contains almost 100 species. Southern Desert Banded Snakes are inoffensive and do not threaten to bite, even when handled. In any other part of the world a small, shovel-nosed desert burrowing snake would be nonvenomous, but in Australia the elapids have occupied this niche.

RELATED SPECIES
The other species in the genus *Simoselaps* are the Northern Desert Banded Snake (*S. anomalus*), which has a glossy black head, and the finely banded West Coast Banded Snake (*S. littoralis*). *Simoselaps* is closely related to the burrowing snakes (*Neelaps*), also from Western Australia.

The Southern Desert Banded Snake is a smooth-scaled snake with a pointed head, a smooth-scaled body, and a short, pointed tail. The head and neck are pale gray, with black flecks, while the body and tail are boldly and broadly banded orange and black.

Actual size

FAMILY	Elapidae: Hydrophiinae
RISK FACTOR	Venomous: venom composition unknown
DISTRIBUTION	Australia: northern and central Australia
ELEVATION	0–2,490 ft (0–760 m) asl
HABITAT	Arid and semiarid woodland, scrubland, grassland, and desert
DIET	Lizards and snakes
REPRODUCTION	Viviparous, with litters of 2–5 neonates
CONSERVATION STATUS	IUCN not listed

ADULT LENGTH
19¾–20¾ in
(500–525 mm)

532

SUTA PUNCTATA
LITTLE SPOTTED SNAKE
(BOULENGER, 1896)

The Little Spotted Snake is a glossy, smooth-scaled snake, olive-brown above, with every scale black-spotted, and white underneath. The head is orange-brown with a series of large, irregular black spots that extend backward onto the nape and neck.

The Little Spotted Snake inhabits northern Australia, from the Kimberley, Western Australia, to Mt. Isa, Queensland, with an isolated population in the Pilbara. It inhabits arid and semiarid habitats from eucalypt woodlands to spinifex grasslands and stony or sandy deserts. The prey of the Little Spotted Snake comprises mostly skinks and small agamids, but it also feeds on blindsnakes (*Anilios*). Although the related, and highly nervous, Curl Snake (*Suta suta*) is considered potentially dangerous to humans, the Little Spotted Snake is not thought capable of causing a serious snakebite. Even so, bites are likely to be very painful and should be avoided. A live-bearer, it produces litters of two to five neonates.

RELATED SPECIES

The genus *Suta* contains three other species, the previously mentioned dangerous Curl or Myall Snake (*S. suta*), from eastern and central Australia, Rosen's Snake (*S. fasciata*) from southern Western Australia, and the Ord Snake (*S. ordensis*) from the Ord River and neighboring drainage systems in northern Western Australia.

Actual size

FAMILY	Elapidae: Hydrophiinae
RISK FACTOR	Venomous: neurotoxins, and possibly hemorrhagins
DISTRIBUTION	New Guinea: Woodlark Island (Milne Bay Province, Papua New Guinea)
ELEVATION	0–260 ft (0–80 m) asl
HABITAT	Rainforests and overgrown creeks
DIET	Earthworms
REPRODUCTION	Believed oviparous, clutch size unknown
CONSERVATION STATUS	IUCN not listed

ADULT LENGTH
27½–31½ in
(700–800 mm)

TOXICOCALAMUS LONGISSIMUS
WOODLARK ISLAND SNAKE
(BOULENGER, 1896)

533

The Woodlark Island Snake was the first species to be described in the endemic New Guinea genus *Toxicocalamus*, which now numbers at least 15 species, with more awaiting description. It is a diurnal inhabitant of leaf litter in rainforest, where it appears to feed exclusively on earthworms. Endemic to Woodlark Island, Milne Bay Province, Papua New Guinea, only 13 specimens are known. Island endemics may be threatened by mining developments on their small islands. The Woodlark Island Snake is the only species in the genus to have had its venom analyzed. It contains the building blocks for some serious neurotoxins and hemorrhagins, components known as three-finger toxins: type-I phospholipase A2, and snake venom metalloproteinases. How these serve a snake that feeds on earthworms, and what they would do to a human, is unknown.

Actual size

The Woodlark Island Snake is slender-bodied, with a round head and a long tail. It is yellow-brown with darker pigment in the center of each scale, giving the appearance of longitudinal stripes. The head is faintly banded, and the prefrontal and preocular scales are fused as a single scale. The arrangement of the head scales is important in identifying members of this genus.

RELATED SPECIES

Toxicocalamus contains species ranging from the shoelace-like Fly River Snake (*T. preussi angusticinctus*) to the stout-bodied Setakwa River Snake (*T. grandis*). Most species are under 3 ft 3 in (1 m), but the recently described Star Mountains Worm-eating Snake (*T. ernstmayri*) achieves 4 ft (1.2 m). *Toxicocalamus longissimus* is one of five Milne Bay island endemics, alongside the Misima Island Snake (*T. misimae*), Sudest Island Snake (*T. mintoni*), Rossel Island Snake (*T. holopelturus*), and Fergusson Island Snake (*T. nigrescens*).

FAMILY	Elapidae: Hydrophiinae
RISK FACTOR	Highly venomous: pre- and postsynaptic neurotoxins, myotoxins, and procoagulants
DISTRIBUTION	Australia: eastern coastal Australia (New South Wales and Queensland)
ELEVATION	0–4,580 ft (0–1,395 m) asl
HABITAT	Coastal rainforest creeks
DIET	Frogs, and small mammals, birds, and lizards
REPRODUCTION	Viviparous, with litters of 5–18 neonates
CONSERVATION STATUS	IUCN not listed

ADULT LENGTH
2 ft 5 in–3 ft 3 in
(0.75–1.0 m)

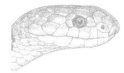

TROPIDECHIS CARINATUS
ROUGH-SCALED SNAKE
(KREFFT, 1863)

The Rough-scaled Snake is a strongly keel-scaled snake with a moderately broad head and large eyes with round pupils. It is generally brown above, with faint darker cross-bars, and yellow to white below. A faint black and orange collar is sometimes present on the nape.

A predator of primarily frogs, but also small mammals, lizards, or birds, the Rough-scaled Snake looks more like a harmless keelback than a dangerous elapid, and several people who have confused the two have not lived to tell the tale. This species possesses highly toxic venom and is easily capable of causing a fatal snakebite. The Rough-scaled Snake occurs in aquatic habitats in coastal rainforest, in northern New South Wales and southern Queensland, with small, isolated populations in southern, coastal Cape York Peninsula, Queensland. It may be active diurnally in cool weather but becomes nocturnal in warm weather. Rough-scaled Snakes produce litters of up to 18 neonates, but more usually they number five to nine.

RELATED SPECIES

Tropidechis carinatus is unusual, though not unique, in being a keel-scaled elapid. It bears a very strong resemblance to, and occurs in sympatry with, the nonvenomous Common Keelback (*Tropidonophis mairii*, page 436), and an incorrect identification could easily have tragic results. *Tropidechis carinatus* is believed closely related to the Tigersnake (*Notechis scutatus*, page 519).

Actual size

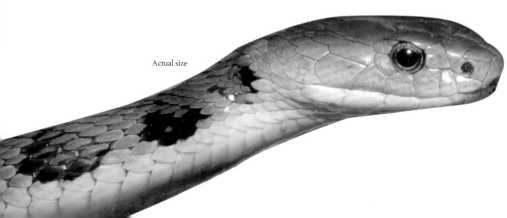

FAMILY	Elapidae: Hydrophiinae
RISK FACTOR	Venomous: venom composition unknown
DISTRIBUTION	Australia: eastern and northeastern Australia
ELEVATION	0–4,350 ft (0–1,325 m) asl
HABITAT	Wet coastal forest, dry woodland, mallee, grassland, semidesert, and sand hills
DIET	Blindsnakes
REPRODUCTION	Oviparous, with clutches of 8–13 eggs
CONSERVATION STATUS	IUCN not listed

ADULT LENGTH
2 ft–2 ft 7 in,
occasionally 3 ft 3 in
(0.6–0.8 m,
occasionally 1.0 m)

VERMICELLA ANNULATA
EASTERN BANDY-BANDY
(GRAY, 1841)

535

Bandy-bandies are so-called because of their alternating black and white banded pattern. The Eastern Bandy-bandy occurs through eastern and northeastern Australia in a wide range of habitats, from wet coastal forests to dry woodland, mallee, grasslands, and sandy desert. It is believe to feed exclusively on blindsnakes (*Anilios*), of which Australia has a large and diverse population. Although venomous, bandy-bandies are inoffensive and reluctant to bite. Their preferred defense involves raising body coils high in the air in a series of stiff loops, while thrashing about continually. It may be that the continual movement of black and white confuses a predator long enough for the bandy-bandy to burrow into the leaf litter and escape.

The Eastern Bandy-bandy is smooth-scaled, and banded black and white on the rounded head, body, and short tail, with bands of each color being the same width. Its main defense involves raising stiff body coils and continually changing its position.

RELATED SPECIES

Four other species of bandy-bandies are recognized: the Intermediate Bandy-bandy (*Vermicella intermedia*) from northwestern Australia; Northern Narrow-banded Bandy-bandy (*V. multifasciata*) from northern Australia; Pilbara Bandy-bandy (*V. snelli*); and Centralian Bandy-bandy (*V. vermiformis*). All these species are from small, localized areas, only *V. annulata* being widespread.

Actual size

FAMILY	Homalopsidae
RISK FACTOR	Rear-fanged, mildly venomous; believed harmless to humans
DISTRIBUTION	Southeast Asia: southern Myanmar, Thailand, Peninsular Malaysia, and Singapore
ELEVATION	Sea-level
HABITAT	River estuaries, mudflats, and possibly mangroves
DIET	Fish
REPRODUCTION	Viviparous, with litters of 1–10 neonates
CONSERVATION STATUS	IUCN Least Concern

ADULT LENGTH
31½ in
(800 mm)

536

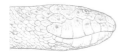

BITIA HYDROIDES
KEEL-BELLIED WATERSNAKE
GRAY, 1842

The dorsal body scales of the Keel-bellied Watersnake are numerous, smooth, and triangular, but non-overlapping, the skin between them being clearly visible in the gaps. The belly scales exhibit a pair of longitudinal ridges that run the length of the body on either side, similar to the ventral scales of arboreal snakes like ratsnakes (such as *Pantherophis*, pages 207–210). However, this is a marine species so the ridges must serve a different purpose. The snake also exhibits typical homalopsid characteristics, including small, dorsolaterally positioned eyes, dorsal valvular nostrils, and tight-fitting lip scales to prevent ingress of water. In common with many other homalopsids it is a rear-fanged venomous species that feeds entirely on small fish, especially gobies. Nothing is known of the effects of its venom on humans.

RELATED SPECIES

Bitia hydroides is most closely related to Gerard's Watersnake (*Gerarda prevostiana*, page 542), the Crab-eating Mangrove Snake (*Fordonia leucobalia*, page 541), and Cantor's Mangrove Snake (*Cantoria violacea*, page 538), but its unusual body scalation makes it unique among homalopsids.

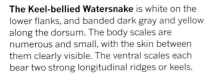

Actual size

The Keel-bellied Watersnake is white on the lower flanks, and banded dark gray and yellow along the dorsum. The body scales are numerous and small, with the skin between them clearly visible. The ventral scales each bear two strong longitudinal ridges or keels.

FAMILY	Homalopsidae
RISK FACTOR	Nonvenomous
DISTRIBUTION	Southeast Asia: Maluku (Indonesia)
ELEVATION	0–2,770 ft (0–845 m) asl
HABITAT	Rainforest, secondary growth, gardens, and plantations
DIET	Earthworms
REPRODUCTION	Viviparous, with litters of up to 5 neonates
CONSERVATION STATUS	IUCN Data Deficient

ADULT LENGTH
19–21 in
(480–535 mm)

BRACHYORRHOS ALBUS
SERAM SHORT-TAILED SNAKE
(LINNAEUS, 1758)

537

Not all homalopsid snakes are marine or freshwater aquatic, and not all are mildly venomous, or even possess fangs. The Seram Short-tailed Snake is a terrestrial, semi-fossorial, and apparently occasionally arboreal snake that inhabits secondary growth, plantations, and gardens, and also presumably pristine rainforest. This snake is frequently found around human habitations and may be diurnal. It is found on the eastern Indonesian island of Seram, in Maluku Province, and on its satellite islands, Ambon, Haruku, Nusa Laut, and Saparau. A population from Pulau Bisa, off Obi Island in North Maluku Province, is also attributed to this species. Short-tailed snakes are believed to prey entirely on earthworms.

The Seram Short-tailed Snake has a cylindrical body with smooth scales, which are almost glossy and iridescent, a strongly pointed head, and a short, rapidly tapering tail. It is uniform light pinkish gray to brown or dark gray above, and uniform white below, except for under the tail, which has brown sutures along the subcaudal scale edges.

RELATED SPECIES

Three other species of short-tailed snakes are recognized: the Buru Short-tailed Snake (*Brachyorrhos gastrotaenius*), Halmahera Short-tailed Snake (*B. wallacei*), and Ternate Short-tailed Snake (*B. raffrayi*), while populations from the Aru Islands and Moratai await study. The related nonvenomous, fangless stout-tailed snakes (*Calamophis*) comprise four species from west New Guinea and the islands of Cenderawasih (formerly Geelvink) Bay.

Actual size

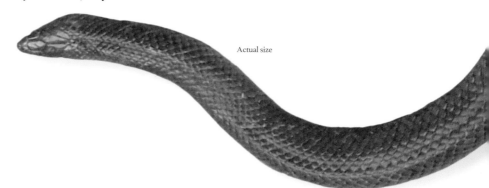

FAMILY	Homalopsidae
RISK FACTOR	Rear-fanged, mildly venomous; believed harmless to humans
DISTRIBUTION	Southeast Asia: southern Myanmar, Thailand, Peninsular Malaysia, Singapore, and Andaman Islands; possibly Borneo, Sumatra, and Timor
ELEVATION	Sea-level
HABITAT	River estuaries, tidal creeks, mudflats, mangrove swamps, and into fresh water
DIET	Crustaceans (shrimps)
REPRODUCTION	Not known, presumed viviparous
CONSERVATION STATUS	IUCN Least Concern

ADULT LENGTH
3 ft 7 in–4 ft 3 in
(1.1–1.3 m)

538

CANTORIA VIOLACEA
CANTOR'S MANGROVE SNAKE
GIRARD, 1858

Cantor's Mangrove Snake is black-brown or purple-brown, the possible origin of the *violacea* part of its name, with a series of narrow, transverse yellowish or whitish bars, which also occur on the head. The underside is gray to yellowish.

Cantor's Mangrove Snake occurs in the Indian-owned Andaman Islands, and along the coastlines of the Malay Peninsula to Singapore. Reports of its presence in Borneo, Sumatra, and Timor are unsubstantiated. It inhabits river mouths, mudflats, and mangrove swamps, and occasionally enters fresh water. By day it shelters in mud-lobster burrows and crab holes, emerging at night to hunt. It preys on pistol shrimps. It is thought that the defense of these shrimps, a loud sound and flash of light created by their pincers, and a chemical they exude, may guide the snake to its prey. Nothing is known of the venom of this snake, although local folklore mentions human fatalities. Theodore Cantor (1809–60) was a Danish physician and naturalist with the British East India Company.

RELATED SPECIES

Cantoria violacea is most closely related to Gerard's Watersnake (*Gerarda prevostiana*, page 542), the Crab-eating Mangrove Snake (*Fordonia leucobalia*, page 541), and the Keel-bellied Watersnake (*Bitia hydroides*, page 536). *Cantoria* is now a monotypic genus, but previously it also contained the extremely rare Trans-Fly Watersnake (*Djokoiskandarus annulatus*).

Actual size

FAMILY	Homalopsidae
RISK FACTOR	Rear-fanged, mildly venomous; harmless to humans
DISTRIBUTION	Southeast Asia: southern Vietnam and Malay Peninsula, to the Philippines, Indonesia, and Timor-Leste
ELEVATION	Sea-level
HABITAT	River estuaries, mudflats, mangroves, and coastal rice paddies
DIET	Fish, and possibly crustaceans, frogs, and tadpoles
REPRODUCTION	Viviparous, but litter size unknown
CONSERVATION STATUS	IUCN not listed

CERBERUS SCHNEIDERII
SCHNEIDER'S BOCKADAM
(SCHLEGEL, 1837)

ADULT LENGTH
Male
3 ft 3 in
(1.0 m)
Female
4 ft
(1.25 m)

539

The origin of the name "Bockadam" is unclear, but the alternative name for these snakes is Dog-faced Watersnake; *Cerberus* was the many-headed dog of hell in Greek mythology. Schneider's Bockadam is a coastal species that inhabits mangrove swamps and estuarine mudflats, but enters fresh water in coastal rice paddies and even crosses roads during heavy rain. It is instantly recognizable by its rough, keel-scaled body and elongate head. The eyes are dorsolateral, enabling the snake to lie submerged with just its eyes and valvular nostrils visible. The prey of Schneider's Bockadam consists primarily of fish, mainly gobies and lizardfish, but crustaceans are also taken, and there is a suspicion that rice paddy-dwelling specimens may take frogs and tadpoles in the absence of fish. Johann Gottlob Schneider (1750–1822) was a German herpetologist.

Schneider's Bockadam has a light or dark brown body, with irregular black cross-bars, and rugose, keeled scales. It has an elongate head, small, dorsolateral eyes, and valvular nostrils.

RELATED SPECIES

Once a single species was recognized, the Asian Bockadam (*Cerberus rynchops*), but this species is now confined to the coasts of South Asia. The Southeast Asian and Indo-Malay Archipelago populations are now a separate species (*C. schneiderii*), as are the northern Australian and southern Papuan populations (*C. australis*). Two species with very limited ranges are also defined, from the Palau Islands (*C. dunsoni*) and the land-locked Lake Buhi, on Luzon, Philippines (*C. microlepis*). This last taxon is listed as Endangered by the IUCN.

Actual size

FAMILY	Homalopsidae
RISK FACTOR	Rear-fanged, mildly venomous; harmless to humans
DISTRIBUTION	Southeast Asia: southern Thailand, Cambodia, and Vietnam
ELEVATION	0–82 ft (0–25 m) asl
HABITAT	Lowland ponds and slow-moving watercourses
DIET	Fish
REPRODUCTION	Viviparous, with litters of 9–13 neonates
CONSERVATION STATUS	IUCN Least Concern

ADULT LENGTH
30¼ in
(770 mm)

540

ERPETON TENTACULATUM
TENTACLED SNAKE
LACÉPÈDE, 1800

The Tentacled Snake is one of the strangest snakes on the planet. It inhabits slow-moving, still, or stagnant habitats across southern Thailand, Cambodia, and Vietnam. The dorsal scales bear shallow keels, while two keels are present in each narrow ventral scale, but its most unusual feature is a pair of fleshy tentacles on either side of the snout. When hunting small fish the snake anchors itself to underwater vegetation by its prehensile tail, and remains motionless in the water, camouflaged by its patterning, shape, and a frequent growth of algae. When a fish approaches, the snake strikes and swallows it in one swift movement. The tentacles have a part to play in prey location, but it is not known how they function. They may detect the vibrations of swimming fish.

RELATED SPECIES

Although a member of the mildly venomous mangrove-, mudflat-, and freshwater-dwelling Homalopsidae, *Erpeton tentaculatum* does not resemble any known living snake. A molecular study suggests it may be a relative of Bocourt's Mudsnake (*Subsessor bocourti*), also from Thailand.

Actual size

The Tentacled Snake is slender-bodied, and may be brown or gray, with every dorsal and ventral scale keeled, but the paired tentacles are this species' most distinctive characteristic.

FAMILY	Homalopsidae
RISK FACTOR	Rear-fanged, mildly venomous; believed harmless to humans
DISTRIBUTION	Asia and Australasia: India to New Guinea, and northern Australia
ELEVATION	Sea-level
HABITAT	River estuaries, mudflats, and mangrove swamps
DIET	Crustaceans (crabs and mud lobsters), and also fish
REPRODUCTION	Viviparous, with litters of 6–15 neonates
CONSERVATION STATUS	IUCN Least Concern

ADULT LENGTH
28 in
(710 mm)

FORDONIA LEUCOBALIA
CRAB-EATING MANGROVE SNAKE
(SCHLEGEL, 1837)

541

Also known as the White-bellied Mangrove Snake, due to its immaculate white underbelly, the Crab-eating Mangrove Snake is one of the most widely distributed members of the Homalopsidae, occurring from the Sunderbans swamps of Bangladesh and West Bengal, India, through the Indo-Malay Archipelago, to New Guinea and northern Australia. A common species, it spends the day hidden inside mud lobster burrows or crab holes, becoming active with the incoming evening tide. Its diet primarily consists of freshly molted crabs, which are grasped, chewed until the snake's rear fangs come into play, and shaken until dismembered, allowing the remainder to be ingested. Small mud lobsters and fish are also reported in this species' diet. Its venom is not considered harmful to humans.

The Crab-eating Mangrove Snake is a variably patterned species that may be dark gray, brown, orange, yellow, or white, either totally, or spotted with black, white, or yellow. However, its belly is always pure white. It has smooth scales, a rounded head, and small eyes.

RELATED SPECIES
The closest relatives of *Fordonia leucobalia* are Gerard's Watersnake (*Gerarda prevostiana*, page 542), Cantor's Mangrove Snake (*Cantoria violacea*, page 538), and the Keel-bellied Watersnake (*Bitia hydroides*, page 536). It is often found in sympatry with one of the bockadams (*Cerberus*, page 539), from which it can be easily distinguished by its smooth rather than strongly keeled scales.

Actual size

FAMILY	Homalopsidae
RISK FACTOR	Rear-fanged, mildly venomous; believed harmless to humans
DISTRIBUTION	Asia: India and Sri Lanka, Myanmar to Cambodia, and Sumatra to the Philippines
ELEVATION	Sea-level
HABITAT	Mangrove swamps, river estuaries, mudflats, and occasionally rocky shores
DIET	Crustaceans (crabs)
REPRODUCTION	Not known, but presumed viviparous
CONSERVATION STATUS	IUCN Least Concern

ADULT LENGTH
20½–22¾ in
(520–580 mm)

542

GERARDA PREVOSTIANA
GERARD'S WATERSNAKE
(EYDOUX & GERVAIS, 1837)

Gerard's Watersnake occurs widely across South and Southeast Asia, from western India to the Philippines, and south to Sumatra. It is an inhabitant of mangrove swamps, tidal estuaries, mudflats, and rocky shores. It shelters inside mud-lobster mounds, hunting at night in muddy pools. Prey consists of freshly molted crabs, which are dismembered before being swallowed, in the same way reported for the larger Crab-eating Mangrove Snake (*Fordonia leucobalia*, page 541). Whether it also feeds on fish has not been established. The venom of Gerard's Watersnake has not been studied, and nor has its reproductive strategy. A specimen was reported to contain "five eggs," but this could indicate oviparity or ovoviviparity (a form of viviparity). Edward Gerard was a taxidermist who worked for the British Museum in the nineteenth century.

RELATED SPECIES

Gerarda prevostiana is closely related to the Crab-eating Mangrove Snake (*Fordonia leucobalia*, page 541), Cantor's Mangrove Snake (*Cantoria violacea*, page 538), and the Keel-bellied Watersnake (*Bitia hydroides*, page 536).

Gerard's Watersnake is dorsally gray or brown with each scale black-edged, and pale yellow, cream, or yellow-gray below, some specimens exhibiting a longitudinal stripe from the gray-sutured yellow lips to the tail. The body scales are smooth and glossy.

Actual size

FAMILY	Homalopsidae
RISK FACTOR	Rear-fanged, mildly venomous; harmless to humans
DISTRIBUTION	Southeast Asia: Thailand, Cambodia, Vietnam, and Peninsular Malaysia
ELEVATION	0–33 ft (0–10 m) asl
HABITAT	Shallow wetlands, reservoirs, irrigation canals, and ponds
DIET	Fish
REPRODUCTION	Viviparous, with litters of 13–33 neonates
CONSERVATION STATUS	IUCN not listed

HOMALOPSIS MERELJCOXI

COX'S MASKED WATERSNAKE

(LINNAEUS, 1758)

ADULT LENGTH
Male
3 ft 5 in
(1.05 m)

Female
4 ft 6 in
(1.38 m)

543

The masked watersnakes of genus *Homalopsis* bear a dark stripe through the eye, reminiscent of a highwayman's mask. Previously most mainland populations were listed as the Puff-faced Watersnake (*H. buccata*), but as part of a recent revision Cox's Masked Watersnake was described for Thai and Cambodian populations with bold dorsal markings and high dorsal scale counts. This population contains the largest known homalopsids, females achieving greater than 4 ft 3 in (1.3 m) total length. Masked watersnakes are freshwater snakes, Cox's Masked Watersnake occurring in natural wetlands and man-made irrigation canals, ponds, and reservoirs. Prey consists of freshwater fish. These snakes are harvested in large and potentially damaging numbers for the skin trade. Merel "Jack" Cox is an American herpetologist specializing in the herpetofauna of Thailand.

Cox's Masked Watersnake is a light brown snake, boldly patterned with dark brown bands, each wider than the interspaces and edged with black. The head is light brown above with a broad, dark brown, mask-like stripe through the eyes and a dark brown V-shaped marking on the dorsum. The undersides are white or yellow.

RELATED SPECIES

The genus *Homalopsis* contains four other species: Hardwick's Spine-bellied Watersnake (*H. hardwickii*, page 508) from northeast India, the Martaban Watersnake (*H. semizonata*) from Myanmar, the Black-bellied Mekong Watersnake (*H. nigroventralis*) from Laos and Cambodia, and the Puff-faced Watersnake (*H. buccata*) from Peninsular Malaysia, Sumatra, Borneo, and Java.

Actual size

FAMILY	Homalopsidae
RISK FACTOR	Rear-fanged, mildly venomous; harmless to humans
DISTRIBUTION	Australasia: northern Australia and southern New Guinea
ELEVATION	Sea-level
HABITAT	River estuaries, mudflats, and mangrove swamps
DIET	Fish, and possibly crabs
REPRODUCTION	Viviparous, with litters of 6–8 neonates
CONSERVATION STATUS	IUCN Least Concern

ADULT LENGTH
17 in
(436 mm)

544

MYRON RICHARDSONII
RICHARDSON'S MANGROVE SNAKE
GRAY, 1849

Sometimes called the Gray Watersnake, because of its usual color, or Gray's Watersnake, after its describer, Richardson's Mangrove Snake occurs on the northern coast of Australia, from the Kimberley to Cape York, through the Torres Strait, and on the southern coast of Papua New Guinea. It is a small snake, and infrequently encountered compared to sympatric species such as the Crab-eating Mangrove Snake (*Fordonia leucobalia*, page 541). This species occurs in the same habitats of mangrove swamps, mudflats, and tidal river estuaries, where it hunts fish, primarily bottom-dwelling or mudflat-inhabiting gobies. Some authors report that this species also feeds on crabs, but that would potentially put it in competition with the larger and more common crab-specialist *Fordonia*.

RELATED SPECIES

Where only one species was previously recognized, now there are three. The Indonesian Aru Island population is now the Aru Mangrove Snake (*Myron karnsi*), while the population around Broome, Western Australia, has also been described as a separate species, the Broome Mangrove Snake (*M. resetari*).

Actual size

Richardson's Mangrove Snake is a polymorphic species; it exists in more than one color pattern. The usual morphotype is pale or dark gray, with dark gray reticulations dorsally and a pale underbelly, but occasional specimens are vivid orange as shown here, with dorsal markings confined to faint dark gray spots. The body scales are smooth, the head relatively long, and the eyes dorsolaterally positioned.

FAMILY	Homalopsidae
RISK FACTOR	Rear-fanged, mildly venomous; harmless to humans
DISTRIBUTION	Australasia: northern Australia and southern New Guinea
ELEVATION	0–920 ft (0–280 m) asl
HABITAT	Freshwater swamps, marshes, billabongs, lagoons, creeks, and drainage ditches
DIET	Fish, frogs, and tadpoles
REPRODUCTION	Viviparous, with litters of 12–15 neonates
CONSERVATION STATUS	IUCN Least Concern

ADULT LENGTH
28–29 in
(710–740 mm)

PSEUDOFERANIA POLYLEPIS

FLY RIVER SMOOTH WATERSNAKE
(FISCHER, 1886)

545

While many homalopsid snakes are coastal marine in distribution, the Fly River Smooth Watersnake is a freshwater inhabitant of a wide variety of slow-moving or still freshwater habitats, from billabongs and lagoons to swamps, marshes, creeks, and the drainage ditches dug to prevent flooding around villages. Prey consists primarily of small freshwater fish, but frogs and their tadpoles are also taken. Once bitten, frogs quickly succumb to the venom, but although bites to humans have been sustained, no symptoms have been observed, and these snakes are considered harmless to humans. In southern Western Province, Papua New Guinea, this snake is hunted and eaten by people from some villages.

The Fly River Smooth Watersnake is olive-green to brown, sometimes almost black, above, and yellow to orange below. A yellow-orange stripe may be present on the side of the head and juveniles are sometimes striped, but many specimens are unicolor above and unmarked. The body scales are smooth, almost glossy, and the dorsolateral eyes are relatively small on the slightly elongate head.

RELATED SPECIES

Pseudoferania polylepis was originally included in the large genus *Enhydris*, but when that genus was revised the monotypic genus *Pseudoferania* was resurrected for this species. Most authors include a second species as a synonym of *P. polylepis*, Macleay's Watersnake (*P. macleayi*), named for the Australian naturalist John William Macleay (1820–91), who collected extensively in southern New Guinea. The internasal scale contacts the loreal scale in *P. polylepis*, but not in *P. macleayi*.

Actual size

FAMILY	Pareatidae
RISK FACTOR	Nonvenomous
DISTRIBUTION	Southeast Asia: Peninsular Malaysia
ELEVATION	3,880–6,730 ft (1,184–2,050 m) asl
HABITAT	Montane tropical rainforest, and cloud and moss forest
DIET	Slugs, and possibly snails
REPRODUCTION	Probably oviparous
CONSERVATION STATUS	IUCN not listed

ADULT LENGTH
19¾–28¾ in
(500–729 mm)

ASTHENODIPSAS LASGALENENSIS
MIRKWOOD FOREST SLUG-SNAKE
LOREDO, WOOD, QUAH, ANUAR, GREER, AHMAD & GRISMER, 2013

The Mirkwood Forest Slug-snake was so named because its cloudy, mossy, upland forest home reminded its discoverers of J. R. R. Tolkien's fictional forest of Mirkwood, Eryn Lasgalen or "Wood of Greenleaves" in his Sindarin language. It is a specialist slug-eater, a member of a worldwide snake guild called "goo-eaters" that is represented in Asia by the Pareatidae. The Mirkwood Forest Slug-eater forages for slugs, and possibly snails, in low tropical vegetation at night. Unlike snakes that feed on bulkier prey, slug-eaters lack a mental groove under the chin, a fold of loose skin that enables the mouth to gape widely. It is probably oviparous, like the Malayan Slug-snake (*Asthenodipsas malaccanus*).

RELATED SPECIES

Asthenodipsas lasgalenensis occurs in sympatry with the Mountain Slug-snake (*A. vertebralis*). Other congeners include the Malayan Slug-snake (*A. malaccanus*), Smooth Slug-snake (*A. laevis*), and Sumatran Slug-snake (*A. tropidonotus*). The related Blunt-headed Slug-snake (*Aplopeltura boa*) occurs throughout Southeast Asia.

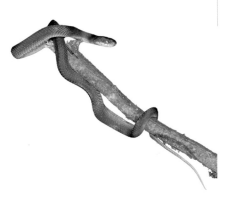

The Mirkwood Forest Slug-snake has a laterally compressed, slender body, a broad, blunt head, smooth scales, and vertically elliptical pupils. It is uniform brown as an adult, and orange to gray as a juvenile, with a patterning of rhomboid bands and a light vertebral stripe. The white lips are dark-edged and the head is brown to black, with dark red or bright orange eyes.

Actual size

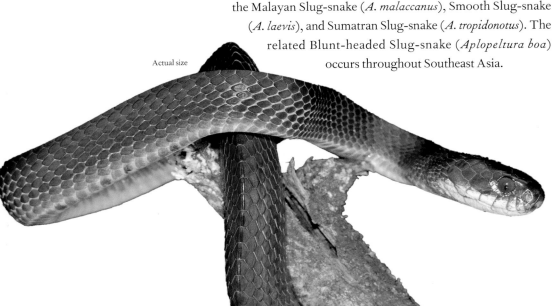

FAMILY	Pareatidae
RISK FACTOR	Nonvenomous
DISTRIBUTION	Southeast Asia: Myanmar, Thailand, Vietnam, Laos, Cambodia, Malaysia, and Indonesia
ELEVATION	1,800–5,840 ft (550–1,780 m) asl
HABITAT	Lowland and submontane forest
DIET	Slugs and snails
REPRODUCTION	Oviparous, with clutches of 3–8 eggs
CONSERVATION STATUS	IUCN Least Concern

ADULT LENGTH
19¾–23¾ in
(500–600 mm)

PAREAS CARINATUS
KEELED SLUG-SNAKE
(BOIE, 1828)

547

The Keeled Slug-snake has two rows of weakly keeled scales on the dorsum, most other slug-eaters being smooth-scaled. It is widely distributed in Southeast Asia, from Myanmar to Bali in Indonesia, at low to moderate elevations. Its preferred habitat is lowland or submontane forest, but it may be found in other habitats. Although called a slug-snake, these snakes will prey on either slugs or snails, moving the lower jaw forward to hook the snail's body with the teeth and then forcibly withdrawing it from the shell. Slug and snail eaters are often referred to as "goo-eaters." This is a widespread "guild" including Asian *Pareas*, *Aplopeltura*, and *Asthenodipsas* (page 546), Neotropical *Dipsas* (pages 273–274), *Sibon* (pages 290–291), and *Sibynomorphus* (page 292), and African *Duberria* (page 384).

RELATED SPECIES

Two subspecies are recognized, the nominate form that occurs through most of the range, and a localized subspecies in Cambodia (*Pareas carinatus unicolor*). The genus *Pareas* contains another 14 species from India, China, Taiwan, and the Ryukyu Islands, to Borneo, several of which occur in sympatry with *P. carinatus*.

The Keeled Slug-eater is an extremely laterally compressed snake with a prehensile tail, a rounded, blunt head, and large, forward-facing eyes with vertically elliptical pupils. Coloration is usually light brown, reddish, or yellowish with fine dark brown bars that fuse with a vertebral stripe.

Actual size

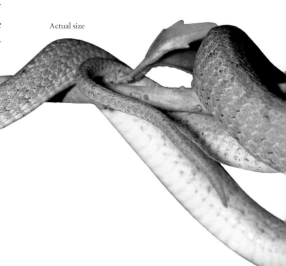

FAMILY	Xenodermatidae
RISK FACTOR	Nonvenomous
DISTRIBUTION	East Asia: central Taiwan and southern Ryukyu Islands
ELEVATION	< 6,560 ft (< 2,000 m) asl
HABITAT	Cool, wet rainforest leaf litter
DIET	Earthworms
REPRODUCTION	Oviparous, clutch size unknown
CONSERVATION STATUS	IUCN Least Concern

ADULT LENGTH
33½ in
(853 mm)

ACHALINUS FORMOSANUS
TAIWAN ODD-SCALED SNAKE
BOULENGER, 1908

548

The Taiwan Odd-scaled Snake is brown above, with a black longitudinal stripe along the back, and has strongly keeled dorsal scales, which are completely fused to the underlying skin. The underside is yellow. The head is elongate, at least twice as long as it is broad, with iridescent scutes.

Snakes in the Xenodermatidae possess scales that are completely fused to the underlying skin, unlike the scales of most other snakes where only the leading edge is embedded in the skin. The odd-scaled snakes (*Achalinus*) are small to moderate-sized, secretive forest snakes that exhibit a semi-fossorial existence, burrowing under rotten logs or into deep leaf litter in search of their earthworm prey. They may be active on the surface after heavy rain, and are then encountered crossing roads. They are very dependent on their cool, wet rainforest habitat and will desiccate and die very quickly when removed from this environment. The Taiwan Odd-scaled Snake occurs on the island of Taiwan and in the southern Ryukyu Islands of Japan.

RELATED SPECIES

There are two subspecies, the nominate form (*Achalinus formosanus formosanus*) from Taiwan, and a subspecies (*A. f. chigirai*) that occurs to the east, on Iriomotejima and Ishigakijima in the southern Ryukyu Islands. *Achalinus formosanus* is most similar to the Taiwan Black Odd-scaled Snake (*A. niger*), the Amami Odd-scaled Snake (*A. werneri*), in the central Ryukyu Islands, and the Japanese Odd-scaled Snake (*A. spinalis*) on mainland Japan. *Achalinus* contains nine species.

Actual size

FAMILY	Xenodermatidae
RISK FACTOR	Nonvenomous
DISTRIBUTION	Southeast Asia: northern Laos
ELEVATION	1,180–2,920 ft (360–890 m) asl
HABITAT	Evergreen forest on rugged karst limestone formations
DIET	Not known
REPRODUCTION	Not known
CONSERVATION STATUS	IUCN not listed

ADULT LENGTH
11¼–14 in
(285–353 mm)

PARAFIMBRIOS LAO
LAOTIAN BEARDED SNAKE
TEYNIÉ, DAVID, LOTTIER, LE, VIDAL & NGUYEN, 2015

549

This monotypic genus was described in 2015, from only two small specimens, collected in evergreen forest on the steep limestone karst escarpments of northern Laos. The Laotian Bearded Snake has a "beard" of small scales along its lower lips, like that of the bearded snakes (*Fimbrios*), but we cannot see this beard of scales in the artwork or photograph. It is also unusual in that it has a double row of dorsal scales, one large and one small, above each ventral scale on the lower flanks, a characteristic only seen in the related Dragon Snake (*Xenodermus javanicus*, page 551). Nothing is known of its natural history, but being closely related to the bearded snakes, it may be terrestrial or semi-aquatic, inhabit cool, wet forest leaf litter along streams, and feed on fish, like Kloss' Bearded Snake (*F. klossi*), amphibians, or earthworms.

The Laotian Bearded Snake is small, with keeled dorsal scales completely fused to the underlying skin, and a rounded head. It is gray, paler below, except for a broad white band around the neck, which may be absent. The mouth is fringed with a beard of small scales.

RELATED SPECIES
Parafimbrios lao is related to the two bearded snakes (*Fimbrios*), Kloss' Bearded Snake (*F. klossi*), and Smith's Bearded Snake (*F. smithi*), from southern Vietnam, Laos, and Cambodia. *Fimbrios* means fringed, a reference to the small labial scales that fringe the lower jaw, the "beard," and *Parafimbrios* means "like *Fimbrios*."

Actual size

FAMILY	Xenodermatidae
RISK FACTOR	Nonvenomous
DISTRIBUTION	Southeast Asia: Malaysian Borneo, (Sabah), and Sarawak, possibly Kalimantan
ELEVATION	2,620–5,910 ft (800–1,800 m) asl
HABITAT	Submontane and montane rainforest, along streams
DIET	Not known
REPRODUCTION	Not known
CONSERVATION STATUS	IUCN Least Concern

ADULT LENGTH
29½ in
(750 mm)

550

STOLICZKIA BORNEENSIS
BORNEO STREAM SNAKE
(BOULENGER, 1899)

The Borneo Stream Snake is only known from a few specimens from Mt. Kinabalu, Mt. Trus Madi, the Crocker Range, in the Malaysian state of Sabah, and Mt. Murud in Sarawak. A row of small scales is present across the head in front of the eyes, a unique arrangement. The body scales are strongly keeled with knobbles on each keel. This poorly known species inhabits montane and submontane rainforest. It is nocturnal and arboreal, but its diet and reproductive strategy are unknown. The genus is named for Ferdinand Stoliczka (1838–74), a Czech herpetologist who worked in Asia.

RELATED SPECIES
Only one other species of *Stoliczkia* is known, the even less well-known Khasi Stream Snake (*S. khasiensis*) from Assam and Meghalaya, northeast India, which lacks the rows of small scales across the head.

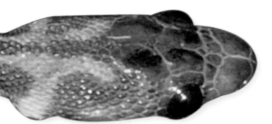

Actual size

The Borneo Stream Snake is an extremely slender snake, laterally compressed with a bulbous head, which is broader than the neck, and a sharply ridged back, protruding eyes, and large, flared nostrils. Almost 30 percent of its length is tail. The dorsal pattern comprises a red-brown or blue-black background, overlain with a black checkerboard pattern. The undersides are brownish and unmarked.

FAMILY	Xenodermatidae
RISK FACTOR	Nonvenomous
DISTRIBUTION	Southeast Asia: Myanmar, Thailand, Malaysia, Indonesia, Java, Sumatra, and Borneo
ELEVATION	1,640–3,610 ft (500–1,100 m) asl
HABITAT	Lowland rainforest and agricultural habitats
DIET	Frogs
REPRODUCTION	Oviparous, with clutches of 2–5 eggs
CONSERVATION STATUS	IUCN Least Concern

ADULT LENGTH
23¾–25½ in
(600–650 mm)

XENODERMUS JAVANICUS
DRAGON SNAKE
REINHARDT, 1836

551

The Dragon Snake is covered in tiny, granular scales, except for five rows of enlarged, keeled scales down the back, three rows at middorsum and a dorsolateral row on either side. These raised scale rows are likely to be the origin of the name "dragon snake" as they present a dragon-like appearance, although other names are more literal, if less poetic, such as the Javan Rough-backed Snake. Although widely distributed, from Myanmar to Borneo and Java, this is a rarely encountered snake, due to its nocturnal activity and secretive, semi-fossorial lifestyle. It spends the day sheltering under logs and emerges at night to hunt frogs in rainforest leaf litter or along streams. It may also be found in areas altered by cultivation, and is known to lay two to five eggs.

The Dragon Snake is long and slender and exhibits a strange dorsal scale arrangement, with three middorsal and two dorsolateral rows of enlarged keeled scales on an otherwise granular-scaled body, and a head covered with granular scales, except for the nasal region. This species is dark gray above, and off-white below.

RELATED SPECIES

This member of the monotypic genus *Xenodermus* lends its name to this curious family of snakes with their scales entirely fused to the underlying skin. Eighteen species in six genera are recognized, four genera being included in this book; only the Indian narrow-headed snakes (*Xylophis*, three species) and bearded snakes (*Fimbrios*, two species) are excluded. *Xenodermus javanicus* is most similar in appearance to the Borneo Stream Snake (*Stoliczkia borneensis*, page 550), but that species lacks the raised dorsal scales.

Actual size

FAMILY	Viperidae: Azemiopinae
RISK FACTOR	Venomous: neurotoxins and anticoagulants
DISTRIBUTION	South-central Asia: southern China (Yunnan), northern Myanmar, and Vietnam
ELEVATION	1,970–6,560 ft (600–2,000 m) asl
HABITAT	Karst and cloud forest, bamboo forest, and fern forest
DIET	Small mammals (rodents and shrews)
REPRODUCTION	Oviparous, with clutches of up to 5 eggs
CONSERVATION STATUS	IUCN Least Concern

ADULT LENGTH
2 ft 4 in–3 ft 3 in
(0.7–1.0 m)

AZEMIOPS FEAE

FEA'S VIPER

BOULENGER, 1888

Fea's Viper is dark gray, with vivid orange bars on its body, and a contrasting white head with a black dorsal arrow-shaped marking, compared to the white head with orange stripes of *Azemiops kharini*. The pupils of the eyes are vertically elliptical like those of a typical viper.

The Azemiopinae, containing only *Azemiops*, is the most primitive subfamily in the Viperidae and considered to be close to the ancestors of the pitvipers, which evolved in south-central Asia. Fea's Viper lacks pits and was originally treated as an elapid, possibly because it possesses smooth body scales and large, regular head scutes, rather than the keeled body scales and granular head scales of most other Asian vipers. In common with most elapids, and in contrast to most vipers, it also lays eggs. It preys primarily on gray shrews that live along the fast-flowing mountain streams in its forested karst landscape. It requires high humidity and low temperatures, and is unable to survive in hot or arid habitats. It is therefore vulnerable to deforestation. Although it is venomous, bites to humans have only produced mild symptoms.

RELATED SPECIES

After 125 years as a single species, *Azemiops feae* was recently split into two species. The western *A. feae* has been called the Black-headed Burmese Viper, due to the broad black arrowhead on top of its head. The White-headed Burmese Viper, from the east of the range, has a pair of orange stripes on top of the head, and was described as a new species, *A. kharini*, to honor the Russian herpetologist and ichthyologist Vladimir Kharin (1957–2013).

Actual size

FAMILY	Viperidae: Crotalinae
RISK FACTOR	Venomous: procoagulants and hemotoxins
DISTRIBUTION	North America: southeastern and eastern USA to northeastern Mexico, excluding Florida Peninsula
ELEVATION	0–1,640 ft (0–500 m), rarely 4,920 ft (1,500 m) asl
HABITAT	Dry deciduous pine–oak or coniferous woodland, especially on rocky slopes, often near water
DIET	Small mammals, birds, lizards, and other snakes
REPRODUCTION	Viviparous, with litters of 2–18 neonates
CONSERVATION STATUS	IUCN Least Concern

AGKISTRODON CONTORTRIX
COPPERHEAD
(LINNAEUS, 1766)

ADULT LENGTH
3 ft 3 in–4 ft 3 in
(1.0–1.3 m)

553

The Copperhead is so named due to its distinctive coppery or reddish head. It is a common inhabitant of rocky, wooded hillsides in the eastern United States, where it may occur alongside the Timber Rattlesnake (*Crotalus horridus*, page 577). Copperheads are often associated with areas of permanent or semi-permanent water. This species is rare in northeast Mexico and absent from the Florida Peninsula. Copperheads warn of their presence by vibrating the tail vigorously on dead leaves, and they also emit a pungent scent when handled. Although prey usually comprises small mammals, from mice to squirrels and young opossums, birds, lizards, and other snakes are also taken. Bites to humans may be relatively common but fatalities are rare, though not unknown. Victims usually experience localized pain, swelling, nausea, headache, dizziness, and general weakness.

The Copperhead is pale tan with a series of orange or brown dumbbell-shaped saddles over the back. The subspecies are distinguished by the width of these saddles, and whether they are dark-edged or not.

RELATED SPECIES
Not to be confused with the unrelated Australian copperheads, *Austrelaps* (page 491), which are elapids, five subspecies of *Agkistrodon contortrix* are recognized from eastern and southeastern USA: the Southern Copperhead (*A. c. contortrix*), Broad-banded Copperhead (*A. c. laticinctus*), Northern Copperhead (*A. c. mokasen*), Osage Copperhead (*A. c. phaeogaster*), and Trans-Pecos Copperhead (*A. c. pictigaster*). This species' closest relative is the Cottonmouth (*A. piscivorus*, page 554).

Actual size

FAMILY	Viperidae: Crotalinae
RISK FACTOR	Highly venomous: procoagulants, hemotoxins, and myotoxins
DISTRIBUTION	North America: southeastern, eastern, and central USA, including Florida Peninsula
ELEVATION	0–2,300 ft (0–700 m) asl
HABITAT	Slow-moving rivers, bayous, lakes, and swamps
DIET	Fish, frogs, small mammals, reptiles, and birds and their eggs
REPRODUCTION	Viviparous, with litters of 4–11 neonates
CONSERVATION STATUS	IUCN Least Concern

ADULT LENGTH
3 ft 3 in–5 ft
(1.0–1.5 m)

AGKISTRODON PISCIVORUS
COTTONMOUTH
(LACÉPÈDE, 1789)

Cottonmouths are usually pale gray or brown with an overlying pattern of darker gray or brown saddles, although the patterning may be obscured in adults, resulting in a uniformly dark snake. The gaping white mouth is a giveaway, and the presence of fangs, heat-sensitive pits, and vertically elliptical pupils distinguish this snake from the harmless watersnakes.

The name Cottonmouth comes from the bright white interior to this snake's mouth, which is displayed when gaping widely in defense. The Cottonmouth's alternative name of Water Moccasin is an indigenous American name distinguishing this species from the related "Highland Moccasin" or Copperhead (*Agkistrodon contortrix*, page 553). The Cottonmouth is unusual in that it is an aquatic pitviper that inhabits a wide variety of watercourses, from slow-moving rivers to mangrove swamps, although it may be encountered in grasslands and woodlands. It exhibits a rather catholic diet, taking a wide variety of fish, amphibians, other reptiles, small mammals, birds and their eggs, and even carrion, although fish and frogs, hunted at night, are the most frequent prey. Bites to humans can be severe, with deaths, and amputations due to tissue death, on record.

RELATED SPECIES

Three subspecies of *Agkistrodon piscivorus* are recognized: the Eastern Cottonmouth (*Agkistrodon piscivorus piscivorus*), Western Cottonmouth (*A. p. leucostoma*), and Florida Cottonmouth (*A. p. conanti*). All three may be confused with the pugnacious, but otherwise harmless, watersnakes of genus *Nerodia* (pages 417–420), such as the Banded Watersnake (*Nerodia fasciata*) or the Diamond-backed Watersnake (*N. rhombifer*, page 420).

Actual size

FAMILY	Viperidae: Crotalinae
RISK FACTOR	Highly venomous: procoagulants, hemorrhagins, and myotoxins
DISTRIBUTION	North America: northwestern Mexico (Tamaulipas, San Luis Potosí, Nuevo León, Hidalgo, and Veracruz)
ELEVATION	0–1,970 ft (0–600 m) asl
HABITAT	Mesquite grassland, thorn forest, and tropical deciduous forest
DIET	Small mammals, frogs, and lizards
REPRODUCTION	Viviparous, with litters of 3–10 neonates
CONSERVATION STATUS	IUCN Least Concern

ADULT LENGTH
23¾–35½ in,
rarely 3 ft 3 in
(600–900 mm,
rarely 1.0 m)

AGKISTRODON TAYLORI
ORNATE CANTIL
BURGER & ROBERTSON, 1951

555

The Ornate Cantil, also known as Taylor's Cantil, is found in northwestern Mexico, widely separated from the ranges of the other former subspecies of the Mexican Cantil (*Agkistrodon bilineatus*). It inhabits primary forest, thorn forest, and rocky karst hillsides, but may also be found in mesquite grasslands. It is crepuscular but also active on overcast days. As a group, the cantils possess more toxic venom than their North American relatives, the Copperhead (*A. contortrix*, page 553) and the Cottonmouth (*A. piscivorus*, page 554). Snakebites to humans can easily be fatal. Juvenile cantils exhibit a yellow tail tip that they use to lure prey within strike range, a technique known as "caudal luring" often used by amphibiophagous (amphibian-eating) snakes. Juveniles prey on frogs or lizards, while adults take small mammals.

The Ornate Cantil is brightly patterned as a juvenile, with light and dark gray saddles with yellow spots on the flanks and five bold facial stripes. With increased maturity the saddles become obscured and almost black, although the yellow spots and facial stripes persist. The Ornate Cantil is very different from the other cantils, which are predominantly brown.

RELATED SPECIES
Agkistrodon taylori was formerly a subspecies of the Mexican Cantil (*A. bilineatus*) from the west coast of Mexico to El Salvador. Two other subspecies are now treated as full species, the Castellana (*A. howardgloydi*) from western coastal Honduras to Costa Rica, and the Yucatán Cantil (*A. russeolus*) from the Yucatán Peninsula of Mexico, and Belize.

Actual size

FAMILY	Viperidae: Crotalinae
RISK FACTOR	Venomous: procoagulants, anticoagulants, and possibly hemorrhagins
DISTRIBUTION	North and Central America: southern Mexico to Panama
ELEVATION	130–5,250 ft (40–1,600 m) asl
HABITAT	Tropical forest, riparian vegetation, pine–oak forest, cloud forest, and wooded savanna
DIET	Small mammals, lizards, and orthopterans
REPRODUCTION	Viviparous, with litters of 13–36 neonates
CONSERVATION STATUS	IUCN not listed

ADULT LENGTH
19¾–35½ in
(500–900 mm)

ATROPOIDES MEXICANUS
CENTRAL AMERICAN JUMPING PITVIPER
(DUMÉRIL, BIBRON & DUMÉRIL, 1854)

The Central American Jumping Pitviper is a short, stout viper with a broad head. Its variable patterning usually consists of a series of vertebral diamond markings on a brown or gray background, with a corresponding series of spots on the flanks. Color pattern and head scalation are important for distinguishing the various species apart.

Despite its scientific name, most of the range of the Central American Jumping Pitviper occurs south of Mexico, as far as Panama. It is the most widely distributed member of its genus. It inhabits forest edges, stream banks, montane pine–oak forest, and cloud forest. Jumping pitvipers are short, stout snakes that form a tight defensive coil with the head in the center, the mouth agape in threat, from which they make sudden powerful strikes that can launch their entire bodies forward, the origin of the "jumping," although the distance that these snakes can strike has been greatly exaggerated. Prey ranges from grasshoppers to lizards and mice. The venom is not as toxic as that of other pitvipers, which is probably why these vipers hold onto their prey following a strike.

RELATED SPECIES

Three other jumping pitvipers also occur from southern Mexico to El Salvador: the Mexican Jumping Pitviper (*Atropoides nummifer*), Guatemalan Jumping Pitviper (*A. occiduus*), and Tuxtlan Jumping Pitviper (*A. olmec*), while Picado's Jumping Pitviper (*A. picadoi*) is found alongside *A. mexicanus* in Costa Rica and Panama. The newly described Indomitable Jumping Pitviper (*A. indomitus*) is limited to a small Honduran mountain range and is listed as Endangered by the IUCN.

Actual size

FAMILY	Viperidae: Crotalinae
RISK FACTOR	Venomous: procoagulants, and possibly hemorrhagins
DISTRIBUTION	Central America: central Costa Rica and Panama
ELEVATION	3,280–9,840+ ft (1,000–3,000+ m) asl
HABITAT	Montane and cloud forest
DIET	Lizards, tree frogs, birds, and small mammals
REPRODUCTION	Viviparous, with litters of 4–8 neonates
CONSERVATION STATUS	IUCN not listed

ADULT LENGTH
19¾–31½ in
(500–800 mm)

BOTHRIECHIS NIGROVIRIDIS
BLACK-SPECKLED PALM-PITVIPER
PETERS, 1862

557

The palm-pitvipers of the Central American genus *Bothriechis* occupy the same niche as the pitvipers of genus *Trimeresurus* (pages 598–602) in Southeast Asia and the bushvipers, genus *Atheris* (pages 605–609), of Africa. All are arboreal predators of arboreal vertebrates, from lizards and tree frogs to birds and rodents, and several are montane species. The Black-speckled Palm-pitviper is a montane species, found in cloud forest and wet forest at elevations from 3,280 to 9,840 ft (1,000–3,000 m). Although it may be relatively common in these habitats, it quickly disappears from disturbed or cultivated areas, and is vulnerable to changes in humidity, temperature, and light levels. Bites to humans cause intense pain, nausea, and breathing difficulties. Human fatalities are unknown, but not impossible.

The Black-speckled Palm-pitviper is emerald green, less often yellow-green, with all scales heavily suffused with black, especially on the dorsum of the body, and black interstitial skin between the scales. Its body is slender and muscular, and the tail is long and prehensile for climbing and anchorage while hunting aloft.

RELATED SPECIES

Of the other ten species of *Bothriechis* only the Side-striped Palm-pitviper (*B. lateralis*), Eyelash Palm-pitviper (*B. schlegelii*, page 558), and recently described Talamancan Palm-pitviper (*B. nubestris*) occur in sympatry with *B. nigroviridis*. The Side-striped Palm-pitviper is green with short yellow and black cross-bars, while the Eyelash Palm-pitviper exhibits distinctive "eyelashes," so both species are easily distinguished from *B. nigroviridis*, but the Talamancan Palm-pitviper requires closer examination. Other members of the genus occur from southern Mexico to Honduras.

Actual size

FAMILY	Viperidae: Crotalinae
RISK FACTOR	Venomous: procoagulants, and possibly hemorrhagins
DISTRIBUTION	North, Central, and South America: southern Mexico to northern South America
ELEVATION	0–8,200 ft (0–2,500 m), rarely > 8,530 ft (2,600 m) asl
HABITAT	Lowland primary and secondary rainforest
DIET	Lizards, tree frogs, birds, and small mammals
REPRODUCTION	Viviparous, with litters of 2–20 neonates
CONSERVATION STATUS	IUCN not listed

ADULT LENGTH
19¾–25½ in,
rarely 31½ in
(500–650 mm,
rarely 800 mm)

BOTHRIECHIS SCHLEGELII
EYELASH PALM-PITVIPER
(BERTHOLD, 1846)

The Eyelash Palm-pitviper is instantly recognizable due to its raised supraciliary "eyelashes." Across its considerable range it exhibits an array of patterns, from a green and brown mottled, lichen-like pattern, to overall green or brown with darker spots, a uniform rust-orange morphotype, and a brilliant golden-yellow morphotype, known as "oropel" in Costa Rica.

The Eyelash Palm-pitviper is one of the best known, most photographed, but also most variable venomous snakes of Latin America. Occurring from Chiapas, southern Mexico, to Colombia and Ecuador, and just entering Venezuela and Peru, it is the most widely distributed member of its genus. The common name originates from a series of short "eyelash" projections above the eye. In lowland forests and along creeks, the Eyelash Palm-pitviper can be very common, especially at night when it hunts small arboreal vertebrates including lizards such as anoles, tree frogs, small mammals, from bats to mice, and even sleeping birds. The Eyelash Palm-pitviper is even reported to be occasionally cannibalistic. Bites to humans are common and fatalities have been recorded on several occasions.

RELATED SPECIES

The genus *Bothriechis* contains 11 species, all of which occur within the range of *B. schlegelii*, but only the Blotched Palm-pitviper (*B. supraciliaris*), from a small area of southwestern Costa Rica near the Panamanian border, exhibits the fleshy eyelashes. The two species are extremely similar, but while *B. schlegelii* usually has markings that cross over the back, those of the Blotched Palm-pitviper are on the upper flanks.

Actual size

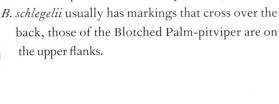

FAMILY	Viperidae: Crotalinae
RISK FACTOR	Venomous: procoagulants, and possibly hemorrhagins
DISTRIBUTION	South America: upper Amazonian Brazil, Colombia, Ecuador, Peru, and Bolivia
ELEVATION	2,620–3,280 ft (800–1,000 m) asl
HABITAT	Lowland tropical rainforest
DIET	Lizards and small mammals
REPRODUCTION	Viviparous, with litters of 3–13 neonates
CONSERVATION STATUS	IUCN not listed

BOTHROCOPHIAS HYOPRORA

AMAZONIAN TOAD-HEADED PITVIPER

(AMARAL, 1935)

ADULT LENGTH
Male
13¾–21 in
(350–530 mm)
Female
19¾–31¾ in
(500–803 mm)

559

The toad-headed pitvipers, genus *Bothrocophias*, are a genus of six fairly stout-bodied, but short, terrestrial pitvipers from the upper Amazon and the Andean slopes of northwestern South America. While several species occur at elevations of 6,560 ft (2,000 m), the Amazonian Toad-headed Pitviper is a lowland species, the most widely distributed member of its genus, occurring through the Amazonian headwaters from Colombia to Bolivia, and eastward into the Amazon Basin, where it occupies the niche of a sedentary sit-and-wait ambusher of terrestrial lizards and rodents. The Amazonian Toad-headed Pitviper is a sexually dimorphic species, females achieving total lengths greater than those of males. Snakebites have caused unconsciousness, edema, and severe hemorrhaging in adult humans, and at least one child fatality.

The Amazonian Toad-headed Pitviper exhibits a slightly turned-up snout tip. It is a short and stout-bodied terrestrial pitviper with patterning comprising alternating dark brown and light brown bands, designed to break up the viper's outline when lying in the leaf litter.

RELATED SPECIES

Bothrocophias hyoprora was once thought related to the Rainforest Hognosed Pitviper (*Porthidium nasutum*, page 593) of Central America and Pacific coastal South America, and was mistakenly placed in the genus *Porthidium*. Both species have distinctive turned-up snout tips. This is more likely a case of convergent evolution, and in most other respects *B. hyoprora* resembles the Small-eyed Toad-headed Pitviper (*B. microphthalmus*), which occupies the Andean slopes to the west of its range.

Actual size

FAMILY	Viperidae: Crotalinae
RISK FACTOR	Highly venomous: procoagulants, and possibly also hemorrhagins, myotoxins, or nephrotoxins
DISTRIBUTION	South America: endemic to Alcatrazes Island, 22 miles (35 km) west of São Paulo, Brazil
ELEVATION	10–875 ft (3–266 m) asl
HABITAT	Atlantic coastal forest
DIET	Centipedes and lizards
REPRODUCTION	Viviparous, with litters of 1–2 neonates
CONSERVATION STATUS	IUCN Critically Endangered

ADULT LENGTH
15¾–19¾ in
(400–500 mm)

560

BOTHROPS ALCATRAZ
ALCATRAZES LANCEHEAD
MARQUES, MARTINS & SAZIMA, 2002

The Alcatrazes Lancehead is a small, slender island species with a pattern consisting of irregular dark brown cross-bands over a paler brown dorsum, which serves to camouflage it in leaf litter. It has relatively large eyes when adult, due to reaching maturity at a smaller size than mainland lanceheads. It is effectively a dwarf island species.

The Alcatrazes Lancehead is endemic to the Ilha de Alcatrazes (½ sq mile / 1.35 sq km) located 22 miles (35 km) west from the coast of São Paulo, Brazil—not to be confused with Alcatraz Island, California ("alcatraz" meaning frigate bird). Alcatrazes Island, Brazil is an island without mammals, the preferred prey of mainland lanceheads. The resident lancehead is a dwarf species that preys primarily on centipedes, but also takes lizards. It is one of the few lanceheads that preys on ectotherms as an adult. The size, diet, and venom composition of the Alcatrazes Lancehead effectively retain the juvenile condition. The small size of adult females also results in small litter sizes, a factor which, together with the island's utilization as an artillery range by the Brazilian Navy, makes this a Critically Endangered species.

RELATED SPECIES

Bothrops alcatraz most resembles the Jararaca (*B. jararaca*, page 568) of mainland Brazil, but being an insular pitviper it may be ecologically closer to the Golden Lancehead (*B. insularis*, page 567) from neighboring Ilha da Queimada Grande. These two species are its closest relatives.

Actual size

FAMILY	Viperidae: Crotalinae
RISK FACTOR	Highly venomous: procoagulants, and possibly also hemorrhagins or cytotoxins
DISTRIBUTION	Southern South America: southern Brazil, Uruguay, Paraguay, and northern Argentina
ELEVATION	0–2,300 ft (0–700 m) asl
HABITAT	Forests, sugarcane plantations, swamps, and grasslands
DIET	Small mammals (rodents and cavies)
REPRODUCTION	Viviparous, with litters of 3–26 neonates
CONSERVATION STATUS	IUCN not listed

ADULT LENGTH
5 ft 2 in–6 ft 7 in
(1.6–2.0 m)

BOTHROPS ALTERNATUS
URUTU
DUMÉRIL, BIBRON & DUMÉRIL, 1854

561

The Urutu is a stout-bodied snake with a broad, lance-shaped head similar to other lanceheads, although the Brazilian name "Urutu" is used for this species. It is found in many lowland habitats in southern Brazil, Uruguay, Paraguay, and northern Argentina, especially those associated with water, such as marshes, swamps, or riverine grasslands. A large snake, the Urutu preys on rodents and cavies such as guinea pigs. It is a medically important species, and a continual threat to agricultural workers because it is frequently encountered in cultivated habitats, including sugarcane plantations, being attracted by the large rodent populations. Bites to humans are rarely fatal, but may result in severe necrosis, and limb loss.

The Urutu is one of the most distinctive species in the genus *Bothrops*, being stout-bodied, large-headed, and boldly marked with a dorsal patterning that resembles a row of old-fashioned telephones, and a series of light-edged dark blotches on the dorsum and sides of the head.

RELATED SPECIES

Bothrops alternatus occurs in sympatry with at least a dozen other species of lanceheads, but can be distinguished from most of them by its distinctive patterning. Other species may exhibit similar patterning, notably the Cotiara (*B. cotiara*), Fonseca's Lancehead (*B. fonsecai*), the giant Jararacussu (*B. jararacussu*), and the Caiçaca (*B. moojeni*), but in none are the markings as boldly defined as in *B. alternatus*.

Actual size

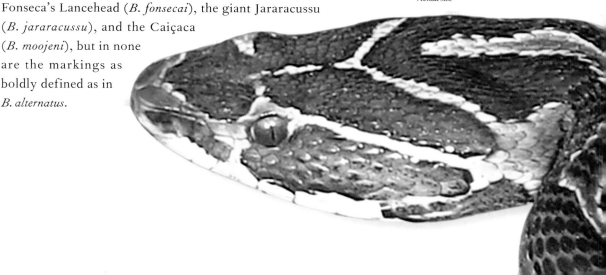

FAMILY	Viperidae: Crotalinae
RISK FACTOR	Highly venomous: procoagulants, and possibly cytotoxins
DISTRIBUTION	Southern South America: Argentina
ELEVATION	0–6,560 ft (0–2,000 m) asl
HABITAT	Sandy grasslands, coastal dunes, salt flats, and pampas
DIET	Lizards, and occasionally frogs
REPRODUCTION	Viviparous, litter size unknown
CONSERVATION STATUS	IUCN not listed

ADULT LENGTH
2 ft 5 in–3 ft 3 in
(0.75–1.0 m)

BOTHROPS AMMODYTOIDES
PATAGONIAN LANCEHEAD
LEYBOLD, 1873

The Patagonian Lancehead is a pastel-colored, slightly stocky snake, with a subdued pattern of spots or blotches on a sandy gray-brown background that blends well with its sandy, grassland habitat. The most significant diagnostic characteristic is its slightly upturned snout.

The Patagonian Lancehead is the southernmost snake species in the world. It is distributed over a vast area in Argentina, from the coastal sand dunes to the Andean foothills at elevations of 6,560 ft (2,000 m), and from Salta Province in the north to Santa Cruz Province in the pampas grasslands of Patagonia, and as far south as 48 degrees south. In this grassland habitat lizards are abundant and constitute its primary prey, especially the iguanine genus *Liolaemus*, but frogs are also taken. This is an instantly recognizable species due to its upturned snout, and in much of its range it is the only pitviper present. Although small, it is believed capable of causing fatal snakebites.

RELATED SPECIES

The only South American pitviper with which *Bothrops ammodytoides* might be confused is the São Paulo Lancehead (*B. itapetiningae*) from southeastern Brazil. This is also a small species with a slightly upturned snout tip, but the ranges of the two species are mutually exclusive.

Actual size

FAMILY	Viperidae: Crotalinae
RISK FACTOR	Highly venomous: procoagulants, hemorrhagins, myotoxins, and cytotoxins
DISTRIBUTION	North, Central, and South America: southern Mexico to Ecuador, northern Peru, Venezuela, and Trinidad
ELEVATION	0–4,920 ft (0–1,500 m), rarely > 8,530 ft (2,600 m) asl
HABITAT	Many habitats, including cultivated areas
DIET	Mammals, birds, lizards, frogs, and centipedes
REPRODUCTION	Viviparous, with litters of 20–90 neonates
CONSERVATION STATUS	IUCN not listed

ADULT LENGTH
6 ft–8 ft 2 in
(1.8–2.5 m)

BOTHROPS ASPER
TERCIOPELO
(GARMAN, 1883)

563

The Terciopelo is found from southeast Mexico, through Central America into Colombia, Ecuador, northern Peru, Venezuela, and Trinidad. It occurs in most habitats, from pristine rainforest to degraded plantations, in lowlands and uplands, and in dry and wet habitats, and often in high densities where the conditions and prey availability are optimal. Prey ranges from centipedes and lizards, for juveniles, to mammals and birds in adults. Cannibalism is also documented. Terciopelos are a major cause of human snakebite morbidity and mortality and it is a much-feared and aggressive species, which will advance on an adversary. At up to 8 ft 2 in (2.5 m) this is one of the largest lanceheads, and although for its length it is relatively slender in build, it has a large and menacing head. Stories of 10 ft (3 m) specimens also exist.

The Terciopelo is a highly variable species, with a background of gray, olive, brown, or even pinkish, and an overlying pattern of dark and light blotches. Some specimens are almost patternless. Some specimens have yellow lips, resulting in the local name "barba amarilla" or "yellow beard."

RELATED SPECIES

Bothrops asper is often confused with the Common Lancehead (*B. atrox*, page 564) from northern South America, of which it was once a subspecies, or the Venezuelan Lancehead (*B. venezuelensis*), with which it occurs in sympatry on the Caribbean coast. The true status of some Venezuelan and Trinidad populations has yet to be determined. The name "fer-de-lance" is inappropriate for this species and belongs instead to the Martinique Lancehead (*B. lanceolatus*).

Actual size

FAMILY	Viperidae: Crotalinae
RISK FACTOR	Highly venomous: procoagulants, hemorrhagins, myotoxins, and possibly also nephrotoxins or cytotoxins
DISTRIBUTION	South America: Amazonia and the Guianas
ELEVATION	0–3,940 ft (0–1,200 m), rarely > 4,920 ft (1,500 m) asl
HABITAT	Lowland tropical rainforest, gallery forest, secondary forest, and cultivated areas
DIET	Mammals, birds, frogs, and lizards
REPRODUCTION	Viviparous, with litters of 8–43 neonates
CONSERVATION STATUS	IUCN not listed

ADULT LENGTH
2 ft 5 in–4 ft
(0.75–1.25 m)

564

BOTHROPS ATROX
COMMON LANCEHEAD
(LINNAEUS, 1758)

The Common Lancehead occurs from Venezuela and the Guianas, through the Amazon Basin to the Andean foothills. Across its range it is arguably the most dangerous snake, being responsible for many serious snakebites. It thrives in many habitats, from pristine tropical rainforest and riverine gallery forest, to degraded secondary growth and cultivation. Although producing smaller litters than its larger relative, and former subspecies, the Terciopelo (*Bothrops asper*, page 563), a gravid female Common Lancehead can quickly populate a newly cleared patch of bush with numerous camouflaged neonates. Adults are terrestrial sit-and-wait ambushers of endotherms, mostly rodents, but juveniles may actively attract prey using their yellow tail tip to lure frogs and lizards within strike range. Juveniles are fairly arboreal, in contrast to the heavier-bodied adults.

RELATED SPECIES

Bothrops atrox occurs in sympatry with several similarly patterned species: Brazil's Lancehead (*B. brazili*), the Caiçaca (*B. moojeni*), and the Jararaca (*B. jararaca*, page 568). Although often referred to as "fer-de-lance," this name is correctly only applied to the Martinique Lancehead (*B. lanceolatus*).

Actual size

The Common Lancehead is patterned with various shaded of brown, with a dark stripe behind the eye. The body patterning can be variable but usually comprises dark dorsal blotches that intersperse with lighter diagonal bars, a pattern also seen in other lancehead species.

FAMILY	Viperidae: Crotalinae
RISK FACTOR	Highly venomous: procoagulants, cytotoxins, possibly myotoxins, and hemorrhagins
DISTRIBUTION	South America: Amazonian Basin, the Guianas, and Brazilian Atlantic coast
ELEVATION	0–3,280 ft (0–1,000 m) asl
HABITAT	Lowland tropical forest
DIET	Frogs, lizards, birds, and small mammals
REPRODUCTION	Viviparous, with litters of 4–16 neonates
CONSERVATION STATUS	IUCN not listed

ADULT LENGTH
2 ft 4 in–3 ft 3 in
(0.7–1.0 m)

BOTHROPS BILINEATUS
TWO-STRIPED FOREST-PITVIPER
(WIED-NEUWIED, 1821)

The Two-striped Forest-pitviper is one of six species of arboreal South American pitvipers that were temporarily placed in their own genus, *Bothriopsis*, in the late twentieth century. They represent the most arboreal members of the genus *Bothrops*. Although it seems generally inoffensive, this species can be considered the second-most dangerous venomous snake in the Amazon, after the Common Lancehead (*Bothrops atrox*, page 564). Deaths have been attributed to even relatively small specimens and its camouflaged coloration and arboreal habits place it on a position to bite high on the body. It inhabits lowland tropical forest, often near water, or edge situations around clearings. Its primary prey consists of frogs or birds, but lizards and rodents are also taken. Both adults and juveniles may caudal lure for prey.

The Two-striped Forest-pitviper is an attractive, slender species with a long, prehensile tail. It is emerald green in color, with or without darker speckling and spots or small brown cross-bars on the dorsum. Some specimens have yellow lips with dark suturing between the scales. This arboreal species easily blends into the Amazonian vegetation.

Actual size

RELATED SPECIES

This is the only green arboreal pitviper within its range, and cannot be confused with any other pitviper, although it may be confused with other arboreal snakes, such as the green parrot snakes (*Leptophis*, page 188). Two subspecies of *Bothrops bilineatus* are recognized, the nominate lower Amazon, Guianas, and Atlantic coastal forest form (*B. b. bilineatus*), and an upper Amazonian forests subspecies (*B. b. smaragdinus*).

FAMILY	Viperidae: Crotalinae
RISK FACTOR	Highly venomous: procoagulants, and possibly also myotoxins, cytotoxins, or hemorrhagins
DISTRIBUTION	West Indies: St. Lucia (Lesser Antilles)
ELEVATION	0–655 ft (0–200 m) asl
HABITAT	Lowland tropical rainforest, rocky stream beds, coastal plains, and plantations
DIET	Small mammals
REPRODUCTION	Viviparous, with litters of up to 37 neonates
CONSERVATION STATUS	IUCN not listed

ADULT LENGTH
3 ft 3 in–6 ft 7 in
(1.0–2.0 m)

BOTHROPS CARIBBAEUS
ST. LUCIA LANCEHEAD
(GARMAN, 1887)

The St. Lucia Lancehead is a fairly variably patterned species. Although most specimens are gray with an overlying dorsal pattern of trapezoidal blotches, and a black stripe behind the eye, others may be brown, reddish, or yellowish. Males are also often darker than females.

One of only two endemic island lanceheads in the Caribbean, the St. Lucia Lancehead is distributed widely across St. Lucia (240 sq miles / 616 sq km), below 655 ft (200 m) asl, excluding the southern third of the island and the arid extreme north. This terrestrial and semi-arboreal lancehead is a lowland rainforest species that also inhabits cocoa or coconut plantations, sheltering inside husk piles during the day and preying on rats and mice. The venom of both the St. Lucia Lancehead and its relative on Martinique (see below) differs from that of South American lanceheads in that it can cause a rapidly fatal arterial thrombosis. There have been attempts to eradicate the lancehead, but ironically this led to the extinction of its only natural predator, the snake-eating St. Lucia Mussurana (*Clelia errabunda*).

RELATED SPECIES
The only other pitviper in the Lesser Antilles, and the closest relative of *Bothrops caribbaeus*, is the neighboring Martinique Lancehead (*B. lanceolatus*), the only species to which the name "fer-de-lance" should be applied. On St. Lucia the only other large snake is the St. Lucia Boa Constrictor (*Boa orophias*), a harmless and endangered island endemic.

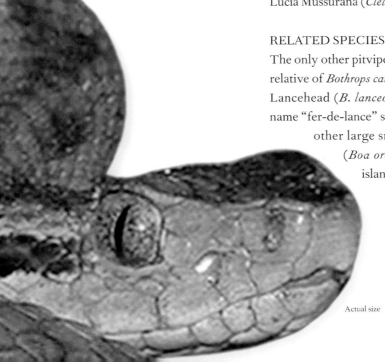

Actual size

FAMILY	Viperidae: Crotalinae
RISK FACTOR	Highly venomous: procoagulants, and possibly also hemorrhagins, myotoxins, or cytotoxins
DISTRIBUTION	South America: Ilha da Queimada Grande (São Paulo, Brazil)
ELEVATION	0–655 ft (0–200 m) asl
HABITAT	Rocky scrub forest
DIET	Primarily passerine birds, occasionally reptiles or centipedes
REPRODUCTION	Viviparous, with litters of 2–10 neonates
CONSERVATION STATUS	IUCN Critically Endangered

ADULT LENGTH
2 ft 4 in–4 ft
(0.7–1.2 m)

BOTHROPS INSULARIS
GOLDEN LANCEHEAD
(AMARAL, 1821)

567

The Golden Lancehead is probably the most venomous snake in the Americas. It inhabits tiny Ilha da Queimada Grande (110 acres/43 hectares), 21 miles (35 km) from the Brazilian mainland. Terrestrial and arboreal, it preys on passerine birds, because there are no mammals on the island. It has long fangs to penetrate plumage, and highly toxic venom, three to five times more toxic than that of any mainland American snake, to subdue prey quickly. It is also recorded to prey on the only other snake present, the White-headed Snail-eater (*Dipsas albifrons*), and centipedes. This lancehead occurs as three "sexes": males, females, and intersex females (females with inactive male genitalia), the males reportedly preferring to mate with intersex females than genuine females. There are an estimated 5,000 lanceheads on this tiny island.

The Golden Lancehead may be orange, golden yellow, or buff in color with faint lighter cross-bands, a patterning that may camouflage it among the orange fruit on the forest floor, where it ambushes feeding birds.

RELATED SPECIES

Bothrops insularis is most closely related to the mainland Jararaca (*B. jararaca*, page 568) and the equally insular Alcatrazes Lancehead (*B. alcatraz*, page 560).

Actual size

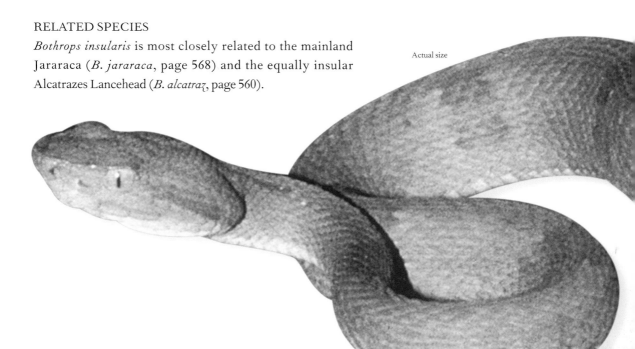

FAMILY	Viperidae: Crotalinae
RISK FACTOR	Highly venomous: procoagulants, hemorrhagins, possibly myotoxins, and cytotoxins
DISTRIBUTION	South America: southeastern Brazil
ELEVATION	0–3,280 ft (0–1,000 m) asl
HABITAT	Tropical deciduous woodland, savanna, and open cultivated habitats
DIET	Small mammals, birds, lizards, frogs, and centipedes
REPRODUCTION	Viviparous, with litters of 18–22 neonates
CONSERVATION STATUS	IUCN Least Concern

ADULT LENGTH
3 ft 3 in–5 ft 2 in
(1.0–1.6 m)

BOTHROPS JARARACA
JARARACA
(WIED-NEUWIED, 1824)

The Jararaca is a very variably patterned species, with ground colors ranging from pale gray to dark brown and a variety of narrow or broad darker saddle markings, which may meet middorsally or alternate on the left and the right. It usually exhibits a black stripe behind the eye.

The Jararaca is a medically important species from Bahia in the Brazilian northeast, to Rio Grande do Sul, southeastern Brazil. It is responsible for many snakebite fatalities, a risk exacerbated by the Jararaca's preference for habitats altered by humans, in which it comes into contact with many poor, itinerant workers. Even those who survive a snakebite may lose limbs and be left disabled, due to the cytotoxic effects of the venom. This is a relatively slender and easily overlooked snake found in the leaf litter of woodland habitats. Adult Jararacas prey primarily on small mammals, but birds and lizards are also commonly taken, while juveniles prefer frogs, birds, and centipedes. The name "Jararaca" is derived from a Tupi Indian word for "large snake."

RELATED SPECIES

The closest relatives of *Bothrops jararaca* are the two island species from Alcatrazes (*B. alcatraz*, page 560) and Queimada Grande (*B. insularis*, page 567). It also resembles the larger Jararacussu (*B. jararacussu*), with which it occurs in sympatry in the south of its range, the Common Lancehead (*B. atrox*, page 564) from the Amazon, and other lancehead species from southeastern Brazil.

Actual size

FAMILY	Viperidae: Crotalinae
RISK FACTOR	Highly venomous: procoagulants, myotoxins, and possibly cytotoxins
DISTRIBUTION	South America: across Amazonia, from the Guianas to the upper Amazon
ELEVATION	0–6,560 ft (0–2,000 m) asl
HABITAT	Lowland primary and secondary rainforest, especially edge situations and canopies
DIET	Tree frogs, lizards, small mammals, and centipedes
REPRODUCTION	Viviparous, with litters of 7–17 neonates
CONSERVATION STATUS	IUCN not listed

ADULT LENGTH
3 ft 3 in–5 ft
(1.0–1.5 m)

BOTHROPS TAENIATUS
SPECKLED FOREST-PITVIPER
WAGLER, 1824

569

The Speckled Forest-pitviper, and several other slender-bodied, prehensile-tailed, arboreal South American pitvipers, were temporarily transferred to the genus *Bothropsis* in the late twentieth century. The distribution of this species is curious, overlying the eastern (Guianan–lower Amazon) and western (upper Amazon) ranges of its relative, the Two-striped Forest-pitviper (*Bothrops bilineatus*, page 565), but with a connecting corridor south of the Amazon. Like its relative, it does not appear to be present in central Amazonia. Snakebites are known for this species, but no fatalities are recorded. However, given the toxicity of the venom of the related Two-striped Forest-pitviper it must be considered highly dangerous. It inhabits lowland rainforest, especially forest edges or canopies, or open secondary growth, where it hunts tree frogs, sleeping lizards, small rodents and opossums, and centipedes at night.

The Speckled Forest-pitviper is slender and camouflaged with mottled browns and greens, the cryptic patterning of some specimens blending superbly with the lichen-covered twigs and vines that festoon the forest edge entanglement around clearings.

RELATED SPECIES

Two subspecies are sometimes recognized, the widely distributed nominate form (*Bothrops taeniatus taeniatus*) and a localized Venezuelan subspecies (*B. t. lichenosus*), known from a single specimen. *Bothrops taeniatus* resembles other cryptically patterned arboreal pitvipers in the Andean part of its range, the Inca Forest-pitviper (*B. chloromelas*) and the Andean Forest-pitviper (*B. pulcher*). It may also be confused with juvenile Common Lanceheads (*B. atrox*, page 564), which are also arboreal.

Actual size

FAMILY	Viperidae: Crotalinae
RISK FACTOR	Highly venomous: procoagulants and hemorrhagins, and secondary cytotoxic effects
DISTRIBUTION	Southeast Asia: Thailand, Vietnam, Cambodia, northern Peninsular Malaysia, and also Indonesia (Java)
ELEVATION	0–4,920 ft (0–1500 m) asl
HABITAT	Dry lowland and hilly habitats
DIET	Small mammals, birds, lizards, and amphibians
REPRODUCTION	Oviparous, with clutches of 10–35 eggs
CONSERVATION STATUS	IUCN not listed

ADULT LENGTH
3 ft 3 in–4 ft 9 in
(1.0–1.45 m)

CALLOSELASMA RHODOSTOMA
MALAYAN PITVIPER
(KUHL, 1824)

The Malayan Pitviper is gray-brown on the sides with a dorsal pattern of alternating dark and light brown cross-bands. A dark postocular stripe contrasts strongly with its white lips. The head is broad and sharply pointed.

The name Malayan Pitviper is inappropriate for this medically important pitviper. Found on dry wooded hillsides in Thailand, Vietnam, and Cambodia, it is absent from tropical rainforests, only entering northwestern Malaysia in Kedah and Perlis states. A relict population (survivor from an earlier period) is also found on Java, Indonesia. The range was continuous during the last glacial period, when sea-levels were lower and dry forest more widespread, but today it is absent from the intervening tropics. As a pitviper it is unusual, possessing smooth scales and laying eggs. It inhabits coffee and rubber plantations in large numbers, inflicting a snakebite toll on workers, causing deaths, through cerebral hemorrhage, and disabling survivors through limb loss. Yet this snake also saves lives, a drug for preventing blood clots after cardiac surgery having been developed from its venom.

RELATED SPECIES

Probably only the harmless Common Mock Viper (*Psammodynastes pulverulentus*, page 397) could be confused with a juvenile *Calloselasma rhodostoma*. Some authorities also recognize a subspecies, *C. r. annamensis*, from southern Cambodia and Vietnam.

Actual size

FAMILY	Viperidae: Crotalinae
RISK FACTOR	Venomous: procoagulants, myotoxins, and possibly hemorrhagins
DISTRIBUTION	North and Central America: southern Mexico to Panama
ELEVATION	4,590–11,500 ft (1,400–3,490 m) asl
HABITAT	Lower and upper montane forest, cloud forest, and meadows
DIET	Small mammals, orthopterans, and lizards
REPRODUCTION	Viviparous, with litters of 2–12 neonates
CONSERVATION STATUS	IUCN not listed

ADULT LENGTH
19¾–32¼ in
(500–820 mm)

CERROPHIDION GODMANI
GODMAN'S MONTANE PITVIPER
(GÜNTHER, 1863)

571

Godman's Montane Pitviper is the most widely distributed of five short, stout members of the genus *Cerrophidion*. It is found in scattered locations from Oaxaca, southern Mexico, to western Panama, at relatively high elevations, where it occupies a variety of habitats including wet and dry montane forest, cloud forest, and open meadows. It is the most widely distributed montane pitviper in Central America. Being short and stout is ideal for snakes at high elevations, where conditions are cool, and prey small and infrequently encountered. Godman's Montane Pitviper preys on small mammals, lizards, and grasshoppers. This is a common species, often seen crawling slowly, or coiled on paths. Snakebites cause gross swelling, intense pain, and nausea, but no fatalities are reported.

Godman's Montane Pitviper is a short, stout pitviper with a broad head and a patterning consisting of three rows of large dark brown or black spots, which may merge over the back to form a zigzag pattern.

RELATED SPECIES

Within genus *Cerrophidion* three species, the Tzotzil Montane Pitviper (*C. tzotzilorum*), Honduran Montane Pitviper (*C. wilsoni*), and Costa Rican Montane Pitviper (*C. sasai*), occur within the range of *C. godmani*. Its range also coincides with four other genera of short, stout pitvipers, *Atropoides* (page 556), *Mixcoatlus* (page 590), *Ophryacus* (page 591), and *Porthidium* (page 593).

Actual size

FAMILY	Viperidae: Crotalinae
RISK FACTOR	Highly venomous: procoagulants, anticoagulants, myotoxins, and hemorrhagins
DISTRIBUTION	North America: southeastern USA (Florida to the Carolinas and Louisiana)
ELEVATION	0–1,640 ft (0–500 m) asl
HABITAT	Coastal hammocks and sand dunes to lowland grassland and woodland, and upland pine forest
DIET	Small mammals and birds
REPRODUCTION	Viviparous, with litters of 7–29 neonates
CONSERVATION STATUS	IUCN Least Concern

ADULT LENGTH
5 ft–8 ft 2 in
(1.5–2.5 m)

CROTALUS ADAMANTEUS
EASTERN DIAMONDBACK RATTLESNAKE
PALISOT DE BEAUVOIS, 1799

The Eastern Diamondback Rattlesnake is the largest rattlesnake species, with historical records of up to around 8 ft 2 in (2.5 m). It inhabits coastal habitats in southeastern USA, is salt-tolerant, and has been recorded swimming strongly between offshore islands, but it is also present in upland pine forests and swampy grasslands. It will shelter from predators or fire in the burrows of nine-banded armadillos or gopher tortoises, and this is where females give birth to their large litters of young. The Eastern Diamondback Rattlesnake is large enough to prey on cottontail rabbits, feral cats, and young turkeys. Such a large rattlesnake is also easily capable of delivering a rapidly fatal human snakebite, which has led to its active persecution, and local extirpations over large areas of its former range.

RELATED SPECIES

Crotalus adamanteus can be distinguished from the Western Diamondback Rattlesnake (*C. atrox*, page 573) by the lack of the black and white banded tail, and from the Timber Rattlesnake (*C. horridus*, page 577) by its bold postocular stripe, which is edged with white. No other large rattlesnakes occur within its range.

Actual size

The Eastern Diamondback Rattlesnake is usually light brown or gray with a bold series of yellow- or white-edged dorsal diamonds along its back, smaller dark spots on the flanks, and a broad black postocular stripe, which is also edged with white or yellow and makes the head very distinctive. Its tail is generally a subdued color, usually yellow-brown.

FAMILY	Viperidae: Crotalinae
RISK FACTOR	Highly venomous: procoagulants and hemorrhagins
DISTRIBUTION	North America: southwestern USA and northern Mexico
ELEVATION	0–4,920 ft (0–1,500 m), rarely > 7,870 (2,400 m) asl
HABITAT	Lowland floodplains, rocky canyons, wooded hillsides, and cactus semidesert
DIET	Small mammals, birds, and lizards
REPRODUCTION	Viviparous, with litters of 6–19 neonates
CONSERVATION STATUS	IUCN Least Concern

ADULT LENGTH
4–6 ft
(1.2–1.8 m)

CROTALUS ATROX

WESTERN DIAMONDBACK RATTLESNAKE

BAIRD & GIRARD, 1863

573

The Western Diamondback Rattlesnake is the archetypical Western movie rattlesnake, and it has been persecuted accordingly, with large, brutal "rattlesnake roundups," organized ostensibly to protect cattle, being partially responsible for its extirpation or rarity across much of its original lowland range, especially in states like Texas and Oklahoma. Western Diamondbacks are found in a wide variety of habitats, from semidesert to rocky arroyos, dry wooded slopes, prairie grassland, and even onto farmland. It has a fairly catholic diet, preying on a range of small mammals to the size of jackrabbits, as well as birds and lizards, the latter being the primary prey of juveniles. The Western Diamondback Rattlesnake is responsible for many serious snakebites across its range.

The Western Diamondback Rattlesnake may be gray, brown, reddish, or yellowish, with a dorsal pattern of large diamonds, and fine white chevrons in the interspaces. The dark postocular strike is not strongly defined but the tail is strongly ringed with black and white, earning the species the colloquial name of "coon-tail rattler" after the raccoon.

RELATED SPECIES

Crotalus atrox occurs in sympatry with several other rattlesnakes, notably the Black-tailed Rattlesnake (*C. molossus*, page 578) and Mohave Rattlesnake (*C. scutulatus*, page 580), as well as some of the smaller montane rattlesnakes along the Mexican border. It is most closely related to the Eastern Diamondback Rattlesnake (*C. adamanteus*, page 572) and Red Diamond Rattlesnake (*C. ruber*). Its black and white banded tail will distinguish it from all these species except the Mohave Rattlesnake, which has narrower black tail rings.

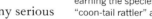
Actual size

FAMILY	Viperidae: Crotalinae
RISK FACTOR	Venomous: venom composition unknown
DISTRIBUTION	North America: Santa Catalina Island (Gulf of California, Mexico)
ELEVATION	0–1,540 ft (0–470 m) asl
HABITAT	Rocky arroyos, and cactus or thorn scrub
DIET	Birds, small mammals, and lizards
REPRODUCTION	Viviparous, with litters of 2–5 neonates
CONSERVATION STATUS	IUCN Critically Endangered

ADULT LENGTH
19¾–27½ in
(500–700 mm)

CROTALUS CATALINENSIS
SANTA CATALINA ISLAND RATTLESNAKE
CLIFF, 1954

The Santa Catalina Island Rattlesnake is a desert-island species with a subtle pastel coloration of ashy gray or pale yellow-brown with faint dorsal saddles. Although it lacks a rattle it will still raise and vibrate its tail in warning, but it makes no sound.

The Santa Catalina Island Rattlesnake is endemic to the 15 sq mile (39 sq km) Santa Catalina Island in the Gulf of California. Its most notable feature is that its tail never develops a rattle, resulting in the alternative name of Rattleless Rattlesnake. Although there are lizards and mice on the island, this slender, agile rattlesnake actively hunts sleeping birds in the thorny bushes of the rocky arroyos at night. A rattle might be a disadvantage, warning roosting birds of the snake's approach. Santa Catalina is uninhabited by humans, but a population of feral cats, which are too large to be prey, poses a threat to the survival of the rattlesnake.

RELATED SPECIES

The closest relatives of *Crotalus catalinensis* are believed to be the Baja California Rattlesnake (*C. enyo*) and the Red Diamond Rattlesnake (*C. ruber*). Other island populations also contain rattleless individuals, most notably the San Esteban Rattlesnake (*C. molossus estebanensis*, page 578).

Actual size

FAMILY	Viperidae: Crotalinae
RISK FACTOR	Venomous: venom composition unknown
DISTRIBUTION	North America: southwestern USA and northwestern Mexico
ELEVATION	0–3,940 ft (0–1,200 m), rarely > 5,910 ft (1,800 m) asl
HABITAT	Sandy desert, and also rocky desert, semidesert, scrub, and dry woodland
DIET	Small mammals, lizards, and occasionally birds or other snakes
REPRODUCTION	Viviparous, with litters of 5–18 neonates
CONSERVATION STATUS	IUCN Least Concern

ADULT LENGTH
19¾ in–31½ in
(500–800 mm)

CROTALUS CERASTES
SIDEWINDER RATTLESNAKE
HALLOWELL, 1854

575

The image of a Sidewinder Rattlesnake moving quickly and diagonally across loose sand, leaving J-shaped tracks in its wake, is familiar to anybody who has watched a documentary about American desert life. This is the most desert-adapted of all the rattlesnakes. Sidewinder Rattlesnakes are small and nocturnal, and they inhabit the deserts of southwestern USA and northwestern Mexico, where they ambush mice, kangaroo rats, and lizards from a position of concealment beneath the sand. Birds and other snakes are also occasional prey. Being of relatively slender build, Sidewinder Rattlesnakes are also capable of climbing into thorny bushes. Although most human snakebites do not result in dangerous symptoms, fatalities have been recorded following the bites of this species.

The Sidewinder Rattlesnake is pastel-colored with desert hues, with a dorsal pattern of alternating dark and light blotches. Its most distinctive characteristic is a raised supraocular horn above each eye, possibly to prevent drifting sand falling over the eye when the snake is concealed under the sand with only its eyes visible.

RELATED SPECIES

There are three subspecies of *Crotalus cerastes*: the Mohave Desert Sidewinder (*C. c. cerastes*), Sonoran Desert Sidewinder (*C. c. cercobombus*), and Colorado Desert Sidewinder (*C. c. lateropens*). Sidewinding is also used as a means of locomotion by unrelated desert-dwelling vipers in other regions, including the Namib Sidewinding Adder (*Bitis peringueyi*, page 616), sand vipers (*Cerastes*, pages 620–621), and MacMahon's Viper (*Eristicophis macmahoni*, page 629). The Baja California Rattlesnake (*C. enyo*) is a close relative of *C. cerastes*.

Actual size

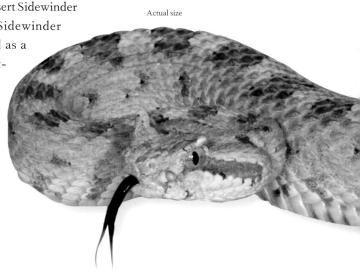

FAMILY	Viperidae: Crotalinae
RISK FACTOR	Highly venomous: variable, with procoagulants, myotoxins, and/or presynaptic neurotoxins
DISTRIBUTION	South America: Venezuela to northern Argentina, and also Aruba Island and islands off the Venezuelan coast
ELEVATION	0–3,280 ft (0–1,000 m), rarely > 5,560 ft (1,695 m) asl
HABITAT	Coastal plains, savanna grasslands, forest edges, dry woodland, semidesert, and arid islands
DIET	Small mammals, birds, and lizards
REPRODUCTION	Viviparous, with litters of 2–20 neonates
CONSERVATION STATUS	IUCN Least Concern

ADULT LENGTH
3 ft 3 in–6 ft
(1.0–1.8 m)

CROTALUS DURISSUS
NEOTROPICAL RATTLESNAKE
LINNAEUS, 1758

The Neotropical Rattlesnake is South America's only rattlesnake, occurring in every mainland country except Chile and Ecuador, and on the southern Caribbean islands of Isla de Margarita, Islas los Testigos, and Aruba. Mainland rattlesnakes inhabit seasonally flooded grasslands, dry woodland, but not rainforests. Mainland individuals may be large, rivaling North American diamondbacks (*Crotalus adamanteus*, page 572, and *C. atrox*, page 573). Rattlesnakes are responsible for nine percent of all serious snakebites in Latin America. Highly variable venom composition means that some populations are highly neurotoxic and others primarily hemotoxic, leading to challenges when treating snakebites and manufacturing and distributing appropriate antivenoms. Antivenom does not exist for all populations.

RELATED SPECIES

Eight subspecies of *Crotalus durissus* are currently recognized: the Guianan Coastal Rattlesnake (*C. d. durissus*); Venezuelan Rattlesnake (*C. d. cumanensis*); Ilha Marajó Rattlesnake (*C. d. marajoensis*); Roraima Rattlesnake (*C. d. ruruima*); South American Rattlesnake (*C. d. terrificus*); Rupununi Rattlesnake (*C. d. trigonicus*); Aruba Island Rattlesnake (*C. d. unicolor*); and Uracoan Rattlesnake (*C. d. vegrandis*). The last two have been treated as full species in the past.

Actual size

The Neotropical Rattlesnake is variable in appearance, although the general pattern is gray or brown with a series of bold diamond marks along the back. A distinctive pair of broad dorsal stripes often extends backward from the head. The Uracoan Rattlesnake (*Crotalus durissus vegrandis*) has an irregular speckled pattern, while the Aruba Island Rattlesnake (*C. d. unicolor*) is a dwarf race with pastel desert coloration and subdued dorsal markings.

FAMILY	Viperidae: Crotalinae
RISK FACTOR	Highly venomous: procoagulants, myotoxins, and hemorrhagins
DISTRIBUTION	North America: eastern and southeastern USA, excluding Florida Peninsula; extinct in Canada
ELEVATION	0–6,560 ft (0–2,000 m) asl
HABITAT	Deciduous upland woodland, especially wooded rocky hillsides
DIET	Small mammals and birds
REPRODUCTION	Viviparous, with litters of 4–14 neonates
CONSERVATION STATUS	IUCN Least Concern, locally endangered and protected

ADULT LENGTH
23¾–31½ in
(600–800 mm)

CROTALUS HORRIDUS
TIMBER RATTLESNAKE
LINNAEUS, 1758

577

Occurring throughout the eastern USA, this species was the first rattlesnake encountered by European colonists and was used as a symbol of defiance against the British during the War of Independence. Today this species is in grave danger, with dens being destroyed, and local extirpations. The last Canadian specimen was killed near Niagara Falls in 1941. Many states now protect the Timber Rattlesnake, but persecution continues. Prey comprises small mammals and birds. As a sedentary sit-and-wait ambusher, the Timber Rattlesnake will spend days waiting for prey. This is also one of the most arboreal of rattlesnakes. Snakebites are extremely serious; the venom is geographically variable, and many fatalities have occurred.

RELATED SPECIES
Crotalus horridus occurs in sympatry with the Eastern Diamondback Rattlesnake (*C. adamanteus*, page 572), Massasauga (*Sistrurus catenatus*, page 597) and Pygmy Rattlesnake (*S. miliarius*). The Canebrake Rattlesnake (*C. h. atricaudatus*) is no longer recognized as a subspecies.

The Timber Rattlesnake is a variably colored species. Northern populations are often dark browns and blacks with their patterning almost obscured, while southern populations, the "canebrakes," may be pinkish gray or yellowish with bold chevron markings and often a broad, brown dorsal stripe. Most marked specimens also exhibit a broad postocular stripe.

Actual size

FAMILY	Viperidae: Crotalinae
RISK FACTOR	Highly venomous: hemorrhagins, and possibly procoagulants
DISTRIBUTION	North America: southern USA and Mexico
ELEVATION	5,580–8,200 ft (1,700–2,500 m) asl
HABITAT	Deciduous pine–oak forest, desert upland, savanna woodland, rocky talus slopes, and even abandoned mines
DIET	Small mammals, and occasionally reptiles or birds
REPRODUCTION	Viviparous, with litters of 3–16 neonates
CONSERVATION STATUS	IUCN Least Concern

ADULT LENGTH
3 ft 3 in–4 ft 3 in
(1.0–1.3 m)

CROTALUS MOLOSSUS
BLACK-TAILED RATTLESNAKE
BAIRD & GIRARD, 1853

The Black-tailed Rattlesnake may be very variable in color, ranging from brown to gray or yellow, with a series of dorsal blotches that may form narrow bands around the body. A black band often runs across the head and through the eyes. The tail is jet black to dark gray, contrasting considerably with the body coloration of yellow specimens.

The Black-tailed Rattlesnake is a widely distributed species, most of its range being south of the Mexican border, where it is found in a wide variety of habitats, from dry woodland to desert uplands and rocky slopes, often in sympatry with other rattlesnake species. Prey is varied, but rodents are the most frequently taken, with birds and lizards also taken on occasion. The Black-tailed Rattlesnake climbs well in low bushes. It has been known to hybridize with the Western Diamondback Rattlesnake (*Crotalus atrox*, page 573). Although it must be considered highly venomous, due to its relatively large size, bites to humans from Black-tailed Rattlesnakes have usually resulted in relatively mild symptoms, with recovery following treatment with antivenom.

RELATED SPECIES

Four subspecies are recognized: the Black-tailed Rattlesnake (*Crotalus molossus molossus*) from southern USA and northern Mexico; Mexican Black-tailed Rattlesnake (*C. m. nigrescens*) from central Mexico; Oaxacan Black-tailed Rattlesnake (*C. m. oaxacus*) from southern Mexico; and San Esteban Rattlesnake (*C. m. estebanensis*), from San Esteban Island in the Gulf of California. Recently the Texan–Chihuahuan population was found to represent a separate species, the Eastern Black-tailed Rattlesnake (*C. ornatus*). Related species include the Mexican West Coast Rattlesnake (*C. basiliscus*) and the Totonacan Rattlesnake (*C. totonacus*), from northeastern Mexico.

Actual size

FAMILY	Viperidae: Crotalinae
RISK FACTOR	Venomous: venom composition unknown
DISTRIBUTION	North America: Central Mexican Plateau
ELEVATION	4,760–8,990 ft (1,450–2,740 m) asl
HABITAT	Grassy upland meadows, pine–oak forest, rocky outcrops, and lava flows
DIET	Small mammals, and birds, frogs, and lizards
REPRODUCTION	Viviparous, with litters of 4–14 neonates
CONSERVATION STATUS	IUCN Least Concern

ADULT LENGTH
2 ft 4 in–3 ft 3 in
(0.7–1.0 m)

CROTALUS POLYSTICTUS
MEXICAN LANCE-HEADED RATTLESNAKE
(COPE, 1865)

The Mexican Lance-headed Rattlesnake is a montane species from the Central Mexican Plateau, and may occur as high as 8,990 ft (2,740 m). The common name originates from its distinctive elongate head. Due to its high-altitude distribution, the Mexican Lance-headed Rattlesnake may be active by day or night, depending on the temperature, and it is especially common in rocky habitats with tall grass, and where mammal burrows afford it an easy refuge from predators, such as birds of prey. It preys on small mammals and occasionally birds, with lizards or frogs also taken. Its habitat is threatened by agricultural development. Only one human snakebite from this species is on record, which resulted in localized swelling and necrosis, and which was treated successfully with antivenom.

RELATED SPECIES

Crotalus polystictus resembles other spotted rattlesnakes, such as the Mexican Dusky Rattlesnake (*C. triseriatus*), with which it occurs in sympatry, or the Twin-spotted Rattlesnake (*C. pricei*) from the Sierra Madre Occidental to the north. Molecular data suggest that *C. polystictus* is a member of a clade of rattlesnakes that also includes the Sidewinder Rattlesnake (*C. cerastes*, page 575) and the Baja California Rattlesnake (*C. enyo*), although neither particularly resembles this species.

The Mexican Lance-headed Rattlesnake is pale to medium gray on the flanks, but brown on the dorsum, with several longitudinal rows of enlarged, irregular, dark gray to black spots, which are especially evident middorsally. This is the origin of its scientific name, *poly* meaning many, and *-stictus* meaning spotted. Its head markings are distinctive, with dark stripes on a pale gray to white background.

Actual size

FAMILY	Viperidae: Crotalinae
RISK FACTOR	Highly venomous: highly variable, with presynaptic neurotoxins, hemorrhagins, cytotoxins, and possibly myotoxins
DISTRIBUTION	North America: southern and southwestern USA, and Mexico
ELEVATION	0–8,200 ft (0–2,500 m) asl
HABITAT	Semiarid grassland, desert, and open scrubland
DIET	Small mammals, and occasionally reptiles or birds
REPRODUCTION	Viviparous, with litters of 2–17 neonates
CONSERVATION STATUS	IUCN Least Concern

580

ADULT LENGTH
3 ft 3 in–4 ft 7 in
(1.0–1.4 m)

CROTALUS SCUTULATUS
MOHAVE RATTLESNAKE
KENNICOTT, 1861

The Mohave Rattlesnake is variable in its coloration. Specimens may be light gray, buff, light brown, or even pale greenish, with a regular series of wide, darker saddles across the back and a dark postocular stripe. The tail is banded black and white, the black bands being narrower than the white bands.

The Mohave Rattlesnake is one of the most dangerous rattlesnakes in North America. Its venom composition and toxicity varies greatly geographically; specimens may look identical but possess completely different venom. Two venom types are found in Arizona specimens. Venom Type A contains a lethal presynaptic neurotoxin called Mohave Toxin that causes death through respiratory paralysis, while Venom Type B is hemorrhagic and cytotoxic, but not neurotoxic. Venom Type A is ten times more toxic than Venom Type B, and is found in Mohave Rattlesnakes across much of the state. The ramifications for antivenom production, and the treatment of snakebites, are considerable. This species preys on mammals, from mice to rabbits, but birds and lizards are also taken.

RELATED SPECIES

Most of the range is occupied by the nominate subspecies (*Crotalus scutulatus scutulatus*), but the southern Mexican population represents a separate subspecies (*C. s. salvini*). *Crotalus scutulatus* occurs in sympatry with other rattlesnake species, and hybridizes with the Western Rattlesnake (*C. viridis*) and South Pacific Rattlesnake (*C. oreganus helleri*). It may be confused with the Western Diamondback Rattlesnake (*C. atrox*, page 573) because both have black and white banded tails, although the black banding is more distinctive in the Western Diamondback.

Actual size

FAMILY	Viperidae: Crotalinae
RISK FACTOR	Venomous: hemorrhagins, and possibly procoagulants
DISTRIBUTION	North America: northern Mexico, just entering southern USA
ELEVATION	4,920–7,550 ft (1,500–2,300 m), rarely 9,190 ft (2,800 m) asl
HABITAT	Deciduous pine–oak or scrub–oak woodland in rocky canyons
DIET	Small mammals, lizards, and birds
REPRODUCTION	Viviparous, with litters of 2–9 neonates
CONSERVATION STATUS	IUCN Least Concern, locally endangered and protected

ADULT LENGTH
Male
21¾–26½ in
(550–670 mm)
Female
17¾–19¾ in
(450–500 mm)

CROTALUS WILLARDI
RIDGE-NOSED RATTLESNAKE
MEEK, 1905

581

The Ridge-nosed Rattlesnake is one of the United States' most protected reptiles. It only enters the country along the Arizona and New Mexico border with Mexico, in mountain ranges known as the Sky Islands. It inhabits wooded, rocky habitats at elevations over 4,920 ft (1,500 m), and feeds on mice, lizards, and small birds. This is a small species, with males exceeding females in length. A conspicuous characteristic of the Arizona Ridge-nosed Rattlesnake is a series of five bold white facial stripes, under the eye, along the lips, and down the chin. It has been suggested that this pattern influenced the war paint of the Apaches who lived in the same mountains, but this story appears to have been largely discounted. Snakebites have so far only resulted in minor envenomings.

The Ridge-nosed Rattlesnake subspecies vary considerably in their coloration and patterning, from pale gray with facial stripes largely absent (*obscurus*), to sandy or red-brown with five bold facial stripes (*willardi*).

RELATED SPECIES

No other rattlesnake resembles *Crotalus willardi*, but five subspecies are recognized. The Arizona Ridge-nosed Rattlesnake (*C. w. willardi*) of Arizona–Chihuahua and New Mexico Ridge-nosed Rattlesnake (*C. w. obscurus*) of New Mexico–Coahuila occur just within the United States. The West Chihuahuan Ridge-nosed Rattlesnake (*C. w. silus*), Del Nido Ridge-nosed Rattlesnake (*C. w. amabilis*), and Southern Ridge-nosed Rattlesnake (*C. w. meridionalis*) occur in the Sierra Madre Occidental of Mexico. Some authorities elevate these subspecies to full species status.

Actual size

FAMILY	Viperidae: Crotalinae
RISK FACTOR	Highly venomous: procoagulants, and possibly also anticoagulants or hemorrhagins
DISTRIBUTION	East Asia: China, Taiwan, and northern Vietnam
ELEVATION	330–4,920 ft (100–1,500 m) asl
HABITAT	Wooded mountains and foothills, in rocky valleys, woodland, and along streams
DIET	Small mammals, lizards, birds, and amphibians
REPRODUCTION	Oviparous, with clutches of 15–26 eggs
CONSERVATION STATUS	IUCN not listed

ADULT LENGTH
3 ft 3 in–5 ft
(1.0–1.5 m)

582

DEINAGKISTRODON ACUTUS
CHINESE COPPERHEAD
(GÜNTHER, 1888)

The Chinese Copperhead has a large, lance-shaped head that terminates in a fleshy, turned-up snout tip. Body patterning consists of a regular series of triangular markings that cross over the back as a narrow dark band. The ground color and markings vary from light brown overlaying pink-brown, to dark black-brown over yellow-brown. A narrow, dark postocular stripe is present.

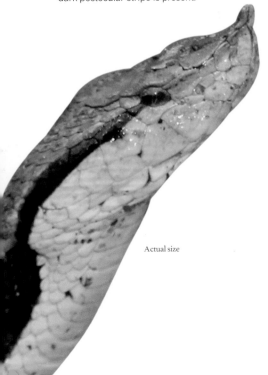

Actual size

The Chinese Copperhead is also known as the Sharp-nosed Viper, Chinese Moccasin, and Hundred-pace Snake, this last name referencing the distance a snakebite victim was thought to be able to walk following the bite. One local Chinese name translates as an even more worrying "five-pacer." Although this is a highly venomous snake capable of causing fatal snakebites, it cannot kill that quickly. It also suffers at the hands of humans, being collected in large numbers for food and traditional medicines. The Chinese Copperhead is relatively unusual in that, like the Malayan Pitviper (*Calloselasma rhodostoma*, page 570), it lays eggs. Prey consists of small mammals, birds, lizards, and frogs. It is an inhabitant of rocky and wooded hillsides in China, Taiwan, and extreme northern Vietnam.

RELATED SPECIES

Deinagkistrodon acutus does not resemble any other snake within its range, but superficially it resembles the Malayan Pitviper (*Calloselasma rhodostoma*, page 570). Molecular studies suggest it may be more closely related to the Mount Kinabalu Pitviper (*Garthius chaseni*, page 583) or Wagler's Temple Pitviper (*Tropidolaemus wagleri*, page 603).

FAMILY	Viperidae: Crotalinae
RISK FACTOR	Venomous: procoagulants, and possibly also anticoagulants or hemorrhagins
DISTRIBUTION	Southeast Asia: Mt. Kinabalu (Sabah, Malaysian Borneo)
ELEVATION	3,000–5,090 ft (915–1,550 m) asl
HABITAT	Submontane forest
DIET	Not known
REPRODUCTION	Oviparous, clutch size unknown
CONSERVATION STATUS	IUCN not listed

ADULT LENGTH
25½ in
(650 mm)

GARTHIUS CHASENI

MOUNT KINABALU PITVIPER

(SMITH, 1931)

583

The Mount Kinabalu Pitviper is known from only a few specimens, mostly collected at altitude on Mt. Kinabalu, with a single specimen from the Crocker Range. Due to its rarity, the natural history of this terrestrial pitviper, described over 80 years ago, is almost entirely a mystery. For some time it was included in the genus *Ovophis*, which gets its name from its oviparous (egg-laying) mode of reproduction. Even its prey preferences are unknown, although it is likely to feed on terrestrial lizards and small mammals. The genus *Garthius* was named in honor of the British herpetologist Garth Leon Underwood (1919–2002), while Frederick Nutter Chasen (1896–1942) was a British zoologist and director of the Raffles Museum.

The Mount Kinabalu Pitviper is a stout-bodied snake that blends well with the leaf litter of its montane rainforest home, being patterned with an irregular design of light and dark brown saddles and bands, and occasional black spots. The head is dark brown, only marked by a black postocular stripe with a pale lower border.

RELATED SPECIES

Garthius chaseni was originally grouped with the stout-bodied, terrestrial Asian mountain pitvipers of genus *Ovophis*, the closest species to Borneo being the Indo-Malayan Mountain Pitviper (*O. convictus*), which occurs in Southeast Asia, including Peninsular Malaysia, and on Sumatra. However, molecular studies suggest a closer relationship with the Chinese Copperhead (*Deinagkistrodon acutus*, page 582).

Actual size

FAMILY	Viperidae: Crotalinae
RISK FACTOR	Venomous: procoagulants, and possibly also anticoagulants or hemorrhagins
DISTRIBUTION	Asia: China, Mongolia, and Russia to Afghanistan, Iran, Uzbekistan, Kazakhstan, and Azerbaijan
ELEVATION	< 9,840 ft (< 3,000 m) asl
HABITAT	Montane forest, alpine meadows, steppe grassland, and semidesert
DIET	Small mammals, birds, and occasionally lizards
REPRODUCTION	Viviparous, with litters of 2–12 neonates
CONSERVATION STATUS	IUCN not listed, but locally protected

ADULT LENGTH
23¾–27 in
(600–690 mm)

584

GLOYDIUS HALYS
HALYS PITVIPER
(PALLAS, 1776)

The Halys Pitviper is a very variable species, with specimens ranging from pale gray to vivid orange. Most are patterned with irregular dark and light cross-bands on the dorsum of the back. A light-edged, dark-centered postocular stripe is usually present.

The Halys Pitviper is the only European pitviper, entering eastern Europe both north and south of the Caspian Sea, in Kazakhstan and Azerbaijan. From the Caspian it occurs through Iran and Afghanistan and eastward to Mongolia, where it is colloquially known as a "Mogoi." Across its range it is common but usually found as individuals rather than in numbers. Its primary prey is small rodents, although birds and, on rare occasions, lizards are taken. The venom is not considered highly toxic and although no deaths are on record following snakebites, caution is recommended. The genus *Gloydius* was named in honor of the American herpetologist Howard Kay Gloyd (1902–78), who devoted much of his life to the study of the *Agkistrodon* complex of pitvipers. "Halys" means "like a chain," a reference to the pitviper's patterning.

RELATED SPECIES

At least five subspecies are recognized: the Siberian Pitviper (*Gloydius h. halys*); Afghan Pitviper (*G. h. boehmei*); Karaganda Pitviper (*G. h. caraganus*); Caucasian Pitviper (*G. h. caucasicus*); and Mongolian Pitviper (*G. h. mogoi*). Head scalation and patterning differences are used to separate the subspecies. Related species include the Chinese Mamushi (*G. blomhoffii*), Amur Pitviper (*G. intermedius*), and Ussuri Mamushi (*G. ussuriensis*).

Actual size

FAMILY	Viperidae: Crotalinae
RISK FACTOR	Venomous: probably procoagulants, possibly myotoxins, neurotoxins, hemorrhagins, and nephrotoxins
DISTRIBUTION	East Asia: Far Eastern Russia, northeastern China, and the Korean Peninsula
ELEVATION	0–4,270 ft (0–1,300 m) asl
HABITAT	Broad-leaved deciduous and needle-leaved coniferous forests, rocky slopes
DIET	Lizards, possibly small mammals and amphibians
REPRODUCTION	Viviparous, litter size unknown
CONSERVATION STATUS	IUCN not listed

ADULT LENGTH
23¾–27½ in
(600–700 mm)

GLOYDIUS INTERMEDIUS
AMUR PITVIPER
(STRAUCH, 1868)

585

The Amur Pitviper inhabits the Amur region of Far Eastern Russia, Manchuria, China, and both North and South Korea, to elevations of 4,270 ft (1,300 m) asl. It inhabits both deciduous and coniferous forest, especially in areas with rocky scree slopes. The inhospitable weather conditions within the range of this species are extreme, leading to periods of hibernation, but being a live-bearing species it is able to breed where eggs would perish. However, as with other members of the genus, females may only reproduce every two years, it taking more than one season of feeding for them to recover their reproductive condition. Human fatalities have been reported following snakebites from members of the genus *Gloydius* and there is no reason to consider this species any less dangerous.

The Amur Pitviper is a fairly stout snake with a typical viperine head and a body pattern consisting of irregular, broad dark to mid-brown bands, over a light brown background, the lower scale rows and venter being pale gray. A dark postocular stripe separates the dark dorsal patterning from the unmarked, off-white lips.

RELATED SPECIES
Related species include the Chinese Mamushi (*Gloydius blomhoffi*), Rock Mamushi (*G. saxatilis*), Shedao Island Pitviper (*G. shedaoensis*, page 586), and Ussuri Mamushi (*G. ussuriensis*). Many of the pitvipers of genus *Gloydius* are similar in appearance, requiring expert examination to determine their identity.

Actual size

FAMILY	Viperidae: Crotalinae
RISK FACTOR	Highly venomous: possibly procoagulants, myotoxins, neurotoxins, nephrotoxins, or hemorrhagins
DISTRIBUTION	East Asia: Shedao Island and Liaoning Province, China
ELEVATION	0–705 ft (0–215 m) asl
HABITAT	Rocky cliffs, and thorn and scrub forest
DIET	Migratory birds
REPRODUCTION	Viviparous, with litters of 1–8 neonates
CONSERVATION STATUS	IUCN Vulnerable

ADULT LENGTH
Male
35¼ in
(895 mm)

Female
39 in
(990 mm)

GLOYDIUS SHEDAOENSIS
SHEDAO ISLAND PITVIPER
(ZHAO, 1979)

586

The Shedao Island Pitviper is generally a gray snake with slightly darker gray or brown cross-bands and bars, these markings extending onto the head, which also exhibits a dark postocular stripe.

The Shedao Island Pitviper is endemic to Shedao ("Snake") Island, a 180-acre (73-hectare) island 8 miles (13 km) from the mainland Liaodong Peninsula, in Liaoning Province, southeastern China, although a recently described subspecies occurs on the mainland. The pitviper is the only reptile on Shedao Island, where it occurs in densities of up to one snake per 11 sq ft (1 sq m), a density potentially greater than that of the Golden Lancehead (*Bothrops insularis*, page 567) on its small Brazilian island. Shedao Island Pitvipers are arboreal and prey on the migratory birds that visit the island every spring and fall. They also hunt prey on the island's rocky cliffs. Birds usually succumb to the snake's bite within one minute. Passerine birds are the preferred prey, but sparrowhawks are occasionally eaten.

RELATED SPECIES

A subspecies (*Gloydius shedaoensis qianshanensis*) has been described from mainland Liaoning Province, while related species include the Chinese Mamushi (*G. blomhoffii*), Amur Pitviper (*G. intermedius*, page 585), Rock Mamushi (*G. saxatilis*), Strauch's Pitviper (*G. strauchi*), and Ussuri Mamushi (*G. ussuriensis*).

Actual size

FAMILY	Viperidae: Crotalinae
RISK FACTOR	Highly venomous: probably procoagulants, and possibly also anticoagulants, hemorrhagins, or nephrotoxins
DISTRIBUTION	South Asia: southern India and Sri Lanka
ELEVATION	985–1,970 ft (300–600 m) asl (India), 0–5,000 ft (0–1,525 m) asl (Sri Lanka)
HABITAT	Wet and dry deciduous forest, evergreen forest, rocky watercourses, plantations, and cultivated areas
DIET	Lizards, small mammals, and amphibians
REPRODUCTION	Viviparous, with litters of 4–18 neonates
CONSERVATION STATUS	IUCN not listed

ADULT LENGTH
11¼–14¾ in,
rarely 19¾ in
(285–375 mm,
rarely 500 mm)

HYPNALE HYPNALE
INDIAN HUMP-NOSED PITVIPER
(MERREM, 1820)

587

The Indian Hump-nosed Pitviper occurs in the Western Ghats of southern India and throughout Sri Lanka, except the arid north. It is a short, stout snake with a triangular head that ends in a slightly upturned "hump-nosed" snout. Generally nocturnal and terrestrial, the Indian Hump-nosed Pitviper is often very common. It shelters in gaps under logs or in leaf litter, from where it ambushes geckos, skinks, mice, and frogs, often waving its tail tip to lure prey within strike range. Being small, sedentary, and cryptically patterned, it is frequently overlooked, and although it does not feature very often in Indian snakebites, it is one of the four most dangerous snakes in Sri Lanka, causing death through renal failure. There is no antivenom available to treat the bites of this species.

RELATED SPECIES

Two related species occur on Sri Lanka, the Scorpion Hump-nosed Pitviper (*Hypnale nepa*) and the Zara Hump-nosed Pitviper (*H. zara*). Wall's Hump-nosed Pitviper (*H. walli*) is now considered a synonym of *H. nepa*. In Sri Lanka the three species may all occur in the same area, but only *H. hypnale* occurs in southern India.

The Indian Hump-nosed Pitviper may be pale brown, orange, or red-brown on the dorsum of the body, with a fine white line along the side of the head, delineating between a paler dorsum and the dark brown sides of the head. The slightly upturned snout distinguishes snakes of this genus and earns them their common name.

Actual size

FAMILY	Viperidae: Crotalinae
RISK FACTOR	Highly venomous: procoagulants, hemorrhagins, possibly anticoagulants, and myotoxins
DISTRIBUTION	Central America: southeastern Costa Rica, including the Osa Peninsula
ELEVATION	3,280–5,250 ft (1,000–1,600 m) asl
HABITAT	Tropical lowland and lower montane forest with 98–180 in (2,500–4,500 mm) annual rainfall
DIET	Mammals (rodents and opossums)
REPRODUCTION	Oviparous, with clutches of 9–16 eggs
CONSERVATION STATUS	IUCN not listed

ADULT LENGTH
6 ft 7 in–8 ft
(2.0–2.4 m)

LACHESIS MELANOCEPHALA
BLACK-HEADED BUSHMASTER
SOLÓRZANO & CERDAS, 1986

The Black-headed Bushmaster is yellow or tan with a series of black rhombic markings along the dorsum of the back. Its most distinctive characteristic is the black cap that completely covers the dorsum of the head and merges with the postocular stripe.

The bushmasters are the largest vipers in the Americas, comparable with the African gaboon vipers (*Bitis gabonica*, page 613, and *B. rhinoceros*). They also possess the longest fangs of any American snake (2–2½ in/50–60 mm), again rivaled only by the gaboon vipers. Bushmasters are the only egg-laying American pitvipers. The Black-headed Bushmaster is a sit-and-wait ambusher of small to medium-sized mammals, from rodents to opossums, often lying motionless in wait for weeks. Bushmasters possess rough skin, every scale being strongly keeled, and the scales along the vertebral column form a distinct ridge, at least anteriorly. The head is rounded and not angular like that of many pitvipers. *Lachesis* was one of the Three Fates of Greek mythology, the goddess who measured the length of the thread of life.

RELATED SPECIES

The Central American Bushmaster (*Lachesis stenophrys*) occurs on the Atlantic versant of Costa Rica and Panama, while the recently described Chocoan Bushmaster (*L. acrochorda*) is found from Panama into Colombia.

Actual size

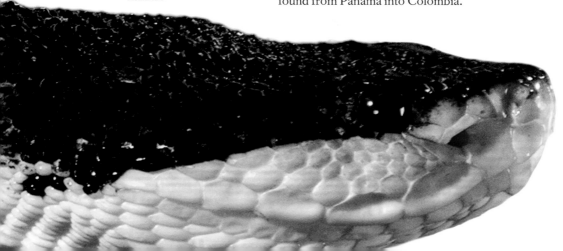

FAMILY	Viperidae: Crotalinae
RISK FACTOR	Highly venomous: procoagulants, hemorrhagins, possibly anticoagulants, and myotoxins
DISTRIBUTION	South America: Amazonia, the Guianas, Trinidad, and Brazilian Atlantic coast
ELEVATION	0–3,280 ft (0–1,000 m), rarely 5,910 ft (1,800 m) asl
HABITAT	Tropical lowland and lower montane forest with 79 in (2,000 mm) to in excess of 157 in (4,000 mm) annual rainfall, and dry forests along rivers; rare in secondary forest
DIET	Mammals (rodents and opossums), and possibly also birds, lizards, or frogs
REPRODUCTION	Oviparous, with clutches of 6–20 eggs
CONSERVATION STATUS	IUCN not listed

ADULT LENGTH
6 ft 7 in–11 ft 10 in
(2.0–3.6 m)

LACHESIS MUTA
SOUTH AMERICAN BUSHMASTER
(LINNAEUS, 1766)

589

The South American Bushmaster occurs throughout Amazonia, from the mouth of the Amazon to the headwaters in Ecuador, Peru, and Bolivia, throughout the Guianas, and on Trinidad. The endangered Brazilian Atlantic coastal population is recognized as a separate subspecies. The South American Bushmaster is much feared in the Amazon, being said to emit a high-pitched whistle, which, if ignored, means certain death. Bites are extremely serious and likely to end fatally unless treated quickly with antivenom. Bushmasters' activity is closely linked to rainfall; they remain hidden in mammal burrows until it rains, and then emerge to ambush mainly rodents, from mice to porcupines, and opossums. Juveniles may take frogs or lizards. The specific name *muta* means "mute," a reference to the curious spiny tip of the bushmaster's tail, hence the old name "mute rattlesnake."

RELATED SPECIES

Two subspecies are recognized: the Amazonian Bushmaster (*Lachesis muta muta*), and the highly endangered Atlantic Coastal Bushmaster (*L. m. rhombeata*). A second South American bushmaster species occurs in the Chocó region of Colombia and Panama, the Chocoan Bushmaster (*L. acrochorda*).

The South American Bushmaster is orange to tan, even reddish, with a series of large, black rhombic markings over its back, which extend into bars on the flanks. The scales are keeled and a raised ridge runs down the vertebrae. A distinctive black postocular stripe is present behind the eye.

Actual size

FAMILY	Viperidae: Crotalinae
RISK FACTOR	Venomous: possibly procoagulants and hemorrhagins
DISTRIBUTION	North America: southern Mexico (Puebla and Oaxaca)
ELEVATION	5,250–7,870 ft (1,600–2,400 m) asl
HABITAT	Dry deciduous forest, montane arid scrub, and high semidesert
DIET	Small mammals and lizards
REPRODUCTION	Viviparous, with litters of 5–8 neonates
CONSERVATION STATUS	IUCN Endangered

ADULT LENGTH
14¾–22¾ in
(375–580 mm)

MIXCOATLUS MELANURUS
BLACK-TAILED
HORNED PITVIPER
(MÜLLER, 1923)

Actual size

The arid upland habitat of the Black-tailed Horned Pitviper has been severely overgrazed by cattle, cleared for farming, and eroded by other agricultural practices. Given that its entire range is probably less than 2,320 sq miles (6,000 sq km), this species may be in danger of extinction. The Black-tailed Horned Pitviper is entirely terrestrial, researchers reporting that is most often encountered sheltering under agaves and cacti in semidesert habitats, or under logs in dry woodland. Due to the low night-time temperatures, the Black-tailed Horned Pitviper is most active in the morning after it has warmed by basking. It then hunts rodents and lizards, such as whiptail lizards.

RELATED SPECIES

Mixcoatlus melanurus was previously included in the genus *Ophryacus* and may be confused with the Mexican Horned Pitviper (*O. undulatus*, page 591). The two can be distinguished by their subcaudal scales, which are undivided in *M. melanurus* but divided in *O. undulatus*. Two hornless *Mixcoatlus* species are known, Barbour's Montane Pitviper (*M. barbouri*) and Brown's Montane Pitviper (*M. browni*), both from Guerrero, southern Mexico.

The Black-tailed Horned Pitviper may be gray or brown with a dorsal pattern of slightly darker blotches, which may coalesce to form a zigzag, and a distinctive black tail. This species also exhibits a broad, raised horn over each eye, which does not curve backward. The tail terminates in a curved spine.

FAMILY	Viperidae: Crotalinae
RISK FACTOR	Venomous: possibly procoagulants and hemorrhagins
DISTRIBUTION	North America: southern Mexico, in the southern Sierra Madre Occidental
ELEVATION	5,910–9,190 ft (1,800–2,800 m) asl
HABITAT	Pine–oak woodland and cloud forest
DIET	Small mammals and lizards
REPRODUCTION	Viviparous, with litters of 3–13 neonates
CONSERVATION STATUS	IUCN Vulnerable

ADULT LENGTH
21¾–27½ in
(550–700 mm)

OPHRYACUS UNDULATUS
MEXICAN HORNED PITVIPER
(JAN, 1859)

591

The Mexican Horned Pitviper occurs at relatively high elevations, in pine–oak or cloud forest habitats in Veracruz, Puebla, Oaxaca, and Guerrero, southern Mexico. It may be terrestrial or arboreal, climbing to heights of 13 ft (4 m) above the ground, even though it does not possess a prehensile tail. Prey comprises either small mammals such as voles, or terrestrial and arboreal lizards, such as alligator lizards and anoles. This is a diurnal species that avoids the cold of the night by hunting during the day. It is most frequently encountered close to mountain streams or in low sparse brush. Nothing is documented regarding snakebites for these small montane pitvipers.

RELATED SPECIES
Today the genus *Ophryacus* contains three species, the other two being the Emerald Horned Pitviper (*O. smaragdinus*), from Veracruz and northern Oaxaca, and the La Soledad Horned Pitviper (*O. sphenophrys*), from southern Oaxaca. The Black-tailed Horned Pitviper (*Mixcoatlus melanurus*, page 590) was formerly also included in *Ophryacus*.

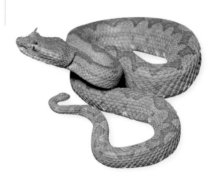

The Mexican Horned Pitviper is gray or brown with a series of irregular dorsal blotches of darker pigment, which may fuse to form a zigzag pattern. It also exhibits curved horns over each eye, although in juveniles the horns may be more wedge-shaped.

Actual size

FAMILY	Viperidae: Crotalinae
RISK FACTOR	Venomous: procoagulants, and possibly anticoagulants or hemorrhagins
DISTRIBUTION	South Asia: southern China, northern India, Nepal, Bangladesh, and Myanmar
ELEVATION	2,300–3,940 ft (700–1,200 m) asl
HABITAT	Montane evergreen forest, including coniferous forest
DIET	Frogs, small mammals, lizards, and birds and their eggs
REPRODUCTION	Oviparous, with clutches of 5–18 eggs
CONSERVATION STATUS	IUCN Least Concern

ADULT LENGTH
3 ft 3 in–4 ft
(1.0–1.2 m)

592

OVOPHIS MONTICOLA
WESTERN MOUNTAIN PITVIPER
(GÜNTHER, 1864)

The Western Mountain Pitviper is a squat snake with a broad, triangular head, small eyes, smooth scales, and a short tail. Its patterning consists of alternating dark and light brown squarish blotches on a gray-brown background, and copper-colored postocular stripes.

The Asian mountain pitviper complex is a complicated group of snakes that look relatively similar and may only be distinguished with certainty, especially in the overlap zones between different species, by using molecular techniques. The Western Mountain Pitviper occurs in evergreen forests in northern India, Nepal, southern China (Yunnan), and into Bangladesh and Myanmar. It is a nocturnal and terrestrial predator of frogs and rodents, but will also take lizards, and birds and their eggs. Unusually for pitvipers, but not uniquely, they are oviparous, and females will guard their eggs. If disturbed the Western Mountain Pitviper will form a coil, flatten its body, and launch repeated short strikes. Little is known regarding the effects of snakebites from this species, although the death of an elderly woman is on record.

RELATED SPECIES

The former subspecies of *Ovophis monticola* have been elevated to specific status, including the Oriental Mountain Pitviper (*O. orientalis*) from China; Taiwan Mountain Pitviper (*O. makazayazaya*); Tonkin Mountain Pitviper (*O. tonkinensis*) from Vietnam and Hainan Island; and Indo-Malayan Mountain Pitviper (*O. convictus*) from Southeast Asia. The most distant *Ovophis* species is the Hime-habu (*O. okinavensis*) in the Ryukyu Islands, a species that was never part of the mountain pitviper complex.

Actual size

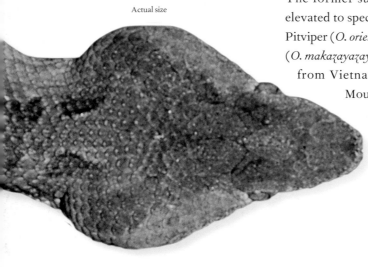

FAMILY	Viperidae: Crotalinae
RISK FACTOR	Venomous: possibly procoagulants or hemorrhagins
DISTRIBUTION	North, Central, and South America: southern Mexico to Ecuador
ELEVATION	0–3,280 ft (0–1,000 m), rarely 5,910 ft (1,800 m) asl
HABITAT	Tropical rainforest and low montane wet forest, and riverine forest in dry regions
DIET	Lizards, and small mammals, birds, or frogs
REPRODUCTION	Viviparous, with litters of 2–15 neonates
CONSERVATION STATUS	IUCN Least Concern

ADULT LENGTH
15¼–25½ in
(400–650 mm)

PORTHIDIUM NASUTUM
RAINFOREST HOGNOSED PITVIPER
(BOCOURT, 1868)

593

The Rainforest Hognosed Pitviper is an inhabitant of wet forests in lowland or low montane locations from southern Mexico to Ecuador. It is also found in riverine forests running through more arid habitats, but does not inhabit the dry forest habitats favored by other members of the genus *Porthidium*. Both diurnal and nocturnal, this pitviper is camouflaged to remain hidden in leaf litter on the forest floor, taking advantage of sunny spots to raise its body temperature. Its primary prey consists of lizards, but mice, birds, and frogs are also eaten, and there are cases of cannibalism and vermivory (earthworm-eating) on record. Hognosed pitvipers are not usually considered dangerous, and although snakebites are known for the Rainforest Hognosed Pitviper, no human fatalities have been recorded.

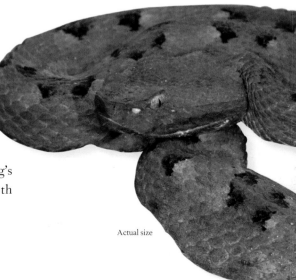

The Rainforest Hognosed Pitviper is a slender species that may be brown or pale blue with a dorsal pattern that may comprise alternating light and dark blotches, occasional black spots, and an orange to red vertebral stripe. The snout is strongly upturned and pointed.

RELATED SPECIES

The genus *Porthidium* currently contains eight other species, the ranges of which all overlap with that of *P. nasutum*, and all of which possess a turned-up snout to some degree. This species may be confused with Dunn's Hognosed Pitviper (*P. dunni*) or the Ujarran Hognosed Pitviper (*P. volcanicum*), both from Oaxaca in southern Mexico, the Slender Hognosed Pitviper (*P. ophryomegas*) from Central America, or Lansberg's Hognosed Pitviper (*P. lansbergii*) from northern South American and Panama.

Actual size

FAMILY	Viperidae: Crotalinae
RISK FACTOR	Highly venomous: hemorrhagins and cytotoxins, and possibly also procoagulants or myotoxins
DISTRIBUTION	East Asia: Japan (Ryukyu Islands, Okinawa and Amami groups)
ELEVATION	0–1,970 ft (0–600 m) asl
HABITAT	Most habitats, from montane forests to sugarcane fields and around human habitations
DIET	Small mammals, and also birds, reptiles, and frogs
REPRODUCTION	Oviparous, with clutches of 5–15 eggs
CONSERVATION STATUS	IUCN not listed

ADULT LENGTH
6 ft 7 in–8 ft,
rarely 10 ft
(2.0–2.4 m,
rarely 3.0 m)

PROTOBOTHROPS FLAVOVIRIDIS
OKINAWA HABU
(HALLOWELL, 1861)

594

The Okinawa Habu is a relatively slender snake with a large, angular, lance-shaped head. Its patterning comprises a reticulated pattern of dark brown over a yellowish to light brown dorsum and pale yellow lower flanks. A dark postocular stripe is present behind the eye.

The Okinawa Habu is one of only three snakes that can be referred to as "problem snakes," and the only one causing problems within its native range, the others being the Brown Treesnake (*Boiga irregularis*, page 146) on Guam, and Burmese python (*Python bivittatus*, page 95) in Florida. The Okinawa Habu is a large pitviper from the mountainous Ryukyu Islands, from Okinawa, Amami-Oshima, and neighboring islands, many of which have large human populations. This species is found everywhere, from pristine montane forest to sugarcane fields. It even enters houses, causing householders to use sticky traps across doors and windows. Snakebites are common, and before antivenom, and measures to control the snakes and their rat prey, 15–24 percent of bites were fatal. Survivors were often left disabled by tissue destruction and necrosis. Today fatalities are rare.

RELATED SPECIES

The northern Ryukyu (Amami) population is recognized as a separate subspecies (*Protobothrops flavoviridis tinkhami*) by some authors. *Protobothrops* contains 14 species, most of which inhabit the Asian mainland, although the Tokara Habu (*P. tokarensis*) and the unrelated Hime-habu (*Ovophis okinavensis*) also occur on the Ryukyu Islands.

Actual size

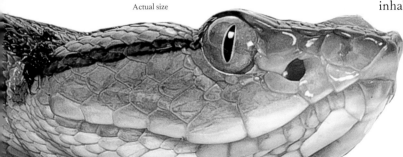

FAMILY	Viperidae: Crotalinae
RISK FACTOR	Highly venomous: procoagulants, and possibly also anticoagulants, myotoxins, or hemorrhagins
DISTRIBUTION	East Asia: China (Mangshan Mountains, Hunan and Guangdong provinces)
ELEVATION	2,620–4,270 ft (800–1,300 m) asl
HABITAT	Subtropical montane forest
DIET	Small mammals and birds
REPRODUCTION	Oviparous, with clutches of 13–21 eggs
CONSERVATION STATUS	IUCN Endangered, CITES Appendix II

ADULT LENGTH
4 ft 7 in–5 ft 7 in
(1.4–1.7 m)

PROTOBOTHROPS MANGSHANENSIS
MANGSHAN PITVIPER
(ZHAO IN ZHAO & CHEN, 1990)

595

The Mangshan Pitviper was originally described in the genus *Ermia*, but this was already allocated to a genus of locusts, so the genus *Zhaoermia* was created. It is now included in *Protobothrops*, based on molecular studies. That such a large pitviper should have evaded discovery until 1990 is surprising, although this is a localized Hunan Province endemic from a single mountain range. Unfortunately, only decades after its discovery this stunning pitviper is considered Endangered because of its limited range (< 116 sq miles/300 sq km), low population numbers, collection for the pet trade and smuggling, and habitat destruction and alteration. The Mangshan Pitviper is arboreal or terrestrial and may be found lying along lichen-covered logs. It is also said to inhabit limestone caves. There are unsubstantiated rumors that it spits venom, but this seems unlikely.

RELATED SPECIES

Although now placed in the genus *Protobothrops*, a genus of 14 species, *P. mangshanensis* may not be closely related to any of its congeners. In appearance it most resembles the variable Jerdon's Pitviper (*P. jerdonii*) from southern China (Yunnan), Myanmar, northeast India, and Nepal.

The Mangshan Pitviper exhibits an amazing pattern that must make it one of the most striking snake species. It is olive with an overlying irregular pattern of brighter greens that even extends onto its broad, blunt head. This is perfect camouflage because it strongly resembles the lichen-covered branches of its montane forest habitat. The tail tip is white, possibly for "caudal luring" of prey.

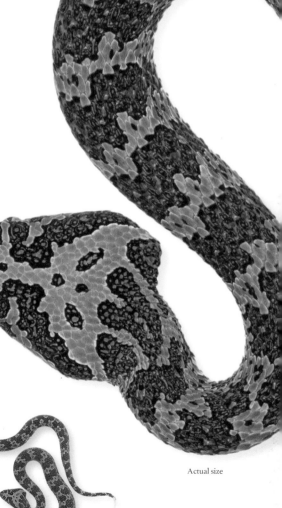

Actual size

FAMILY	Viperidae: Crotalinae
RISK FACTOR	Venomous: hemorrhagins, and possibly anticoagulants
DISTRIBUTION	Southeast Asia: Vietnam and Laos
ELEVATION	655–1,970 ft (200–600 m) asl
HABITAT	Tropical semi-evergreen forest in karst limestone
DIET	Unknown
REPRODUCTION	Oviparous, with clutches of 3–4 eggs
CONSERVATION STATUS	IUCN Endangered

ADULT LENGTH
3 ft 3 in–4 ft 3 in
(1.0–1.3 m)

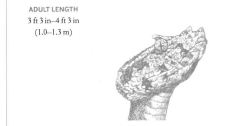

PROTOBOTHROPS SIEVERSORUM
THREE HORNED-SCALED PITVIPER
(ZEIGLER, HERRMANN, DAVID, ORLOV & PAUWELS, 2000)

The Three Horned-scaled Pitviper is gray-brown with an alternating series of dark and light brown squarish blotches along its back, the dark blotches sometimes coalescing to form a zigzag. Its body is slender, and its most distinctive characteristic is the three elevated horns over each of its eyes.

The Three Horned-scaled Pitviper was originally described in the monotypic genus *Triceratolepidophis*, named for its three-horned appearance, but based on molecular studies it was then transferred to the large Asian genus *Protobothrops*. It is a slender, arboreal pitviper, only known from the karst limestone of the Annamite Mountains, on the Vietnam–Laos border. It inhabits semi-evergreen tropical forest in an area gravely threatened by illegal logging and slash-and-burn agriculture. Collection for the illegal pet trade must also be considered a potential threat, as is the case with the much larger Mangshan Pitviper (*P. mangshanensis*, page 595). Although captive specimens have been induced to feed on rodents, its natural diet is undocumented.

RELATED SPECIES

Although now included in genus *Protobothrops*, *P. sieversorum* may not be closely related to any of its congeners. However, it does bear a striking resemblance to the Fi-Si-Pan Horned Pitviper (*P. cornutus*), which occurs in a number of localities within Vietnam and southern China and which is also slender and gray-brown, with fleshy hornlike projections over the eyes.

Actual size

FAMILY	Viperidae: Crotalinae
RISK FACTOR	Venomous: procoagulants and hemorrhagins
DISTRIBUTION	North America: Canada, USA, and Mexico
ELEVATION	0–4,920 ft (0–1,500 m) asl
HABITAT	Low-lying grasslands, meadows, swamps, and bogs in the northeast, to prairie and desert grasslands in the south
DIET	Small mammals, lizards, snakes, frogs, and centipedes
REPRODUCTION	Viviparous, with litters of 5–20 neonates
CONSERVATION STATUS	IUCN Least Concern

ADULT LENGTH
19¾–37½ in
(500–950 mm)

SISTRURUS CATENATUS
MASSASAUGA
(RAFINESQUE, 1818)

597

Also known as the Swamp Rattlesnake, the Massasauga is a relatively aquatic species, the only other aquatic American pitviper being by the Cottonmouth (*Agkistrodon piscivorus*, page 554). Although it is found from the Great Lakes region of Canada to the Rio Grande in Mexico, the bulk of the Massasauga's fragmented range is in the United States. Its habitats are usually open grasslands, prairie, or swamps, or alongside permanent watercourses in drier areas. Its diet varies across its range, adults of western populations preferring rodents, with juveniles taking lizards, while all age classes take frogs in northern populations. Snakes and centipedes are also occasional prey items. Snakebites are relatively common but deaths are rare, although the related Pygmy Rattlesnake (*Sistrurus miliarius*) is capable of causing fatalities.

The Massasauga is a gray to olive-brown snake with a dorsal pattern of darker gray or brown blotches middorsally and a corresponding series of smaller spots along the flanks. A dark postocular stripe, edged white below, is present behind the eye.

RELATED SPECIES

Three subspecies of the Massasauga have been recognized: the Eastern Massasauga (*Sistrurus catenatus catenatus*) in the northeast, Desert Massasauga (*S. c. edwardsii*) in the south, and Western Massasauga from the central US states, although this is now treated as a full species (*S. tergeminus*) by some authors. The closest relatives to *S. catenatus* are the three subspecies of the Pygmy Rattlesnake (*S. miliarius*).

Actual size

FAMILY	Viperidae: Crotalinae
RISK FACTOR	Venomous: procoagulants, and possibly hemorrhagins
DISTRIBUTION	Southeast Asia: Indonesia (eastern Java, Bali, and Lesser Sunda Islands; Lombok to Timor and Wetar)
ELEVATION	0–3,940 ft (0–1,200 m) asl
HABITAT	Tropical wet or dry forest, low montane forest, flooded grasslands, rice paddies, and coastal scrub
DIET	Small mammals, frogs, birds, and lizards
REPRODUCTION	Viviparous, with litters of up to 17 neonates
CONSERVATION STATUS	IUCN Least Concern

ADULT LENGTH
Male
23¾ in
(600 mm)

Female
32 in
(810 mm)

598

TRIMERESURUS INSULARIS
LESSER SUNDAS PITVIPER
KRAMER, 1977

In the late twentieth century the large pitviper genus *Trimeresurus* was the subject of considerable taxonomic change, with the species being split between eight genera. The current situation is that all 50 species are returned to *Trimeresurus* with the eight genera reduced to the status of subgenera. Although the Lesser Sundas Pitviper is extremely agile and possesses a prehensile tail, suggesting a highly arboreal existence, this snake may just as easily be encountered on or near the ground and often hunts frogs, lying in ambush beside small pools in shallowly flooded grassland or woodland. Small mammals, birds, and lizards also feature in its diet. Large specimens are capable of causing human fatalities, with several being documented on Timor where this species is especially common.

The Lesser Sundas Pitviper occurs in three color morphotypes. The most frequently encountered is the typical green morphotype, with yellow-green lips, that occurs through much of the range. Some specimens from the Komodo Islands are cyan or bluish in color while those from Wetar and eastern Timor are bright yellow, that morphotype occurring alongside green specimens in Timor.

RELATED SPECIES

There are many green pitvipers in Southeast Asia, but *Trimeresurus insularis* is most closely related to, and was formerly a subspecies of, the White-lipped Pitviper (*T. albolabris*) of mainland Southeast Asia, southern Sumatra, and western Java. Another closely related former subspecies is the Nepalese Pitviper (*T. septentrionalis*). These species, together with the Mangrove Pitviper (*T. purpureomaculatus*, page 602), are now placed in the subgenus *Trimeresurus* of genus *Trimeresurus*.

Actual size

FAMILY	Viperidae: Crotalinae
RISK FACTOR	Venomous: possibly myotoxins, otherwise unknown
DISTRIBUTION	Southeast Asia: Philippines (Batanes Islands)
ELEVATION	0–3,280 ft (0–1,000 m) asl
HABITAT	Seasonal tropical rainforest on volcanic slopes subject to typhoons
DIET	Small mammals
REPRODUCTION	Oviparous, clutch size unknown
CONSERVATION STATUS	IUCN Data Deficient

ADULT LENGTH
27½–34 in
(700–865 mm)

TRIMERESURUS MCGREGORI
BATAN PITVIPER
TAYLOR, 1919

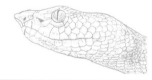

The Batan Pitviper is endemic to the Batanes Islands, the northernmost inhabited archipelago of the Philippines, where it is known to occur on Batan Island (13½ sq miles/35 sq km) and where it inhabits dense forests growing on the island's volcanic slopes. Also known as McGregor's Pitviper, it was named in honor of Richard Crittenden McGregor (1871–1936), an Australian-born American ornithologist who collected the holotype, and was bitten in the process. He suffered a swollen arm for several days but little pain. Little is known about the natural history of the Batan Pitviper, although it is reported to feed on small mammals and is also oviparous, like other species in the subgenus *Parias*.

The Batan Pitviper occurs as two color morphotypes, a bright yellow morphotype, which may have dark gray spots and bars dorsally, and a white to pale gray morphotype. Locals claim the yellow snakes are arboreal and the white snakes are terrestrial.

RELATED SPECIES
Trimeresurus mcgregori was once treated as a subspecies of the Philippine Pitviper (*T. flavomaculatus*). These two species, and related species, were placed in the genus *Parias*, but this taxon is now reduced to the status of a subgenus of *Trimeresurus*. Related species include Hagen's Pitviper (*T. hageni*) from Indonesia, Malcolm's Pitviper (*T. malcolmi*) from Borneo, the Sumatran Pitviper (*T. sumatranus*), and the rare Schultze's Pitviper (*T. schultzei*) from Palawan.

Actual size

FAMILY	Viperidae: Crotalinae
RISK FACTOR	Venomous: venom composition unknown
DISTRIBUTION	Southeast Asia: Nepal, Bhutan, northeast India, Bangladesh, Myanmar, northeast Thailand, and northern Laos and Vietnam
ELEVATION	1,970–6,560 ft (600–2,000 m) asl
HABITAT	Montane and submontane tropical forest, tea plantations, streamside vegetation, and bamboo forest
DIET	Birds, frogs, lizards, and small mammals
REPRODUCTION	Viviparous, with litters of 7–15 neonates
CONSERVATION STATUS	IUCN not listed

600

ADULT LENGTH
2 ft 4 in–3 ft 5 in
(0.7–1.05 m)

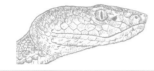

TRIMERESURUS POPEIORUM
POPE'S BAMBOO PITVIPER
SMITH, 1937

Pope's Bamboo Pitviper is a slender, bright green or cyan pitviper with a paler green underside. The iris of the eye is reddish, and a red postocular stripe continues from the eye to the angle of the jaw. A pair of fine white and red longitudinal stripes runs along the lower flanks to the red-brown prehensile tail.

Named in honor of the eminent American herpetologist Clifford Hillhouse Pope (1899–1974), Pope's Bamboo Pitviper is one of the best known of all Asian green pitvipers, many other species being misidentified as this species by the inexperienced. It used to occupy a wide range but its former subspecies have now all been elevated to specific status. It is an inhabitant of dense vegetation in submontane, montane, and bamboo forests, often in the lower vegetation levels, scrub, and bushes, and along streams. It hunts birds, lizards, frogs, and small mammals at night. This species comes into contact with people in tea plantations but, although venomous, it is small, relatively inoffensive, and not considered dangerous to humans, especially compared to related pitvipers within its range.

RELATED SPECIES

The closest relatives of *Trimeresurus popeiorum* are the Thai Peninsula Bamboo Pitviper (*T. fucatus*) and Cameron Highlands Pitviper (*T. nebularis*) from Peninsular Malaysia; Tioman Bamboo Pitviper (*T. buniana*) from Tioman Island; Barat Bamboo Pitviper (*T. barati*) and Toba Bamboo Pitviper (*T. toba*) from Sumatra; and Sabah Bamboo Pitviper (*T. sabahi*) from Borneo. All are included in the subgenus *Popeia*.

Actual size

FAMILY	Viperidae: Crotalinae
RISK FACTOR	Venomous: venom composition unknown
DISTRIBUTION	Southeast Asia: Indonesia (Java, Sumatra, and Natuna and Mentawai islands)
ELEVATION	1,640–5,250 ft (500–1,600 m) asl
HABITAT	Low montane forest and lowland rainforest, bamboo forest, and tea and coffee plantations
DIET	Small mammals, frogs, birds, and lizards
REPRODUCTION	Viviparous, with litters of 7–33 neonates
CONSERVATION STATUS	IUCN Least Concern

ADULT LENGTH
21¾–28¼ in,
rarely 36¼ in
(550–720 mm,
rarely 920 mm)

TRIMERESURUS PUNICEUS

JAVANESE FLAT-NOSED PALM-PITVIPER

(BOIE, 1827)

601

The relatively small, cryptically patterned Javanese Flat-nosed Palm-pitviper blends into the vegetation or the leaf litter of low montane forests or plantations. It is rarely encountered, either because it is so well camouflaged or because population densities are low. Males are primarily terrestrial, while females are more arboreal, but both will climb with agility in low vegetation and may be encountered in tea or coffee plantations where they obviously pose a risk to plantation workers. Bites are reported to cause intense pain and extensive swelling, severe symptoms from a relatively small snake, and although it is not known if deaths have occurred, they must be considered a possibility. Prey comprises small vertebrates, especially mice, but also lizards, birds, and frogs, which are hunted at night.

RELATED SPECIES

The closest relatives of *Trimeresurus puniceus* are the Sumatran Palm-pitviper (*T. andalasensis*); Bornean Palm-pitviper (*T. borneensis*); Siberut Palm-pitviper (*T. brongersmai*) from Siberut, Mentawai Islands; Truong Son Palm-pitviper (*T. truongsonensis*) from Vietnam; and Wirot's Palm-pitviper (*T. wiroti*) from the Malay Peninsula. All belong to the subgenus *Craspedocephalus*.

The Javanese Flat-nosed Palm-pitviper is gray, brown, or yellow-brown with a pattern comprising irregular darker and lighter blotches or saddles, which contrast strongly with the background color in males, less so in females. The saddles may themselves be speckled with dark or light flecks. The head is very obviously flattened.

Actual size

FAMILY	Viperidae: Crotalinae
RISK FACTOR	Venomous: procoagulants, and possibly also anticoagulants, hemorrhagins, or cardiotoxins
DISTRIBUTION	Southeast Asia: Myanmar, Thailand, Peninsular Malaysia, Singapore, and Indonesia (Sumatra)
ELEVATION	Sea-level
HABITAT	Mangrove swamps and coastal forests
DIET	Small mammals, frogs, and lizards
REPRODUCTION	Viviparous, with litters of 7–15 neonates
CONSERVATION STATUS	IUCN not listed

ADULT LENGTH
2 ft 4 in–3 ft 6 in
(0.7–1.07 m)

TRIMERESURUS PURPUREOMACULATUS
MANGROVE PITVIPER
(GRAY, 1832)

The Mangrove Pitviper is very variable in coloration. Specimens may be mottled green and brown with pale cross-bands one scale wide, bright orange with slightly darker dorsal markings, or dark gray without any patterning. In most specimens the undersides are white, often with yellow, green, or brown spotting. The head is often large, rounded, and stocky, and the tail is long and prehensile.

The Mangrove Pitviper, sometimes called the Shore Pitviper or Moonlight Pitviper, is a coastal species that inhabits coastal swamp forest and mangrove swamps in the Irrawaddy Delta, Myanmar, the coast of the Malay Peninsula and Singapore, and western Sumatra, Indonesia. It is a highly arboreal species that climbs well, although males are more slenderly built than females. The Mangrove Pitviper is a large and impressive snake that can launch a long and far-reaching strike from a branch. Its prey consists of small mammals, frogs, and lizards, which are hunted in the evening or at night. The Mangrove Pitviper has a reputation for fast and far-reaching strikes, and snakebites may have serious results, with child fatalities on record.

RELATED SPECIES

Trimeresurus purpureomaculatus is related to the Andaman Pitviper (*T. andersoni*), which used to be a subspecies. Other relatives include the White-lipped Pitviper (*T. albolabris*) of mainland Southeast Asia, the Nepalese Pitviper (*T. septentrionalis*), and the Lesser Sundas Pitviper (*T. insularis*, page 598). These species are included in the subgenus *Trimeresurus*, with the Hon Son Pitviper (*T. honsonensis*), Kanburi Pitviper (*T. kanburiensis*), Big-eyed Pitviper (*T. macrops*), and Beautiful Pitviper (*T. venustus*).

Actual size

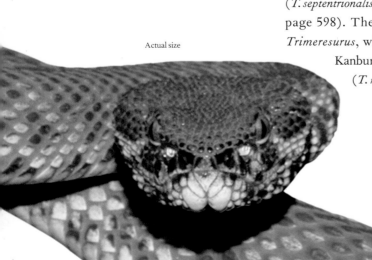

FAMILY	Viperidae: Crotalinae
RISK FACTOR	Venomous: procoagulants, and possibly also anticoagulants, hemorrhagins, or cytotoxins
DISTRIBUTION	Southeast Asia: peninsular Thailand and Malaysia, Indonesia (Sumatra), and neighboring islands
ELEVATION	0–3,940 ft (0–1,200 m) asl
HABITAT	Lowland tropical forests, coastal forest, and mangroves
DIET	Birds, small mammals, lizards, and frogs
REPRODUCTION	Viviparous, with litters of 15–41 neonates
CONSERVATION STATUS	IUCN Least Concern

TROPIDOLAEMUS WAGLERI

WAGLER'S TEMPLE PITVIPER

(BOIE, 1827)

ADULT LENGTH
Male
20½ in
(520 mm)
Female
36¼ in
(920 mm)

603

This is the famous pitviper from the Penang Snake Temple, where dozens of lazy snakes may be seen draped on branches, supposedly soothed into docility by the burning incense. In truth it is a fairly placid snake that is often slow to strike. Female temple pitvipers are impressive snakes with huge heads that terminate in a slightly forward-facing, pointed snout. This species feeds on birds and small mammals, although juveniles and smaller males may feed on frogs and lizards. It usually inhabits lowland tropical forests and will also enter mangrove forest. Highly arboreal, it is rarely encountered on the ground. Large females may produce over 40 neonates at one birthing.

RELATED SPECIES

Tropidolaemus wagleri was originally recognized as a single widely distributed species, but it is now confined to the Malay Peninsula and Sumatran populations. Separated from *T. wagleri* were the Southern Philippine Temple Pitviper (*T. philippensis*), Northern Philippine Temple Pitviper (*T. subannulatus*), which also occurs in Borneo, and Broad-banded Temple Pitviper (*T. laticinctus*) of Sulawesi, Indonesia. Hutton's Pitviper (*T. huttoni*) is the only other member of the genus, known only from two juvenile specimens collected in the Wavy Mountains, southern India, in the 1940s.

Actual size

Wagler's Temple Pitviper is sexually dimorphic and dichromatic. Males and juveniles are small and green, with a reddish stripe through the eye (see illustration above right) and yellow bands on the body. Females are much stouter and longer, black above with yellow bands around the body and yellow lips, and pale yellow to white below.

FAMILY	Viperidae: Viperinae
RISK FACTOR	Venomous: venom composition unknown
DISTRIBUTION	East Africa: Tanzania (Udzungwa and Ukinga mountains)
ELEVATION	5,580–6,230 ft (1,700–1,900 m) asl
HABITAT	Montane bush and bamboo thickets, and tea plantations
DIET	Earthworms, slugs, and possibly frogs
REPRODUCTION	Probably oviparous, with clutches of up to 10 eggs
CONSERVATION STATUS	IUCN Vulnerable

ADULT LENGTH
13¾–14 in
(350–360 mm)

604

ATHERIS BARBOURI
UDZUNGWA MOUNTAIN VIPER
LOVERIDGE, 1930

Also known as the Uzungwe Bushviper, Worm-eating Bushviper, or Barbour's Short-headed Bushviper, the Udzungwa Mountain Viper has been included in its own genus, *Adenorhinus*. It is one of the least-known venomous snakes of Africa, its entire range being two small Tanzanian mountain ranges around 5,910 ft (1,800 m) asl, where it inhabits thick bush. It is extremely vulnerable to habitat disturbance. The Udzungwa Mountain Viper is a terrestrial leaf-litter species, but it may also be semi-fossorial. It is the only viper in the world known to feed primarily on soft-bodied earthworms and slugs, although adults may also take frogs. It is thought to be active by day, after rain. Females may be oviparous. Nothing is known of its venom and no bites are recorded.

The Udzungwa Mountain Viper has rough, keeled body scales and a very short head, which is also covered in rough, keeled scales. They are pale brown with dorsal and lateral rows of large, dark brown blotch markings. The head is unpatterned.

RELATED SPECIES
Atheris barbouri is unlikely to be confused with any other bushvipers. It is believed most closely related to the Usambara Forest Bushviper (*Atheris ceratophora*, page 605), with which it occurs in sympatry.

Actual size

FAMILY	Viperidae: Viperinae
RISK FACTOR	Venomous: procoagulants, and possibly hemorrhagins
DISTRIBUTION	East Africa: Tanzania (Usambara, Udzungwa, and Uluguru mountains)
ELEVATION	2,300–6,560 ft (700–2,000 m) asl
HABITAT	Low montane forest, bush, and woodland
DIET	Frogs, and possibly small mammals or lizards
REPRODUCTION	Viviparous, litter size unknown
CONSERVATION STATUS	IUCN Vulnerable

ADULT LENGTH
15¾–21¾ in
(400–550 mm)

ATHERIS CERATOPHORA
USAMBARA FOREST BUSHVIPER
WERNER, 1896

The Usambara Forest Bushviper is a highly localized viper that inhabits the Eastern Arc Mountains, the Usambara, Udzungwa, and possibly Uluguru ranges, where it occurs in low montane forest, on the forest floor or 3 ft 3 in (1 m) off the ground. An alternative name is Horned Bushviper, a reference to a clump of one to three raised horns over each of its eyes, this being the only arboreal viper in Africa with horns. All bushvipers have strongly keeled scales. Nocturnal or crepuscular, it feeds on small reed frogs, but may also prey on small mammals or lizards. A live-bearer, its litter size is unrecorded. Snakebites have caused intense pain and severe necrosis. There is no antivenom available to treat bites by any of the bushvipers.

The Usambara Forest Bushviper may be quite variably patterned, being uniform yellow, brown, olive, or black, or overlaid with darker zigzag markings. Its most distinguishing characteristic is the elevated horns over each eye, and it is the only arboreal African viper so adorned.

RELATED SPECIES

Atheris ceratophora is the closest relative of the curious Udzungwa Mountain Viper (*Atheris barbouri*, page 604). If the 18 species of bushvipers are the African equivalent of Asian bamboo pitvipers (*Trimeresurus*, pages 598–602), or American palm-pitvipers (*Bothriechis*, pages 557–558), then the horn-bearing *Atheris ceratophora* most resembles the Eyelash Palm-pitviper (*Bothriechis schlegelii*, page 558).

Actual size

FAMILY	Viperidae: Viperinae
RISK FACTOR	Venomous: procoagulants, and possibly hemorrhagins
DISTRIBUTION	West Africa: Senegal and The Gambia to Côte d'Ivoire and Ghana
ELEVATION	0–1,970 ft (0–600 m) asl
HABITAT	Lowland tropical forests and dense bush
DIET	Small mammals
REPRODUCTION	Viviparous, with litters of 6–9 neonates
CONSERVATION STATUS	IUCN Least Concern

ADULT LENGTH
19¾–27½ in
(500–700 mm)

ATHERIS CHLORECHIS
WESTERN BUSHVIPER
(PEL, 1852)

606

The Western Bushviper is generally dark or light green with occasional faint yellow spots, a pale green to bluish belly, a white throat, and a yellow tail tip. The iris of the eye is the same color as the rest of the head. Although adults are green, newborn neonates are tan, taking on the green pigmentation within 24 hours of birth.

Actual size

Most bushvipers occur in East and Central African rainforests and mountain ranges, but the Western Bushviper is one of only two West African species. Records from Nigeria and Gabon may be based on misidentifications. Its primary habitat is lowland tropical rainforest, where the bushviper may climb 6 ft 7 in (2 m) from the ground. It climbs well and possesses a long, prehensile tail. Small mammals are known to be included in the diet of this viper but whether it also takes birds, lizards, or frogs is unknown. The Western Bushviper is a relatively large species that may be capable of delivering a serious bite, yet no antivenom exists to treat bushviper bites.

RELATED SPECIES

Atheris chlorechis resembles the Variable Bushviper (*Atheris squamigera*, page 609), from which it can be distinguished by its much finer and less rugose scalation, especially on the head and snout. In Côte d'Ivoire it may be found in sympatry with the Tai Hairy Bushviper (*A. hirsuta*), a much more rugose, brown species. Its resemblance to the arboreal, green pitvipers of Southeast Asia or tropical America (*Trimeresurus*, pages 598–602 and *Bothriechis*, pages 557–558) is remarkable.

FAMILY	Viperidae: Viperinae
RISK FACTOR	Venomous: procoagulants, and possibly hemorrhagins
DISTRIBUTION	Central and East Africa: Democratic Republic of the Congo, Uganda, Rwanda, Kenya, and Tanzania
ELEVATION	2,950–7,870 ft (900–2,400 m) asl
HABITAT	Rainforest, thorn forest, and papyrus and reedbeds along rivers
DIET	Lizards and frogs, and possibly small mammals, young birds, or snails
REPRODUCTION	Viviparous, with litters of 2–12 neonates
CONSERVATION STATUS	IUCN not listed

ADULT LENGTH
23¾–27½ in
(600–700 mm)

ATHERIS HISPIDA
ROUGH-SCALED BUSHVIPER
LAURENT, 1955

607

The Rough-scaled Bushviper has an extremely fragmented distribution, occurring in low montane rainforest localities in the countries surrounding Lake Victoria, and westward into the Democratic Republic of the Congo. Also known as the Spiny Bushviper, this species has extremely rugose, keeled scales on its head and body that curve outward to exacerbate its spiky, almost fir-cone appearance. A highly arboreal and agile species, the Rough-scaled Bushviper climbs well and may perch 6–10 ft (2–3 m) above the ground, atop foliage or reeds alongside rivers. This bushviper is nocturnal and will hunt on the ground, with prey comprising lizards and frogs. A snail was found in the gut of the holotype, and nestling birds and small mammals are also considered possible prey. There is no antivenom to treat bushviper snakebites.

The Rough-scaled Bushviper has extremely rugose scales that curve outward to present a spiky appearance, as well as large eyes on a relatively short head, and a long, prehensile tail. It is sexually dichromatic, males usually being yellow-green while females are olive-brown, both with irregular chevron markings on the head and body.

RELATED SPECIES
Atheris hispida could be confused with the Tai Hairy Bushviper (*A. hirsuta*) from Côte d'Ivoire, West Africa, but their ranges are mutually exclusive. In East Africa it is most likely to be confused with the extremely variable Variable Bushviper (*A. squamigera*, page 609), with which it occurs in sympatry, although *A. hispida* has much more rugose scales and occupies drier habitats than its congener.

Actual size

FAMILY	Viperidae: Viperinae
RISK FACTOR	Venomous: procoagulants, and possibly hemorrhagins
DISTRIBUTION	East Africa: Uganda, Rwanda, Burundi, Tanzania, and the Democratic Republic of the Congo
ELEVATION	3,280–9,190 ft (1,000–2,800 m) asl
HABITAT	Papyrus and swamp reedbeds, elephant grass plains, and montane forest
DIET	Lizards, small mammals, and frogs
REPRODUCTION	Viviparous, with litters of 4–13 neonates
CONSERVATION STATUS	IUCN not listed

ADULT LENGTH
27½–29½ in
(700–750 mm)

ATHERIS NITSCHEI
GREAT LAKES BUSHVIPER
TORNIER, 1902

The Great Lakes Bushviper can be a stunning snake, being bright green with deep black markings on the head and the body, and pale or yellow-green undersides. Juveniles are brown, changing to the adult livery when they are three to four months old.

The Great Lakes Bushviper, or Sedge Bushviper, occurs in five Rift Valley countries, at medium to high elevations, from Lake Albert, south around Lake Edward, Lake Kivu and the western shores of Lake Victoria, to the northern shores of Lake Tanganyika. Highly arboreal, it is found up to 10 ft (3 m) above the ground, in papyrus and reedbeds, tall elephant grass, and in small trees. Although nocturnal, it will bask during the day. It hunts small terrestrial mammals but also takes frogs and lizards, including chameleons, which are presumably captured aloft. Prey may also be attracted within strike range by caudal luring. This is a relatively large bushviper and since there is no antivenom available it may be capable of delivering a serious snakebite.

RELATED SPECIES

The Rungwe Bushviper (*Atheris rungweensis*) from the eastern shore of Lake Tanganyika, Tanzania, and southern Lake Tanganyika to Lake Malawi, Malawi, was formerly a southern subspecies of *A. nitschei*. Green specimens of the widely distributed Variable Bushviper (*A. squamigera*, page 609) may be confused with *A. nitschei* in the north of its range, but that species lacks black markings.

Actual size

FAMILY	Viperidae: Viperinae
RISK FACTOR	Venomous: procoagulants, and possibly hemorrhagins
DISTRIBUTION	West, Central, and East Africa: Ghana to Angola, Uganda, and Kenya
ELEVATION	2,300–5,580 ft (700–1,700 m) asl
HABITAT	Tropical rainforests, forest clearings, and reedbeds
DIET	Small mammals, lizards, frogs, and other snakes
REPRODUCTION	Viviparous, with litters of 7–9 neonates
CONSERVATION STATUS	IUCN not listed

ADULT LENGTH
19¾–31½ in
(500–800 mm)

ATHERIS SQUAMIGERA
VARIABLE BUSHVIPER
HALLOWELL, 1854

609

One of the largest, and certainly the most widely distributed, of African bushvipers, the Variable Bushviper is found in rainforest habitats across West, Central, and East Africa, from Ghana and Togo in the west, to Uganda and Kenya in the east, and Angola in the south. It also inhabits reedbeds. The Variable Bushviper is a slender, agile, and arboreal species, which will climb to 20 ft (6 m) above the ground. Prey consists of small rodents, lizards, frogs, and other snakes. An adult human fatality is recorded for this species but no antivenom exists to treat bushviper bites, and polyvalent African antivenom and blood transfusions failed to save the victim, who experienced gross swelling, incoagulable blood, and hypertension until his death six days post-bite. These snakes are dangerous.

The Variable Bushviper is not always green in color, adults varying from yellow to gray or red. The scales are often tipped with yellow, forming irregular bands around the body. Juveniles are yellow-green and take on the adult coloration at three to four months. The iris of the eye often contrasts with the body color.

RELATED SPECIES
The Rough-scaled Bushviper (*Atheris hispida*, page 607) occurs in sympatry with *A. squamigera*, but it is easily distinguished by its much more rugose scales. Two species were elevated from within *A. squamigera*, the Mayombe Bushviper (*Atheris anisolepis*) and Broadley's Bushviper (*A. broadleyi*). Further species may yet be identified within the wide-ranging *A. squamigera*.

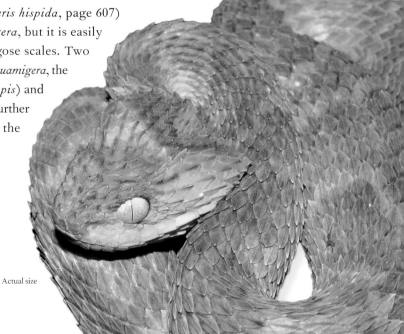

Actual size

FAMILY	Viperidae: Viperinae
RISK FACTOR	Venomous: hemorrhagins and cytotoxins, and possibly also procoagulants, anticoagulants, or nephrotoxins
DISTRIBUTION	Africa: Senegal to Somalia and south to South Africa; also southern Morocco and southwestern Arabian Peninsula, Yemen
ELEVATION	0–11,500 ft (0–3,500 m) asl
HABITAT	Savanna, dry woodland, arid scrub, and all habitats except rainforest, true desert, and high mountains
DIET	Mammals, birds, lizards, snakes, frogs, and even tortoises
REPRODUCTION	Viviparous, with litters of 20–60, rarely up to 156, neonates
CONSERVATION STATUS	IUCN not listed

ADULT LENGTH
3 ft–6 ft 3 in
(0.9–1.9 m)

BITIS ARIETANS
PUFF ADDER
MERREM, 1820

610

The Puff Adder is one of the largest, most recognizable, and widely distributed African snakes, occurring throughout sub-Saharan Africa, from Senegal to Somalia, south to the Cape, except in rainforests, true desert, and high mountains. Populations also exist in southern Morocco and the southwestern Arabian Peninsula. Puff Adders feed on a wide variety of vertebrate prey, the strangest being a young leopard tortoise. Females produce many offspring, the record being 156. Adult Puff Adders are so heavy that they use the "caterpillar crawl" method, moving in a straight line driven by waves of intercostal muscular contractions. Juveniles may be arboreal and even adults can swim. Fatal snakebites are documented, but more bites lead to amputation due to necrosis.

RELATED SPECIES

The Somalia population is treated as a subspecies (*Bitis arietans somalica*). Molecular studies suggest that *B. arietans* may be a species complex containing several cryptic species.

Actual size

The Puff Adder is a large, stout-bodied viper with a broad, rounded head. Its coloration may be variable, from pale yellow to dark yellow, browns, and reds, but this color is overlaid by a series of distinctive, dark, backward-facing chevrons.

FAMILY	Viperidae: Viperinae
RISK FACTOR	Highly venomous: neurotoxins, and possibly also myotoxins, procoagulants, anticoagulants, hemorrhagins, nephrotoxins, cardiotoxins, or cytotoxins
DISTRIBUTION	Southern Africa: South Africa, Lesotho, Swaziland, Zimbabwe, and Mozambique
ELEVATION	0–9,840 ft (0–3,000 m) asl
HABITAT	Rocky outcrops with grassy slopes, and montane fynbos
DIET	Lizards, small mammals, and amphibians
REPRODUCTION	Viviparous, with litters of 5–16 neonates
CONSERVATION STATUS	IUCN Least Concern

ADULT LENGTH
19¾–23¾ in
(500–600 mm)

BITIS ATROPOS
BERG ADDER
(LINNAEUS, 1758)

611

Berg Adders have a fragmented distribution, from the Cape of Good Hope, where they may be found near sea-level, to the Drakensberg and Lebombo mountains, where they may occur at 9,840 ft (3,000 m) asl. Other populations are found on the Zimbabwe–Mozambique border. Most populations are montane; this is a species adapted to living in relatively high-elevation rocky grassland and fynbos, feeding on lizards, which are killed by its neurotoxic venom. Adults may move to feeding on small mammals, and amphibians are also taken. Neurotoxic venom is unusual for African vipers, and no antivenom is available for Berg Adder bites, so snakebites may be serious in their outcome.

The Berg Adder has a dorsal background color that ranges from sandy brown to dark gray, overlain by four rows of small, triangular, dark brown or gray blotches, two on the back and one on each flank. The dorsal markings extend onto the head as an arrowhead, which may be surrounded by a pale line that extends the length of the body as a pair of dorsolateral stripes.

RELATED SPECIES
Two subspecies have been recognized, the nominate form (*Bitis atropos atropos*) throughout most of the range, and another in northeast South Africa (*Bitis a. unicolor*). It is suggested that this latter population may deserve specific status, and the Zimbabwe population may represent a third species. The patterning of *B. atropos* is similar to that of several harmless snakes, notably the Spotted Skaapsteker (*Psammophylax rhombeatus*) and the Rhombic Egg-eater (*Dasypeltis scabra*), both of which may derive some protection from their mimicry.

Actual size

FAMILY	Viperidae: Viperinae
RISK FACTOR	Venomous: presynaptic neurotoxins, and myotoxins
DISTRIBUTION	Southwest Africa: Namibia, Botswana, and western South Africa
ELEVATION	985–5,250 ft (300–1,600 m) asl
HABITAT	Arid semidesert, sandveld, and rocky outcrops
DIET	Lizards, small mammals, and frogs
REPRODUCTION	Viviparous, with litters of 5–20 neonates
CONSERVATION STATUS	IUCN not listed

ADULT LENGTH
11¾–20¼ in
(300–515 mm)

BITIS CAUDALIS
HORNED ADDER
(SMITH, 1839)

612

The Horned Adder is a small but widely distributed viper of arid, sandy, or rocky habitats across Namibia, Botswana, and South Africa. Its cryptic pastel-colored patterning helps it blend in almost invisibly on the ground, while a curved horn over each eye may shade the eye in bright sun, or prevent wind-blown sand from covering it when waiting in ambush for prey. Although lizards are its main prey, this small viper also takes frogs and mice. The venom is a presynaptic neurotoxin, presumably to deal with lizards, but it also contains a myotoxin, affecting neuromuscular function. Bites to humans are usually mild, and there are no fatalities recorded.

The Horned Adder exhibits a pastel-colored rhombic pattern comprising a variety of browns, grays, buffs, creams, and pale blues that break up the outline of the snake. There can be considerable variation between different individual adders, and males are more vividly patterned than females. The iris of the eye is gray or brown, depending on the background color of the neighboring scales.

RELATED SPECIES

Bitis caudalis is most closely related to the Namaqua Dwarf Adder (*B. schneideri*) and Namib Sidewinding Adder (*B. peringueyi*, page 616), but it is also similar in appearance to the Many-horned Adder (*B. cornuta*) of coastal Namibia to South Africa. Not all Many-horned Adders are horned, but those that are can be distinguished from *B. caudalis* by the presence of a tuft of irregular horns, rather than a single horn over each eye.

Actual size

FAMILY	Viperidae: Viperinae
RISK FACTOR	Highly venomous: procoagulants, hemotoxins, and possibly also anticoagulants, cardiotoxins, or cytotoxins
DISTRIBUTION	Central and southeast Africa: Nigeria to Uganda, south to Zambia, Tanzania, Mozambique, and South Africa
ELEVATION	0–4,920 ft (0–1,500 m) asl
HABITAT	Primary and secondary forest clearings
DIET	Mammals (mice and rats to antelopes), and birds
REPRODUCTION	Viviparous, with litters of 6–45 neonates
CONSERVATION STATUS	IUCN not listed, locally protected in South Africa

ADULT LENGTH
3 ft 3 in–5 ft
(1.0–1.5 m)

BITIS GABONICA
EAST AFRICAN GABOON VIPER
DUMÉRIL, BIBRON & DUMÉRIL, 1854

613

The East African Gaboon Viper is one of Africa's largest vipers, its length exceeded only by the West African Gaboon Viper (*Bitis rhinoceros*) and the Puff Adder (*B. arietans*, page 610). Its fangs may be the longest known, at over 2 in (50 mm). Camouflaged, it lies motionless in ambush in sun-dappled leaf litter, before delivering a rapid "stab and release" bite. Prey ranges from grass mice and pouched rats to small antelopes and birds. The East African Gaboon Viper inhabits tropical forest clearings and plantations in Nigeria and Central Africa, with isolated populations in coastal Tanzania, Mozambique, and South Africa, where the species is totally protected. Bites to humans are rare and fatalities unrecorded, but possible following untreated bites.

The East African Gaboon Viper is a stunning snake, with its geometric Persian carpet pattern of pastel browns, grays, pinks, purples, and creams, its broad head patterned like a leaf, even down to the midvein, and its pale cream or gray eyes. This is a species that disappears easily in leaf litter due to its amazing color pattern.

RELATED SPECIES

This species is easily distinguished from the West African Gaboon Viper (*Bitis rhinoceros*), which occurs from Guinea to Ghana, by the presence of dark teardrop markings under each eye, and the absence of the pointed horns on the snout that earn the West African species its scientific name *rhinoceros*. The two gaboon viper species are ecologically separated by the forest–savanna mosaic of the Togo Gap, west of Nigeria.

Actual size

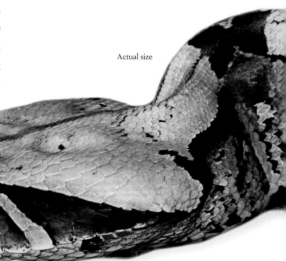

FAMILY	Viperidae: Viperinae
RISK FACTOR	Highly venomous: composition unknown, possibly hemorrhagins
DISTRIBUTION	West and Central Africa: Ghana to Angola, Uganda, and western Kenya
ELEVATION	0–8,200 ft (0–2,500 m) asl
HABITAT	Tropical rainforests, riverine forest, woodland, grassland, swamps, and plantations
DIET	Small mammals, frogs, and fish
REPRODUCTION	Viviparous, with litters of 15–40 neonates
CONSERVATION STATUS	IUCN not listed

ADULT LENGTH
19¾–31½ in
(500–800 mm)

BITIS NASICORNIS
RIVER JACK
(SHAW, 1802)

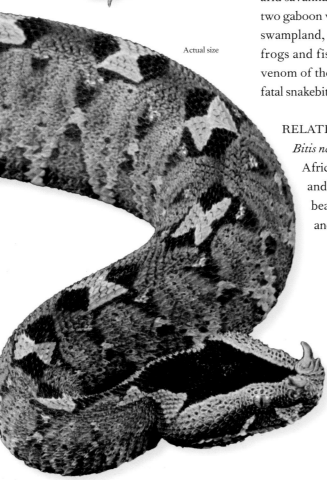

Actual size

The River Jack is also known as the Rhinoceros Viper, a name that may be confused with the West African Gaboon Viper (*Bitis rhinoceros*). It occurs in tropical rainforest and riverine forest, where its cryptic, geometrical patterning blends with the forest-floor leaf litter. There are two blocks of distribution, in the rainforests of Central and West Africa, separated by the arid savanna–woodland Togo Gap, which also separates the two gaboon viper species. River Jacks also occur in grassland, swampland, and cultivation. Rodents are the main prey, but frogs and fish are also reported. Little is known about the venom of the River Jack; bites are rare, but there has been a fatal snakebite in captivity.

RELATED SPECIES

Bitis nasicornis is similar in appearance to both the West African and East African Gaboon Vipers (*B. rhinoceros* and *B. gabonica*, page 613), the former species also bearing nasal horns. Hybrids between *B. nasicornis* and gaboon vipers are also known in the wild.

The River Jack is patterned with a complex design of pinks, blues, browns, and blacks, the pattern beginning as a black arrowhead on the viper's head, then continuing as a series of alternating geometric shapes on the back, with corresponding triangular markings on the flanks. The snout bears a large pair of impressive curved horns.

FAMILY	Viperidae: Viperinae
RISK FACTOR	Venomous: venom composition unknown
DISTRIBUTION	East Africa: Ethiopian Highlands
ELEVATION	6,560–9,840 ft (2,000–3,000 m) asl
HABITAT	Montane grassland and coffee plantations
DIET	Probably small mammals
REPRODUCTION	Viviparous, with up to 16 neonates
CONSERVATION STATUS	IUCN not listed

ADULT LENGTH
29½ in
(750 mm)

BITIS PARVIOCULA
ETHIOPIAN MOUNTAIN VIPER
BÖHME, 1976

615

The rare Ethiopian Mountain Viper's natural history is poorly documented. Confined to high elevations in the Ethiopian Highlands, it is believed to feed on small mammals such as rodents, which it takes in captivity. It has been found in three localities in the Highlands, and is associated with high-elevation grasslands, but may also occur in coffee plantations. Whether it is more widely distributed is not currently known. Captive specimens have produced up to 16 neonates. There are no snakebites on record and nothing is known about its venom. Unfortunately this species is much in demand for the pet trade and overcollecting may represent a more serious threat to its continued survival than outright persecution.

RELATED SPECIES

Bitis parviocula is tentatively placed in the subgenus *Macrocerastes* with the other large African vipers, the East African Gaboon Viper (*B. gabonica*, page 613), West African Gaboon Viper (*B. rhinoceros*), and River Jack (*B. nasicornis*, page 614). Another species of large Ethiopian viper, the Bale Mountain Adder (*B. harenna*), was described in subgenus *Macrocerastes* as recently as 2016.

The Ethiopian Mountain Viper is brown to olive with a dorsal pattern of yellow crosses surrounded by a circle of black, and smaller yellow and black markings on the lower flanks. The head bears a yellow V-shaped marking separating the brown dorsum of the head from the brown pigment of the cheeks. A pale stripe passes from the eye to the lip.

Actual size

FAMILY	Viperidae: Viperinae
RISK FACTOR	Venomous: venom composition unknown
DISTRIBUTION	Southwest Africa: Namibia and southern Angola
ELEVATION	Near sea-level
HABITAT	Namib Desert and coastal dunes
DIET	Lizards
REPRODUCTION	Viviparous, with litters of 3–10 neonates
CONSERVATION STATUS	IUCN Least Concern

ADULT LENGTH
8–9¾ in,
rarely 12¾ in
(200–250 mm,
rarely 325 mm)

BITIS PERINGUEYI
NAMIB SIDEWINDING ADDER
(BOULENGER, 1888)

Actual size

Also known as Peringuey's Adder, this is an extremely desert-adapted snake from the Namib Desert and coastal dunes of Namibia and southern Angola. It moves across the loose shifting sand using the same diagonal sidewinding locomotion as Sidewinder Rattlesnakes (*Crotalus cerastes*, page 575) in the United States, the Arabian Horned Viper (*Cerastes gasperettii*, page 620), and MacMahon's Viper (*Eristicophis macmahoni*, page 629) in Pakistan, Afghanistan, and Iran. It will shuffle down into the sand, leaving only its dorsally positioned eyes visible. Because rainfall in the Namib is rare, the viper obtains much of its water from its prey, sand lizards and barking geckos, but additional water may be obtained by lapping condensed fog from its scales. Snakebites are rare and symptoms reported are mild.

RELATED SPECIES

Bitis peringueyi is related to the Horned Adder (*B. caudalis*, page 612), but it more closely resembles some of the other southwest African desert-dwelling dwarf vipers, such as the Namaqua Dwarf Adder (*B. schneideri*) and Desert Mountain Adder (*B. xeropaga*). All these small vipers are threatened by illegal collection for the pet trade.

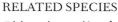

The Namib Sidewinding Adder is cryptically patterned with pastel desert hues, pinks, yellows, reds, and browns, but it has a black tail tip, which can be used to lure prey within strike range. It has dorsally positioned eyes, an adaptation for living in loose sand.

FAMILY	Viperidae: Viperinae
RISK FACTOR	Venomous: venom composition unknown
DISTRIBUTION	East Africa: Kenya
ELEVATION	> 4,920 ft (> 1,500 m) asl
HABITAT	Montane grasslands and scrub, and also valley woodland
DIET	Small mammals and lizards
REPRODUCTION	Viviparous, with litters of 7–12 neonates
CONSERVATION STATUS	IUCN not listed

ADULT LENGTH
15¾–19¾ in
(400–500 mm)

BITIS WORTHINGTONI
KENYAN HORNED VIPER
PARKER, 1932

617

The Kenyan Horned Viper is endemic to a small portion of the Rift Valley in south-central Kenya, at elevations above 4,920 ft (1,500 m) asl. The climate is cold, probably forcing the viper to adopt a more diurnal existence, although some authors consider it to be nocturnally active in the chill of the night. It also hibernates during the winter period. The habitat of this viper consists of high-elevation grassland and scrub along rocky escarpments, where it hunts lizards and small mammals, but it also occurs in more protected valleys in *Acacia* woodlands. The Kenyan Horned Viper is reportedly a slow-moving snake that reacts angrily if disturbed. Nothing is known of its venom, and there are no snakebites on record.

The Kenyan Horned Viper is a gray snake with a dorsal pattern of black triangular markings that form butterfly patterns along its back, and a corresponding series of small black squares on the flanks, with a pale flash separating the dorsal from the lateral black marks and forming a punctuated dorsolateral stripe. The most distinguishing feature is the high raised horns over the eyes.

RELATED SPECIES

This member of the genus *Bitis* is unlikely to be confused with any of its congeners, and it is placed in its own subgenus, *Keniabitis*. Neither do any other snakes in East Africa bear a resemblance to *B. worthingtoni*.

Actual size

FAMILY	Viperidae: Viperinae
RISK FACTOR	Venomous: venom composition unknown
DISTRIBUTION	Southeastern Africa: Tanzania to KwaZulu-Natal, South Africa
ELEVATION	0–5,910 ft (0–1,800 m) asl
HABITAT	Riverine grasslands, coastal thicket, and dry savanna
DIET	Frogs and toads, and also small mammals
REPRODUCTION	Oviparous, with clutches of 3–9 eggs
CONSERVATION STATUS	IUCN not listed

ADULT LENGTH
11¾–19¾ in
(300–500 mm)

618

CAUSUS DEFILIPPII
SNOUTED
NIGHT ADDER
(JAN, 1863)

The Snouted Night Adder is so-called because it is the only night adder species with a sharply upturned snout. Patterning is usually gray or brown with a series of broad, alternating light and dark brown rhombic blotches along the dorsum of the back. It has a dark V-shaped marking on the back of the head and dark postocular stripes.

Night adders have smooth scales that feel velvety to the touch, and round pupils, unlike typical vipers, which possess keeled scales and vertically elliptical pupils. The Snouted Night Adder belies its common name, being also active by day, especially after heavy rain. It is terrestrial, occasionally semi-arboreal, and most commonly encountered in grasslands fringing rivers or lakes. It also inhabits arid savannas and coastal thicket. The primary prey of night adders are frogs and toads, but even though this is the smallest night adder, it is also capable of taking small mammals. When disturbed it will puff up its body and hiss loudly, before elevating its body and making a vigorous strike. Snakebites cause intense pain, swelling, and fever. There is no antivenom available for night adder bites.

RELATED SPECIES

Causus defilippii may be confused with the Common Egg-eater (*Dasypeltis scabra*), which also exhibits a bold V-shaped marking on the back of the head. The turned-up snout distinguishes *C. defilippii* from the sympatric Rhombic Night Adder (*C. rhombeatus*).

Actual size

FAMILY	Viperidae: Viperinae
RISK FACTOR	Venomous: venom composition unknown
DISTRIBUTION	West and Central Africa: Senegal to Ethiopia, south to northern Angola
ELEVATION	0–2,300 ft (0–700 m) asl
HABITAT	Tropical forest, savanna, and semidesert
DIET	Frogs, toads, and small mammals
REPRODUCTION	Oviparous, with clutches of 6–20 eggs
CONSERVATION STATUS	IUCN not listed

ADULT LENGTH
11¾–27½ in
(300–700 mm)

CAUSUS MACULATUS
WEST AFRICAN NIGHT ADDER
(HALLOWELL, 1842)

619

The West African Night Adder occurs in many habitats, from tropical forest to moist savanna and arid semidesert, from Senegal to Ethiopia and south to northern Angola. There is also a record from southern Mauritania, a desert country. This species has round pupils and smooth, velvety scales, in contrast to the typical viper arrangement of vertically elliptical pupils and rough, keeled scales. Although generally terrestrial it can climb, and is commonly encountered around human habitations during the day, hunting frogs and toads. Small mammals are also taken. Night adders often give away their hiding places by hissing loudly. They will puff themselves up and strike rapidly. Snakebites cause pain, swelling, fever, and other mild symptoms, but no deaths are recorded. There is no antivenom available to treat snakebites.

The West African Night Adder is stout-bodied and usually brown, gray, or greenish, with a series of dark rhombic markings down the center of its back, which are not edged with white as they are in the similarly patterned Rhombic Night Adder (*Causus rhombeatus*). There is a fine black V-shaped marking on the back of its head.

RELATED SPECIES

Causus maculatus is similar in appearance to the Rhombic Night Adder (*C. rhombeatus*) of East Africa, with which it occurs in sympatry in the east of its range. It also occurs in sympatry with the Forest Night Adder (*C. lichtensteinii*), but that species is easily distinguished, being slender and green.

Actual size

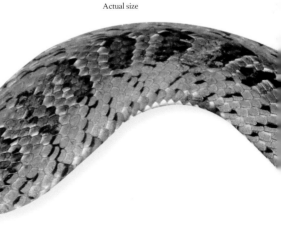

FAMILY	Viperidae: Viperinae
RISK FACTOR	Venomous: procoagulants and cytotoxins, and possibly hemorrhagins
DISTRIBUTION	Arabia and Middle East: Saudi Arabia, Bahrain, Qatar, UAE, Oman, Yemen, Israel, Jordan, Iraq, and Iran
ELEVATION	0–4,920 ft (0–1,500 m) asl
HABITAT	Sandy desert and vegetated or stony semidesert
DIET	Lizards, and also small mammals and birds
REPRODUCTION	Oviparous, with clutches of 4–20 eggs
CONSERVATION STATUS	IUCN Least Concern

ADULT LENGTH
17¾–33 in
(450–840 mm)

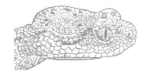

CERASTES GASPERETTII
ARABIAN HORNED VIPER
LEVITON & ANDERSON, 1967

The Arabian Horned Viper is more often hornless than horned. Its body is colored with desert hues of sandy yellow or pink, with a dorsal pattern of dark brown irregular bars and a brown postocular stripe behind the eye.

Actual size

The Arabian Horned Viper may be horned or hornless (see illustration above left), the hornless form being more frequently encountered. It inhabits the Arabian Peninsula, Iraq, and Iran, with an isolated population in the Arava Valley of Jordan and Israel. Its habitats include sandy desert and stony semidesert. On dunes it burrows into the sand during the day, emerging to hunt prey such as lizards at night. The viper leaves J-shaped tracks in the sand as it sidewinds over the dunes, and when in ambush position it can be found just under the surface. When alarmed, the Arabian Horned Viper will saw-scale like a carpet viper (*Echis*, pages 624–628) and launch a rapid strike. Snakebites cause pain, swelling, nausea, and necrosis, and may prove fatal. The species is named for the American herpetologist, and Arabian specialist, John Gasperetti (1920–2001).

RELATED SPECIES

Cerastes gasperettii is similar in appearance to the Sahara Horned Viper (*C. cerastes*) of North Africa, a subspecies of which (*C. c. hoofieni*) occurs in the south-western Arabian Peninsula. In the rest of its range it is only likely to be confused with the relatively rare, rocky desert-dwelling Persian False Horned Viper (*Pseudocerastes persicus*, page 636). A subspecies (*C. g. mendelssohni*) occurs in Israel and Jordan.

FAMILY	Viperidae: Viperinae
RISK FACTOR	Venomous: procoagulants, cytotoxins, and possibly hemorrhagins
DISTRIBUTION	North Africa: Western Sahara and Mauritania to Sinai (Egypt) and Israel
ELEVATION	Near sea-level
HABITAT	Sandy desert and dunes
DIET	Small mammals and lizards
REPRODUCTION	Uncertain, one female laid 8 eggs which hatched within hours
CONSERVATION STATUS	IUCN Least Concern

ADULT LENGTH
9¾–21 in
(250–530 mm)

CERASTES VIPERA
SAHARA SAND VIPER
(LINNAEUS, 1758)

The Sahara Sand Viper is poorly known despite its range encompassing every North African country except Sudan. It inhabits sandy desert, where it ambushes lizards, such as fringe-toed lizards, or small mammals, using its black tail to lure them within strike range, while it lies hidden with only its dorsolaterally positioned eyes visible. Disturbed Sahara Sand Vipers form their bodies into concentric-moving curves and create a rasping, saw-scaling sound similar to that produced by congeners and the saw-scale and carpet vipers of genus *Echis* (pages 624–628). Although this is a small species, bites from larger relatives have caused pain, swelling, vomiting, and necrosis.

The Sahara Sand Viper is a pastel, desert hue-colored snake, being shades of yellow, cream, or sandy pink, with a contrasting black tail, which is used for caudal luring. The eyes are positioned dorsolaterally.

RELATED SPECIES
The larger Sahara Horned Viper (*Cerastes cerastes*) and *C. vipera* occupy mutually exclusive ranges, the Sahara Horned Viper preferring rocky desert and oasis habitats. *Cerastes vipera* is hornless, while the Sahara Horned Viper and Arabian Horned Viper (*C. gasperettii*, page 620) may be horned or hornless. The recently described and related Tunisian Sand Viper (*C. boehmei*) possesses a tuft of scales over each eye.

Actual size

FAMILY	Viperidae: Viperinae
RISK FACTOR	Highly venomous: procoagulants, myotoxins, hemorrhagins, and possibly neurotoxins
DISTRIBUTION	Middle East: Syria, Israel, Jordan, and Lebanon
ELEVATION	66–5,250 ft (20–1,600 m) asl
HABITAT	Mediterranean oak woodland, rocky slopes, and cultivated gardens, fields, and olive groves
DIET	Small mammals, lizards, and birds
REPRODUCTION	Oviparous, with clutches of 5–25 eggs
CONSERVATION STATUS	IUCN Least Concern

ADULT LENGTH
4–5 ft
(1.2–1.5 m)

DABOIA PALAESTINAE
PALESTINE VIPER
(WERNER, 1938)

The Palestine Viper is gray, olive, red, or yellow, overlaid with a dorsal pattern of dark brown, rhombic-shaped blotches, which often coalesce to form a zigzag, and a series of brown blotches on the flanks. The head bears a dark V-shaped marking on its dorsum and dark postocular and subocular stripes.

The Palestine Viper is a large snake, achieving lengths up to 5 ft (1.5 m). The activity of this species depends on the season. It is more diurnal in winter, becoming nocturnal in summer, and although a terrestrial species, it climbs into bushes and small trees to bask or avoid the heat of the day. Adult Palestine Vipers prey on rodents and birds, and also take chameleons. Juveniles prey on lizards. This is a Mediterranean oak woodland species that also inhabits rocky slopes, but with many habitats being altered by humans it is also likely to occur in cultivated olive groves or gardens. Given the size of this viper and the potency of the venom of some of its relatives, snakebites should be considered extremely serious medical emergencies.

RELATED SPECIES

The genus *Daboia* contains four other species, the two Russell's vipers (*D. russelii*, page 623, and *D. siamensis*) in South and Southeast Asia, the Desert Viper (*D. deserti*) in Tunisia and Libya, and the Moorish Viper (*D. mauritanica*) in Morocco, Algeria, and Tunisia.

Actual size

FAMILY	Viperidae: Viperinae
RISK FACTOR	Highly venomous: presynaptic neurotoxins, procoagulants, myotoxins (Sri Lanka), hemorrhagins, and possibly nephrotoxins or cytotoxins
DISTRIBUTION	South Asia: India, Pakistan, Nepal, Bangladesh, and Sri Lanka
ELEVATION	0–9,040 ft (0–2,755 m) asl
HABITAT	Woodland, forest edge, scrub, gardens, rice paddies, and grassland
DIET	Small mammals and lizards
REPRODUCTION	Viviparous, with litters of 20–40 neonates
CONSERVATION STATUS	IUCN not listed, CITES Appendix III (India)

ADULT LENGTH
4–5 ft 2 in
(1.2–1.6 m)

DABOIA RUSSELII
SOUTH ASIAN RUSSELL'S VIPER
(SHAW & NODDER, 1797)

623

Patrick Russell (1726–1805) was a Scottish surgeon for the East India Company who studied Indian snakes and snakebites. The South Asian or Western Russell's Viper is one of the most dangerous snakes in the world. It occurs in most habitats, from sea-level to almost 9,200 ft (2,800 m) asl, often in high densities. Russell's vipers prey on rats, although juveniles also feed on lizards. This species is especially common in rice paddies, where it bites workers during the harvest. When disturbed, Russell's vipers issue a loud, slow, threatening hiss, puffing up the body and making rapid, far-reaching strikes. Without antivenom, death is a strong possibility, through cerebral hemorrhage or renal failure. Thousands of people die annually in South Asia. The venom varies across its range, impacting on the treatment of snakebite victims.

The South Asian Russell's Viper is a stunningly beautiful snake. It is orange or pale brown to red-brown dorsally with a bold pattern comprising three rows of dark red-brown lozenge shapes, each with a bold black edge, the dorsal row often fusing to form a punctuated zigzag.

RELATED SPECIES

A second species, the Southeast Asian or Eastern Russell's Viper (*Daboia siamensis*), has a fragmented distribution through Southeast and East Asia, occurring in isolated pockets in Myanmar, Thailand, southern China, Taiwan, and Indonesia (Flores and the Komodo Islands). At one time each of these populations was treated as separate subspecies of *D. russelii*.

Actual size

FAMILY	Viperidae: Viperinae
RISK FACTOR	Highly venomous: procoagulants, hemorrhagins, and possibly also anticoagulants, nephrotoxins, or cytotoxins
DISTRIBUTION	South and western Asia, and Arabia: India and Sri Lanka to Afghanistan, the Caspian states, and Iran; also Qatar, UAE, and Oman
ELEVATION	0–6,560 ft (0–2,000 m) asl
HABITAT	Rocky valleys and hills, gravel plains, scrub habitats, around old buildings, and almost anywhere
DIET	Small mammals, lizards, frogs, small snakes, and invertebrates, including scorpions
REPRODUCTION	Oviparous or viviparous, with clutches or litters of 2–23 eggs or neonates
CONSERVATION STATUS	IUCN not listed

ADULT LENGTH
19¾–31½ in
(500–800 mm)

ECHIS CARINATUS
COMMON SAW-SCALE VIPER
(SCHNEIDER, 1801)

624

The Common Saw-scale Viper is a brown snake with an irregular rhombic pattern. The eye of Arabian specimens is bright orange. This species' classic defensive posture of coiling into concentric curves and moving backward enables it to saw-scale a warning, move away from the threat, and watch its enemy at the same time.

Actual size

The Common Saw-scale Viper may be small but it is one of the most dangerous snakes in the world. It occurs in very high densities at night, in areas where people walk barefoot and sustain snakebites; 20 percent of untreated bites terminate fatally, due to cerebral hemorrhage. This is a widely distributed Asian species, its common name originating from the warning sawing sound created when the serrated keels of its obliquely arranged lateral body scales rasp together. This behavior is common to all *Echis*, although Afro-Arabian species are called "carpet vipers." Prey includes mice and lizards, but some populations specialize in scorpions and the venom of Common Saw-scale Vipers varies according to their prey preferences. Some populations lay eggs, while others are live-bearers.

RELATED SPECIES

The Indian population is the nominate subspecies (*Echis carinatus carinatus*). The Sri Lankan population was allocated separate status (*E. c. sinhaleyus*), but most authors place it in the nominate subspecies. The Central Asian population, from Pakistan to Uzbekistan, is sometimes recognized as a subspecies (*E. c. multisquamatus*), as is the Astola Island population, Pakistan (*E. c. astolae*). One widely distributed subspecies is the Sindh Saw-scale Viper (*E. c. sochureki*) from Pakistan, Iran, Qatar, UAE, and Oman.

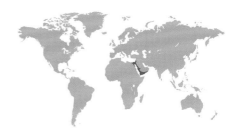

FAMILY	Viperidae: Viperinae
RISK FACTOR	Highly venomous: procoagulants, hemorrhagins, and possibly also anticoagulants, nephrotoxins, or cytotoxins
DISTRIBUTION	Arabia and Middle East: eastern Egypt, Israel, Jordan, western Saudi Arabia, Yemen, and Dhofar, southern Oman
ELEVATION	0–8,530 ft (0–2,600 m) asl
HABITAT	Dry rocky habitats, around agricultural areas
DIET	Small mammals, lizards, toads, birds, and invertebrates, including scorpions
REPRODUCTION	Oviparous, with clutches of 6–10 eggs
CONSERVATION STATUS	IUCN not listed

ADULT LENGTH
19¾–31½ in
(500–800 mm)

ECHIS COLORATUS
PAINTED CARPET VIPER
GÜNTHER, 1878

The Painted Carpet Viper occurs from Egypt to Israel and south along the Red Sea coast of Saudi Arabia to Yemen and the Dhofar region of southern Oman. Also known as Burton's Carpet Viper, it inhabits rocky rather than sandy terrain, and often occurs alongside humans in areas altered by irrigation or agriculture. It exhibits a catholic diet, feeding on rodents, lizards, toads, and large invertebrates, including scorpions. Observers have also noticed a particular hunting tactic that may indicate a specialization to feed on birds. The snake climbs into low vegetation overhanging small pools of water, and waits with its head down, possibly for birds coming to drink. This has also been observed in the Milos Viper (*Macrovipera schweizeri*, page 631). The Painted Carpet Viper is highly venomous, and snakebites are serious.

RELATED SPECIES

The most closely related species to *Echis coloratus* is the Oman Carpet Viper (*E. omanensis*, page 627) from Musandam (northern Oman) and UAE. The Dead Sea population of *E. coloratus*, which has larger eyes, has been proposed as a separate subspecies (*E. c. terraesanctae*).

The Painted Carpet Viper has a much more bulbous and less angular head than the Common Saw-scale Viper (*E. carinatus*, page 624). Its body color ranges from yellow to brown or light gray laterally, and darker brown dorsally, with patterning comprising a longitudinal row of dark-edged, square, or rhomboid blotches of the same light color as the flanks.

Actual size

FAMILY	Viperidae: Viperinae
RISK FACTOR	Highly venomous: procoagulants, hemorrhagins, and possibly also anticoagulants, nephrotoxins, or cytotoxins
DISTRIBUTION	West Africa: Senegal and Guinea-Bissau to Chad, and south to the coast of Togo and Benin
ELEVATION	0–3,280 ft (0–1,000 m) asl
HABITAT	Arid savanna grasslands and dry woodland
DIET	Small mammals, lizards, birds, amphibians, and invertebrates
REPRODUCTION	Oviparous, with clutches of 6–20 eggs
CONSERVATION STATUS	IUCN not listed

ADULT LENGTH
15¾–23¾ in
(400–600 mm)

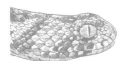

626

ECHIS OCELLATUS
WEST AFRICAN CARPET VIPER
STEMMLER, 1970

The West African Carpet Viper is a stout snake with a short, rounded head and a patterning comprising three longitudinal rows of irregular dark brown to black blotches on a light brown background, the lateral dark spots exhibiting distinctive white centers that immediately distinguish this species.

The West African Carpet Viper has been called the most dangerous snake in Africa. It occurs through an area of high human populations, including much of Nigeria, Africa's most populous country, and it probably causes more serious bites than much-feared species like mambas or cobras. The ramifications of snakebites from this species are extremely serious, death through cerebral hemorrhage being a strong possibility for untreated cases. Although an inhabitant of arid savannas, the West African Carpet Viper is also known to enter dry woodland, where it will bite people collecting firewood, especially as it may occur in very high densities in favorable habitats. This species is a fairly generalist feeder, taking a wide variety of small vertebrates and large invertebrates as prey.

RELATED SPECIES

The White-bellied Carpet Viper (*Echis leucogaster*), identified by its immaculate white belly, inhabits the arid Sahel countries, from Mauritania to Niger, north of the range of *E. ocellatus*, and it also occurs in Morocco and Western Sahara, while the tiny, white-spotted Mali Carpet Viper (*E. jogeri*) is endemic to central Mali.

Actual size

FAMILY	Viperidae: Viperinae
RISK FACTOR	Highly venomous: procoagulants, hemorrhagins, and possibly also anticoagulants, nephrotoxins, or cytotoxins
DISTRIBUTION	Arabia: northern Oman and UAE
ELEVATION	0–2,950 ft (0–900 m) asl
HABITAT	Rocky wadis and irrigated gardens
DIET	Small mammals, lizards, birds, and toads
REPRODUCTION	Oviparous, with clutches of 6–10 eggs
CONSERVATION STATUS	IUCN Least Concern

ADULT LENGTH
19¾–27½ in
(500–700 mm)

ECHIS OMANENSIS
OMAN CARPET VIPER
BABOCSAY, 2004

627

The Oman Carpet Viper is confined to the Arabian Peninsula, in eastern UAE and northern Oman, where it inhabits rocky wadis and hillsides, and also occurs in areas irrigated for gardens. It avoids sandy habitats. It reaches elevations of 2,950 ft (900 m) in the Musandam Peninsula of northern Oman. Prey items include small rodents, lizards, and possibly birds or toads, which are ambushed, often around small pools of water. As is the case with all carpet vipers, this is a highly dangerous snake, which will saw-scale a warning and strike rapidly if ignored. Snakebites may have extremely serious consequences, including death if antivenom is not administered.

The Oman Carpet Viper is a slender viper with a bulbous head. It may be light gray or tan with a narrow to broad dark dorsum, punctuated by a series of dark-edged spots, blotches, or bars of the lighter lateral ground color. It also possesses a dark postocular stripe.

RELATED SPECIES
Echis omanensis was, until recently, treated as a northern population of the Painted Carpet Viper (*E. coloratus*, page 625), to which it bears a strong resemblance. The two species are closely related, but separated by a vast sandy desert into which neither ventures. The Sindh Saw-scale Viper (*E. carinatus sochureki*, page 624) occurs on gravel plain habitats close to the rocky wadis of the Oman Carpet Viper, but the two are unlikely to be confused.

Actual size

FAMILY	Viperidae: Viperinae
RISK FACTOR	Highly venomous: procoagulants, hemorrhagins, and possibly also anticoagulants, nephrotoxins, or cytotoxins
DISTRIBUTION	North and northeast Africa, and Arabia: Algeria to Egypt, south to Kenya, and western Saudi Arabia to Yemen
ELEVATION	0–5,580 ft (0–1,700 m) asl
HABITAT	Arid savanna, desert, and semidesert
DIET	Small mammals, lizards, and large invertebrates
REPRODUCTION	Oviparous, with clutches of 4–6 eggs
CONSERVATION STATUS	IUCN Least Concern

ADULT LENGTH
19¾–23¾ in
(500–600 mm)

ECHIS PYRAMIDUM
NORTHEAST AFRICAN CARPET VIPER
(GEOFFROY SAINT-HILAIRE, 1827)

The Northeast African Carpet Viper is usually gray to gray-brown, darker on the dorsum, with a longitudinal pattern of white spots with black interspaces on the back, and white inverted chevrons over black spots on the flanks.

The Northeast African Carpet Viper has an extremely punctuated distribution. The main range includes both southern coasts of the Red Sea, from Egypt, into Sudan, and south to the Horn of Africa, Somalia, the Saudi Arabian coast, and the coast of Yemen. Isolated populations are found in Egypt, Ethiopia, Kenya, Algeria, and Libya. Different populations take different prey, from mice to lizards, or even scorpions, solifugids, and centipedes. Population densities can be very high; for example, 7,000 specimens were collected in one 2,510 sq mile (6,500 sq km) Kenyan region in four months. The toothless, harmless egg-eating snakes (*Dasypeltis*, page 157) obtain protection by mimicking the patterning and defensive behavior of these dangerous saw-scaling vipers.

RELATED SPECIES

Several subspecies may be recognized, from northeast Kenya (*Echis pyramidum aliaborri*), northwest Kenya and Ethiopia (*E. p. leakeyi*), and Egypt to Algeria (*E. p. lucidus*), with the large Red Sea populations allocated to the nominate form (*E. p. pyramidum*). Other species in the same area include the Somali Carpet Viper (*E. hughesi*), the Dhofar Carpet Viper (*E. khosatzkii*), and the Big-headed Carpet Viper (*E. megalocephalus*), from Nokra Island, in the Red Sea. The White-bellied Carpet Viper (*E. leucogaster*) is also a member of this species group.

Actual size

FAMILY	Viperidae: Viperinae
RISK FACTOR	Highly venomous: procoagulants, and possibly neurotoxins
DISTRIBUTION	Western Asia: Pakistan, Afghanistan, and Iran
ELEVATION	2,490–4,300 ft (760–1,310 m) asl
HABITAT	Sandy desert, especially sand dunes
DIET	Small mammals, lizards, and birds
REPRODUCTION	Oviparous, with clutches of up to 12 eggs
CONSERVATION STATUS	IUCN not listed

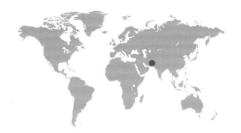

ADULT LENGTH
23¾–27½ in
(600–700 mm)

ERISTICOPHIS MACMAHONI
MACMAHON'S VIPER
ALCOCK & FINN, 1897

629

The "Asian sidewinder," MacMahon's Viper is a desert-adapted species that sidewinds like the Sidewinder Rattlesnake (*Crotalus cerastes*, page 575), sand and horned vipers (*Cerastes*, pages 620–621), and Namib Sidewinding Adder (*Bitis peringueyi*, page 616). A rare species, from the Balochistan deserts of the Pakistan–Afghanistan–Iran border, it inhabits sandy dunes. It ambushes rodents, lizards, and small birds. Although a terrestrial species, it has a prehensile tail and climbs low shrubs. Its long head accommodates large, valvular nostrils, and an arrangement of enlarged nasal scales, which prevent ingress of blown sand. The small, tight-fitting scales of the lips also prevent the entry of sand. Bites from this species are rare but its venom is considered as toxic as that of the carpet vipers (*Echis*, pages 624–628). Human fatalities are known.

MacMahon's Viper is light sandy brown with small white and black spots scattered along its dorsolateral surface and irregular dark brown spots on the dorsum. Also present is a white and brown postocular stripe. The enlarged nostrils and nasal scales are prominent on the head.

RELATED SPECIES

The closest relatives of MacMahon's Viper are believed to the false horned vipers (*Pseudocerastes*, pages 636–637), which also have some of the same desert adaptations to avoid ingress of blown sand.

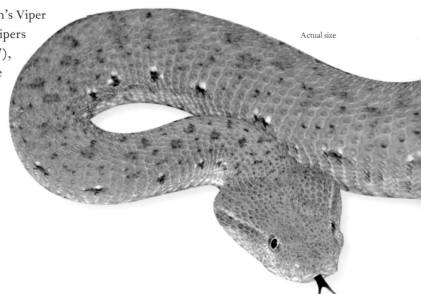

Actual size

FAMILY	Viperidae: Viperinae
RISK FACTOR	Highly venomous: procoagulants, and possibly also hemorrhagins, myotoxins, or cytotoxins
DISTRIBUTION	Middle East, Central Asia, and North Africa: Cyprus, Israel, and Turkey to India and Tajikistan; also Tunisia
ELEVATION	0–8,200 ft (0–2,500 m) asl
HABITAT	Vegetated rocky montane hillsides and valleys
DIET	Small mammals, lizards, and birds and their eggs
REPRODUCTION	Oviparous or viviparous, with up to 35 eggs or neonates
CONSERVATION STATUS	IUCN not listed

ADULT LENGTH
5 ft–6 ft 7 in
(1.5–2.0 m)

630

MACROVIPERA LEBETINA
LEVANT VIPER
(LINNAEUS, 1758)

The Levant Viper is a stocky snake with a large, blunt head. Its coloration tends toward being unicolor, either sandy tan or gray, with patterning limited to an alternating checkerboard of darker square spots on the dorsum.

Also known as the Blunt-nosed Viper, the Levant Viper is a large and dangerous snake. It prefers vegetated rocky hills and valleys, but is also found in human-altered habitats, which brings it into contact with humans and domesticated herds. It inflects serious snakebites across its extensive range; even camels have been killed by this snake. Prey ranges from rodents to lizards, birds, and birds' eggs, and different subspecies or populations vary in whether they are live-bearing or egg-laying. Although generally nocturnal or crepuscular, the Levant Viper may be active by day in overcast weather, and despite being a heavy-bodied terrestrial species, it is also able to climb into low bushes. The Cypriot and Caucasian populations represent the only European populations of this species.

RELATED SPECIES

Most vipers that were considered closely related to *Macrovipera lebetina* were moved to genus *Daboia*, leaving only its former subspecies, the Milos Viper (*Macrovipera schweizeri*, page 631). *Macrovipera lebetina* contains at least five subspecies, the nominate subspecies being confined to Cyprus. The other subspecies occur from Israel to Afghanistan (*M. l. obtusa*); northern Afghanistan to Kashmir, India (*M. l. cernovi*); in Uzbekistan and Central Asia (*M. l. turanica*); and in Algeria (*M. l. transmediterranea*).

Actual size

FAMILY	Viperidae: Viperinae
RISK FACTOR	Highly venomous: procoagulants, and possibly also hemorrhagins, myotoxins, or cytotoxins
DISTRIBUTION	Europe: Greece (Cyclades)
ELEVATION	0–2,455 ft (0–748 m) asl
HABITAT	Vegetated rocky hills, rocky meadows, and dry stone walls
DIET	Birds and lizards
REPRODUCTION	Oviparous, with clutches of up to 10 eggs
CONSERVATION STATUS	IUCN Endangered

ADULT LENGTH
19¾–27½ in,
rarely 3 ft 3 in
(500–700 mm,
rarely 1.0 m)

MACROVIPERA SCHWEIZERI
MILOS VIPER
(WERNER, 1935)

631

The Milos Viper, or Cyclades Viper, is endemic to four small islands in the western Cyclades (Kimolos, Polyaigos, Sifnos, and Milos), of which Milos, at 62 sq miles (160 sq km), is the largest. Its habitats include rocky meadows and dry stone walls. The Milos Viper is diurnal in winter and nocturnal in summer, and is most active when migratory passerine birds are arriving. It climbs into low bushes around standing water and ambushes thirsty birds coming to drink. Juveniles prey on lizards. Female Milos Vipers produce up to ten eggs, but take two years to get back into breeding condition, due to the seasonality of prey. This is Europe's rarest viper, being threatened by mining, fire, roads, and poaching for the reptile trade. Hans Schweizer (1891–1975) was a herpetologist who specialized in vipers.

RELATED SPECIES
The only close relative of *Macrovipera schweizeri* is the Levant Viper (*M. lebetina*, page 630), of which it was once a subspecies, the nearest population being on Cyprus. The Sifnos population is sometimes treated as a subspecies (*M. schweizeri siphnensis*).

The Milos Viper is similar in appearance to the Levant Viper (*Macrovipera lebetina*, page 630), albeit considerably smaller. It may be orange, reddish, tan, or gray with a pattern of obscure, darker square markings on the back. It is stout-bodied with a blunt head.

Actual size

FAMILY	Viperidae: Viperinae
RISK FACTOR	Venomous: procoagulants and hemorrhagins
DISTRIBUTION	East Africa: Kenya (Mt. Kenya and Aberdare Range)
ELEVATION	8,860–12,500 ft (2,700–3,800 m) asl
HABITAT	Open montane moorland
DIET	Lizards, frogs, and small mammals
REPRODUCTION	Viviparous, with litters of 1–3 neonates
CONSERVATION STATUS	IUCN not listed

ADULT LENGTH
11¾–13¾ in
(300–350 mm)

MONTATHERIS HINDII
KENYA MOUNTAIN VIPER
(BOULENGER, 1910)

One of the smallest vipers in the world, and one with a very limited range, the Kenya Mountain Viper is only found in two localities, on Mt. Kenya and the nearby Aberdare Range, between 8,860 and 12,500 ft (2,700 and 3,800 m) asl. Living above the tree-line, this tiny viper inhabits windswept moorland, sheltering from the cold, rain, and birds of prey in tussock grass clumps, only emerging in sunlight to bask. Its prey ranges from skinks and chameleons to frogs and mice. Females produce up to three neonates. The entire range of this potentially threatened snake may be less than a few football pitches, yet it has no international protection. Its range does lie within two national parks, which affords some local protection, but no Kenya Mountain Vipers have been seen for several years.

The Kenya Mountain Viper is brown with a double series of dark brown or black triangle markings along its back, arranged with the points touching or almost touching middorsally, and a series of similar but less distinct markings along the flanks. A fine pair of yellowish stripes runs the length of the body, separating the dorsal from the lateral markings.

RELATED SPECIES
Montatheris hindii was previously included in the bushviper genus *Atheris*, which comprises arboreal species. Other terrestrial species removed from *Atheris* include the Floodplain Viper (*Proatheris superciliaris*, page 635).

Actual size

FAMILY	Viperidae: Viperinae
RISK FACTOR	Highly venomous: procoagulants, hemorrhagins, and possibly neurotoxins
DISTRIBUTION	Western Asia: Turkey, Armenia, Azerbaijan, and Iran
ELEVATION	3,610–8,860 ft (1,100–2,700 m) asl
HABITAT	North-facing mountain slopes, juniper forest, and steppe
DIET	Small mammals, lizards, birds, and insects
REPRODUCTION	Viviparous, with litters of 3–11 neonates
CONSERVATION STATUS	IUCN Near Threatened

MONTIVIPERA RADDEI

ARMENIAN MOUNTAIN VIPER

(BOETTGER, 1890)

ADULT LENGTH
Male
2 ft 7 in–3 ft 3 in
(0.8–1.0 m)

Female
2 ft–2 ft 4 in
(0.6–0.7 m)

633

The Armenian Mountain Viper is endemic to the southern Caucasus, in northeastern Turkey, northwestern Iran, Armenia, and Azerbaijan, where it inhabits rocky, wooded slopes, meadows, montane steppe, juniper forests, and cultivation. Diurnal or crepuscular during the spring and summer, it hibernates through the extreme winter. It is reported that both juveniles and adults feed on insects in the spring, bulking up on rodents and lizards during the summer months. Birds are only occasionally taken. The female Armenian Mountain Viper is much smaller than the male, in contrast to some other vipers. The limited range of this viper is affected by land-use change, and specimens are poached for the reptile trade. It is highly venomous and capable of delivering a fatal snakebite.

The Armenian Mountain Viper is light or dark gray with a dorsal pattern comprising a longitudinal series of large orange spots, which may have black edges and may also be fused to form an irregular zigzag pattern. A dark postocular stripe is present behind the eye.

RELATED SPECIES

The most closely related species to *Montivipera raddei* are believed to be the other Iranian species, the Iranian Mountain Viper (*M. albicornuta*) and Latifi's Viper (*M. latifii*). Two subspecies are recognized, the northern *M. r. raddei* and the southern *M. r. kurdistanica*. The genus *Montivipera* contains nine species.

Actual size

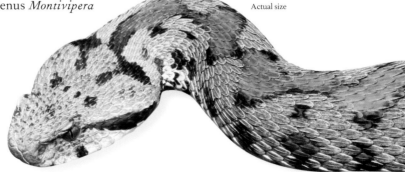

FAMILY	Viperidae: Viperinae
RISK FACTOR	Highly venomous: probably procoagulants and hemorrhagins, and possibly also neurotoxins
DISTRIBUTION	Western Asia and southeast Europe: Turkey and Greece (Thrace)
ELEVATION	0–8,860 ft (0–2,700 m) asl
HABITAT	Rocky and well-vegetated hillsides and valleys
DIET	Small mammals, lizards, birds, and invertebrates
REPRODUCTION	Viviparous, with litters of 2–10 neonates
CONSERVATION STATUS	IUCN not listed

ADULT LENGTH
Male
2 ft 7 in–3 ft 3 in
(0.8–1.0 m)

Female
3 ft–4 ft 3 in
(0.9–1.3 m)

634

MONTIVIPERA XANTHINA
OTTOMAN VIPER
(GRAY, 1849)

The Ottoman Viper resembles a large "adder," being pale brown or light gray with a bold dorsal pattern of darker brown or gray blotches, which often fuse to form a zigzag pattern. The head bears two angular spots on the nape and a postocular stripe behind the eye.

The Ottoman Viper is the only member of the mountain viper genus *Montivipera* to occur in Europe, being distributed from Turkey, across the Bosphorus into Thracian Greece, and also on many of the islands of the Aegean. It is the largest venomous snake in Europe, with female specimens frequently exceeding 3 ft 3 in (1 m) in length. Preferred habitats include rocky wooded slopes and meadows, especially those close to watercourses with dense vegetation for cover. In some localities population densities are considerable, even in areas altered by humans. Adults prey on rodents, lizards, and birds, while juveniles take lizards and also feed on locusts and centipedes. Large specimens may easily deliver a fatal snakebite.

RELATED SPECIES

Montivipera xanthina is the largest member of the genus *Montivipera*, but eight other species are described, distributed across western Asia. The Bulgardagh Viper (*M. bulgardaghica*) and Turkish Mountain Viper (*M. albizona*) inhabit Turkey; Wagner's Viper (*M. wagneri*) straddles the Turkey–Iran border; the Lebanon Viper (*M. bornmuelleri*) occurs in the Lebanon, Israel, and Syria; the Armenian Mountain Viper (*M. raddei*, page 633) is found in the Caucasus; and the Iranian Mountain Viper (*M. albicornuta*), Kuhrang Mountain Viper (*M. kuhrangica*), and Latifi's Viper (*M. latifii*) occur in Iran.

Actual size

FAMILY	Viperidae: Viperinae
RISK FACTOR	Highly venomous: procoagulants, and possibly hemorrhagins
DISTRIBUTION	East Africa: southern Tanzania, southern Malawi, and northern Mozambique
ELEVATION	0–2,620 ft (0–800 m) asl
HABITAT	Low-lying floodplains, grasslands, and coastal lowlands
DIET	Frogs and toads, and small mammals on occasion
REPRODUCTION	Viviparous, with litters of 3–16 neonates
CONSERVATION STATUS	IUCN not listed

ADULT LENGTH

Male
19¾–21¾ in
(500–550 mm)

Female
19¾–23¾ in
(500–600 mm)

635

PROATHERIS SUPERCILIARIS
FLOODPLAIN VIPER
(PETERS, 1855)

Also known as the Lowland Swamp Viper, this viper has a punctuated distribution, occurring in floodplains and low-lying grasslands at either end of Lake Malawi, in Tanzania and Malawi, and along the Mozambique coast. The Floodplain Viper is a crepuscular species that hunts frogs and toads, but occasionally also rodents. If disturbed it adopts a defensive posture not dissimilar to that of carpet vipers (*Echis*, pages 624–628), rubbing its scales together and hissing loudly, followed by a rapid strike if the warning is unheeded. Snakebites are reported to be very painful and, although no deaths are reported, this species is thought capable of delivering a fatal snakebite. No antivenom exists to treat its bite.

The Floodplain Viper is stout-bodied, with a rounded head and strongly keeled scales. It is blue-gray, with patterning comprising a series of black oval blotches down the back, which may coalesce into a zigzag, and a corresponding series of dark blotches on the flanks, separated from the dorsal markings by short white flashes. Smaller yellow-brown blotches may be present in the interspaces. The head is boldly marked with alternating black and white oblique chevrons.

RELATED SPECIES
Proatheris superciliaris was formerly included in the arboreal bushviper genus *Atheris* (pages 605–609). Another terrestrial viper removed from genus *Atheris* is the Kenya Mountain Viper (*Montatheris hindii*, page 632).

Actual size

FAMILY	Viperidae: Viperinae
RISK FACTOR	Highly venomous: probably neurotoxins, otherwise unknown
DISTRIBUTION	Western Asia and Arabia: Iran, Afghanistan, Pakistan, Turkey, Oman, and UAE
ELEVATION	1,100–7,220 ft (335–2,200 m) asl
HABITAT	Rocky desert and semidesert
DIET	Small mammals, lizards, birds, and invertebrates
REPRODUCTION	Oviparous, with clutches of 10–21 eggs
CONSERVATION STATUS	IUCN not listed

ADULT LENGTH
2 ft 4 in–3 ft 3 in
(0.7–1.0 m)

PSEUDOCERASTES PERSICUS
PERSIAN FALSE HORNED VIPER
(DUMÉRIL, BIBRON & DUMÉRIL, 1854)

The Persian False Horned Viper is pale gray-brown with scattered and alternating brown spots along its back. The "horns" over the eyes comprise composite clumps of small scales rather than actual horns.

False horned vipers do not possess single-scale horns, like horned vipers (*Cerastes*, pages 620–621), but rather a clump of elevated scales that together form a composite elevated ridge over the eye. The Persian False Horned Viper is distributed across Iran, Afghanistan, and Pakistan, with an isolated population in the Zagros Mountains, on the Turkish–Iranian border, and another on the rugged Musandam Peninsula of Oman and UAE, where it is especially rare. Whether either of these isolated populations are distinct and separate species has yet to be determined. The habitat of this viper comprises rocky desert and semidesert, and it is active at night. Prey includes rodents, lizards, including geckos and agamids, occasionally small birds, and large invertebrates. Snakebites from this species may be serious.

RELATED SPECIES

The two other members of genus *Pseudocerastes* are Field's False Horned Viper (*P. fieldi*), in the northern Arabian Peninsula, and the Iranian Spider-tailed Viper (*P. urarachnoides*, page 637), in western Iran. The closest species to *Pseudocerastes* is the monotypic MacMahon's Viper (*Eristicophis macmahoni*, page 629) from the Afghanistan–Pakistan–Iran border.

Actual size

FAMILY	Viperidae: Viperinae
RISK FACTOR	Venomous: venom composition unknown
DISTRIBUTION	Western Asia: western Iran
ELEVATION	655–985 ft (200–300 m) asl
HABITAT	Rocky and well-vegetated hillsides and valleys
DIET	Birds, and possibly small mammals
REPRODUCTION	Oviparous, clutch size unknown
CONSERVATION STATUS	IUCN not listed

ADULT LENGTH
23¾–31½ in
(600–800 mm)

PSEUDOCERASTES URARACHNOIDES
IRANIAN SPIDER-TAILED VIPER
BOSTANCHI, ANDERSON, KAMI & PAPENFUSS, 2006

637

Known only from a few specimens from the Zagros Mountains, western Iran, this is one of the strangest of snakes, and certainly one of the most startling of recent herpetological discoveries. The Iranian Spider-tailed Viper possesses a tail that is shaped like a spider, the name *urarachnoides* coming from *ura*, meaning "tail," *arachno*, meaning "spider," and *oides*, meaning "like." The tail bears a number of projections that resemble spiders' legs, protruding from a bulbous tip, shaped like a spider's body, which is used as a lure to attract prey within strike range. Although a bird specialist, this viper may also take shrews. The scales of the head and body are much more strongly keeled than those of its congeners, and this may serve to cryptically break up the outline of this species.

RELATED SPECIES
Pseudocerastes urarachnoides is related to the Persian False Horned Viper (*P. persicus*, page 636) and Field's False Horned Viper (*P. fieldi*), while the genus *Pseudocerastes* is related to genus *Eristicophis*.

The Iranian Spider-tailed Viper is pale gray with a series of gray-brown blotches across its back. The scales of the body and head are extremely rugose and serve to break up the snake's outline. Its most striking characteristic is the curious spider-like tail tip.

Actual size

FAMILY	Viperidae: Viperinae
RISK FACTOR	Highly venomous: presynaptic neurotoxins, and probably also procoagulants and hemorrhagins
DISTRIBUTION	Southeast Europe: Greece, the Balkans, and European Turkey, to northern Italy and Austria
ELEVATION	0–8,200 ft (0–2,500 m) asl
HABITAT	Open rocky hillsides, scree slopes, dry stone walls, and vineyards
DIET	Small mammals, birds, and lizards
REPRODUCTION	Viviparous, with litters of 3–18 neonates
CONSERVATION STATUS	IUCN Least Concern

ADULT LENGTH
Male
31½–35½ in
(800–900 mm)

Female
23¾–27½ in
(600–700 mm)

VIPERA AMMODYTES
NOSE-HORNED VIPER
(LINNAEUS, 1758)

The Nose-horned Viper is sexually dichromatic, males being gray with black or dark gray zigzag markings while females are pale tan to brown with a dark brown zigzag. A V-shaped marking on the back of the head may be bold or indistinct. In some specimens the zigzag is broad and resembles a longitudinal wavy stripe.

Also known as the European Sand Viper, and distributed through southeastern Europe from Austria to Turkey, this species is easily recognized by the fleshy, hornlike protuberance on its snout, which faces forward in the nominate form but is erect in the other two subspecies. This characteristic is only seen in one other European viper, Lataste's Viper (*Vipera latastei*, page 642) of the Iberian Peninsula. Its habitats range from open rocky hillsides to cultivated areas like vineyards and dry stone walls, although in the summer the viper may move into woodland. Prey consists of mice, ground-nesting birds, and lizards. Snakebites to humans are extremely serious; this is the most dangerous European viper, with fatalities on record.

RELATED SPECIES

Vipera ammodytes belongs to the *Vipera* subgenus, along with the Asp Viper (*V. aspis*, page 639), Lataste's Viper (*V. latastei*, page 642), and Dwarf Atlas Mountain Viper (*V. monticola*, page 643). The Transcaucasian Sand Viper (*V. transcaucasiana*), of northeastern Turkey, is a close relative and former subspecies. Three subspecies are recognized, from the Balkans to Austria and Italy (*V. a. ammodytes*); in Bulgaria and Romania (*V. a. meridionalis*); and in Greece, Albania, and European Turkey (*V. a. montandoni*).

Actual size

FAMILY	Viperidae: Viperinae
RISK FACTOR	Venomous: probably procoagulants, hemorrhagins, and possibly neurotoxins
DISTRIBUTION	Western Europe: France, Italy, Switzerland, and northeastern Spain
ELEVATION	0–9,610 ft (0–2,930 m) asl
HABITAT	Woodland, upland meadows, gardens, and coastal habitats
DIET	Small mammals, lizards, and birds
REPRODUCTION	Viviparous, with litters of 2–12 neonates
CONSERVATION STATUS	IUCN Least Concern

VIPERA ASPIS
ASP VIPER
(LINNAEUS, 1758)

ADULT LENGTH
Male
27½–33½ in
(700–850 mm)

Female
23¾–27½ in
(600–700 mm)

639

Asp Vipers are relatively small, blunt-headed vipers with a slightly upturned snout. This species is found in a wide variety of habitats, from woodland to open meadows, to altitudes of almost 9,840 ft (3,000 m) in the Alps and Pyrenees. It feeds on rodents, lizards, and occasional small birds. Unlike the Northern Adder (*Vipera berus*, page 640), the Asp Viper hibernates singly and does not use communal hibernacula (overwintering dens). The range of the Asp Viper overlaps that of the Northern Adder, Seoane's Viper (*V. seoanei*, page 644), and Lataste's Viper (*V. latastei*, page 642), but the species are ecologically allopatric (ocurring in separate habitats). Although the Asp Viper is small, serious snakebites have occurred.

RELATED SPECIES

Vipera aspis is part of the *Vipera subgenus*, with the Nose-horned Viper (*V. ammodytes*, page 638), Lataste's Viper (*V. latastei*, page 642), and Dwarf Atlas Mountain Viper (*V. monticola*, page 643). Five subspecies are recognized, from central and southern France (*V. a. aspis*); the Swiss Alps (*V. a. atra*); northern Italy (*V. a. francisciredi*); southern Italy and Sicily (*V. a. hugyi*); and southwestern France and the Pyrenees (*V. a. ʒinnikeri*). The harmless Viperine Watersnake (*Natrix maura*, page 416) resembles *V. aspis*.

The Asp Viper demonstrates sexual dichromatism. Males are gray with dark gray to black markings, and females brown with dark brown markings. Females of some montane populations (*V. a. aspis*) are melanistic. Many specimens exhibit a bold zigzag, but in some this pattern may be reduced to narrow cross-bars. A V-shaped head marking may be present or absent.

Actual size

FAMILY	Viperidae: Viperinae
RISK FACTOR	Venomous: probably procoagulants and hemorrhagins, and possibly neurotoxins
DISTRIBUTION	Europe, and Central and East Asia: Scotland to Scandinavia, and east to Sakhalin Island (Russia)
ELEVATION	0–9,840 ft (0–3,000 m) asl
HABITAT	Heathland, woodland, cultivated land, and coastal cliffs
DIET	Small mammals, lizards, and occasionally birds
REPRODUCTION	Viviparous, with litters of 3–20 neonates
CONSERVATION STATUS	IUCN not listed

ADULT LENGTH
Male
19¾–23¾ in
(500–600 mm)

Female
19¾–31½ in,
(500–800 mm),
occasionally 3 ft 3 in (1.0 m)
in Scandinavia

VIPERA BERUS
NORTHERN ADDER
(LINNAEUS, 1758)

The Northern Adder is the most widely distributed terrestrial snake in the world. In the British Isles it occurs through England, Wales, and Scotland, and on some Scottish islands, but not the Isle of Man or Ireland. In Europe it occurs from northern France to Scandinavia and the Kola Peninsula of Siberia, 125 miles (200 km) north of the Arctic Circle, and eastward through Kazakhstan and Mongolia, to far eastern Russia and Sakhalin Island. Scotland to Sakhalin is 4,794 miles (7,715 km). Its habitats vary greatly, from heathland to woodland and cultivated land. Adults take voles and mice, while juveniles feed on lacertid lizards and slow worms. Birds are also occasional prey. Snakebites are common but rarely fatal, only 12 deaths occurring in the UK during the entire twentieth century, the last in 1975.

The Northern Adder is a sexually dichromatic species, males being light gray with a black zigzag, and females brown with a dark brown zigzag, both sexes bearing a V-shaped marking on the head. Some females, especially from high latitudes or altitudes, are melanistic and patternless; being black enables them to warm quickly when basking.

RELATED SPECIES

Vipera berus belongs to the *Pelias* subgenus, with Seoane's Viper (*V. seoanei*, page 644) and the Caucasian Viper (*V. kaznakovi*, page 641). Most of this species' range is occupied by the nominate subspecies. Other subspecies occur in the Balkans (*V. b. bosniensis*), and far-eastern Russia and Sakhalin Island (*V. b. sachalinensis*). *Vipera berus* may be confused with the harmless Viperine Watersnake (*Natrix maura*, page 416).

Actual size

FAMILY	Viperidae: Viperinae
RISK FACTOR	Venomous: venom composition unknown
DISTRIBUTION	Eastern Europe and Asia Minor: Georgia, Armenia, and northeast Turkey
ELEVATION	1,440–7,870 ft (440–2,400 m) asl
HABITAT	Woodlands, wooded slopes, and meadows near water
DIET	Small mammals, lizards, and invertebrates
REPRODUCTION	Viviparous, with litters of 3–5 neonates
CONSERVATION STATUS	IUCN Vulnerable

ADULT LENGTH
19¾–27½ in
(500–700 mm)

VIPERA KAZNAKOVI
CAUCASIAN VIPER
NIKOLSKY, 1909

641

A group of closely related viper species with localized upland ranges, the *kaznakovi* group, is found in the Caucasus, between the Black and Caspian Seas. The Caucasian Viper occurs around the Black Sea's eastern coast, in Georgia and Armenia, and possibly as far south as northeastern Turkey, where it inhabits wooded slopes and meadows. Adults prey on mice and lizards, while juveniles take invertebrates. The range of the Caucasian Viper is small, and threats to this species include overcollection for the reptile trade, and development for tourism and agriculture. Other members of the *kaznakovi* group are listed as Endangered or Critically Endangered by the IUCN. Aleksandr Kaznakov was a Russian naturalist, and director of the Caucasus Museum.

The Caucasian Viper can be a stunning snake. The adult ground color is orange, red, or pink, with jet-black markings on the head and body that form a broad zigzag or which may almost totally obscure the background color, leaving the orange as spots that contrast with the black surrounds. In juveniles the patterning may be more obvious but under a red wash.

RELATED SPECIES

Vipera kaznakovi is a member of the *Pelias* subgenus with the Northern Adder (*V. berus*, page 640) of Eurasia, and Seoane's Viper (*V. seoanei*, page 644) of Iberia. Its other relatives are several other small vipers with isolated ranges in the Caucasus: Darevsky's Viper (*V. darevskii*), Dinnik's Viper (*V. dinniki*), the Western Caucasian Viper (*V. magnifica*), Orlov's Viper (*V. orlovi*), and the Black Sea Viper (*V. pontica*), and Nikolsky's Adder (*V. nikolskii*) in Ukraine.

Actual size

FAMILY	Viperidae: Viperinae
RISK FACTOR	Venomous: possibly procoagulants, hemorrhagins, or neurotoxins
DISTRIBUTION	Southwest Europe and northwest Africa: Spain, Portugal, Morocco, Algeria, and possibly Tunisia
ELEVATION	0–9,120 ft (0–2,780 m) asl
HABITAT	Arid rocky, vegetated hillsides, scree slopes, cork oak woodland, hedgerows, dry stone walls, and coastal dunes
DIET	Small mammals, lizards, and birds
REPRODUCTION	Viviparous, with litters of 3–15 neonates
CONSERVATION STATUS	IUCN Vulnerable

ADULT LENGTH
19¾–27½ in
(500–700 mm)

VIPERA LATASTEI
LATASTE'S VIPER
BOSCA, 1878

Lataste's Viper may be gray, brown, or reddish with a bold, black-edged zigzag pattern down the dorsum of the back, and alternating spots on the flanks, although in some specimens the markings may be indistinct. The nasal projection aids in the identification of this species.

Actual size

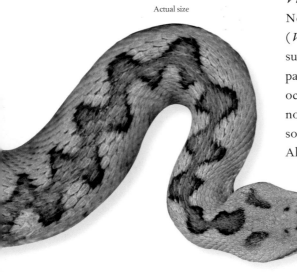

Lataste's Viper, also sometimes called the Snub-nosed Viper, is a small snake with a fleshy nasal projection similar to that seen in the Nose-horned Viper (*Vipera ammodytes*, page 638) of southeastern Europe. It inhabits rocky, wooded slopes and cork oak woodland in the Iberian Peninsula and northwest Africa, and is threatened in parts of its range by habitat alteration, such as the planting of conifers, as well as active persecution, collection, and males being killed on roads. With a preference for rocky habitats it also inhabits dry stone walls, and has been found in coastal sand dunes. Snakebites from this species may be serious. Vital-Fernand Lataste (1847–1934) was a French herpetologist.

RELATED SPECIES

Vipera latastei is a member of the *Vipera* subgenus with the Nose-horned Viper (*V. ammodytes*, page 638), Asp Viper (*V. aspis*, page 639), and its closest relative and former subspecies, the Dwarf Atlas Mountain Viper (*V. monticola*, page 643). Two subspecies are recognized, the nominate form occurring through most of Spain and Portugal except the north, and a southern subspecies (*V. l. gaditana*) found in southwest Iberia, and in the Atlas Mountains of Morocco, Algeria, and possibly western Tunisia.

FAMILY	Viperidae: Viperinae
RISK FACTOR	Venomous: venom composition unknown
DISTRIBUTION	Northwest Africa: High Atlas of Morocco
ELEVATION	3,940–4,270 ft (1,200–1,300 m), possibly 13,100 ft (4,000 m) asl
HABITAT	Alpine thorn-cushion habitats
DIET	Lizards and insects
REPRODUCTION	Viviparous, with litters of 2–3 neonates
CONSERVATION STATUS	IUCN Near Threatened

ADULT LENGTH
11¾–13¾ in
(300–350 mm)

VIPERA MONTICOLA

DWARF ATLAS MOUNTAIN VIPER

SAINT GIRONS, 1953

643

The Dwarf Atlas Mountain Viper is one of the smallest of vipers, only achieving up to 13¾ in (350 mm) in total length. It inhabits the High Atlas Mountains of Morocco at elevations over 3,940 ft (1,200 m), occurring in dense thorn-cushion and tussock grass growths, and feeding on lizards and insects. It may even occur as high as 13,100 ft (4,000 m). Activity is diurnal and winter hibernation essential. The entire range of the Dwarf Atlas Mountain Viper is less than 7,720 sq miles (20,000 sq km), and it is threatened by the clearance and removal of woodland for firewood. Population recovery is slow due to the low litter size of two to three neonates, and an irregular breeding cycle resulting from the inhospitable high-elevation climate. Snakebites are unknown for this species.

The Dwarf Atlas Mountain Viper is a gray to gray-brown snake with the typical viper marking of a black zigzag, although this marking may be narrow and appear more as a series of cross-bars linked by a longitudinal stripe. A former subspecies of Lataste's Viper (*Vipera latastei*, page 642), this species has a small fleshy horn on the snout tip.

RELATED SPECIES

Vipera monticola is a member of the *Vipera* subgenus and is most closely related to Lataste's Viper (*V. latastei*, page 642), of which it was once a subspecies. It is especially closely related to the southern subspecies (*V. l. gaditana*), which also occurs in the Atlas Mountains.

Actual size

FAMILY	Viperidae: Viperinae
RISK FACTOR	Venomous: possibly procoagulants, hemorrhagins, or neurotoxins
DISTRIBUTION	Southwestern Europe: northern Spain and Portugal
ELEVATION	0–6,230 ft (0–1,900 m) asl
HABITAT	Woodland, rocky hillsides with cover, and agricultural areas in walls or hedgerows
DIET	Small mammals, lizards, and birds
REPRODUCTION	Viviparous, with litters of 2–10 neonates
CONSERVATION STATUS	IUCN Least Concern

ADULT LENGTH
19¾–23¾ in
(500–600 mm)

VIPERA SEOANEI
SEOANE'S VIPER
LATASTE, 1879

Seoane's Viper is not a snake of open or rocky terrain like its congeners, preferring habitats with dense protective cover such as woodland or hedgerows or walls around meadows, especially at lower elevations. Snakes at higher elevations are found in scrubby, rocky alpine habitats. The primary prey taken by adult vipers are small mammals such as mice or voles, although ground-nesting birds or lizards are also taken. Juveniles prey more on lizards, including slow worms. Melanistic specimens comprise a relatively large portion of its populations, especially among females, where being black enables a gravid female to warm up when basking more quickly than a normal zigzag-patterned female. Viktor Seoane (1832–1900) was a Galician amateur naturalist.

Seoane's Viper usually resembles a small Northern Adder (*Vipera berus*, page 640), with its brown or gray coloration and darker zigzag markings. Fully black melanistic specimens are fairly common and many individuals exhibit a broad dorsal stripe rather than a zigzag.

RELATED SPECIES

Vipera seoanei belongs to the *Pelias* subgenus, with the Northern Adder (*V. berus*, page 640), of which it was originally a subspecies, and the Caucasian Viper (*V. kaznakovi*, page 641). Two subspecies are recognized, the nominate form occupying the northern and northwestern coast of Spain and Portugal, and another subspecies in the Cantabrian Mountains of Spain (*V. s. cantabrica*).

Actual size

FAMILY	Viperidae: Viperinae
RISK FACTOR	Venomous: venom composition unknown
DISTRIBUTION	Europe: France, Italy, the Balkans, Greece, Hungary, Romania, Bulgaria, and Moldova
ELEVATION	820–9,840 ft (250–3,000 m) asl
HABITAT	Grassland and meadows to juniper woodland
DIET	Insects, lizards, and small mammals
REPRODUCTION	Viviparous, with litters of 3–8 neonates
CONSERVATION STATUS	IUCN Vulnerable, CITES Appendix I

ADULT LENGTH
19¾–23¾ in
(500–600 mm)

VIPERA URSINII

ORSINI'S MEADOW VIPER

(BONAPARTE, 1835)

645

Orsini's Meadow Viper was named for the Italian naturalist Antonio Orsini (1788–1870), by Charles Bonaparte (1803–57), the naturalist nephew of Napoleon Bonaparte. Europe's smallest viper, it has a fragmented distribution from southern France to the Black Sea coast of Moldova, in a wide array of different habitats, from lowland grasslands to upland juniper woodland and alpine meadows. It is also known as the Karst Viper, due to its presence on limestone hillsides. It feeds on locusts and other large insects, and occasionally small vertebrates. The IUCN list it as Vulnerable, but the Austrian population is already believed extinct, while the Hungarian populations are small and extremely vulnerable. Threats include overgrazing by sheep, and habitat degradation by pigs. Some populations are so small that zoos have stepped in with captive breeding programs.

Orsini's Meadow Viper is a petite species that may be variably patterned across its range and across subspecies, from unicolor to a typical adder zigzag pattern on a gray or brown background.

RELATED SPECIES

Vipera ursinii is in the subgenus *Pelias*, and four subspecies are recognized, from southern France and central Italy (*V. u. ursinii*); Croatia to Macedonia (*V. u. macrops*); Hungary and Romania (*V. u. rakosinensis*); and Romania, Bulgaria, and Moldova (*V. u. moldavica*). The Greek Meadow Viper, a former subspecies, has been elevated to specific status as *V. gracea*. Close relatives are the Steppe Viper (*V. renardi*) from Romania and Ukraine to China, and Lotiev's Viper (*V. lotievi*) from the Caucasus.

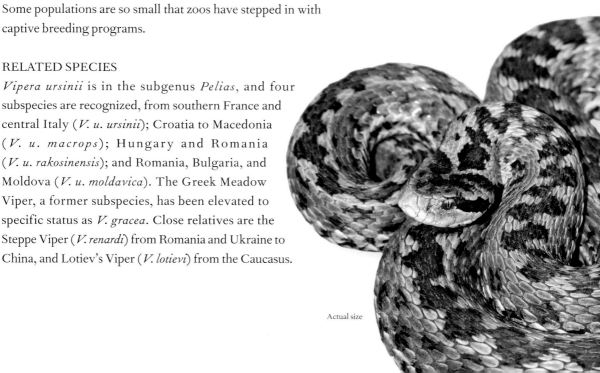

Actual size

GLOSSARY

anthropogenic Caused by the agencies of man.

anticoagulant A component of snake venom that prevents coagulation (clotting) of the blood leading to prolonged bleeding. *See also* Procoagulant.

arboreal Being adapted to move and live in the trees.

asl Above sea level.

autotomy The voluntary shedding of the tail as a defensive strategy, with the ability to regenerate the lost tail. *See also* Pseudautotomy.

basal The most primitive or ancestral member of a phylogenetic tree of related organisms.

bsl Below sea level.

bufotoxins Toxins found in the parotoid glands of toxic toads.

Caatinga A form of desert vegetation comprising arid scrubland and thorn forest, found in Brazil.

canthus rostralis A distinctive ridge that runs along the side of the head from the eye to above the nostril, often seen in vipers.

carinate, bicarinate, tricarinate Keeled body scales may bear one, two, or three keels.

caudal lure/luring Some young snakes, especially pitvipers and pythons, will slowly move the contrastingly colored tail tip to tempt potential prey within strike range.

Cerrado A form of savanna comprising woodland and grassland, found in Brazil.

Chaco A dry, hot lowland habitat found in southeastern Brazil, Paraguay, and northern Argentina.

clade A group of organisms comprising all the evolutionary descendants of a common ancestor. It may not necessarily have a name.

cloaca (adj. cloacal) The common genital-excretory opening of reptiles and birds.

cloacal plate Either entire or divided, the scale covering the cloaca.

congeneric Two or more species belonging to the same genus.

conspecific Two or more individuals belonging to the same species.

convergent evolution When unrelated organisms in different geographic areas evolve to occupy the same niche, the result being that the organisms often resemble one another, a classic case being the Green Tree Python (*Morelia viridis,* page 94) and Emerald Treeboa (*Corallus caninus,* page 108). Sometimes called parallel evolution.

crepuscular Active during dusk and dawn. *See also* Diurnal.

cryptic species A species that is well camouflaged to blend in with its background.

cryptozoic Inhabiting secretive, hidden habitats such as leaf litter or under stones.

dentary A tooth-bearing bone of the lower jaw of reptiles.

dichromatism Males and females are different colors.

dimorphism Males and females are different shapes or possess different adornments.

dipterocarp A type of tall tree primary forest dominant in Southeast Asia.

diurnal Active during the day, opposite of nocturnal. *See also* Crepuscular.

dorsal (n. dorsum) Pertaining to the upper surface of the body or head (opposite of ventral, venter).

dorsolateral On the upper flanks or sides of the body.

Duvernoy's gland The venom producing glands of a rear-fanged snake, named for the French anatomist F. M. Duvernoy.

endemic Only found in a single defined location, i.e. an island, mountain, or country.

endotherm (adj. endothermic) An animal that maintains its body temperature metabolically, i.e. a mammal or bird.

estivate Passing the dry season in a state of dormancy, the equivalent of hibernation or brumation in the cold.

extirpation Localized extinction.

fossorial Living in the soil or sand, or in the subsoil leaf litter. *See also* Semi-fossorial.

frontal scale A large dorsal head scale located between the eyes (see diagram on page 17).

fynbos A specific habitat comprising evergreen scrub and heathland found in the Cape of South Africa, from the Afrikaans for "fine bush."

genus (pl. genera, adj. generic); also subgenus The taxonomic category between family and species that contains a number of similar, presumably related, species. Some large genera are split into subgenera. Written in italics with a large case initial letter.

Gondwana, Gondwanan Pertaining to the southern super-continent following the break-up of Pangaea in Mesozoic times. Comprising South America, Africa, Madagascar, the Seychelles, Arabia, India, Australia, southern New Guinea, New Zealand, and Antarctica.

hemipenes (sing. hemipenis) The paired male sex organs of snakes and lizards.

hemolytic A component of venom that damages or destroys the red blood corpuscles.

hemorrhagin A component of snake venom that causes breakdown of blood vessels resulting in leakage of blood into the tissues.

hibernaculum (pl. hibernacula) A communal winter denning retreat used by certain snakes such as the Northern Adder (*Vipera berus,* page 640) or Timber Rattlesnake (*Crotalus horridus,* page 577).

holotype The single name-bearing specimen used to describe a species, usually housed in a natural history museum. Related specimens from the same location are known as paratypes.

incertae sedis A term that means "of uncertain placement," indicating that a species' position in a family has not been adequately determined.

internasal The scales on the dorsum of the head, between the left and right nasal scales (see diagram on page 17).

interspace The area of background color between bands or patches of patterning.

interstitial The skin between the scales on a snake or lizard.

IUCN International Union for the Conservation of Nature.

karoo A South African semi-desert habitat.

keeled scales Keeled scales are rough because they bear one or more raised ridges (opposite of smooth scales).

loreal scale The scale located between the preocular and nasal scales (see diagram on page 17).

labials, infralabials, supralabials Of the lip scales: supralabials—the upper lip scales, infralabials—the lower lip scales; many pythons and boas have heat-sensitive pits in their supralabials and infralabials (see diagram on page 17).

macrostomatan Large-mouthed. The Macrostomata is a clade of snakes capable of preying on animals wider than their own heads, i.e. above the level of the pipesnakes and shieldtails.

maxillary The tooth- or fang-bearing bone in the upper jaw of snakes.

melanistic Where all patterning is obscured by black pigment, commonly seen in female snakes living in cooler conditions because they will warm more quickly when basking.

mental groove A groove containing a fold of loose skin down the throat of most advanced snakes that enables the articulated lower jaws to be opened widely to engulf large prey.

mental scale The anteriormost scale of the lower jaw, located between and in front of the first infralabials (see diagram on page 17). *See also* Rostral.

mesic A habitat with a moderate supply of water. *See also* Xeric.

monotypic (taxonomic) A genus containing a single species or a family containing a single genus.

morph/morphotype A phase or form of patterning that differs from the normal pattern. Some species exhibit several different morphotypes in nature.

morphology The study of variation in the body shape, size, coloration, or patterning of organisms.

MYA Million years ago.

necrosis Tissue death and destruction, possibly caused by a cytotoxic venom. May lead to amputation and limb loss.

neonate A newly born offspring of a viviparous (live-bearing) species.

nominate subspecies The subspecies that bears the same name as the species, i.e. *Vipera berus berus*, *Natrix natrix natrix*.

ontogenetic change Any change in coloration, patterning, or morphology that occurs with maturity from juvenile to adult, i.e. the change from yellow or orange to green in Green Tree Pythons or Emerald Treeboas.

ophiophagous Eating snakes, i.e. the King Cobra (*Ophiophagus hannah,* page 480). This is not cannibalism unless the snake eats its own species.

orthopteran Grasshoppers, locusts, and katydids.

oviparous The reproduction strategy involving laying eggs (opposite of viviparous).

parietals A pair of enlarged dorsal scales on the rear of the head behind the frontal scale (see diagram on page 17).

parthenogenesis When a female gives birth or lays eggs without contact with a male. There is only one obligate parthenogenetic snake, the Brahminy Blindsnake (*Indotyphlops braminus*, page 57), but there have been many cases of facultative parthenogenesis when females of normally sexual species have reproduced without a mate, usually in captivity. Obligate parthenogenes are all-female species that are always parthenogenetic, facultative parthenogenes are sexual species which have occasionally reproduced parthenogenetically.

perianthropic Living alongside man.

pheromone (adj. pheromonal) A chemical substance released into the environment by an animal to attract a mate.

phylogenetic Relating to the evolutionary relationships and diversity of organisms.

piscivore/piscivorous Feeding on fish.

polymorphic Appearing in several different morphotypes or phases.

polyvalent antivenom An antivenom developed to treat the bites of a number of unrelated venomous snakes found within the same geographical area. Opposite of monovalent antivenom—antivenom produced to treat bites from one species or group of closely related species.

postocular The scale or scales directly in front of and in contact with the eye (opposite of preocular). See diagram on page 17.

prefrontals A pair of dorsal head scales behind the internasals but in front of the frontal scale (see diagram on page 17).

procoagulant A component of snake venom that causes coagulation (clotting) of the blood. Procoagulant venoms eventually cause prolonged bleeding by using up all the clotting factor. *See also* Anticoagulant.

psammophilous (n. psammophile) Living in sandy environments such as deserts.

pseudautotomy/caudal pseudautotomy The voluntary loss of the tail without the ability to regenerate a new tail (lizards can regenerate their tails while autotomizing but snakes do not).

reticulate (n. reticulation) Patterning that resembles a network of lines and blotches.

riparian Relating to the bank of a river, stream, lake, or wetland.

rostral The anteriormost scale on the dorsum of the head, located between and in front of the first supralabials (see diagram on page 17). S*ee also* Mental Scale.

rugose Rough, as in keeled scales.

scutes The large regular scales of the head.

semi-fossorial Living and moving through the leaf litter or under logs. *See also* Fossorial.

sister species, sister taxon/taxa A pair of closely related species.

species A taxonomic rank below the level of genus which is written as a binomial in italics with only the generic part of the name receiving a large case initial letter, i.e. *Vipera berus.*

species complex An often widely distributed species thought to contain a number of cryptic species.

species group A group of closely related species.

specific status When a subspecies is considered distinct enough to be elevated to full species.

subcaudal The scales on the underside of the tail, usually paired, occasionally single.

subspecies A taxonomic rank below the level of species that is written as a trinomial, i.e. *Vipera berus berus*.

647

supraciliary Small scales around the eye which may be elevated into small horns in some species i.e. the Eyelash Palm-pitviper (*Bothriechis schlegelii*, page 558).

supraocular A large scale located above the eye (see diagram on page 17).

supralabial *See* Labials.

sympatry (adj. sympatric) Where two species occupy the same location and habitat.

synonym/synonymy (adj. synonymized) When a species is considered no longer valid and is subsumed into another species.

taxon (pl. taxa) Any named group of organisms at any rank, a species, genus, family.

thanatosis A defensive strategy whereby an animal "plays dead" to avoid predation.

tubercle (adj. tuberculate) A soft, raised protuberance of the skin, often of sensory nature.

648

type genus The genus for which a family was named, i.e. *Coluber* for Colubridae.

type species The first described species within a particular genus.

type locality The locality where the holotype of a particular species was collected.

ventral (n. venter) Pertaining to the lower surface of the body or head (opposite of dorsum, dorsal).

versant The land sloping down from a mountain range.

vestigial A body part that is no longer used, which has become reduced in size due to evolution.

vicariant (n. vicariance) When two populations of a taxonomic group of organisms are separated by a geographical barrier such as a river, mountain range, or ocean, resulting in them evolving differently.

viviparous The reproduction strategy involving giving birth to neonates (opposite of oviparity).

xeric A dry habitat with little or no water. *See also* Mesic.

RESOURCES

The following is a selection of useful books, field guides, and web sites currently available for those with an interest in snakes or the wider field of herpetology.

BOOKS (General)

Aldridge, R.D. & D.M. Sever. *Reproductive Biology and Phylogeny of Snakes*. CRC Press, 2011

Campbell, J.A. & E.D. Brodie. *Biology of the Pitvipers*. Selva Publishing, 1992

Dreslik, M.J., W.K. Hayes, S.J. Beaupre & S.P. Mackessy. *The Biology of the Rattlesnakes II*. ECO Publishing, 2017

Gower, D., K. Farrett & P. Stafford. *Snakes*. Natural History Museum, 2012

Greene, H.W. *Snakes: The Evolution of Mystery in Nature*. University of California Press, 1997

Hayes, W.K., M.D. Caldwell, K.R. Beaman & S.P. Bush. *The Biology of the Rattlesnakes*. Loma Linda University Press, 2008

Henderson, R.W. & R. Powell. *Biology of the Boas and Pythons*. Eagle Mountain Press, 2007

Lilywhite, H.B. *How Snakes Work: Structure, Function and Behavior of the World's Snakes*. Oxford University Press, 2014

McDiarmid, R.W., M.S. Foster, G. Guyer, J.W. Gibbons & N. Chernott. *Reptile Biodiversity: Standard Methods for Inventory and Monitoring*. University of California Press, 2012

O'Shea, M. *Venomous Snakes of the World*. New Holland/ Princeton University Press, 2005

O'Shea, M. *Boas and Pythons of the World*. New Holland/ Princeton University Press, 2007

Pough, F.H., R.M. Andrews, M.L. Crump, A.H. Savitsky, K.D. Wells & M.C. Bradley. *Herpetology* (4th edition). Sinauer Publishing, 2016

Schuett, G.W., M. Höggren, M.E. Douglas & H.W. Green. *Biology of the Vipers*. Eagle Mountain Press, 2001

Vitt, L.J. & J.P. Caldwell. *Herpetology: An Introductory Biology of Amphibians and Reptiles*. Academic Press, 2014

Wallach, V., K.L. Williams & J. Boundy. *Snakes of the World: A Catalogue of Living and Extinct Species*. CRC Press. 2014

Zug, G.R. & C.H. Ernst. *Snakes in Question*. Smithsonian Institution Press, 1996

FIELD GUIDES

North America
(state field guides not listed)

Ernst, C.H. & E.M. Ernst. *Snakes of the United States and Canada*. Smithsonian Books, 2003

Heimes, P. *Herpetofauna Mexicana, Volume 1: Snakes of Mexico*. Chimaira, 2016

Powell, R. & R. Conant. *Peterson Field Guide to the Reptiles and Amphibians of Eastern and Central North America* (4th edition). Houghton Mifflin Harcourt, 2016

Stebbins, R.C. & S.M. McGinnis. *Peterson Field Guide to the Reptiles and Amphibians of Western North America* (4th edition). Houghton Mifflin Harcourt, 2018

Tennant, A. & R.D. Bartlett. *Snakes of North America: Western Regions*. Gulf Publishing, 2000

Tennant, A. & R.D. Bartlett. *Snakes of North America: Eastern & Central Regions*. Gulf Publishing, 2000

Central and South America, and the West Indies (national guides not listed)

Bartlett, R.D. & P.P. Bartlett. *Reptiles and Amphibians of the Amazon: An Ecotourist's Guide*. University of Florida Press, 2002

Campbell, J.A. & W.W. Lamar. *Venomous Reptiles of the Western Hemisphere* (2 vols.). Comstock Cornell, 2004

Crother, B.I. *Caribbean Amphibians and Reptiles*. Academic Press, 1999

Köhler, G. & L.D. Wilson. *Reptiles of Central America*. Herpeton Verlag, 2003

Savage, J.M. *The Amphibians and Reptiles of Costa Rica*. University of Chicago Press, 2002

Schwartz, A. & R.W. Henderson. *Amphibians and Reptiles of the West Indies: Descriptions, Distributions, and Natural History*. University of Florida Press, 1991

Europe (national field guides not listed)

Arnold, E.N. & D.W. Ovenden. *Reptiles and Amphibians of Europe*. Harper Collins, 2004

Beebee, T. & R. Griffiths. *Amphibians and Reptiles of Europe*. HarperCollins, 2000

Kreiner, G. *The Snakes of Europe: All species west of the Caucasus Mountains*. ECO Publishing/Chimaira, 2007

Speybroeck, J., W. Beukema, B. Bok, J. van der Voort & I. Velikov. *Field Guide to the Reptiles and Amphibians of Britain and Europe*. Bloomsbury Press, 2016

Africa and Madagascar (national guides not listed)

Branch, B. *Field Guide to Snakes and Other Reptiles of Southern Africa*. Struik, 1998

Branch, B. *Pocket Guide: Snakes and Reptiles of South Africa*. Struik, 2016.

Geniez, P. *Snakes of Europe, North Africa and the Middle East: A Photographic Guide*. Princeton University Press, 2018

Glaw, F. & M. Vences. *A Field Guide to the Amphibians and Reptiles of Madagascar* (3rd edition). Vences & Glaw Verlag, 1994

Henkel, F-W. & W. Schmidt. *The Amphibians and Reptiles of Madagascar, the Mascarenes, the Seychelles and the Comoros Islands*. Krieger Publishing, 2000

Howell, K., S. Spawls, H. Hinkel & M. Menegon. *Field Guide to East African Reptiles* (2nd edition). Bloomsbury, 2017

Marais, J. *A Complete Guide to the Snakes of Southern Africa* (2nd edition). Struik, 2004

Asia and Arabia (national guides not listed)

Chan-ard, T., J.W.R. Parr & J. Nabhitabhata. *A Field Guide to the Reptiles of Thailand*. Oxford University Press, 2015

Da Silva, A. *Colour Guide to the Snakes of Sri Lanka*. R&A Publishing, 1990

Das, I. *A Naturalist's Guide to the Snakes of South-East Asia*. John Beaufoy Publishing, 2012

Das, I. *A Field Guide to the Reptiles of South-East Asia*. Bloomsbury, 2015

Egan, D. *Snakes of Arabia: A Field Guide to the Snakes of the Arabian Peninsula and its Shores*. Motivate Publishing, 2008

Steubing, R.B., R.F. Inger & B. Lardner. *A Field Guide to the Snakes of Borneo* (2nd edition). Natural History Books (Borneo), 2014

Whitaker R. & A. Captain. *Snakes of India: A Field Guide*. Draco Books, 2008

Australasia and Oceania (state field guides not listed)

Cogger H.G. *Reptiles and Amphibians of Australia* (7th edition). CSIRO Publishing, 2014

McCoy, M. *Reptiles of the Solomon Islands*. Pensoft, 2006

Mirtschin, P., A.R. Rasmussen & S.A. Weinstein. *Australia's Dangerous Snakes: Identification, Biology and Envenoming*. CSIRO Publishing, 2017

O'Shea, M. *A Guide to the Snakes of Papua New Guinea*. Independent Publishing, 1996 (a much expanded and revised edition covering all of New Guinea is in preparation.)

Swan, S.K. & G. Swan. *A Complete Guide to the Reptiles of Australia* (5th edition). Reed/New Holland, 2017

649

HERPETOLOGICAL SOCIETIES

Society for the Study of Reptiles and Amphibians (SSAR)
ssarherps.org

American Society of Ichthyologists and Herpetologists (ASIH)
www.asih.org

Herpetologists' League (HL)
herpetologistsleague.org

Societas Europaea Herpetologica (SEH)
seh-herpetology.org

Australian Herpetological Society (AHS)
www.ahs.org.au

Herpetological Association of Africa (HAA)
www.africanherpetology.org

British Journal of Herpetology (BHS)
www.thebhs.org

Deutsche Gesellschaft für Herpetologie und Terrarienkunde (DGHT)
www.dght.de/startseite

Herpetological Conservation Trust (HCT)
www.arc-trust.org

European Snake Society
www.snakesociety.nl/index-e.htm

International Herpetological Society (IHS)
www.ihs-web.org.uk

USEFUL WEB SITES

World Congress of Herpetology (WCH)
www.worldcongressofherpetology.org

Reptile Database
reptile-database.reptarium.cz (use Advanced Search facility**)**

International Herpetological Symposium (IHS)
www.internationalherpetologicalsymposium.com

International Union for the Conservation of Nature (IUCN) Red List of Threatened Species
www.iucnredlist.org

Convention on International Trade in Endangered Species of Fauna and Flora (CITES)
www.cites.org

World Association of Zoos and Aquariums (WAZA)
www.waza.org

INDEX *of* COMMON NAMES

651

INDEX *of* SCIENTIFIC NAMES

653

654

655

INDEX *of* TAXONOMIC GROUPS

–idea = family; **–inae** = subfamily; **–idia** = higher taxa

ACKNOWLEDGMENTS

AUTHOR ACKNOWLEDGMENTS

A big thank you to all the photographers who have provided their excellent images for this book, and thanks to all my fellow herpetologists, zoologists, and film-makers who have shared the ups and downs of herping trips and fieldwork, all around the world. There are too many great friends and colleagues to list but you all know who you are. Thanks also to the Ivy Press team, Kate Shanahan, Caroline Earle, Alison Stevens, Liz Drewitt, Susi Bailey, Ginny Zeal, and David Anstey, who helped make this project such an enjoyable one. And thanks to my partner Bina Mistry, for keeping me fed and watered at my desk during the 30 months it took to write this book, and for her helpful comments on the text and images.